D1756220

WITHDRAWN

011274

POSTHARVEST
PHYSIOLOGY
of
PERISHABLE
PLANT PRODUCTS

POSTHARVEST PHYSIOLOGY OF PERISHABLE PLANT PRODUCTS

———**Stanley J. Kays**———

University of Georgia, Athens

An **avi** Book
Published by Van Nostrand Reinhold
New York

AN AVI BOOK
(AVI is an imprint of Van Nostrand Reinhold)

Copyright © 1991 by Van Nostrand Reinhold
Library of Congress Catalog Card Number 90-44213
ISBN 0-442-23912-2

Printed in the United States of America.

Van Nostrand Reinhold
115 Fifth Avenue
New York, New York 10003

Chapman and Hall
2-6 Boundary Row
London, SE1 8HN, England

Thomas Nelson Australia
102 Dodds Street
South Melbourne 3205
Victoria, Australia

Nelson Canada
1120 Birchmount Road
Scarborough, Ontario MIK 5G4, Canada

16   15   14   13   12   11   10   9   8   7   6   5   4   3   2   1

Library of Congress Cataloging-in-Publication Data

Kays, Stanley J.
    Postharvest physiology of perishable plant products/
    Stanley J. Kays.
        p.   cm.
    "An AVI Book"—T.p. verso.
    Includes bibliographical references.
    ISBN 0-442-23912-2
    1. Plant products—Postharvest physiology.   I. Title.
SB130.K38   1991
631.5'6—dc20   90-44213
                                                                      CIP

To Professor William Raymond Kays and Charlotte Loretta Kays
for a lifetime of contributions to horticulture

CONTENTS

PREFACE

There are two primary reasons for offering a broad overview of postharvest physiology. First, many critical functions that are operative during plant growth shift after harvest when the input of energy, water, and other essential requisites ceases in most products. Consequently, postharvest physiology differs substantially from what is covered in a typical plant physiology text. Second, the value of the majority of live agricultural plant products approximately doubles between harvest and retail sales. The total cost of losses occurring late in the production–harvest–marketing sequence are substantially greater than those incurred during the production phase. Likewise, inputs essential to prevent or minimize these losses are often only a minute fraction of the overall costs for the product. Therefore, a better understanding of the functional processes after harvest makes both biological and economic sense.

This book focuses on the functional processes controlling physical and chemical changes in live plants and plant products after harvest. The objective is to provide a concise overview of the theoretical principles and processes governing these changes. Examples include agronomic crops, fruits, nuts, vegetables, flowers, woody ornamentals, seeds, and other forms of plant propagules and turf grasses, with individual examples ranging from intact plants to pollen. Emphasis is placed on the basic postharvest physiological principles rather than on detailing optimum storage and handling conditions for individual crops. The latter information can be found in several excellent reference books cited in the text. A solid understanding of the basic processes operating should provide the reader with an understanding of the rationale for specific recommendations for commodities and with the potential to anticipate appropriate conditions for lesser known or new products for which handling and storage recommendations are not yet available.

This book is intended to be a useful overview reference for professionals in agronomy, botany, horticulture, forestry, food science, pest management, and agriculture engineering. It can also serve as a suitable textbook for junior/senior-level undergraduates and/or first year graduate students in these academic disciplines. Due to the diverse backgrounds of undergraduate students taking a course in postharvest physiology, the text is written so that only freshman botany and chemistry are essential prerequisites. The text can complement a plant physiology course in the student's normal curriculum.

There is generally more material provided than can be covered in a typical semester course. Based on the varied background of the students, instructors may choose to select certain portions of the text and substitute material from their own field of interest. For example, chapter 2 describes the structure of individual products from distinct organs to the subcellular level. If this information is covered in prerequisite courses, other information may be emphasized. Sections from the text not covered in a course provide a valuable reference source for the students.

References are presented at the end of each chapter and are separated into two groups: research papers cited in the text, and books and reviews for additional reading. The research papers (cited in the text as superscript numerals) represent only a small selection of the published information available. It would be impossible to include all references and still provide a readable text. Rather, specific references are included to provide examples of a particular concept. It is hoped that handling the references in this manner will enhance the reader's access to the literature and interest in pursuing the subject further.

I am grateful to Drs. R. C. Herner, R. E. Paull, M. Knee, and R. L. Shewfelt for reviewing the entire manuscript and to Drs. L. Waters, N. H. Banks, H. Y. Wetzstein, R. J. Romani, B. E. Michel, and S. E. Kays for critically reviewing various chapters. I would also like to acknowledge the scientists that have made their photographs and drawings available and the many contributions to the text made by Dr. Dave Simons. Special thanks goes to R. A. Walkup for her assistance in producing many of the illustrations and reviewing and coordinating many aspects of the finalization of the text.

ANATOMICAL ABBREVIATIONS

C	crystal	O	osmiophilic globule
Ch	chromosome (s)	P	plastid
Cs	cristae	Pc	precipitated compounds
CW	cell wall	Pe	pistil
CW_1	primary cell wall	PE	plastid envelope
CW_2	secondary cell wall	PM	plasma membrane
E	epidermal cell	PP	protoplastid
ER	endoplasmic reticulum	R	ribosome
G (Gr)	granum	S	stroma
GB	golgi body	Sb	starch body
LB	lipid body	Se	sepal
M	mitochondria	Ssc	substomatal cavity
Me	membrane	St	stamen
Ml	middle lamella	T	thylakoid
Mt	microtubule	Us	upper surface
N	nucleus	V	vacuole
Ne	nuclear envelope	Vb	vascular bundle
NM	nuclear membrane	Ve	vesicle
Nu	nucleolus		

SCIENCE AND PRACTICE OF POSTHARVEST PLANT PHYSIOLOGY

NATURE OF THE DISCIPLINE

The handling, storage, and marketing of plants and plant parts is one of the major preoccupations of human societies. It is an integral component of the human food supply chain and is therefore as diverse as are the cultures and the foods used. This diversity is increased by the use of perishable plants and their parts for decorative purposes and for modification of the environment.

A major concern with the handling of perishable plant material is the maintenance of condition; therefore it is necessary to consider not only the nature of that plant material but also the technological and economic aspects associated with getting the material to the consumer.

Postharvest physiology is the division of plant physiology dealing with functional processes in plant material after it has been harvested. Postharvest physiology is concerned with plants or plant parts that are handled and marketed in the living state including seeds, fruits, vegetables, cut flowers and foliage, nursery products, turf, vegetative propagules, and edible fungi. Some of these products are intact plants but the majority are isolated plant parts. Included in the range of products are all the various plant organs, and there are examples of all stages of development from germinating seed and juvenile shoots to mature plants, storage organs, and dormant seeds.

Postharvest physiology deals with the time period from harvest or removal of the plant from its normal growing environment to the time of ultimate utilization, deterioration, or death. When preharvest and harvesting factors have a direct influence on postharvest responses, they are also considered to be vital components of the complete postharvest picture.

Just what is meant by utilization varies with the product. For seeds and cuttings used for propagation and for transplants, utilization is planting. Utilization of cut flowers and foliage is displaying and maintaining them for decorative purposes. Continued development may be desired during utilization while senescence and death typically occur toward the end of their useful life. However, for the majority of products, utilization is the time of death due to consumption or processing. Unfortunately, much of the harvested plant material never reaches this point of utilization but is discarded because of deterioration due to senescence, stress responses, pathogen activity, insect attack, or mechanical damage.

FUNDAMENTAL NATURE OF PERISHABLE PLANT PRODUCTS

The most important characteristic of perishable plant products is that they are alive and therefore continue to function metabolically. However, their metabolism is not identical with that of the parent plant growing in its original environment, since the harvested product is under varying degrees of stress.

In the case of severed plant parts, harvesting, packaging, and handling interferes with or eliminates entirely some of the essential requisites for plant growth. Water and mineral nutrient supply from the soil is eliminated instantaneously as is, in most cases, the flow of carbon and energy from photosynthesis. Exposure to light is changed and in many cases virtually eliminated and the availability of oxygen and concentration of carbon dioxide is altered. The process of harvesting often results in substantial wounding; packaging and transport can cause further mechanical damage. The gravitational orientation of the harvested produce is commonly altered. It is frequently subjected to physical pressure, to a substantially altered temperature regime, and to an undesirable gaseous environment. Overall, living plant material is typically subjected to very harsh treatment during its postharvest life.

In the case of intact plants, such as containerized nursery products and transplants, the stresses incurred may be less extreme; however, the same factors are involved and the plants respond in a similar manner. It is useful to look at the handling of a typical product such as lettuce (*Lactuca sativa*, L.)* and to recognize what is being done at each stage. To even better appreciate the effects of these practices, a comparison can be drawn between plants and humans. A fundamental characteristic of all living organisms is that they respond in a multitude of interacting ways to counteract the effects of stresses to which they are exposed to maintain as near as possible a homeostatic condition within the organism. If the stress is so severe that it exceeds the physical or physiological tolerance of the organism, then death occurs. In this respect plants and humans react similarly and therefore can be compared.

In the case of the lettuce, a substantial wound is inflicted by severing the stem. In so doing, the supply of water from the roots is eliminated while the detached leafy portions of the plant continue to lose water. Unless this water loss from the leaves and wounded tissue is inhibited, the inevitable consequence is loss of turgor and wilting.

Harvest also eliminates the supply of mineral nutrients essential for normal metabolic activity. The harvested product is then dependent on recycling of those nutrients already present.

During handling of a head of lettuce, whether during harvesting and mar-

---

* Due to the diverse range of common names used worldwide for agricultural products, the botanical name is listed following the first usage of a common name in each chapter. *Hortus Third*[3] is the authority for the systematic nomenclature when applicable.

keting or in a retail store, leaf breakage is common. Trimming off outer leaves results in more wounds and additional breakage of the leaves occurs when the heads are packed tightly into containers. Some heads are placed upright whereas others are turned upside down, a distinctly abnormal gravitational orientation. Packing into the container effectively eliminates photosynthetic light and therefore the plant's external source of energy. In the confined environment within the container, gas exchange is restricted.

While the lettuce head is injured and its circumstance is radically altered in contrast to the preharvest state, it is still alive and continues to respond, metabolically adjusting to its new set of conditions. In fact, its rate of respiration is increased due to the treatment it has received. This situation in turn may more rapidly deplete the oxygen concentration within the container and increase that of carbon dioxide. Both of these changes will precipitate further changes in the metabolism of the lettuce. Injury also results in production of ethylene by the tissue, which if allowed to accumulate in the confined space of the container will accelerate the rate of senescence of the lettuce itself.

Respiration results in the release of energy as heat, causing the temperature of the lettuce to rise. In turn, deterioration processes, such as water loss, senescence, and rate of growth of pathogens, are accelerated. The latter may find an ideal environment with the rising temperature, high humidity, and ready access to the host tissue through extensive wounds. In addition, during transport to the consumer the harvested produce is subject to physical pressure, vibration, bruising, and frequently to temperatures and humidities that accelerate senescence. Finally the product is presented in the retail market to the consumer as "farm-fresh" produce.

Despite this apparently severe treatment, perishable produce must be harvested and moved through some handling and transportation system to its site of utilization while maintaining its live status and optimum condition. There is an inherent conflict between the requirements of human societies and the biological nature of the harvested perishable produce. Harvesting is a very effective way to kill a plant part, yet living plant produce must be moved from sites of production to sites of consumption that are commonly distant from each other. In the less extreme case, containerized plants cannot be moved from their growing environment to the suboptimal environments of transport vehicles and of most marketing situations without some deterioration of the plants occurring.

As a direct consequence of this inherent conflict between the need to harvest plant parts, which precipitates their death, and the need to keep them alive, compromises must be accepted. Compromises are an essential element at each level of postharvest handling of perishable plant products. They may take the form of compromises in temperature to minimize metabolic activity while avoiding chilling injury, or in oxygen concentration to minimize aerobic respiration yet avoid anaerobic respiration, or in the tightness of packing to minimize pressure damage while avoiding vibration damage, and so on.

Understanding the nature of the harvested product and the effects of handling practices is essential for arriving at the most appropriate compromises to maintain optimum condition of the produce. There is unfortunately no fixed recipe or solution for each product; rather the most appropriate practices must be worked out by the individual operator for each particular situation, taking into account physiological, physical, personnel, and economic factors.

EVOLUTION AND HISTORY OF
POSTHARVEST STORAGE

An evolutionary analysis of storage can be approached from several different biological positions. In some cases, animals store food internally. An extreme example can be seen in the hibernation of bears during the winter. In this case, luxury consumption of food is made during times of abundance, and the food is stored in the form of fat to be utilized as an energy source during hibernation. In many cases, however, food is stored external to the animal and typically, at least from a historical context, in a nonprocessed and readily perishable form.

A surprisingly large number of animal and insect species utilize some form of food storage. For some, the act of storage represents an instinctive behavioral process; however, for many others, storage is a learned response. Squirrels and chipmunks store food in underground caches for use during the winter months. Several large feline carnivores store fresh kills under protected conditions. For example, leopards often place for later use unconsumed portions of animals in crotches or branches of trees as protection from scavengers. Ants such as the genus *Pogonomyrmex* collect and store seeds in their underground nests. Other forms of harvester ants have evolved an even more complex system where harvested leaves are used to culture fungi in specialized underground storage areas of the nest.

Only with humans has storage evolved to a highly complex level and even this practice is a very recent development on an evolutionary time scale. Storage was no doubt utilized by early humanoids other than *Homo sapiens;* however, this subject has been little explored. As current knowledge of the evolution of man is analyzed, it becomes apparent that development of the ability to store food represented an extremely important step in the evolutionary process. Surprisingly, this topic has been virtually ignored to date.

The following questions are examples of the many that remain unanswered: When did man or the predecessor of man first start storing food? What impact did this storage have on survival potential, mobility, delineation of roles between sexes, changes in diet, sedimentation of populations, the evolution of agriculture? This deceptively simple process of food storage has been intimately associated with the evolutionary development of mankind.

Prehistoric Period

Man evolved slowly over many millennia. During the last 14–16 million years, however, major changes leading to modern man occurred (fig. 1-1). During this period, the predecessor of man moved from the edge of the tropical rainforest to the savanna. The ability of certain animals to function on two feet (bipedalism) rather than all fours evolved and some evolved over thousands of years to become hominids. *Ramapithecus,* the earliest direct ancestor of man, is known to have existed 14–15 million years ago.

The reason for this move from the tropical rainforest to the savanna is not known; however, the consequences were monumental. Food availability changed radically due to the seasonal variability of the savanna. (Alternating wet and dry seasons are essential elements in the maintenance of a grass-

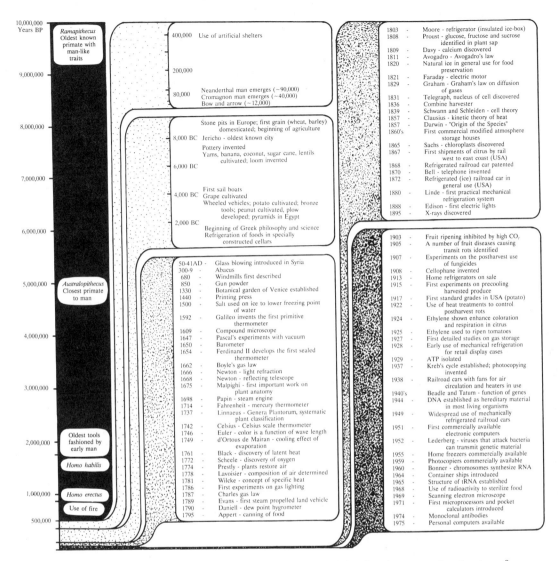

Figure 1-1. Chronology of the emergence of man and the development of technology.[9]

land.) Food acquisition became more critical with "feast and famine" cycles in food availability being common. Selection conditions for developing the ability to store food, even in the most primitive sense of the word, were operative. Many anthropologists argue that during this period important changes in diet occurred with the introduction of seeds into the diet to supplement fruit, roots, grubs, and meat. Grains and nuts store well whereas fruits and meat rot quickly. This ability to store grain and to ration it as needed helped even out the fluctuations in food supply.

More distinct differences in roles between males and females probably began to evolve when the predecessor of man entered the savanna. Males increasingly concentrated on hunting while females gathered fruits, nuts, seeds, and other vegetable material. It is interesting to note that vegetable products typically represented substantially more than 50% of the diet (estimates of up to 80% have been made). Hence, women in hunter-gatherer societies were the first to be involved in postharvest handling of plant material as the acquisition and storage of vegetable material was within their sphere of responsibility.

It is not certain how distinctly the roles of males and females had delineated or to what level storage had evolved by the time of *Australopithecus*, 3–4 million years B.P. (before the present). There is at present no evidence that food-gathering containers were used by *Australopithecus*; however, materials such as rawhide that might have been used are themselves highly perishable. It is fairly certain that food gathering had evolved to a point where more food was collected than was required for immediate consumption, which in itself represents a form of storage in that it prevents field losses due to shattering of seed heads or loss to other organisms.

Modern man developed slowly over a 5-million-year period, generally evolving in a sequence from *Australopithecus* to the genus *Homo* and to the species-level stages of *H. habilis* (approximately 2 million years ago) and *H. erectus* (approximately 0.3 million years ago). *H. sapiens* subsequently evolved from Neanderthal (90,000 years ago) through Cro-Magnon (30–35,000 years ago) to modern man (10,000 years ago).* Very little is known at present concerning the development of storage practices during the prehistoric period of man's evolution. During this period researchers can only speculate, based on diet, the level of complexity of the tools being used, and the degree of social organization as to the probable role and level of development of storage practices. The earliest written records discovered to date are in the form of cave paintings from the Cro-Magnon period approximately 35,000 years ago.

Around 10,000 years ago the last major glacial period was drawing to a close, marking the end of a long series of extreme climatic changes. Diverse changes in the flora began to occur at the end of the ice age and these were largely completed by 8,000 years B.P. These changes set the stage for the first steps toward agriculture that occurred in an area of the Near East called the Fertile Crescent.

The impetus resulting in the development of agriculture is not known; however, the impact of agriculture on the development of civilization was monumental.[5] Man was no longer forced to adapt to their environment; instead they began to alter and shape it. The door was opened for the development of all of the complex societies and civilizations that have followed.

Storage must have played a critical role in the development of agriculture. Seeds of wheat (*Triticum aestivum*, L.) and other early annual domesticates had to be stored in a manner that would maintain their viability until the next growing season. In addition, grain to be utilized for food had to be stored in such a way as to minimize losses. Rotting due to excess moisture and losses due to insects and rodents—typical problems confronted by early farmers— were well documented during the Roman era.

While the practice of storage must have evolved during development of hunter-gatherer societies, with the onset of agriculture its importance became

---

* There is no existing evidence that each successive stage leads directly to the next. For example, *Ramapithecus* could represent a side branch on the evolutionary tree, eventually becoming extinct, without actually leading to *Australopithecus*.

much more critical. Technological advances such as pottery that could be used as storage vessels began to occur and this progress in turn affected the subsequent development of agriculture. The sequence of plant domestication in the Near East underscores the importance of storage and its subsequent effect on agricultural development. Wheat was domesticated at least 9,000 years ago, followed by chickpeas (*Cicer arietinum*, L.), lentils (*Lens culinaris*, Medic.), and fava beans (*Vicia faba*, L.) during the next millennium (fig. 1-1). All were crops that were relatively stable in storage if handled properly. Later species to be domesticated in the centuries following around 6,000 years ago were almonds (*Prunus dulcis*, (Mill.) Webb), pistachios (*Pistacia vera*, L.), walnuts (*Juglans regia*, L.), figs (*Ficus carica*, L.), dates (*Phoenix dactylifera*, L.), apricots (*Prunus armeniaca*, L.), grapes (*Vitis vinifera*, L.), and olives (*Olea europaea*, L.). Again, all were crops that could be stored successfully for extended periods in natural or processed forms.

Early Advances in Storage Technology

Advances in technology enhanced the ability of humans to successfully store harvested plant products. Containers made of plant and animal material were probably in widespread use during the last ice age; however, due to their perishable nature none are known to have survived. Basket making was a highly developed art by 7,000 years B.C. Early baskets were used for gathering seeds, nuts, fruits, and other vegetable material and for storage until their subsequent use.

The ability to make pots of fired clay was an extremely significant advance in storage technology and its appearance coinciding with the first domestication of grain-bearing plants cannot be considered mere chance. For the first time, it allowed stored products to be sealed in containers that would significantly decrease the chances of loss (fig. 1-2). By 6,500 years ago pottery had become common and specialized craftsmen made a wide range of pots for various uses. Without successful storage of surplus food, however, labor specialization could not have occurred and the development of arts and crafts would have stagnated.

Another early technique developed for the storage of grain was underground pits and silos (fig. 1-3). These are known to have been used by pre-Neolithic societies in the Middle East 9,000 to 11,000 years ago. Early pits or silos were small and shallow but with time the use of larger pits developed.[7,22] By the time of the Roman era, the silo was one of the major means of long-term storage of grain and it continued to be one of the main storage facilities used until the early nineteenth century. Even today grain may be stored in pits in the major grain-producing and exporting countries of the world. Pits are also currently used in traditional societies for storage of a wide range of vegetable and even fruit products.

The earliest record of modified atmosphere storage of plant products comes from the Roman era (fig. 1-4). Varro[30] gave a detailed account of the construction of underground pits for storage of grain and of sealing them after they were filled. He also stressed the dangers of entering them too quickly after opening. During storage, respiration of the grain decreased the oxygen concentration and elevated the carbon dioxide level tremendously, decreasing the chances of losses due to insects and rodents. Entry of the silos before

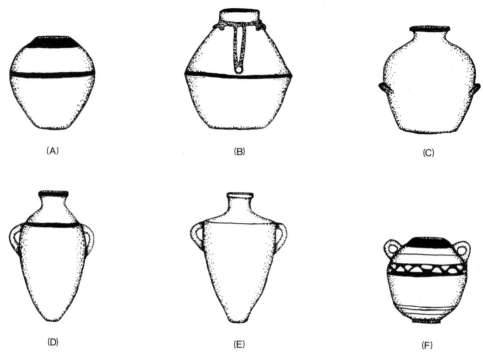

Figure 1-2. Examples of Palestinian pottery from several archeological periods. *A* and *B:* the Chalcolithic period; *C:* early Bronze age; *D:* middle Bronze age; *E:* late Bronze age; *F:* Byzantine period *(after Gonen[11]).* Large vessels with wide mouths were routinely used for larger foodstuffs and those with narrow mouths held liquids and small grains. Use of tightly fitted or sealed lids created modified environments within the containers and eliminated access by herbivores.

adequate aeration could be fatal, so the Romans devised the technique of lowering a burning lamp into the silo to detect "foul air." As is true today, a number of techniques and variations of greater or lesser utility were advocated for storage of grain in the Roman era. Pliny recognized the importance of harvesting a crop for storage at the proper stage of ripeness and discussed various techniques used.[23] He relates that "some people tell us to hang up a toad by one of its longer legs at the threshold of the barn before carrying the corn into it." Even today misconceptions and superstitions sometimes influence how harvested products are stored.

Processing of plant products was also developed to facilitate storage. Domesticated fruits such as figs, grapes, and apricots and many fruits gathered from the wild were dried (fig. 1-5). In some instances, grains and nuts were parched, and grains and grapes were often fermented. The net effect was to enhance the storage potential of the products.

Developments Leading to Modern Storage Technology

The importance of storage temperature, storage gas atmosphere, product moisture management, transportation, storage facilities, and storage insects and pathogens was appreciated to varying degrees by the time of the Roman

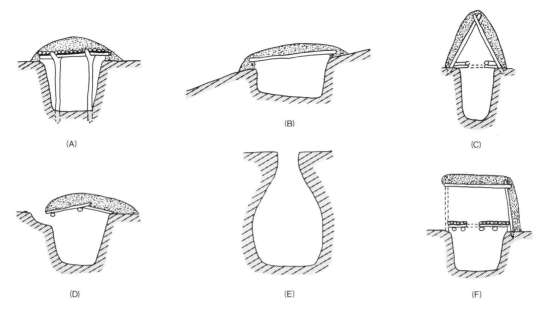

Figure 1-3. A wide range of relatively simple storage structures has been utilized by man over the centuries. Storage pits have been described from the later stages of the Paleolithic period (Shanidar cave).[25, 26] Storage pits are found with increasing frequency at sites dating from 9000–7000 B.C. and some display distinct technological improvements (e.g., plastered walls).[20] Taylor[28] suggests that the Neolithic Revolution—the time period when man began to make the transition from a hunter-gatherer subsistence—could with equal justification be called the "storage revolution." Storage facilities can be separated into aboveground and below-ground structures. Below-ground structures were often preferred due to the simplicity of construction and greater security, especially during more turbulent periods, since the store can be more easily hidden. The drawings are examples of potato (*A, B, D*) and grain (*C, E, F*) storage pits.[7]

Figure 1-4. Structures, some of which have undergone little technological improvement since their use 8,000–10,000 years B.P., remain widely used by farmers in many areas of the world today. For example, this photograph shows a traditional potato (*Solanum tuberosum,* L.) store currently used by farmers in the Puma of southern Peru.[24]

(A)

(B)

Figure 1-5. Storing the harvest *(A)* and a scribe *(B)* checking the storage of raisins *(Vitis vinifera,* L.). These were recorded in hieroglyphics found in the Egyptian tomb of Beni Hasan dating from around 2,500 years B.P. *(from Newberry[18]).*

Empire. While little of the basic principles governing these factors was understood, many were utilized to enhance the reliability and quality of stored products.

Storage Temperature

By the beginning of the Roman Empire, the importance of refrigeration was known but not widely utilized for harvested products. Varro[30] discusses the undesirable effect caused by respiratory heating during prolonged storage of olives and in grain stored before being sufficiently mature. The importance of cool temperatures for the storage of apples was also well known. During this period, wealthy citizens had cellars packed with ice or snow that were used for the preservation of luxury foods. Caves and pits provided areas that were much cooler than outside temperatures and were also used.

As early as 1550, the effect of salt and other chemicals on lowering the freezing point of water had been established* and was used by the Roman aristocracy to cool wine and drinking water. Salt was to play an important role in the refrigerated transport of fruits and vegetables well into the twentieth century.

---

* Reportedly discovered by the Spanish physician Blasius Villafrance.

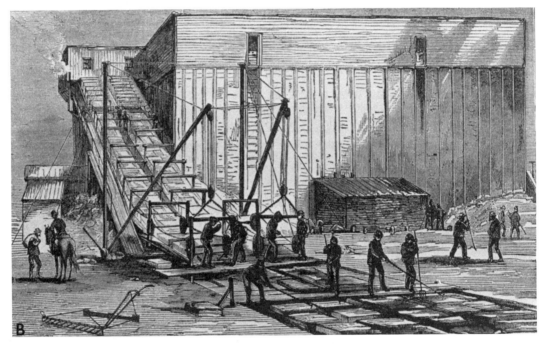

Figure 1-6. Ice gathering *(A)* and storage *(B)* on the Hudson River near New York in the late 1800s.[2]

Improvements and utilization of temperature management for harvested plant products progressed slowly until the Industrial Revolution.[12] In 1803, Thomas Moore invented the refrigerator, which consisted of an insulated icebox.[16] By 1820, ice collected during the winter months in northern areas of the temperate zone came into general use for the preservation of foods (fig. 1-6). A thriving ice business developed within the United States, and by 1834, shipments were being made to Havana, the West Indies, Rio de Janeiro, Ceylon, Calcutta, Bombay, Madras, Batavia, Manila, Singapore, Mauritius, and Australia.

Ice was utilized for refrigeration in specially constructed railway cars as early as 1868. The cars were insulated and galvanized-iron tanks held a salt and ice mixture for cooling.[27] The latter were subsequently replaced with a

V-shaped box containing around 3,000 pounds of ice that transversed the length of the car. By 1872, the refrigerated railroad car came into general use.

The availability and price of natural ice varied considerably from year to year. Cold winters meant cheap ice and mild winters, such as in 1890 in the United States, resulted in a much reduced supply and higher prices. The need for a mechanical means of refrigeration was apparent. A machine for refrigerating air by the evaporation of water in a vacuum was patented in 1755 by a Scotsman named Cullen. However, it was not until 1859 that a machine, using ether, was commercially marketed by a firm in Australia. The ammonia compression machine developed by Professor Carl Linde of the University of Munich opened the way for the large-scale production of ice. By 1889, there were more than 200 commercial ice plants in the United States, and by 1909, around 2,000.

Early use of refrigeration machinery was primarily for the production of ice rather than the direct cooling of storage houses. By the late 1800s, interest in the use of refrigerated warehouses for the commercial storage of fruit developed.[26] In 1876, a storage house cooled with barrels of ice was constructed by the Western Cold Storage Company in Chicago for the preservation of fruit. An estimate made in 1901 indicated that 600 commercial establishments storaged fruit and other produce.

Use of mechanical refrigeration spread rapidly, and by 1928, the use of cold temperature display cases in retail food stores was common. Refrigeration had become an indispensable part of the food distribution system in many countries.

Low-temperature stress, both preharvest and postharvest, was also a matter of concern during Roman times. Columellia[30] recommended the burning of piles of chaff around vineyards and fields to prevent injury to the crop when frost threatened. Fruit rooms in the 1700s and onward typically had a small stove that was used to elevate the temperature when absolutely essential. Much of the postharvest research around the beginning of this century was directed toward determining the optimum temperature for storage of various plant produce. It was commonly believed that temperatures as low as 0°C were undesirable for fruits such as apples (*Malus sylvestris,* Mill.) and pears (*Pyrus communis,* L.) and that higher temperatures (2°C) were essential. This belief was shown, however, not to be the case; quality could be maintained substantially longer for many cultivars at the lower temperatures.

The inability to control storage room temperatures precisely created problems that are seldom encountered with the controlled mechanical refrigeration systems used today. From the Roman times until well into the twentieth century, cool dry conditions were recommended for fruit storage and caused serious losses in fruit quality due to desiccation. Dry conditions were recommended because the inability to precisely control the temperature of the storage room resulted in significant temperature fluctuations. When a moist, warm period followed cold weather, the fruit temperature was lower than the air temperature and water condensed on the surface of the fruit. If these conditions were allowed to remain, they had disastrous consequences due to stimulated pathogen activity. It was, therefore, recommended that fruit rooms be kept dry and be aired only when the internal and external temperatures were alike. Techniques that would dry the air (e.g., fire, $CaCl_2$) were also often employed. Accurate control of storage temperature in modern storage rooms has to a large extent eliminated this problem. However, with traditional storage systems and occasionally with mechanical failure of modern refrigeration systems, the problems encountered with imprecise control of temperature are dramatically illustrated.

Alteration of the Storage Gas Atmosphere

The effect of sealing grain in underground silos on the storage gas atmosphere was documented during Roman times and probably practiced prior to that period. The primary benefit, the control of insects and rodents, as well as the dangers to humans, were recognized by the Romans. Cato[30] recommended storing grapes in clay pots, which would have also resulted in a modified atmosphere. Various elaborations of this basic technique, resulting in a modified storage gas atmosphere, have been described since that period. For example, R. Brookes[6] gave the following description in 1763 of a fruit storage technique "communicated to the public by the Chevalier *Southwell,* and which has been used in France with success":

> As many expedients have beeen tried among us, for preferving fruit frefh all the year, I fhall beg leave to give one communicated to the public by the Chevalier *Southwell,* and which has been ufed in *France* with fuecefs. Take of Salt Petre one pound, of Bole Armenic two pounds, of common Sand well freed from its earthy parts, four pounds, and mix all together ; after this let the fruit be gathered with the hand before it be thorough ripe, each fruit being handled only by the ftalk ; lay them regularly, and in order, in a large wide mouthed glafs veffel ; then cover the top of the glafs with an oiled paper, and carrying it into a dry place, fet it in a box filled all round to about four inches thicknefs, with the aforefaid preparations, fo that no part of the glafs veffel fhall appear, being buried in a manner in the prepared Nitre ; and at the end of the year fuch fruits may be taken out as beautiful as they were when firft put in.

The first scientific studies on modified storage atmospheres were conducted in 1819 by Jacques Berard[4] in France. Berard demonstrated that harvested fruits utilize oxygen and liberate carbon dioxide during storage and that fruits placed in containers devoid of oxygen did not ripen. Benjamin Nyce in the 1860s built and operated a fruit storage warehouse in Cleveland, Ohio that utilized the basic principles of modified atmosphere storage.[9] Unfortunately, little interest was generated among fruit storage workers of that time and the idea did not take hold.

During the early part of the twentieth century, the effect of high CO_2 on inhibiting the ripening of apples and softening of peaches was reported. In 1927, Franklin Kidd and Cyril West published their studies on the utilization of storage gas atmosphere to control ripening and longevity of apples.[13] An essential element of gas atmosphere storage that developed during this period was the necessity to accurately control the gas composition of the storage atmosphere. Although the term *controlled atmosphere storage* was not utilized until 1941, the requirement of control was apparent. Research during this period concentrated on optimum storage gas atmospheres for various products and their cultivars and methods for constructing and operating controlled atmosphere storage rooms.

Commercial utilization of controlled atmosphere storage began in England in 1929 when a grower stored around 30 tons of apples in 10% CO_2. As the news of the success of this storage technique spread, it stimulated research and utilization in many countries. By 1989, 2.6 billion pounds of apples were stored under controlled atmosphere conditions in the United States alone.

SIGNIFICANCE OF POSTHARVEST PLANT PHYSIOLOGY AND HANDLING OF PERISHABLE PLANT PRODUCTS

When plants or plant parts are used by humans, whether for food, for aesthetic purposes, or for modifying the environment, there is always a postharvest component. Just how important this postharvest component is in the overall business of satisfying human requirements for plant products varies widely. It is influenced by the nature of the product itself, particularly its perishability, the intended use of the product, the environmental and handling conditions to which the product is exposed, the relative abundance of the product at the time, the culture of the society, and socioeconomic factors. In all cases the importance of the postharvest component is increased with the time lapse between harvest and use. The greater the time lapse for a product the more important is the postharvest component.

Factors Influencing the Postharvest Component

There is a very wide range in the degree of perishability of plants and plant products. Dry seeds such as cereal grains and pulses are perhaps the least perishable of the major food items. They are comparatively low in moisture content, on the order of 3.4–15%, and are protected from excessive moisture loss, microbial infection, and mechanical damage by specialized tissues such as the seed coat. Under appropriate conditions many species of seed can be maintained in good condition for years with relatively simple storage facilities and treatments. However, the prolonged storage time adds to the significance of the postharvest component. In contrast, juvenile products such as okra (*Abelmoschus esculentus,* (L.) Moench.) or lettuce have little protection, are harvested when they are actively growing, and often contain 85% or more water. In addition, they have a high metabolic rate, they have little protection against water loss or microbial infection, and they are very susceptible to mechanical damage. Under the best of conditions, these plant parts can be maintained for a few weeks; under ambient conditions, they last only a few days.

Differences in postharvest perishability probably delineate the separation of agronomic and horticultural crops better than any other single characteristic. Most agronomic products tend to be relatively stable in contrast with most horticultural products, which are highly perishable. This situation unfortunately has tended to deemphasize the importance given to the postharvest period of agronomic crops even though extensive losses do occur.

The way in which harvested products are to be utilized also affects how they can be handled during the postharvest period. For example, produce that is to be used for fresh market is handled quite differently from that being processed. Produce for the fresh market commonly passes through several stages or operations, each requiring time and handling, and is often transported over long distances. Therefore, much care and effort is needed to maintain its condition. In contrast, fresh produce for processing is usually produced close to the processing plant. It is harvested and moved quickly with a minimum of handling to that point. The produce has similar physiological and deteriorative characteristics in both cases, but the time required between harvesting and processing limits the extent of deterioration and therefore the benefits to be gained from intensive postharvest care.

In many instances, particularly in subsistence agriculture situations, perishable crops are used by the grower or sold and exchanged locally over short distances in a short time span. In such cases, the postharvest requirements are minimized, but not eliminated. An understanding of the physiology of the produce and of the causes of deterioration can lead to simple and inexpensive changes in postharvest handling practices that can greatly reduce losses.

In the more mechanized societies, the trend is for the production of perishable products to shift further and further away from the major markets. The spread of cities over nearby production centers and the subsequent increase in land costs are common problems that force production to more distant sites. Mechanization of production is an important factor in minimizing costs in many countries. For it to be economical, large tracts of land not usually available close to the markets are required. In addition, the demands from consumers for year-round supplies of major products encourages production to shift location with seasonal changes of climate. Hence, winter production of summer-growing crops can be achieved in subtropical regions but the harvested produce must then be transported back to the markets in the temperate regions. In such cases, careful attention to the postharvest requirements of the crops is essential and a substantial cost factor is added to these products.

The importance attached to the postharvest handling of produce is also influenced by current supply/demand situations. When supply is high in relation to demand, the price paid is lower. This can result in less effort and expense during marketing and the acceptance of greater losses of product.

Finally, cultural and socioeconomic factors have a major influence on the importance attached to the postharvest component. Traditional postharvest practices have evolved over long periods of time and are intimately associated with the local culture and the structure of the society. Although these practices have withstood the test of time, our current understanding of postharvest behavior may offer opportunities for improvement. Acceptance of these changes by traditional societies, and by the more developed ones, may be difficult to achieve. Clear demonstration of the economic or social value to those involved in production and marketing may be required before they are prepared to place a greater emphasis on the postharvest care of their products.

Governments may also exert a substantial influence on the emphasis placed on postharvest care of produce through policies relating to national food supply goals. Policies on transportation and communications systems, support programs, and taxation greatly affect how products are handled after harvest. Government policy may not necessarily improve postharvest practices or reduce postharvest losses. Changes in postharvest practices may have undesirable secondary effects in other areas such as employment and distribution of population.

Postharvest Losses

A postharvest loss is any change in the quantity or quality of a product after harvest that prevents or alters its intended use or decreases its value. Postharvest losses vary greatly in kind (e.g., ranging from losses in volume to subtle losses in quality), magnitude, and where in the postharvest handling system they occur.[21] Likewise, individual crops differ greatly in their susceptibility to loss.

How important are postharvest losses in the overall production, marketing, and utilization scheme of agricultural plant products? Accurate estimates of net losses simply are not available. Although a detailed cost analysis can be made for an individual crop at a specific time and place (e.g., within an individual village or production brigade), these measurements do not extrapolate accurately to an all-encompassing province, country, or worldwide estimate for the crop. Losses in one village may be radically different from another just down the road. Likewise losses one year within an individual village may differ tremendously with the next. Ideally a summation of losses for each crop at the local, state, national, and finally international level are needed for an accurate assessment. A very general estimate of the potential importance of postharvest losses can be seen in the increase in value of a product after harvest, that is, the distribution of costs between production and marketing segments of the overall system. While it is often difficult to obtain precise data on the actual costs of production and marketing of plant products from the grower right through to consumer, general estimates can be made from production and marketing statistics together with estimates of the general profit margins involved (table 1-1). Based on these estimates it is apparent that approximately 50% of the retail value of fresh produce is accrued after harvest.

Another way to assess the significance of the postharvest component is to look at the extent of losses that occur during this phase. Once again a meaningful determination of these losses is extremely difficult to obtain since they are specific to both location and time. Thus, at present it is impossible to substantiate statistically these losses on a national or international basis. However, authoritative groups such as the National Research Council of the United States and the Food and Agriculture Organization of the United Nations have made estimates for planning purposes in developing countries. Their conservative estimates are that the minimum overall loss of the more durable cereal grains and legumes is around 10% while that for more perishable food products is approximately 20%. Reported losses for individual crops are often much higher, ranging from 0.5% to 100% (tables 1-2 and 1-3).

Table 1-1. Estimated Distribution of Total Costs Between Production and Postharvest Costs as Percentage of Cost to Consumer

Crop	Production %	Postharvest %
Snap bean (*Phaseolus vulgaris*, L.)	53.7	46.3
Lima bean (*Phaseolus lunatus*, L.)	61.5	38.5
Okra (*Abelmoschus esculentus*, (L.) Moench.)	50.6	49.4
Sweetpotato (*Ipomoea batatas*, (L.) Lam.)	50.5	49.5
Sweet corn (*Zea mays* var. *rugosa*, Bonaf.)	49.0	51.0
Cabbage (*Brassica oleracea*, L. Capatita group)	46.7	53.3
Apple (*Malus sylvestris*, Mill.)	44.4	55.6
Peach (*Prunus persica*, (L.) Batsch.)	53.1	46.9
Karume azalea (*Rhododendron* spp.)	45.5	54.5
Pin oak (*Quercus palustris*, Muenchh.)	33.3	66.7
Juniper (*Juniperus* spp.)	46.0	54.0

Note: Data derived in part from Cooperative Extension Service, University of Georgia publications, 1981–1982.

Table 1-2. Reported Production and Loss Figures in Less-Developed Countries

Commodity	Production (1,000 tonnes)	Estimated Losses (% of total crop)
Roots/Tubers		
Carrots (*Daucus carota*, L.)	557	44
Potatoes (*Solanum tuberosum*, L.)	26,909	5–40
Sweetpotatoes (*Ipomoea batatas*, (L.) Lam.)	17,630	35–95
Yams (*Dioscorea* spp.)	20,000	10–60
Cassava (*Manihot esculenta*, Crantz.)	103,486	10–25
Vegetables		
Onions (*Allium cepa*, L.)	6,474	16–35
Tomatoes (*Lycopersicon esculentum*, Mill.)	12,755	5–50
Plantain (*Musa* spp.)	18,301	35–100
Cabbage (*Brassica oleracea*, L. Capatita group)	3,036	37
Cauliflower (*Brassica oleracea*, L. Botrytis group)	916	49
Lettuce (*Lactuca sativa*, L.)	—	62
Fruits		
Bananas (*Musa* spp.)	36,898	20–80
Papayas (*Carica papaya*, L.)	931	40–100
Avocados (*Persea americana*, Mill.)	1,020	43
Peaches (*Prunus persica*, (L.) Batsch.) apricots (*Prunus armeniaca*, L.) nectarines (*Prunus persica* var. *nucipersica*, (Suckow) Schneid.)	1,831	28
Citrus (*Citrus* spp.)	22,040	20–95
Grapes (*Vitis vinifera*, L.)	12,720	27
Raisins (*Vitis vinifera*, L.)	475	20–95
Apples (*Malus sylvestris*, Mill.)	3,677	14

Source: From NSF.[19]

Table 1-3. Postharvest Losses of Wheat, Rice, and Maize

Crop	Country	Storage Method	Cause of Loss	Storage Period (months)	% Loss	References
Wheat (*Triticum aestivum*, L.)	India	Gunny sacks	Insects	6–12	28–50	14
	India	Bulk	Insects	—	3.5	31
	Egypt	Stack	—	12	36–48	1
Rice (*Oryza sativa*, L.)	Sri Lanka, Nepal, Malaysia, Vietnam	—	Insects	30	30	15
	West Africa	Farm storage	—	12	2–10	19
Maize (*Zea mays*, L.)	United States	—	Insects	12	0.5	29
	Tanzania	—	—	—	20–100	10
	India	Farm storage	—	9	2–11	14

Postharvest losses in developed countries are generally considered to be less than in the developing regions of the world but reliable estimates of these are not readily available. A USDA study[29] that dealt with isolated segments of the total postharvest component and with a variable range of products reported losses in durable products such as grain, dried fruits, and nuts due to insects and other storage problems to be 3.62%. Losses of 5.42% of fruits and 10.3% of vegetables during transit, unloading, and retail marketing were also reported. These figures do not take into account postharvest losses on the farm; during sorting, grading, packaging, storage, and wholesale marketing; or at the consumer end of the chain.

The importance of the large number of stages, each commonly under the control of a different organization, in the marketing of perishable plant products is illustrated in figure 1-7. While each operator along the chain may be prepared to accept some loss, the cumulative effect of all these losses can be substantial. The rubbish bin is a major consumer of produce.

While complete physical loss of produce is the most dramatic example, much of the total loss is due to reduced quality and nutritional value. Postharvest care, therefore, is as much concerned with maintenance of quality as with quantity. It should be applied to all produce, not only to the top-quality material. The aim is to maintain as much as possible of the existing quality, whatever that may be. It can be argued that even greater care is needed for produce that has already started to deteriorate than for top-quality material, since the established deterioration sequence must be inhibited (e.g., preventing water loss from wounded tissue and preventing infection of the wound as well as maintaining physiological condition) and the existing condition maintained.

The full significance of quality loss may not be immediately obvious as in the case of storage of seed or other propagules. Seed that has been allowed to deteriorate during storage may still be viable, but if planted, its rate of establishment and subsequent yield will be impaired. Thus what may super-

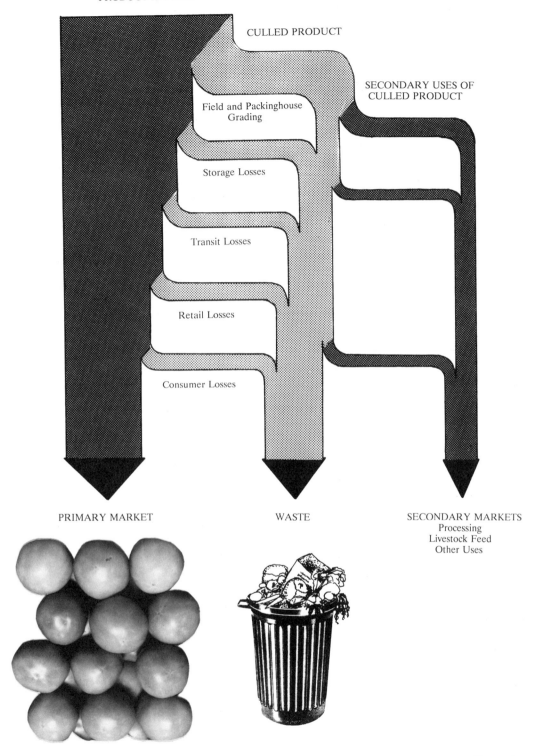

Figure 1-7. A flow chart of postharvest losses for tomatoes (*Lycopersicon esculentum*, Mill.) as they move toward final utilization (data from Campbell[8]). Inferior grades that would be discarded in some countries often remain in the food chain in developing countries. These find alternate markets that have different quality tolerances.[17]

ficially appear to be a small postharvest loss in quality of the seed is in fact greatly amplified into a serious loss in yield of the crop.

Although it is not possible to eliminate losses, the extent of these losses can be reduced through better understanding of the nature of harvested produce and use of appropriate technology. There is, however, a cost associated with the use of many postharvest techniques, but the cost of increasing the usable supply of a particular product by reducing losses is considerably less than the cost of producing more to compensate for postharvest losses. One example often quoted is that the total energy cost of good grain storage practices is about 1% of the energy cost of producing that grain. Other advantages gained through using improved postharvest practices to increase usable supply, in contrast to increased production, are fewer adverse effects on health, safety, and the environment.

REFERENCES

1. Abou-Nasr, S., H. S. Salama, I. I. Ismail, and S. A. Salem. 1973. Ecological studies on insects infesting wheat grains in Egypt. *Z. fuer Angola Entomol.* **73**:203–212.
2. Anon. 1875. Gathering and storing ice. *Sci. Am.* **32**(March 27):195.
3. Bailey Hortorium. 1976. *Hortus Third, a Concise Dictionary of Plants Cultivated in the United States and Canada.* Macmillian, N.Y., 1290p.
4. Berard, J. E. 1821. Memoire sur la maturation des fruits. *Ann. Chim. Phys.* **16**:152–183, 225–251.
5. Braidwood, R. J., and C. A. Reed. 1957. The achievement and early consequences of food production: A consideration of the archeological and natural-historical evidence. *Cold Springs Harbor Symp. Quant. Biol.* **22**:19–31.
6. Brookes, R. 1763. *The Natural History of Vegetables.* The Bible and Sun, London, p. xii.
7. Buttler, W. 1938. Pits and pit-dwellings in Southeast Europe. *Antiquity* **10**:25–36.
8. Campbell, D. T. 1988. A systems analysis of postharvest injury of fresh fruits and vegetables. Master's thesis, University of Georgia, 165p.
9. Dalrymple, D. G. 1969. The development of an agricultural technology: Controlled atmosphere storage of fruit. *Technol. and Cult.* **10**:35–48.
10. FAO. 1981. Food loss prevention in perishable crops. *FAO Agric. Ser. Bull. 43,* 72p.
11. Gonen, R. 1973. *Ancient Pottery.* Cassell, London, 95p.
12. Goosman, J. C. 1924–1927. History of refrigeration. *Ice and Refrig.* **66**:297, 446, 541; **67**:33, 110, 181, 227, 428; **68**:70, 135, 335, 413, 478; **69**:99, 149, 203, 267, 372; **70**:123, 197, 312, 503, 612; **71**:81.
13. Kidd, F., and C. West. 1927. A relation between the concentration of oxygen and carbon dioxide in the atmosphere, rate of respiration, and length of storage of apples. *Food Invest. Board Rep. London for 1925, 1926,* pp. 41–42.
14. Khare, B. P. 1972. Insect pests of stored grain and their control in Uttar Pradesh. *G. B. Pant Univ. Agric. Technol. Res. Bull. No. 5.* Dept. Entomol., Coll. Agric. Exp. Stn., Patnagar, U.P., 153p.
15. Lever, R. J. A. W. 1971. Losses in rice and coconut due to insect pests. *World Crops* **23**:66–67.
16. Moore, Thomas. 1803. *Essay on the Most Eligible Construction of Ice-Houses; Also, a Description of the Newly Invented Machine Called the Refrigerator.* Bonsal and Niles, Baltimore, 28p.

17. Mukai, M. K. 1987. Postharvest research in a developing country: A view from Brazil. *HortScience* **22:**7–9.
18. Newberry, P. E. 1893. *Beni Hasan. Part I*. Kegan, Paul Trench, Triibner and Co., London, 85p. and plates.
19. NSF. 1978. *Report of the Steering Committee for Study on Postharvest Food Losses in Developing Countries*. National Research Council, National Science Foundation, Washington, D.C., 206p.
20. Perrot, J. 1966. Le gisement natufiende Mallaha (Èynan), Israël. *L'Anthropologie* **70:**437–484.
21. Pierson, T. R., J. W. Allen, and E. W. McLaughlin. 1982. Produce losses. *Mich. State Univ., Agric. Econ. Rep. 422,* 48p.
22. Puleston, D. E. 1971. An experimental approach to the function of classic Maya chultuns. *Am. Antiq.* **36:**322–335.
23. Rackham, H. 1938. *Pliny Natural History,* vol. 5. Harvard University Press, Cambridge, Mass., 544p.
24. Rhoades, R., M. Benavides, J. Recharte, E. Schmidt, and R. Booth. 1988. Traditional potato storage in Peru: Farmer's knowledge and practices. International Potato Center, Lima, Peru. *Potatoes Food Syst. Res. Ser. Rep.* **4:**1–67p.
25. Solecki, R. S. 1964. "Shanidar" Cave, a late Pleistocene site in Northern Iraq. *6th Int. Congr. Quaternary Rep.* **4:**413–423.
26. Solecki, R. S. 1964. "Zawi Chemi Shanidar," a post-Pleistocene village site in northern Iraq. *6th Int. Congr. Quaternary Rep.* **4:**405–412.
27. Taylor, W. A. 1900. The influence of refrigeration on the fruit industry. *USDA Yearb. Agric. 1900,* pp. 561–580.
28. Taylor, W. W. 1973. Storage and the Neolithic Revolution. In *Estudios Dedicados al Prof. Dr. Luis Pericot*, University of Barcelona, Spain, pp. 194–197.
29. USDA. 1965. Loss in agriculture. *USDA Agric. Res. Ser., Agric. Handb. No. 291,* 120p.
30. Varro, M. Terentius. 1800. *The Three Books of M. Torentina Varro Concerning Agriculture*. T. Owen (trans.). Oxford, 257p.
31. Wilson, H. R., A. Singh, O. S. Bindra, and T. R. Evertt. 1970. *Rural Wheat Storage in Ludhiana District, Punjab*. Ford Foundation, New Delhi, 32p.

ADDITIONAL READINGS

Bellingham, C. 1604. *The Frviterers Secrets*. R. B., London, 28p.
Birx, H. J. 1988. *Human Evolution*. Charles C Thomas, Springfield, Ill., 359p.
Bynum, W. F., E. J. Browne, and R. Porter (eds.). 1981. *Dictionary of the History of Science*. Macmillan, London, 494p.
Campbell, B. G. 1988. *Humankind Emerging*. Scott, Foresman, and Co., Boston, 552p.
Cohen, M. N. 1977. *The Food Crisis in Prehistory: Overpopulation and the Origins of Agriculture*. Yale University Press, New Haven, Conn., 341p.
Cooper, E. 1972. *A History of Pottery*. Longman, London, 276p.
FAO. 1977. *Analysis of FAO Survey of Postharvest Crop Losses in Developing Countries* (AGPP/MISC/27). Food an Agricultural Organization of the United Nations, Rome.
FAO. 1981. Food loss prevention in perishable crops. *FAO Agric. Serv. Bull. 43,* 72p.
Gonen, R. 1973. *Ancient Pottery*. Cassell, London, 95p.
Harlan, J. R. 1976. *Origins of African Plant Domestication*. Aldine, Chicago, 498p.
Hellemans, A., and B. Bunch. 1988. *The Timetables of Science*. Simon and Schuster, New York, 656p.

Ho, P.-T. 1975. *The Cradle of the East: An Inquiry into the Indigenous Origins of Technique and Ideas of Neolithic and Early Historic China, 5,000–1,000* BC. University of Chicago Press, Chicago, 400p.

NSF. 1978. *Report of the Steering Committee for Study on Postharvest Food Losses in Developing Countries*. National Research Council, National Academy of Sciences. Washington, D.C., 206p.

Partridge, M. 1974. *Farm Tools Through the Ages*. New York Graphic Society, Boston, 240p.

Reed, C. A. 1977. *Origins of Agriculture*. Aldine, Chicago, 1,013p.

Sankalia, H. D. 1964. *Stone Age Tools: Their Techniques, Names and Probable Functions*. Poona, India, 114p.

Spier, R. F. G. 1970. *From the Hand of Man, Primitive and Preindustrial Technologies*. Houghton Mifflin Co., Boston, 159p.

Varro, M. Terentius. 1800. *The Three Books of M. Torentina Varro Concerning Agriculture*, T. Owen (trans.). The University Press, Oxford, England, 257p.

Walters, H. B. 1905. *History of Ancient Pottery*, 2 vols. John Murray, London.

White, K. D. 1967. *Agricultural Implements of the Roman World*. Cambridge University Press, Cambridge, Mass., 232p.

2

NATURE AND STRUCTURE OF HARVESTED PRODUCTS

How plants and plant products are handled after harvest and the changes that occur within them during the postharvest period are strongly influenced by their basic structure. For example, plant parts that function in nature as storage organs behave after harvest in many ways that are distinctively different than structurally dissimilar parts such as leaves or flowers. As a consequence, it is desirable to be familiar with the structure not only of the general product but with the tissues and cells that comprise it. This chapter describes general morphological groupings of harvested products, the tissues that are aggregated to form these products, and the structure of the cells that make up these tissues.

One extremely important concept of plant morphology is that structure, whether at the organ, tissue, or the cellular or subcellular level, is not fixed but is in a state of transition. Changes in structure are especially important during the postharvest period. A structural unit once formed is eventually destined to be degraded and recycled back to carbon dioxide. Marked changes in structure occur with eventual use of the product (e.g., consumption of a food product) or loss due to pathogen invasion. Although less dramatic, distinct and important structural alterations occur in plant products during storage and marketing. Formation of fiber cells, loss of epicuticular waxes during handling, and chloroplast degradation or transformation are but a few of many changes that can occur. At the cellular and subcellular level, very pronounced changes occur and these intensify as the product approaches senescence, or as in the case with seeds and intact plants, during the beginning of renewed growth. It is of paramount importance, therefore, that we view organs, tissues, cells, and subcellular bodies of harvested products as structural units in a state of change. Decreasing this rate of change is in most cases an essential requisite for successful extended storage of the product.

GROUPING OF HARVESTED PRODUCTS BASED ON MORPHOLOGY

The range of plant products that are harvested and used by mankind is vast and the characteristic responses after harvest are so varied that some system of grouping or classification is necessary. Our interest in postharvest physiology stems from the practical requirement of getting perishable plant products from producers to consumers and maintaining the desired supply and quality throughout the year. The classification most useful for this purpose is one that brings together products with similar environmental requirements after harvest for maintenance of quality or those that are susceptible to chilling or other types of injury. Although this system is extremely useful to the practical operator, it does not help us to understand the nature of the harvested produce. Likewise, classification according to the product's use, that is, fruits, vegetables, florist items, nursery products, agricultural seed products, does not aid in understanding those products and there is often considerable overlap of the categories. For example, foliage is not only used as a florist item but also as vegetative cuttings and edible vegetables. Their characteristic postharvest behavior is very similar, irrespective of the end use. Fruits such as tomato are used as vegetables but still ripen like other fruits. Understanding the nature of the product, rather than for what it is to be used, is more useful and allows prediction of the likely response of harvested products.

Botanical classification into family, genus, and species is of only limited value with regard to postharvest handling.* Whole plants or any of the separate organs from that plant may have vastly different characteristics and behavior, yet all are from the same botanical specimen.

A classification according to plant part and stage of development is a satisfactory alternative. It allows an understanding of the nature of the harvested product, and therefore a means of predicting behavior. Both physical and physiological characteristics are indicated. It should be recognized that classifications are systems devised by scientists and applied to the subject of their choice to serve a particular purpose. In this case, the intent is to classify harvested plant products in order to reduce their number for easy management and to be able to predict the behavior and handling requirements of products for which there is no specific information available. Because of the very nature of biological systems and the inherent variation both within and between species, as well as the wide range of uses of plant products, no classification system can be perfect. For example, the storage temperature requirements for sweetpotato (*Ipomoea batatas,* (L.) Lam.) and beetroot (*Beta vulgaris,* L.), both root crops, differ substantially. As a consequence, scientists must be prepared to allow for the exceptions and to refine or alter the classification for specific purposes.

Classifying according to plant part and knowing the characteristics of those parts makes it easier to understand the current and potential processes that may be operative during the postharvest period.

---

* Botanical classification can, however, be useful in tracing the evolution of certain traits.

Intact Plants

Whole plants are considered to be harvested when they are removed from the production environment. They retain both the shoot and root system that may or may not still be associated with soil or growing media. These whole plants should have little or no harvest injury and maximum capacity to continue or to recommence growth and development. They commonly have access to all the normal requirements for plant growth, that is, water, mineral nutrients, oxygen, carbon dioxide, and light (energy), and are susceptible to all the influences on plant growth: physical, chemical, and biological.

Germinated seed such as bean or alfalfa sprouts (*Medicago sativa,* L.) have the parent seed as an energy source and do not require light. They have a very high metabolic rate and release a substantial amount of heat. In the juvenile shoot there has been little development of a cuticle and the roots are exposed. The seedlings are therefore very susceptible to water loss and to mechanical damage.

Bare root seedlings and rooted cuttings are physiologically more developed than germinated seeds, but are also very subject to water loss and mechanical damage. These young plants do not have stored reserves, as do germinated seeds, so they are dependent on either continued photosynthesis or minimized metabolic rate for maintenance of quality. More mature whole plants can tolerate harsher conditions as they are generally less brittle, have a better-developed cuticle, and have accumulated some stored reserves. The extreme is seen in dormant woody plants where metabolic rate is suppressed and leaves and active feeder roots are absent, so susceptibility to water loss, mechanical damage, and microbial infection is greatly reduced.

It is easier to maintain the quality of intact plants marketed in containers. The root systems are not exposed and as a consequence are less subject to stress and mechanical damage. The medium around the plant roots provides a reservoir of mineral nutrients and particularly water. However, the medium only increases the volume of water available to the plant. Water will be depleted unless it is periodically replenished or the plant is kept in an environment where loss does not occur. Very often during retail sales, containerized plants are subjected to water stresses of sufficient magnitude to result in quality losses.

Tissue-cultured plants are a special case of containerized intact plants. Due to the special environment under which they are produced, they are particularly delicate, but at the same time, they are relatively well protected within the microenvironment of their containers. Their compact size and controlled environment often enable them to be shipped through normal mail and parcel delivery systems that are not accustomed to handling live material. Special precautions, however, are necessary to maintain orientation, to protect from temperature extremes, and to avoid prolonged delays.

Detached Plant Parts

Plant parts utilized by humans come from virtually every portion of the plant. This fact is perhaps best illustrated by vegetables (fig. 2-1), which range from immature flowers to storage roots.

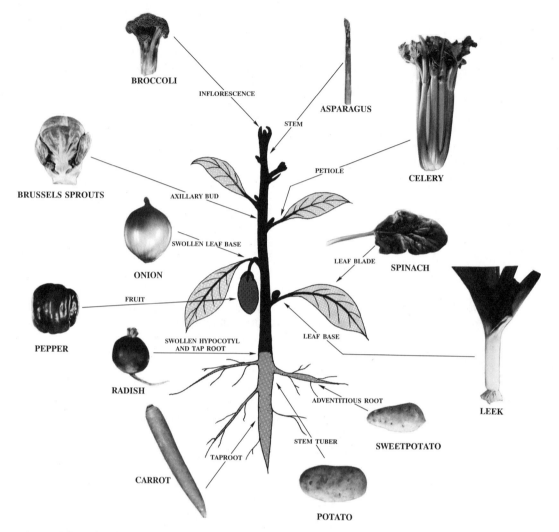

Figure 2-1. Plant parts utilized by man are derived from virtually every portion of the plant. This diagram illustrates examples of various parts from a cross-section of plants used as vegetables.

Aboveground Structures

LEAVES

Leaves, widely consumed as foods, are also utilized for ornamental purposes. In some cases, they may also represent propagules for asexually reproducing plants. Morphologically the leaves of most dicotyledonous plants are comprised of a leaf blade, the thin flattened portion of the leaf, and a petiole that attaches the blade to the stem. Many leafy crops have intermediate to large petioles [e.g., spinach (*Spinacea oleracea*, L.), collards (*Brassica oleracea*, L. Acephala group), Ceylon spinach (*Basella rubra*, L.)], while others have only sessile (having the blade attached directly to the stem) or very much reduced petioles [e.g., Chinese cabbage (*Brassica rapa*, L. Pekinensis group), lettuce (*Lactuca sativa*, L.), cabbage (*Brassica ole-*

racea, L. Capitata group)]. Several species have been developed in which the leaf petiole is the product of interest [e.g., celery (*Apium graveolens* var. *dulce,* (Mill.) Pers.), rhubarb (*Rheum rhaponticum,* L.)]. However, these tend to have significantly different postharvest responses and as a consequence are covered separately.

While attached to the plant, the primary function of leaves is the acquisition of carbon through photosynthesis. Leaves also control, to a large extent, transpiration by the plant, which helps to regulate their temperature. After harvest, these functions are seldom operative. The leaf loses its potential to acquire additional carbon (energy) and is cut off from its supply of transpirational water. Hence the energy required by the cells of the leaf for the maintenance of life processes must now come from recycled carbon found within. Leaves of most species do not act as long-term carbon storage sites. This lack of energy reserves decreases their potential postharvest life expectancy. As a consequence, leaves are stored under conditions that will minimize the rate of utilization of these limited energy reserves.

Water loss from harvested leaves is an additional limitation to long-term storage. With detachment from the plant, the leaves can no longer replenish water lost through transpiration. The leaf responds with closure of the stomata, the primary avenue of water loss. Closure greatly decreases the rate of water loss but does not eliminate it.

The exterior surface of leaves of some species (e.g., cabbage and collards) is covered with a relatively thick waxy cuticle that helps to decrease water loss. The rate of water loss, therefore, is modulated by both the nature of the product and the environmental conditions in which it is stored. Thus, leafy products are usually stored under conditions of high relative humidity and reduced temperature to minimize the loss of water.

Another method that is used to help increase the longevity of leafy products is harvesting the leaves and stem intact. Although the stem may not be consumed, it acts as a reservoir for both water and carbon, helping to increase the life expectancy of the leaves.

PETIOLES

Petioles of dicotyledonous leaves function on intact plants as conduits for the transport of photosynthates from the leaf to sites of utilization and for the transport of water and nutrients from the root system to the leaf. The petiole also provides support and positioning for the leaf within the aerial canopy of the plant. In a few cases, the petiole acts as a site for the storage of photosynthetic carbon.

Several plants have well-developed fleshy petioles that are primary morphological components. Celery, rhubarb, and pak-choi (*Brassica rapa,* L. Chinensis group) are utilized either exclusively or largely for their edible petioles. These petioles tend to have more stored energy reserves than leaves. They also lose water less readily due to a smaller surface-to-volume ratio and fewer disruptions in the continuity of the surface. Thus products such as celery and rhubarb have a considerably greater potential storage duration than leaves. Nevertheless, losses of water and carbon remain of critical importance and storage conditions very similar to those for leaves are utilized for petioles.

STEMS, SHOOTS, AND SPIKES

Postharvest products in which the stem is a primary or essential component can be separated into two general classes: (1) Products such as asparagus (*Asparagus officinalis*, L.), coba (*Zizania latifolia*, Turcz.), sugarcane (*Saccharum officinarum*, L.), lettuce (*Lactuca sativa*, L.), kohlrabi (*B. oleracea*, L. Gongylodes group), tsatsai (*B. juncea* var. *tsatsai*, Makino), and bamboo shoots (*Bambusa* spp. and others) are utilized almost exclusively for their stem tissue even though rudimentary leaves may be present. (2) With many ornamental foliages and floral spikes, the stem represents an essential part of the product although it is generally considered as secondary to the leaves and/or flowers present. Common examples would be gladiolus spikes or any one of many species used for cut foliage. The former class tends to be largely meristematic or composed of very young tissues and is metabolically highly active. These species typically can continue to take up water and elongate if placed on moist pads or in shallow pans of water. Exceptions would be kohlrabi and tsatasi, which are formed at the base of the stem and are more mature and metabolically less active. Products such as ornamental asparagus (*Asparagus asparagoides*, Wight) and floral spikes have developed leaves and/or flowers. These also have relatively high metabolic rates and can continue to take up water if handled properly.

Several stem products (e.g., gladiolus (*Gladiolus* spp.) and flowering ginger (*Alpinia purpurata*, (Vieill.) K. Schum.) spikes or asparagus spears) exhibit strong geotropic responses after harvest. If stored horizontally, the apical portion of the stem will elongate upward, producing a bent product of diminished quality.

Optimum storage conditions vary in this relatively diverse group; however, moist, cool conditions predominate. Holding the cut base of the stem in water is either desirable or essential for many stem crops.

FLOWERS

Flowers are compressed shoots made up of specialized foliar parts that are adapted for reproduction. A significant number of harvested plant products are, in fact, flowers. Flowers represent a diverse group varying in size, structure, longevity, and use. Uses range from aesthetic appeal to articles of food (e.g., cauliflower (*B. oleracea*, L. Botrytis group), broccoli (*B. oleracea*, L. Italica group), lily blossoms (*Lilium* spp.)).

Flowers are made up of young, diverse, metabolically active tissues, typically with little stored carbon. There are distinct limits on potential longevity when detached from the parent plant, even when held under optimum conditions. In fact, while attached to the parent plant, many of the flowers' structural components display only a brief functional existence. Anthesis or flower opening represents a very short period in the overall sequence from flower initiation to seed maturation. Almost invariably, floral products are highly perishable, seldom lasting more than a few weeks after harvest.

From a handling and storage perspective, flowers can be separated into two primary groups: those that are detached from the parent plant at harvest and those that remain attached. Most floral products fall into the former classification. As with other detached plant parts, after harvest they are unable to fix additional carbon through photosynthesis* nor are they able to transport in photosynthate from adjacent leaves on the parent plant. With many floral products, attached stems represent a reserve of carbon and water that can in part be utilized by the flower. In many cases, this reserve greatly extends the potential longevity of the individual flowers. Thus flower crops are typically made up of much more than just the floral tissues. Attached stems and leaves represent important components and these structures often strongly influence the postharvest behavior of the flower.

Some flowers are marketed attached to the parent plant [e.g., potted *Chrysanthemum × morifolium,* Ramat. or azalea (*Rhododendron* spp.)]. The flower or flowers may represent the sole reason for purchase or their presence may simply enhance the attractiveness of the plant, which is the article of primary interest. Both advantages and disadvantages are realized during the postharvest period by having the flowers attached to the parent plant. While the flowers benefit from a continued supply of photosynthate, water, and minerals, storage conditions are generally dictated by what is best for the entire plant and not for the flowers alone. Optimum storage temperature for the plant may not necessarily coincide with the optimum storage temperature of the flower.

Individual flowers are borne on a stem or flower stalk, the structure and arrangement of which varies widely between species. They may be found as a solitary flower or on spikes, umbels, panicles, and other variations. Structurally a complete flower is made up of sepals, petals, stamens, and pistil borne on a receptacle. Imperfect flowers lack one or more of these floral parts.

The sepals, from the Greek word for covering, are leaflike scales that make up the outermost part of the flower. Although leaflike, the sepals are structurally not as complex as leaves. Usually they are green, although on some species they may have the same coloration as the petals. Together the sepals of a single flower are called the calyx.

The petals (the Greek word for flower leaves) are also modified leaves, and as with the sepals, they are of greater structural simplicity than an actual leaf. The petals of the flower, especially those utilized for their ornamental appeal, are generally brightly colored and showy. Collectively the petals and sepals make up the perianth.

Interior to the petals is the pollen-bearing part of the flower, the stamen. While this male floral part is occasionally found singly, usually there are multiple stamen within a flower. Together, the stamen within a single flower make up the androecium. The upper portion of the stamen, which produces the pollen, is the anther. It is supported on a slender stalk called the filament.

The female portion of the flower is the pistil. Structurally it is comprised of three major parts: the stigma, which is the apical tip of the pistil that acts as the receptive surface for the pollen; the style, which is an elongated column of tissue connecting the stigma with the ovary; and the ovary, which is at the base and is the reproductive organ of the flower. Within the ovary are individual ovules attached to the placenta that develop into seeds when

---

* Photosynthesis can occur under the appropriate conditions; however, the rate is exceedingly low.

fertilized. The pistil may contain a single carpel or several carpels fused together. Collectively the carpels make up the gynoecium.

FRUITS

A fruit is a matured ovary plus associated parts; the term includes both the fleshy fruits of commerce such as bananas (*Musa* spp.) and apples (*Malus sylvestris*, Mill.)—products normally associated with the term "fruit"—and dry fruits such as seeds and nuts. Many of the vegetables that people consume are, in fact, botanically fruits (e.g., tomato (*Lycopersicon esculentum*, Mill.), squash (*Cucurbita* spp.), melons, and peas (*Pisum sativum*, L.)). It is evident, therefore, that the term *fruit* encompasses a very broad range in morphological, biochemical, and physiological variation.

Fleshy Fruits

The actual morphological part of fleshy fruits that is consumed varies widely (fig. 2-2). The fleshy portion can be derived from either the pistil or from parts

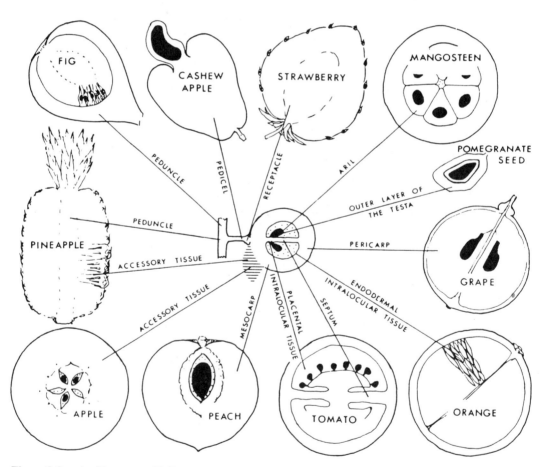

Figure 2-2. A wide range of inflorescence structures can become fleshy and make up the edible portion of fruits. The diagram illustrates the origin of the fleshy portion of a number of common fruits (*from Coombe[4]*).

other than the pistil (accessory parts). The ovary wall, which develops into the pericarp, can be substantial in some fruits. It is subdivided into three regions: the exocarp, mesocarp, and endocarp (from the outside inward). Practically all parts of the floral structure can develop, in various species, into the fleshy part of the fruit. In the peach (*Prunus persica,* Batsch) the edible portion is principally mesocarp tissue. In many species the accessory tissues of the fruit dominate over the carpellary tissue, making up a majority of the fruit's volume. For example, with the apple the edible portion is accessory tissue, whereas in the strawberry (*Fragaria* × *Ananassa,* Duchesne) it is largely receptacle tissue. Because of this diversity in morphological makeup of various fruits, an equally wide range in chemical composition between various types of fruits and in biochemical changes within these tissues after harvest should be anticipated.

Fleshy fruits may be divided into several subclasses based on their morphology. These include the (1) berry, a pulpy fruit from a single pistil with one or more carpels that develop several to many seeds [e.g., tomato, papaya (*Carica papaya,* L.)]; (2) hesperidium, fruits with carpels containing a leathery rind with inner pulp juice sacs or vesicles (e.g., orange, (*Citrus sinensis,* (L.) Osbeck), lime (*C. aurantifolia,* Swingle), grapefruit (*C.* × *paradisi,* Macf.), lemon (*C. limon,* (L.) Burm.f.)); (3) pepo, fruits derived from an inferior ovary that develops from multiple carpels bearing many seeds (e.g., squash, cucumber (*Cucumis sativus,* L.), melons (*Cucumis melo,* L.)); (4) drupe, a simple fruit derived from a single carpel where the mesocarp tissue becomes thick and fleshy (e.g., peach, plum (*Prunus americana,* Marsh.), cherry (*Prunus avium,* L.), olive (*Olea europaea,* L.)); and (5) pome, simple fruit comprised of several carpels, the edible portion of which is made up of accessory tissue (e.g., apple, pear). Thus mesocarp, aril, peduncle, pedicel, receptacle, testa, placental, endodermal, and other tissues may, in respective fruits, represent all or a major portion of the fleshy edible portion of the fruit (fig. 2-2). As a consequence, classification of fruits on a strict morphological basis is not overly useful after harvest in that handling and storage requirements do not follow the same classification.

From a postharvest context, it is more useful to separate fleshy fruits into those that have the potential to ripen after harvest (climacteric fruits) and those that must be ripe when gathered (nonclimacteric fruits). Climacteric fruits such as apples, tomatoes, pears (*Prunus communis,* L.), and bananas typically have much more flexibility in the rate at which they may be marketed. For example, it is possible to store mature, unripe apples for as long as 9 months before eventual ripening. This flexibility greatly alters how we approach the postharvest handling, storage, and marketing of these products. Most nonclimacteric fruits are ripe at harvest and, although there are notable exceptions to their potential storage duration (e.g., dates (*Phoenix dactylifera,* L.) and citrus), many have relatively short storage lives. As a consequence, these fruits tend to be marketed relatively quickly.

Dried Fruits

With dried fruits, the fruit wall is sclerenchymatous and dry at maturity. Agronomic food crops such as wheat, rice, and soybeans or seeds of many other species are examples of dried fruits or ovules of dried fruits. In contrast to the fleshy parenchymatous tissue of succulent fleshy fruits, the fruit wall of these species is dry at maturity. In a number of dried fruit species, the integuments are completely fused to the ovary wall such that the fruit and seed are one entity (e.g., lettuce). Likewise, they may also have accessory parts such as bracts or glumes that remain attached. Many dried fruits are dehiscent with the fruit wall splitting open at maturity. Most dehiscent fruits contain multiple ovules. Thus dehiscence represents a means for dispersal of the individual seeds. Dried fruits, both dehiscent and indehiscent, can be divided into several subclasses based on fruit morphology. The following are subclasses of dehiscent fruits:

1. *Capsule:* The fruit is formed from two or more united carpels, each of which contains many seeds. Common examples of capsules would be the fruits of the poppy (*Papaver* spp.), iris (*Iris* spp.), and Brazil nut (*Bertholletia excelsa*, Humb. & Bonpl.).
2. *Follicle:* The fruit is formed from a single carpel that splits open only along the front suture at maturity. Follicles are found widely in floral crops such as the peony (*Paeonia lactiflora*, Pall.), *Delphinium* spp., and columbine (*Aquilegia* spp.).
3. *Silique:* The fruit is formed from two long halves that are separated longitudinally. Examples would be fruits in the Brassicaceae family such as those of the radish (*Raphanus sativus*, L.) and mustard (*Brassica juncea*, (L.) Czern.).
4. *Legume:* The fruit is formed from a single carpel that splits along two sides when mature. Examples would be soybean (*Glycine max*, (L.) Merr.) and the common bean (*Phaseolus vulgaris*, L.).

Indehiscent fruits typically contain a single ovule and do not split open upon reaching maturity. The following are subclasses of indehiscent fruits:

1. *Achene:* It is a simple, thin-walled fruit containing one seed and in which the seed coat is free, attached to the pericarp at only one point. Examples of species having achenes are the strawberry, sunflower (*Helianthus annuus*, L.), and *Clematis* spp. Although the seedlike portion of the strawberry is botanically a dried fruit, it is treated after harvest as a fleshy fruit in that the fleshy receptacle is the edible portion of interest.
2. *Caryopsis:* It is a one-seeded fruit in which the thin pericarp and the seed coat are adherent in the fruit. Rice (*Oryza sativa*, L.), wheat (*Triticum aestivum*, L.), and barley (*Hordeum vulgare*, L.) are plants with a caryopsis fruit.
3. *Nut:* With nuts, the single seed is enclosed within a thick, hardened, outer pericarp. Examples of species having nuts would be the walnut (*Juglans regia*, L.), pecan (*Carya illinoensis*, (Wang.) C. Koch), and acorn (*Quercus* spp.).
4. *Samara:* They are one- or two-seeded fruits that possess a wing-like appendage formed from the ovary wall. The winged seeds of the ash (*Fraxinus* spp.), maple (*Acer* spp.), and elm (*Ulmus* spp.) are examples of plants with this type of fruit.
5. *Schizocarp:* The fruit is formed from two or more carpels that split upon reaching maturity, yielding one-seeded carpels. The fruits of many of the Umbelliferae (e.g., carrot (*Daucus carota*, L.), parsley (*Petroselinum crispum*, (Mill.) Nym.)) are schizocarps.

Mature ovules or seeds contained within the ripened dried fruit represent the primary article of interest for agriculturists. In most instances the seeds are removed from the fruit during harvest or before storage. Seeds are comprised of an embryo, endosperm, and testa.

The embryo consists of an axis at one end that is the root meristem and at the other are the cotyledon(s) and shoot meristem. The level of structural complexity varies widely, for example, lateral shoot or adventitious root primordia may be present. The endosperm represents the storage tissue of the seed. Not all seeds have endosperms, although those utilized in agriculture almost exclusively have well-developed endosperms. Endosperm structure varies widely, for example, in cell wall thickness and amount and nature of stored materials. The principal storage component in most seeds is starch with lower amounts of protein and lipids. In oil seed crops and some nuts, lipids are the primary storage component. In many seeds [e.g., Poaceae (Gramineae)], the outer layers of cotyledon cells are important sites for the production of the enzymes required for the remobilization of the stored material during germination. The testa or seed coat represents a protective barrier for the seed. Its structure (e.g., thickness), composition, and physical properties vary considerably.

Like fleshy fruits, dried fruits and seeds begin a progressive, irreversible series of deteriorative changes after maturation and harvest. While it is not possible to stop the process of deterioration, we can, through proper storage conditions, greatly decrease the rate. One important postharvest consideration influencing the way in which the product is handled is the eventual use of the dried fruit or individual seed. Seeds that are to be used for reproduction typically have more stringent storage requirements than seeds that are to be processed into foods or other products. Deteriorative processes in seed stocks result in losses of seedling vigor and subsequently in the ability of the seed to germinate. In most texts, seeds that will no longer germinate due to deterioration are considered dead. In many cases, however, the majority of the cells may be alive. While no longer useful for reproduction, these seeds remain, in many cases, useful for processing. It is also possible for seed to lose its functional utility as a raw product for processing (due to postharvest deterioration) without complete loss of the ability to germinate. An example would be when low levels of lipid peroxidation cause rancidity in an oil seed crop destined for human consumption causing the quality to be impaired (e.g., pecans). Postharvest peroxidation of seed lipids, however, normally is thought to be a primary factor in storage deterioration of seed germination potential.[13,20]

In general there are four critical factors that affect the rate of postharvest losses of dried fruits and seeds. These are the nature of the fruit or seed, its moisture content, and the temperature and oxygen concentration of the storage environment. Therefore, proper handling and storage can greatly extend the functional utility of these products.

OTHER STRUCTURES

Mushrooms, members of the Ascomycetes and Basidiomycetes, represent another form of detached aboveground structures. Some, such as truffles, may be formed below-ground. Unlike seed-producing Angiosperms, mushrooms reproduce from spores. The actual mushroom of commerce is the fruiting body of the organism. It is comprised of three distinct parts: the pileus or umbrella-like cap; the lamellae or delicate spore-forming gills at the base of the pileus; and the stipe, the stalk on which the pileus is held (fig. 2-3). The fruiting body may also be thin, ear-shaped, and gelatinous or have various other structural variations.

Mushrooms sold undried as a fresh product have a relatively short storage and marketing potential. After harvest, the fruiting body continues to develop with the pileus, initially in the form of a tight button or closed cap, opening to expose the gills. With spore shed the fruiting body begins to deteriorate with development of off-odors. Mushrooms are also subject to quality losses after harvest due to breakage, bruising, and discoloration. As the pileus opens, the mushroom becomes much more susceptible to mechanical damage.

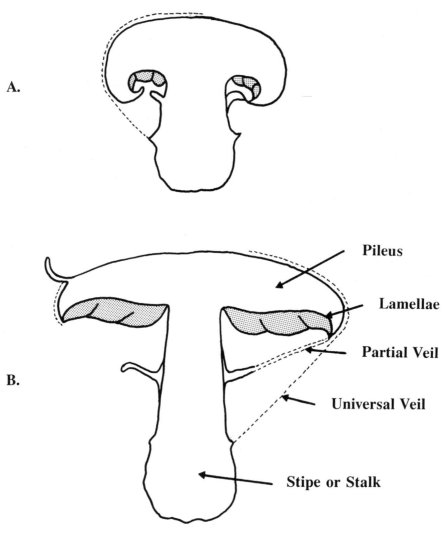

Figure 2-3. Schematic presentation of an immature *(A)* and mature *(B)* carpophore of *Agaricus bitorquis*. Individual parts are the pileus or cap, lamellae or gills, partial veil having separated from the annulus, universal veil, and stipe or stalk *(after Singer[17])*.

Below-Ground Structures

Subterranean storage organs are specialized structures in which products of photosynthesis accumulate and serve to maintain the plant during periods of environmental stress such as winter or drought. These underground structures do not normally contain chlorophyll but derive the energy necessary for a continued low rate of metabolism from their stored reserves. Commonly buds in these storage organs or in parent tissue associated with them are dormant when harvested, and therefore have a suppressed rate of metabolism. While it is desirable for this dormancy to be maintained, it is not always possible. If suitable environmental conditions prevail, active growth commences at the expense of the storage reserves.

ROOTS

Roots are modified to form storage organs in several crops. They are characteristically swollen structures that may contain reserves, primarily of starch and sugars.

The edible radish is partly root and partly hypocotyl tissue. The secondary xylem parenchyma continues to grow and divide and forms the bulk of the radish. There is relatively little stored reserve in the radish; it is mainly cellulose and water. Small amounts of glucose, fructose, and starch may accumulate under different cultural conditions. Similarly, the carrot storage root is actually formed from hypocotyl and taproot tissue. The cortex sloughs off and secondary growth results in a central xylem core surrounded by phloem and pericycle tissue. In all these tissues, there is massive development of parenchyma in which starch is stored. The fleshy root of beet (*Beta vulgaris,* L.) is also of hypocotyl-root origin but in this case secondary development is different. A series of cambia develop outside the primary vascular core, each producing strands of xylem and phloem embedded in parenchyma. This structure results in the concentric ring pattern characteristic of cross sections of beet roots. Starch and sugars accumulate in the parenchyma cells.

The sweetpotato storage organ is an adventitious root in which an anomalous secondary growth occurs. There is a large proportion of parenchyma in the primary xylem and cambia develop in this tissue. These produce xylem, phloem, and many storage parenchyma cells in the tissue that was originally xylem. The normal phloem surrounds this unusual development of the xylem, as does the periderm that originated in the pericycle. During secondary growth of these storage roots the cortex and epidermis do not continue to grow but are split by the expanding tissues of the secondary stele and eventually slough off. Periderm arises in the pericycle during secondary growth and forms a protective tissue to replace the epidermis. Its protective capacity is due to the accumulation of layers of suberin, a fatty acid substance, and wax in the cell walls.

This capacity to form a protective periderm is not only important in the normal development of storage roots but also in wound healing of roots, particularly sweetpotato, and of modified stems such as the potato (*Solanum tuberosum,* L.) tuber. The exposed cells are first sealed with suberin and other fatty materials. High humidity, proper temperature, and adequate aeration are necessary for sealing of the wound, providing an environment that stimulates cell division and formation of a periderm. If excessive moisture is present suberization may be inhibited and callus tissue may be formed instead.

RHIZOMES AND TUBERS

While rhizomes and tubers are underground structures with the superficial appearance of roots, they are anatomically stems. They arise from lateral buds near the base of the main stem and grow predominantly horizontally through the soil. They have nodes and internodes with leaves, sometimes reduced to scale leaves, at the nodes. Buds, often referred to as eyes, form in the axil of leaves. These buds may elongate into shoots and adventitious roots may develop from the stem tissue. Rhizomes may be somewhat enlarged and function as storage organs. In some species, they are used for vegetative propagation of the plant, for example, many of the members of the *Iris* genus. For others, the rhizomes may be utilized for both propagation and consumption as in ginger (*Zingiber officinale*, Roscoe) and lotus (*Nelumbo nucifera*, Gaertn.).

Potato tubers form as swellings at the end of short rhizomes. In potato, the swelling begins by division of the parenchyma cells in the pith followed by division in the cortex and vascular regions. Vascular elements become separated by parenchyma tissue. Starch begins to accumulate in the cortex and later in the deeper tissues of the vascular region and the pith. The epidermis is replaced by a suberized periderm derived from the subepidermal layer. Numerous lenticels are formed by the production of loose masses of cells under the stomates of the original epidermis. Under favorable conditions the lenticels proliferate and rupture the epidermis and are evident as small white dots on the surface of the tuber. Tissues of the tuber retain their capacity to form periderm even after harvest, provided satisfactory humidity and temperature conditions are maintained. This capacity to heal periderm that is damaged during harvest and handling or even to heal cuts across the tuber in preparation of sets (seed pieces) for propagation is a very important characteristic. Without the formation of this protective layer, the damaged tuber would be subject to excessive water loss and microbial infection.

Yams (*Dioscorea* spp.) are another very important food crop usually regarded as tubers. However, Onwueme[12] points out some important differences, particularly the lack of scale leaves, nodes, or buds. The so-called tubers do not arise from stem tissue but rather as outgrowths of the hypocotyl. While the tubers do not have preformed buds they are able to form buds from a layer of meristematic tissue just below the surface. Once dormancy is broken these buds grow and can be the cause of major postharvest losses. Tubers are covered by several layers of cork (phellem) that arise from successive cork cambia (phellogen). Periderm formation may continue after harvest and is essential for wound healing.

BULBS

Bulbs, as occur in onion (*Allium cepa*, L.), tulip (*Tulipa* spp.), and hyacinth (*Muscari* spp.) are underground buds in which the stem is reduced to a plate with very short internodes and the sheathing leaf bases are swollen to form a storage organ. At horticultural maturity, the above-ground parts of leaves shrivel and die but the swollen leaf bases remain alive. There is no anatomical distinction between these two parts and therefore no formation of a protective layer such as occurs in abscission zones or periderms. Therefore, adequate curing or drying of the neck and outer leaf bases is needed if quality is to be maintained during the postharvest period.

Bulbs, as storage organs, have a dormant period during which their metabolic rate is low and keeping quality is good. However, they contain intact shoots and are therefore capable of growth if dormancy is broken and suitable environmental conditions exist.

CORMS

Corms are short, thickened, underground stems. When dormancy is broken, the terminal bud or lateral buds grow into new plants with adventitious roots arising from the base of parent corms and from the new shoots. The dry leaf bases and hollow flower stem of *Gladiolus* spp. tend to remain more or less attached to the corm and afford protection from water loss and mechanical damage.

The taro (*Colocasia esculenta*, (L.) Schott), on the other hand, does not retain its leaf bases but has a well-developed periderm on the corm. Secondary corms, or cormels, arise from lateral buds on the corm. The Chinese water chestnut (*Eleocharis dulcis*, Trin.) forms corms at the end of slender rhizomes. These corms have a scaly, brown periderm.

NON-STORAGE ORGANS

Products included in this category are primarily used as propagation material. To be effective as such, they must contain some stored reserves to supply energy and nutrients for the early development of the new plant.

Root Cuttings

While a number of species can be propagated from root cuttings (e.g., Peony roots, *Paeonia* spp.), there is relatively little use of this technique as other propagation systems are generally less labor intensive and are quicker.

Root cuttings are commonly taken during the dormant stage so they do not display a high level of metabolic activity. Typically, root cuttings are thickened secondary roots with a well-developed periderm. There are often no obvious anatomical features that can be used to determine the orientation of root cuttings but polarity is maintained with shoots being produced from the proximal end and roots from the distal end. It is, therefore, important to use a handling system that identifies the ends. One such system is to cut the proximal end straight and the distal end slanted.

Root cuttings of bramble fruits (*Rubus* spp.) are occasionally marketed commercially.

Crowns

Crown is a general term referring to the junction of shoot and root systems of a plant. Those marketed in the horticultural trade are most commonly herbaceous perennials and consist of numerous branches from which adventitious roots arise. Crowns are divided by cutting, preferably during the dormant period, and are marketed before or soon after the onset of active growth. Some plants, such as asparagus, develop thick storage roots as well.

Crowns with young shoots and active roots are susceptible to mechanical damage and the cut surfaces are subject to microbial infection and water loss. Stored reserves are present in the mature stems and roots and the metabolic rate is comparatively low.

TISSUE TYPES

The structure of harvested products can be further subdivided into five general tissue types of which the products are composed. These include the dermal, ground, vascular, support, and meristematic tissues. Dermal tissues form the interface between the harvested product and its external environment and as a consequence are extremely important after harvest. They often have a dominant influence on gas exchange (water vapor, oxygen, and carbon dioxide) and the resistance of the product to physical and pathological damage during handling and storage. With some postharvest products, the dermal tissues are also important components of the products' visual appeal. The luster imparted by the epicuticular waxes on an apple or the presence of desirable pigmentation in the surface cells are examples.

Ground tissues make up the bulk of many edible products, especially fleshy products such as roots, tubers, seeds, and many fruits. These often act as storage sites for carbon; however, they have many other functions. Maintenance of their characteristic texture, composition, flavor, and other properties after harvest is especially important. Vascular tissues are responsible for movement of water, minerals, and organic compounds throughout the plant. Their function is especially critical during growth; however, in detached plant parts their primary role is diminished in importance. With some products, for example, carrots, the vascular tissues make up a significant portion of the edible product. Support tissues such as collenchyma and sclerenchyma give structural support to the plant. Collenchyma tissues are found widely within plants, whereas sclerenchyma tissues, being much more lignified and ridged, are less common in most edible products. Lastly, meristematic tissues are comprised of cells that have the capacity for active cell division. As a consequence, they are of particular importance during growth.

Dermal Tissues

The dermal tissues cover the outer surface of the plant or plant part and constitute its interface with the surrounding environment. The two principal types of dermal tissues are the epidermis and the periderm.

Epidermis

Epidermal cells covering the surface of the plant are quite variable in size, shape, and function. Most are relatively unspecialized, although scattered throughout these unspecialized cells are often highly specialized cells or groups of cells such as stomates, trichomes, nectaries, hydathodes, and various other glands.

The epidermis is typically only one layer of cells in thickness, although with some species, parts of the plant may have a multilayered epidermis. Epidermal cells are typically tabular in shape and vary in thickness with species and location on the plant. The cells are alive, metabolically active, and may contain specialized organs such as chloroplasts and pigments within the vacuoles. Complex fatty substances called cutin are found intermeshed within the outer cell wall and on the outer surface of the epidermis making up the cuticle (fig. 2-4). The cuticle is found on all aboveground surfaces and often on significant portions of the root system. The surface of the waxy cuticle may be seen as flat plates or as rods or filaments protruding outward from the surface (fig. 2-5).

STOMATA

Stomata are specialized openings in the epidermis that facilitate the bidirectional exchange of gases (water vapor, carbon dioxide, oxygen, etc.). Although varying widely in appearance (fig. 2-6), stomata are comprised of

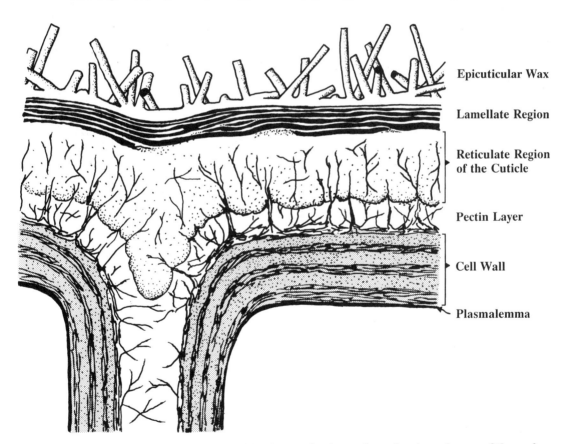

Epicuticular Wax

Lamellate Region

Reticulate Region of the Cuticle

Pectin Layer

Cell Wall

Plasmalemma

Figure 2-4. Representation of a plant cuticle. Successive layers from the plasmalemma of the surface cell outward are the cell wall and the pectin layer of the middle lamella; the reticulate region of the cuticle in which cutin and wax are transversed by cellulose fibrils; the lamellate region where there are separate layers of cutin and wax; and the epicuticular wax *(after Juniper and Jeffree[8])*.

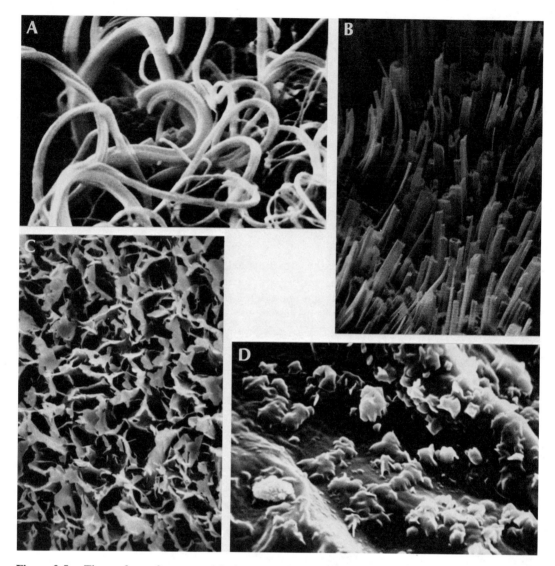

Figure 2-5. The surface of many aerial plant parts is covered with a layer of epicuticular wax. The deposition and continuity of the wax varies widely from long coiled cylindrical rods *(A)*, to long tubules *(B)*, to fringed plates *(C)*, to irregularly shaped plates *(D)* *(from Baker[1])*.

two specialized cells, called the guard cells, which through changes in their internal pressure alter the size of the stomatal opening. Below the guard cells is the substomatal chamber (fig. 2-7).

Stomata are found on most aerial portions of the plant; however, they are most abundant on leaves. The number of stomata per unit surface area of leaf varies widely among species, cultivar, and environmental conditions under which the plants are grown. Typically the lower surface of the leaf (fig. 2-7) has the greatest number of stomata; in some cases, they are found only on the lower surfaces.

Stomatal aperture is altered by light, carbon dioxide concentration, and plant water status. The mechanics of opening and closing of the stomatal aperture appear to be largely controlled by the movement of potassium ions

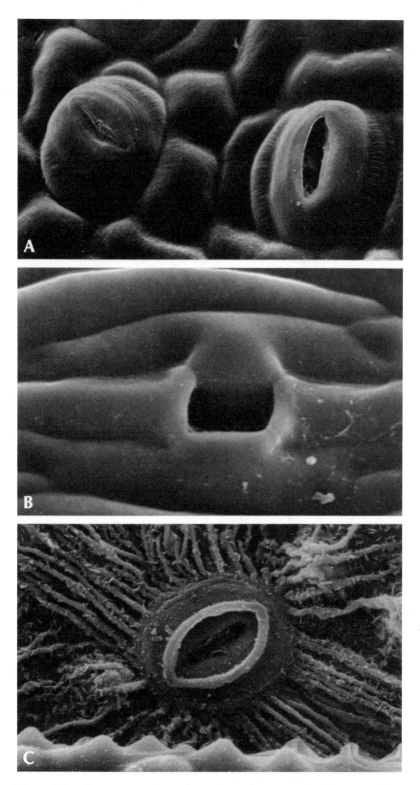

Figure 2-6. Stomata seen from the surface of *(A) Liquidambar styraciflua,* L.; *(B) Pinus caribaea,* Morelet. × *P. palastris,* Mill. hybrid; and *(C) Cornus florida,* L. *(Photographs courtesy of H. Y. Wetzstein.)*

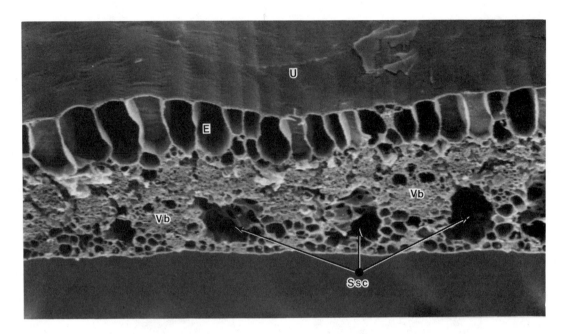

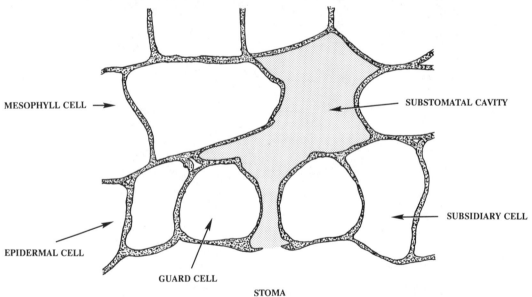

MESOPHYLL CELL →

SUBSTOMATAL CAVITY

SUBSIDIARY CELL

EPIDERMAL CELL

GUARD CELL

STOMA

Figure 2-7. *Top:* An electron micrograph of a *Cyperus* spp. leaf displaying stomata with large substomatal cavities (Ssc) on the lower surface (U, upper surface of the leaf; E, epidermal cell; Vb, vascular bundle). *Bottom:* Representation of a lemon leaf stomate in cross section. *(Electron micrograph courtesy of H. Y. Wetzstein.)*

between the guard cells and their adjacent neighboring cells. When the potassium ion concentration in the guard cells is high, the stomate is open; when it is low, closure occurs.

Stomatal closure represents a means of conserving water within the tissue. When plant parts are severed at harvest from the parent plant, the supply of water from the root system is eliminated and the stomata typically decrease their aperture markedly. During the postharvest period, stomata also represent potential sites for entry of some pathogenic fungi. These fungi bypass the plants' surface defense mechanisms (chiefly the cuticle and epicuticular waxes) and gain rapid entry into the interior.

TRICHOMES

Trichomes represent another type of specialized epidermal cells that may be found on virtually every part of the plant. These extend outward from the surface, greatly expanding the surface area (fig. 2-8). As a class, trichomes are highly variable in structure, ranging from glandular to nonglandular, single-celled to multicelled. The four primary groups include hairs on aerial plant parts, scales, water vesicles, and root hairs. The surface fuzz on peach fruits is an example of trichomes.

Root hairs greatly expand the water and nutrient-absorbing surface of the root system as well as anchoring the elongating root from which they are borne. With aerial trichomes, their role is less well defined. In some species, they represent one means in which the plant combats insect predation. The specialized hooked trichomes of some species trap insects, thus providing a

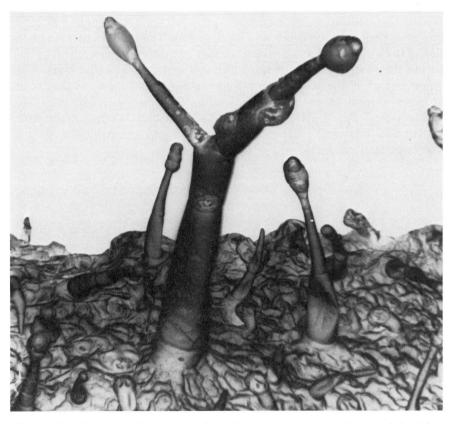

Figure 2-8. Surface trichomes are found in various shapes, sizes, and densities, depending on species and position on the plant. The electron micrograph displays large trichomes on the adaxial surface of a *Nicotiana tabacum,* L. leaf. *(Electron micrograph courtesy of R. F. Severson.)*

physical barrier (fig. 2-9). Trichomes may contain a number of chemicals that attract or repel specific insects. In addition, trichomes have been implicated as sites for the sequestering and/or secreting of certain chemicals. Excess salts taken up by some species are removed via the trichomes. Trichomes are also known to exert a pronounced effect on the boundary layer of air around the organ affecting the exchange of gases.

During the postharvest period, trichomes broken during harvesting or handling provide primary entry sites for pathogens. Operations such as de-fuzzing peaches (trichome removal), as a consequence, can greatly decrease the life expectancy of the product if appropriate treatments are not utilized.

NECTARIES

Nectaries are multicellular surface glands found on flowers and other aerial plant parts that secrete sugars and certain other organic compounds (fig. 2-10). Their position on flowers varies with species. They may be found on the petals, sepals, stamen, ovaries, or receptacle. Sugars are one of the primary groups of organic compounds secreted; their presence facilitates pollination by insects.

HYDATHODES

Hydathodes are much more complex than just modified epidermal cells; however, their function is in many ways similar. They represent a modification of both the vascular and ground tissues along the margins of leaves that permit the passive release of water through surface pores (fig. 2-11). Guttation, which is the discharge of water in the liquid state, causes small droplets to form on the leaves and occurs through the hydathodes. The surface pore or opening is a stomate that is not capable of altering its aperture. Hydathodes represent potential sites for water loss and pathogen entry in harvested leafy products.

Periderm

On plant parts such as roots and stems that increase in thickness due to secondary growth, the epidermis is replaced by a protective tissue called the periderm. Periderm is also formed in response to wounding in most species. When roots are damaged during harvesting, wound periderm forms over the wound surfaces, decreasing the risk of pathogen entry.

The periderm is composed of three tissue types: (1) the phellogen or cork cambium from which the other cells of the periderm arise; (2) the phellem or cork, which is the protective tissue or the exterior surface; and (3) the phelloderm, which is tissue found interior to the cambium (fig. 2-12). The phellem or cork cells are not alive when mature and typically have suberized cell walls. They are tightly arranged, having virtually no intercellular space between adjacent cells, and are found in layers of varying cell number and thickness.

Some harvested products such as sweetpotato roots or white potato tubers must be held under conditions favorable for wound periderm formation prior to storage. Warm temperature and high relative humidity (i.e., 7 days at 29°C,

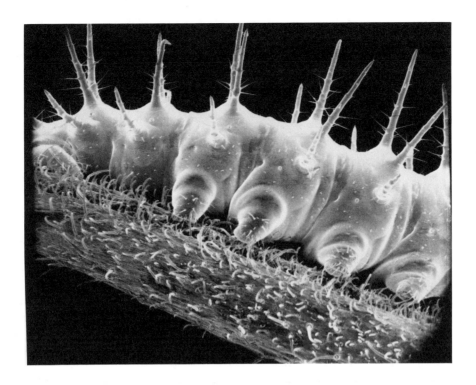

Figure 2-9. Trichomes may in some species function as defense structures, discouraging insect herbivores. The photographs show the trapping of second *(top)* and third *(bottom)* instar larvae of *Heliconius melpomene* by the hooked trichomes on the leaves and petioles of *Passiflora adenopoda*. Damage to trichomes during or after harvest, however, provides entry sites for postharvest pathogens *(from Gilbert[7]).*

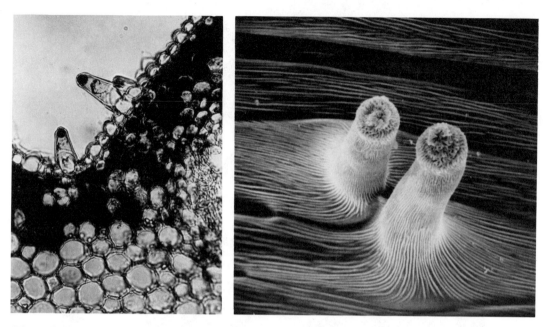

Figure 2-10. Electron micrographs of nectaries from the flower of *Tropaeolum majus,* L. On some species nectaries may be found on other organs (e.g., leaves, bracts, sepals). Their postharvest importance lies largely in their potential role as sites for the entry of pathogens. *(Electron micrographs by Rachmilevitz and Fahn.[14])*

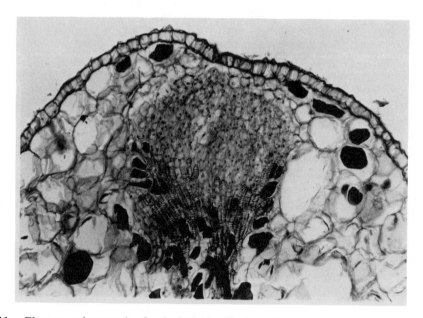

Figure 2-11. Electron micrograph of a hydathode. Hydathodes represent one of several types of specialized secretory structures. They secrete water brought to the leaf periphery by tracheids. The water then moves through an area of loosely packed parenchyma cells called the epithem, exiting through modified stomata (pores), no longer capable of closing. Secretion of water, called guttation, results in the formation of droplets along the border of the leaf. *(Electron micrograph courtesy of A. Fahn.[6])*

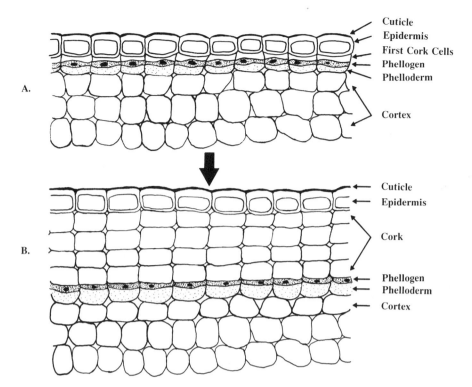

Figure 2-12. When plant parts increase in thickness due to secondary growth, the epidermis is replaced by periderm *(A)* composed of the phellogen, the phellen or cork, and the phelloderm. The phellogen divides *(B)*, forming successive layers of cork cells that act as a protective barrier *(redrawn from Kramer and Kozlowski[9])*.

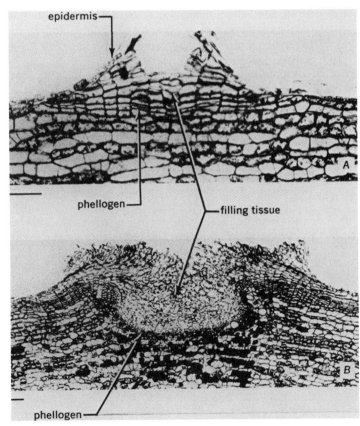

Figure 2-13. Cross sections of young *(top)* and old *(bottom)* lenticels from avocado *(Persea americana,* Mill.) stems. Note the phellogen at the base of the lenticels and the extensive filling of the older lenticel *(bottom) (from Esau[5])*.

90–95% RH for sweetpotato roots) favor rapid wound periderm formation, thus decreasing water loss and pathogen invasion during storage.

Interspersed on the periderm of many species are lenticels comprised of groups of loosely packed cells having substantial intercellular space (fig. 2-13). These appear to be present to facilitate the diffusion of gases into and out of the plant. Unlike stomata, lenticels are not capable of closure and as a consequence their presence enhances the potential for water loss from the product after harvest.

Lenticels range in size from extremely small to up to 1 cm in diameter depending on the species and location on the plant. They are often seen as groups of cells with a more vertical orientation, protruding above the surface of the periderm (fig. 2-13).

Ground Tissue

Parenchyma cells form the ground tissue of most postharvest products. In fleshy fruits and roots and in seeds they act as storage sites for carbohydrates, lipids, or proteins and make up the bulk of the edible portion. In leaf tissue, parenchyma cells have numerous chloroplasts and have a photosynthetic function. Parenchyma cells may also act as secretory cells and can resume meristematic activity in response to wounding.

Parenchyma cells have numerous sides or facets and are highly variable in shape. The number of sides ranges from approximately 9 to 20 or more. Within a single mass of relatively homogenous parenchyma cells, both the number of sides and the actual size of the individual cells often varies. In fruit, roots, and tubers there is considerable intercellular space, whereas in seeds the parenchyma cells are much more compacted. In some aquatic species, the parenchyma cells, called aerenchyma, are very loosely packed, facilitating diffusion of gases.

The characteristics of individual parenchyma cells are to a large extent dependent on the function of the tissue and its composition. Photosynthetic parenchyma in the mesophyll of leaves are often referred to as chlorenchyma due to the abundance of chlorophyll. Storage parenchyma cells exhibit characteristics that are in part dependent on the organic compounds sequestered within their specialized plastids. Parenchyma cells of tubers, roots, and some fruits contain amyloplasts that store starch. The parenchyma cells in flowers contain chromoplasts with various colorful pigments.

Parenchyma cells typically have relatively thin primary cell walls although in some seeds these walls may be rather thick. This general lack of structural rigidity of the wall makes the parenchyma cells rapidly lose their shape and textural properties when water is lost. In many postharvest products, the actual percentage of the total water present that must be lost before significant alterations occur in the tissues' physical properties is often relatively small.

Support Tissue

Collenchyma

Collenchyma cells are in many ways similar to parenchyma cells; however, they have thickened cell walls and provide structural support for the plant (fig. 2-14). The cells are strong and flexible with nonlignified cell walls. The walls are primary in nature but considerably thicker than those found surrounding parenchyma cells. In addition, collenchyma cells tend to be more elongated in shape than parenchyma cells. The walls of collenchyma cells are much more pliable than their structural counterparts, sclerenchyma cells (fig. 2-15). In addition, they remain metabolically active and have the ability to degrade much of the wall if induced to resume meristematic activity.

The walls are composed primarily of cellulose, pectins, and hemicelluloses but are not lignified. Thickening of the wall occurs as the cell grows. Mechanical stress, caused by wind, increases the extent to which the walls thicken. In older parts of the plant, collenchyma cells may form secondary cell walls, becoming sclerenchyma.

As a support tissue, collenchyma is found in the aerial parts of the plant and not in the roots. The cells generally are located just below the surface of leaves, petioles, and herbaceous stems. The strands of cells found in the edible petioles of celery are collenchyma cells and vascular tissue.

Sclerenchyma

Sclerenchyma cells lend hardness and structural rigidity to plants and plant parts. They have lignified secondary cell walls formed after the cells have completed expansion (fig. 2-15). At maturity many are nonliving and no longer contain protoplasts. The cells are found throughout the plant but seldom in the extensive homogenous masses in which parenchyma and collenchyma cells are found. Rather sclerenchyma cells tend to be found individually or in small clusters interspersed among other cell types. Their shape is highly variable but is still used as the basis for the two general classes of sclerenchyma. Sclereids tend to be shorter and more compact than fibers, which typically are quite elongated. Sclereids are highly variable in shape; some are compact and more or less regular, whereas others may be highly branched. They are often found in layers or clusters in epidermal, ground, and vascular tissues of stems, leaves, seed, and some fruits. The stone cells in pears are an example of sclereids (fig. 2-16).

Fibers are long sclerenchyma cells (fig. 2-17) varying from quite short to as long as 250 mm in the ramine (*Boehmeria nieva,* (L.) Gaud.) fibers of commerce. They function as support elements in nonelongating plant parts, especially stems and the leaves and fruit of some species. The formation of fiber cells can occur after harvest in some products, decreasing their acceptability, for example, in asparagus and okra.

Vascular Tissue

Vascular tissues provide the conduits for the movement of water and nutrients throughout the plant. Of the tissues found in the plant, vascular tissue is the most complex, being composed of several types of cells. The xylem and phloem are the two general groups of vascular tissues. Within the xylem, water, minerals, and some organic compounds from the root system move upward throughout the plant. Carbohydrates (chiefly sucrose) and to a much lesser extent other organic compounds formed in the leaves or apical meristem are transported both acropetally and basipetally in the phloem.

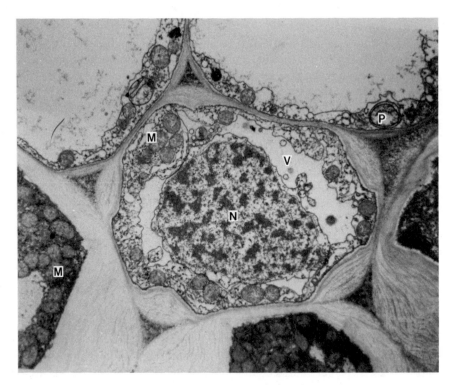

Figure 2-14. Electron micrograph of a collenchyma cell. Particularly noticeable are the much thickened primary cell walls that lend structural support to the tissue and the presence of protoplasm containing a central nucleus (N), vacuole (V), and a number of mitochondria (M). The cell wall is nonlignified, containing large amounts of pectin and water. *(Electron micrograph ×8,000 from Ledbetter and Porter.[10])*

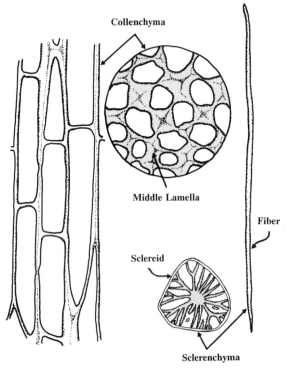

Collenchyma

Middle Lamella

Fiber

Sclereid

Sclerenchyma

Figure 2-15. Supporting cells of the collenchyma (left) and sclerenchyma (right) type are often elongated. Sclereids, one form of sclerenchyma, tend to be shorter and compact. Collenchyma cells have irregularly thickened primary cell walls and differ from sclerenchyma cells in that the latter usually have well-developed secondary cell walls *(after Esau[5])*.

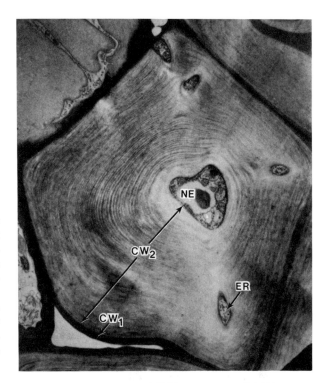

Figure 2-16. Stone cells such as those found in pear (*Pyrus communis,* L.) fruit represent an example of a sclereid. Extensive secondary cell wall (CW_2) surrounds the cytoplasm, which contains mitochondria, plastids, and nuclear envelope (NE). The primary cell wall (CW_1) is found external to the outer layers of the secondary wall (ER, endoplasmic reticulum). (*Electron micrograph ×4,800 courtesy of Ledbetter and Porter.[10]*)

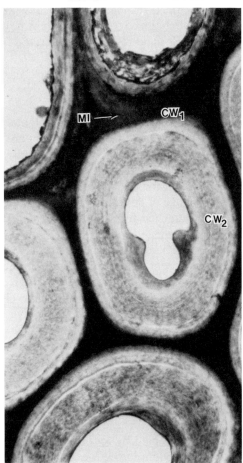

Figure 2-17. Fiber cells seen in cross section. The central fiber has an empty central cavity surrounded by a thick secondary cell wall (CW_2) that is actually composed of three layers. Adjacent to the exterior layer of the secondary cell wall is the primary cell wall (CW_1), followed by the middle lamella (Ml). (*Electron micrograph ×8,000 courtesy of Ledbetter and Porter.[10]*)

Xylem

The xylem functions primarily as a tissue for the conduction of water, but it also has storage and support functions. This diversity in role is in part due to the occurrence of several types of cells making up the xylem. Tracheids and vessel members are the water-conducting cells (fig. 2-18). Tracheids are elongated cells, tapering toward the ends, with secondary cell walls that impart structural support to the plant. At maturity they are nonliving. Water moves through openings, called pits, in the sides of the cells into adjacent pits of neighboring tracheids.

Vessel members, also elongated cells but with flattened porous ends, are joined end to end, forming vessels. These may be as much as a meter or greater in length in some species. Vessels appear to be a more complex form of water-conducting cells found in angiosperms. They appear to have evolved from tracheids during the evolutionary development of the angiosperm.

Associated with tracheids and vessel members are parenchyma cells that act as storage sites. These cells may contain starch, lipids, tannins, or other material and are particularly important in the secondary xylem of woody perennials. Sclereid and fiber cells are also part of the xylem, providing further structural support.

The xylem may be of either primary or secondary origin. Primary xylem is formed during the initial development of the plant, arising from the pro-cambium. Secondary xylem develops during the secondary thickening of stems and roots, and is derived from the vascular cambium.

Phloem

The phloem is the photosynthate-conducting tissue of vascular plants. Carbohydrates formed in the leaves are transported both acropetally to the growing tip of the plant and basipetally toward the root system. The directional allocation depends on the position and strength of competing sinks within the plant for photosynthate, the position of the individual leaf on the plant, its stage of development, time of day, and other factors. In plant parts that are detached from the parent plant at harvest, phloem transport is essentially eliminated.

Phloem, like the xylem, is composed of several types of cells; however, phloem tends to be less sclerified. Photosynthate and other organic compounds are transported through specialized elongated cells called sieve elements (fig. 2-19). While sieve elements are much like the tracheids of the xylem, they differ in being alive at maturity, although they do not contain a nucleus or vacuole. Individual cells are joined end to end, forming sieve tubes. At the ends of each cell is a porous plate called the sieve area or sieve plate that allows, and may in part control, the movement of material from one cell to the next. The remainder of the cell wall is variable in thickness and may also have perforated regions.

Associated with a sieve element are one or more parenchymatous companion cells. These appear to partially control the anucleate sieve element and are joined by an interconnecting membrane system. Companion cells also function during the loading of photosynthate into the phloem and the sub-

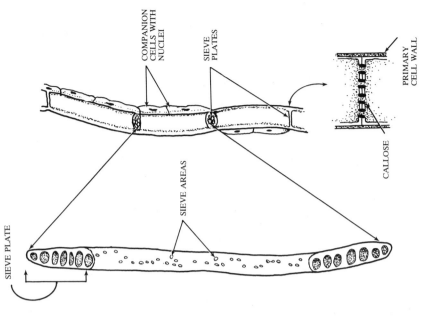

Figure 2-18. Water and minerals are moved throughout the plant in specialized xylem cells: tracheids and vessel elements. Tracheids have tapered end walls and represent a more primitive form of xylem than vessel elements. The geometry of the secondary cell wall varies from spiral to extensive. When there is an extensive secondary wall, numerous pits are generally present through which water moves from tracheid to tracheid. Vessel elements, however, are found end to end with water moving through their perforated end walls (*redrawn from I. P. Ting*[18]).

Figure 2-19. Photosynthates and other organic constituents move through the phloem cells. Illustrated at the left is an elongated sieve-tube element with sieve plates on the end walls formed from groups of sieve areas. Sieve areas are also found on the side walls through which some lateral transport takes place. At the right are three sieve-tube cells attached end to end, forming an elongated conduit.

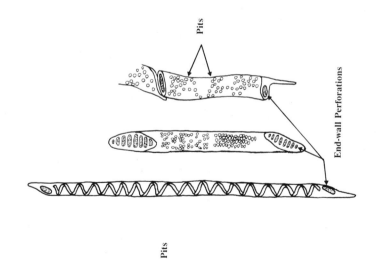

Tracheids

Vessel Elements

Pits

Pits

End-wall Perforations

COMPANION CELLS WITH NUCLEI

SIEVE PLATES

SIEVE AREAS

SIEVE PLATE

PRIMARY CELL WALL

CALLOSE

sequent unloading upon arrival at the sink site. These parenchyma cells may also act as storage sites for an array of organic compounds.

Fibers (sclerenchyma cells) are also associated with the phloem as are sclereids. These provide structural support and rigidity to the system.

Phloem, like the xylem, can be divided into two general classes based on origin. The primary phloem is derived from the procambium and the secondary phloem arises from the vascular cambium during secondary thickening.

Meristematic Tissue

Meristematic tissues are comprised of groups of cells that retain the ability for cell division. Their primary function is in protoplasmic synthesis and the formation of new cells. Meristematic cells are typically small with a thin primary wall and few vacuoles (fig. 2-20). Some of these cells undergo division, forming new cells.

Apical meristematic tissue is found in the growing tip of shoots and roots. It gives rise to the primary growth and structure of the plant. Lateral meristems give rise to the secondary growth of tubers, storage roots, and woody stems. A third type of meristematic tissue is the intercalary meristem found in the growing stems of monocots.

When plant parts are decapitated at harvest, there is generally little activity in the meristematic tissues. Some tissues, however, can and do recycle nutrients and water into these cells, resulting in growth. After extended cold storage, the apical portion of the stem of cabbage will resume growth. With intact plants during storage, conditions are generally selected to minimize growth, and as a consequence, meristematic activity is also repressed. Maintenance of the meristematic tissue in a healthy condition in intact plants, however, is essential. With improper storage, the apical meristems can readily die, decreasing both the quality of the product and the rate at which it recovers upon subsequent planting.

CELLULAR STRUCTURE

Cells are the structural units of living organisms, aggregations of which form tissues. Plant cells vary widely in size, organization, function, and response after harvest. They differ from animal cells by the presence of a rigid cell wall and a large central vacuole. Interior to the cell wall is the plasma membrane or plasmalemma that separates the interior of the cell and its contents, the protoplasm, from the cell's external environment. Much of the cell's energy transfer and synthetic and catabolic reactions occur within the cytoplasm, as do its information storage, processing, and transfer system. Within the cytoplasm of eukaryotic cells (organisms other than bacteria and blue-green algae) are numerous organelles and cytoplasmic structures. These provide a means of compartmentalization of areas within the cell that have specific functions. Organelles such as the nucleus, mitochondria, plastids, microbodies, and dictyosomes and cytoplasmic structures such as microtubules, ribosomes, and the endoplasmic reticulum are found within the cytoplasm (fig. 2-21). While the number of various organelles differs with cell type, age,

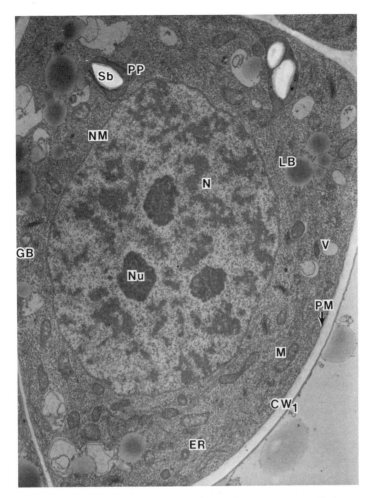

Figure 2-20. Meristematic cells are characteristically small, have only a few small vacuoles (V) and a relatively thin primary cell wall (CW₁). (N, nucleus; Nu, nucleoli; NM, nuclear membrane; PM, plasma membrane; PP, protoplastid; ER, endoplasmic reticulum; LB, lipid body; M, mitochondrion; Sb, starch body; GB, Golgi bodies). *(Electron micrograph courtesy of H. Amerson.)*

and location within the plant, a general example of the relative number per cell is presented in table 2-1.

The response of plant products after harvest is a function of the collective responses of these subcellular organelles. An understanding of the structure and function of cells and their individual components provides the basis for a more thorough understanding of postharvest alterations occurring in the product.

Cell Wall

The cell wall provides the mechanical support for the plant and individual plant parts. Rigidity and structure in a harvested product are due to a large

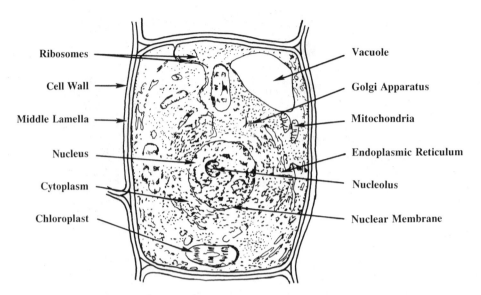

Figure 2-21. Diagram of the structures within an individual plant cell.

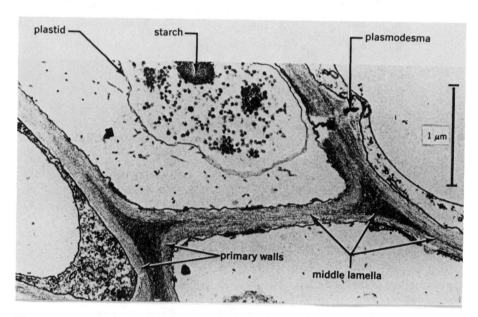

Figure 2-22. The primary cell walls of phloem tissue from *Nelumbo nucifera* Gaertn. Note the middle lamella, found externally to the primary cell wall, and the absence of a secondary cell wall (*from Esau*[5]).

extent to the presence of cell walls. The cell's primary wall is formed from the cytoplasm during cell division. Its largest single component is cellulose that forms the wall's general architecture (fig. 2-22). The primary cell wall also contains hemicellulose, pectin, both structural and nonstructural proteins, water, and other organic and inorganic substances. Epidermal cells, for example, may also contain waxes. Secondary cell walls, formed interior to

Table 2-1. Comparison of Size, Number, and Force Required to Precipitate Cellular Constituents

	Diameter	*No. per Cell*	*Force to Precipitate (g)*
Nucleus	5–20 μm	1	100
Chloroplasts	5–20 μm	20–200	1,000
Mitochondria	1–5 μm	500–2,000	10,000
Ribosomes	250 Å	5–50 × 10^5	105,000
Enzymes	20–100 Å	5–50 × 10^8	

the primary cell wall, provide much greater rigidity to the cell due to the presence of lignin. The secondary cell walls also contain cellulose and hemicelluloses, but very little pectin. Most edible plant products do not have a large number of cells with secondary cell walls.

In the primary cell wall, cellulose molecules, composed of long chains of glucose subunits, occur in orderly strands called micelles (fig. 2-23). Groups of micelles are arranged into microfibrils that make up the cellulose framework of the wall. This cellulose component is quite stable, and as a consequence, there is little alteration after harvest in the cellulose structure of the cell wall.

Pectins are found extensively in the middle lamella, the area between neighboring cell walls, and act as cementing agents holding the adjacent walls together. They are also interspersed in the cellulose microfibrils along with hemicellulose. After harvest or during ripening many pectins are solubilized, which alters the strength of the cell-to-cell structural framework of the tissue, and as a consequence, the texture of the product. Alterations in the pectin composition of the middle lamella are thought to play an important role in the softening process in products such as apple or pear fruit.

Cellular Membranes

The cell is separated from its surrounding environment by a thin membrane (approximately 7.5 nm wide) found just interior to the cell wall. The plasmalemma or plasma membrane, as other cellular membranes, is composed of a viscous lipid bilayer (fig. 2-24). Interspersed on the surface and within the membrane are a number of other molecules (e.g., proteins) that carry out an array of critical functions. For example, some sites on the membrane control the specificity of transport of molecules across the membrane, while others catalyze specific reactions. Thus, the membrane is more than just a barrier; it is a dynamic functional system.

A number of other membranes are found within the cell (fig. 2-25). Organelles such as the nucleus, mitochondria, and plastids are bound by two membranes. The vacuole and microbodies, on the other hand, are enclosed by a single membrane. Also found throughout the cytoplasm is the endoplasmic reticulum. It forms a continuous membrane system that functions as a reactive surface for many biochemical reactions and in compartmentalization of certain compounds. Ribosomes, which carry out the assembly of new proteins, are often attached to the endoplasmic reticulum.

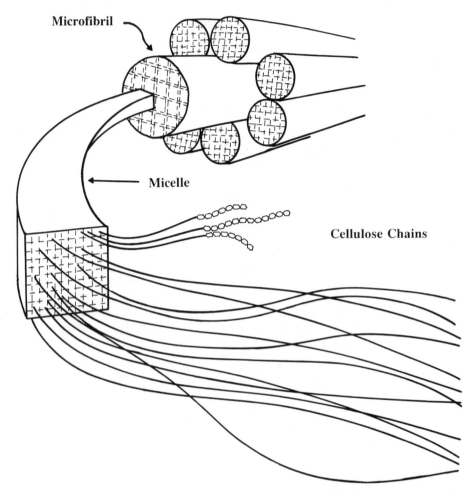

Microfibril

Micelle

Cellulose Chains

Figure 2-23. The microfibril ultrastructure of a primary cell wall. Illustrated are the cellulose chain micelles that make up individual microfibrils *(after Bonner and Galston[3]).*

Cell membranes are composed of lipids, especially phospholipids and glycolipids, and proteins, both lipoproteins and nonconjugated proteins. Their composition imparts a flexible, fluid-like structure under normal ambient conditions. At low temperatures fluidity can be greatly reduced, especially in certain areas of the membrane. This reduction is thought to impair the function of the membrane, leading to several of the physiological disorders found with chilling injury.

Cytoplasm

The cytoplasm is the viscous matrix that envelops all of the more differentiated parts and organelles of the protoplasm. It is the cellular mass interior to and including the plasma membrane and exterior to the vacuolar membrane, the tonoplast. Individual organelles (e.g., nucleus, mitochondria, plastids),

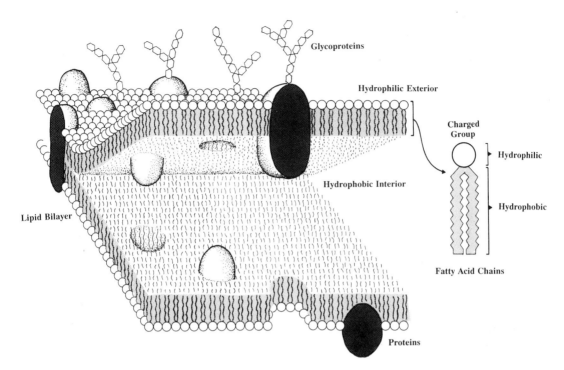

Figure 2-24. Plant membrane bilayer separated as would be visualized using freeze fracture preparation. Present is the lipid bilayer with hydrophilic head groups and hydrophobic tails, proteins, and glycoproteins. The structure of a common membrane phospholipid illustrates the hydrophilic head group and the hydrophobic tail *(after Wolfe[21])*.

while found in the cytoplasm, are not considered part of it. The cytoplasm contains proteins, carbohydrates, amino acids, lipids, nucleic acids, and other substances that are water soluble. Generally, nonorganelle structures such as microtubules, ribosomes, and the endoplasmic reticulum are also considered part of the cytoplasm. Although under certain conditions the cytoplasm can assume a gel-like structure, typically it is viscous and can move within the cell. This movement, called protoplasmic streaming, is quite substantial in some types of cells.

Nucleus

Nearly all living cells of agricultural plant products are uninucleate. Exceptions would be mature sieve tube cells that no longer contain a nucleus and certain specialized cells that are multinucleate. The nucleus represents the primary repository of genetic information within the cell and thus functions in the replication of this information during cell division, and more importantly from a postharvest context, in the control of protein synthesis. Specific enzymes essential for ripening and other postharvest changes are assembled in the cytoplasm. The nucleus is delimited by two porous membranes separated by a perinuclear space (fig. 2-26).

Interior to the nuclear membrane is the nuclear matrix or nucleoplasm that

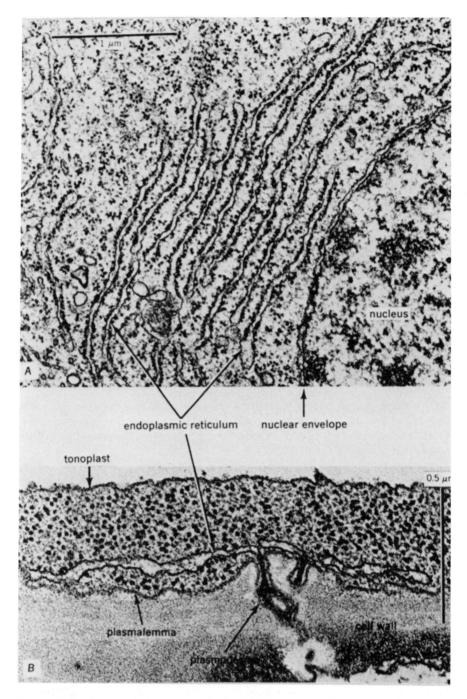

Figure 2-25. Electron micrograph of leaf cells of *Nicotiana tabacum,* L. *(A)* and *Beta vulgaris,* L. *(B)* displaying the endoplasmic reticulum, plasmalemma, nuclear envelope, and tonoplast membrane *(from Esau[5]).*

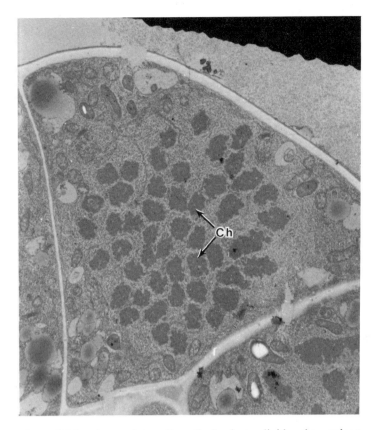

Figure 2-26. As meristematic cells begin to divide, the nuclear membrane disappears and the chromatin becomes organized into chromosomes (Ch, dark masses in the center of the above *Pinus taeda,* L. cell). *(Electron micrograph courtesy of H. Amerson.)*

contains deoxyribonucleic acid (DNA), ribonucleic acid (RNA), nucleic acids, proteins, lipids, and other substances. Also found in the nuclear matrix is a dark spherical body, the nucleolus. It functions in the storage of RNA and the synthesis of ribosomal RNA. The chromosomes are also found here; however, with the exception of during cell division, they are not in a tightly coiled state but are found as a matrix of DNA and protein called chromatin. These associated proteins, a significant number of which are histones, act as repressors and activators of transcription, impart three-dimensional structure, aid in protection, and have various other functions.

Mitochondria

Mitochondria are small, elongate, occasionally spherical structures, 1 μm to 5 μm long. The number of mitochondria per cell varies with cell type, age, and species, ranging from a few hundred to several thousand. Their primary function is in cellular respiration. Within the mitochondria occur the energy conversion processes of the tricarboxylic acid cycle and the electron transport system. As a consequence, the mitochondria are of critical importance in the recycling of stored energy after harvest.

Mitochondria, like the nucleus, are enclosed within a double membrane. The outer membrane is relatively porous while the inner membrane contains numerous tubular folds or extensions called cristae (fig. 2-27). The energy transfer proteins of the electron transport system are situated on the surface of cristae. Found free within the mitochondrial matrix are many of the enzymes of the tricarboxylic cycle (some, e.g., succinic dehydrogenase, are located in the inner membrane). Also located in this aqueous portion of the mitochondria are RNA and DNA that control the synthesis of certain mitochondrial enzymes. The presence of genetic information has led to the belief by some scientists that the mitochondria were originally autonomous organisms that became associated with eukaryotic cells early in the evolution of life.

Increases in the number of mitochondria within a cell appear to begin with an increase in their size, which is followed by division of the mitochondria into two separate organelles.

Plastids

Plant cells contain a distinct group of organelles called plastids. As with the nucleus and mitochondria, these organelles are enclosed within a double membrane. Plastids are found with differing form, size, and function. The three principal types are chloroplasts (plastids containing chlorophyll and are concerned with photosynthesis), chromoplasts (plastids containing other pigments such as carotene or lycopene in red tomato fruit), and the nonpigmented leucoplasts. Many leucoplasts act as storage sites for specific types of carbon compounds within the cell and are prevalent in a wide range of harvested products. Specialized leucoplasts include amyloplasts that contain starch, elaioplasts that contain fats, and aleuroplasts that contain proteins. Much of the research on plastids has focused on chloroplasts; however, the structure, physiology, and biochemistry of these specialized leucoplasts are of considerable interest in postharvest biology in that they represent sites of stored energy.

Plastids arise from proplastids inherited with the cytoplasm in newly formed cells. These proplastids appear to divide and differentiate into the various types of plastids, depending on the nature of the cell in which they exist. Their final form is not static, however, in that there is often considerable interconversion between types of plastids (fig. 2-28). In many cases, pronounced changes are associated with major physiological events such as fruit ripening or senescence.

Contained within a plastid such as a chloroplast are proteins, lipids, starch grains, DNA, RNA, and various organic compounds. There is also a complex inner lamellar membrane system of varying levels of complexity and an embedding matrix called the stroma. In chloroplasts, this membrane system is composed of grana, seen as flattened thylakoids stacked in the shape of a cylinder, and frets, an inner-connecting membrane system transversing the stroma between individual grana (fig. 2-29). In the stroma the photosynthetic

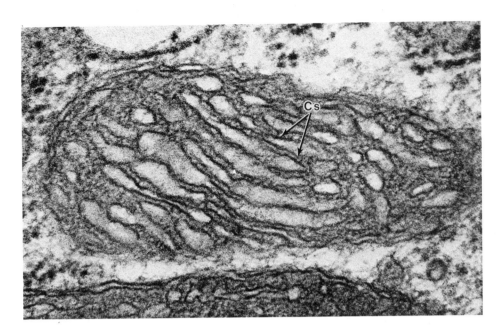

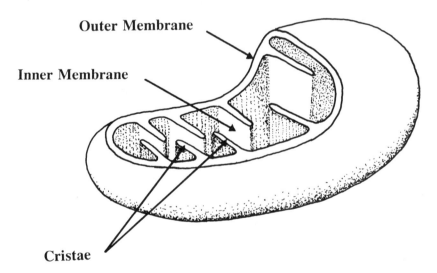

Figure 2-27. Electron micrograph of a mitochondrion showing the outer membrane and cristae (Cs) and diagram of what is thought to be the arrangement of the double membrane forming the outer membrane, inner membrane, and cristae. *(Electron micrograph ×200,000 courtesy of W. W. Thomson; drawing after Ting.[18])*

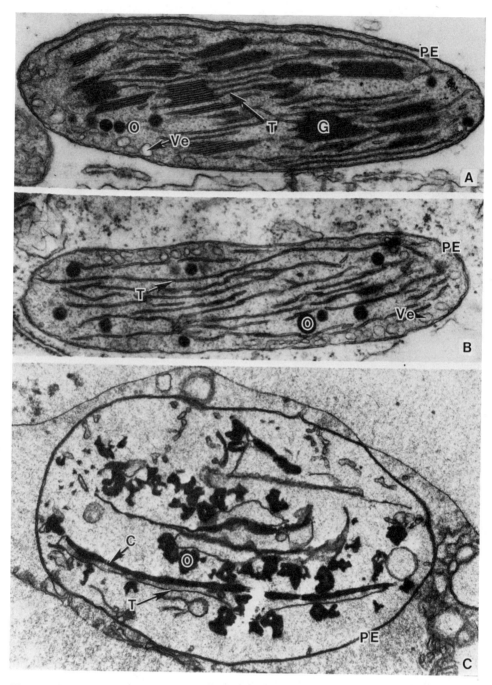

Figure 2-28. The transition of an elongated chloroplast *(A)* from tomato fruit with distinct granal and stomatal thylakoids, to a metamorphosing chloroplast *(B)* with only a few vestiges of granal thylakoids remaining, and finally a chromoplast *(C)* containing lycopene crystals (C, crystal; G, granum; O, osmiophilic globule; PE, plastid envelope; T, thylakoid; Ve, vesicle). *(Electron micrographs (A) ×40,700 (B) ×38,000 and (C) ×20,100 courtesy of Rosso.[16])*

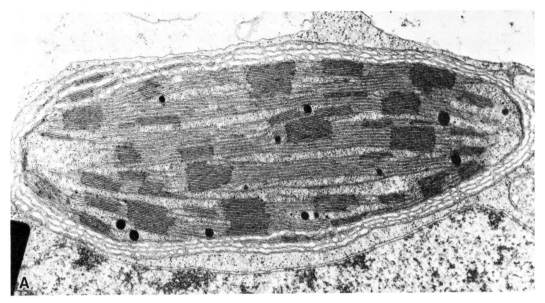

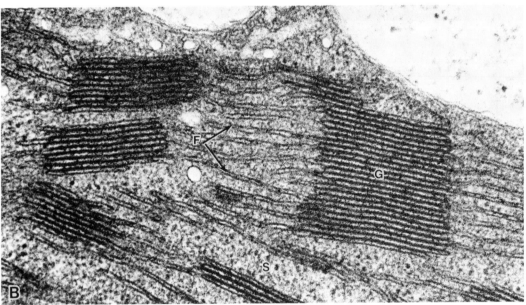

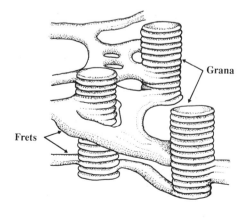

Grana

Frets

Figure 2-29. Electron micrograph of a chloroplast *(A)* with enlarged view of a granal region *(B)* (G, grana; S, stroma; F, fret). The drawing shows the arrangement of the grana and fret structure within the granal region. *(Electron micrographs (A) ×62,600 and (B) ×40,000 courtesy of W. W. Thompson; drawing after Bidwell.[2])*

CO_2 fixation reactions occur, while the photosynthetic photosystems are found on the thylakoid membranes.

Microbodies

Microbodies are small (0.5–1.5 μm), spherical organelles (fig. 2-30) found in a variety of plant species and types of cells. There may be from 500 to 2,000 microbodies in a single cell. They are bound by a single membrane and contain specific enzymes. The enzymes present and the function of the microbody vary depending on its biochemical type. Microbodies that are associated with chloroplasts and function in glycolate oxidation during photorespiration are called peroxisomes. Those present in oil-rich seeds and that function in the conversion of lipids to sugars during germination (glyoxylate cycle) are called glyoxysomes.

Microbodies appear to be formed from invaginations on the smooth endoplasmic reticulum. Their internal matrix is generally amorphous although crystalline substances may be present.

Vacuole

In mature cells, the vacuole is seen as the central feature within the cytoplasm (fig. 2-31). Young meristematic cells have numerous small vacuoles, which may with time enlarge and coalesce into the large central vacuole of the mature cell. Structurally, the vacuole is bound by a single membrane, called the tonoplast, which exhibits differential permeability to various molecules. Thus the movement of many molecules into and out of the vacuole is closely controlled. Contained within the vacuole are a diverse array of possible compounds. For example, the vacuole may contain sugars, organic acids, amino acids, proteins, tannins, calcium oxalate, anthocyanins, phenolics, alkaloids, gums, and other compounds. These may be dissolved in the aqueous medium, found as crystals (fig. 2-31), or congealed into distinct bodies.

The vacuole has three primary functions within the cell. It acts as a disposal site for "waste" material from the cytoplasm. This function is essential since most plant cells have very little potential to excrete unwanted substances external to the cell. As a consequence the vacuole functions as a repository for these substances. The vacuole also functions in the maintenance of the turgor pressure of the cell. Absorption of water by the vacuole provides the outward force that contributes to the shape and texture of the cell. Loss of turgor pressure after harvest diminishes product quality.

Finally the vacuole contains a large number of hydrolytic enzymes, for example, proteases, lipases, nucleases, and phosphatases. Under normal conditions these enzymes act in part in the recycling of compounds from the cytoplasm. However, when the tonoplast is ruptured due to injury or senescence, these hydrolytic enzymes are released into the cytoplasm. Here they attack a wide range of cellular constituents, accelerating the rate of disorganization and death of the cell. Conversely, some of the enzymes are present in the cytoplasm and the substrate is released from the vacuole. The discoloration reactions occurring after cells sustain mechanical injury (e.g., bruis-

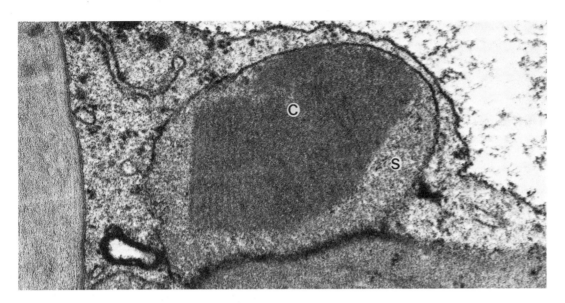

Figure 2-30. Electron micrograph of a microbody from a *Citrus* mesophyll cell (S, stroma; C, crystalline inclusion of protein, possibly the enzyme catalase). *(Electron micrograph ×150,000 courtesy of W. W. Thompson.)*

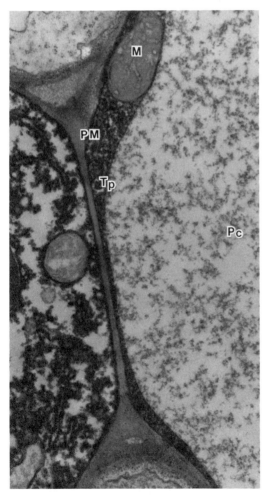

Figure 2-31. Electron micrograph of the central vacuole in a mature *Liquidambar styraciflua*, L. cell. Within the vacuole are precipitated components (Pc) (PM, plasma membrane; Tp, tonoplast; M, mitochondria). *(Electron micrograph courtesy of H. Y. Wetzstein.)*

ing) are the result of the action of phenoloxidase on phenols released from the vacuole into the cytoplasm.

Dictyosomes

Dictyosomes are small (1–3 μm in diameter by 0.5 μm thick), subcellular organelles composed of stacks of flat, circular cisternae, each enclosed by a single membrane (fig. 2-32). It is not clear whether the dictyosomes (also known as Golgi bodies) are in fact free organelles or are simply organellelike inclusions on the endoplasmic reticulum with which they are associated. The number of cisternae per dictyosome and number of dictyosomes per cell vary. There may be from several to as many as 20 cisternae in a single dictyosome and from a small number to more than 100 dictyosomes per cell depending on cell type. Cells that are actively secreting compounds from the cytoplasm, for example, during cell wall development or mucilages from the tips of roots, tend to have an abundance of dictyosomes. As a consequence, the function of dictyosomes is believed to be in the secretion of cellular compounds, chiefly polysaccharides, outside of the plasma membrane. This process is accomplished through the formation of small spherical vesicles of polysaccharides by the cisternae. The vesicles migrate to the plasma membrane where the vesicle membrane and the plasma membrane fuse, with the contents of the vesicle being released to the exterior.

Ribosomes

Ribosomes are small bodies (.017–.025 μm in diameter) that are the sites of protein synthesis within the cell (fig. 2-33). They are found in the cytoplasm both associated with the endoplasmic reticulum and free, dispersed singly or in small groups. Ribosomes are also found in the nuclei, plastids, and mitochondria. These, however, appear to be distinct from those found in the cytoplasm. Because of the relatively short life expectancy of many proteins, and the large number required by the cell, many ribosomes are needed. Estimates of the number of ribosomes per cell range from 500,000 to 5,000,000.

When a messenger RNA, carrying the code for a specific protein from the nucleus, unites with several ribosomes, the resulting complex is called a polysome or polyribosome. Transfer RNAs within the cytoplasm bind to amino acids, bringing these basic building blocks of proteins to the polysome for incorporation into the new protein molecule.

Microtubules

Microtubules appear as hollow tubes approximately 24 nm in diameter and several μm in length (fig. 2-34). They are composed of 10–13 protein fibrils. These subunits are largely composed of a globular protein called tubulin.

Microtubules appear to have several functions within the cell. During cell division, microtubules form the spindle fibers that control chromosome mi-

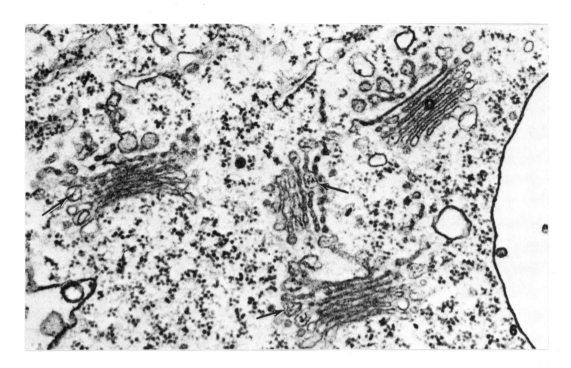

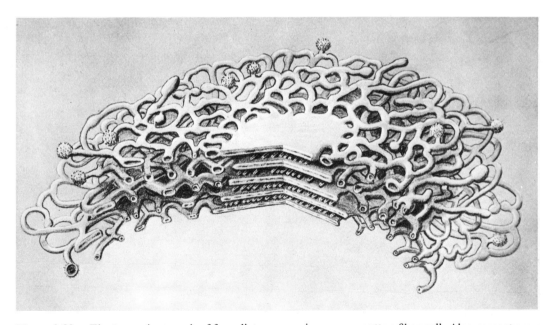

Figure 2-32. Electron micrograph of four dictyosomes in a young cotton fiber cell. Also present are a large number of ribosomes. The drawing represents what is thought to be the three-dimensional structure of a dictyosome. *(Electron micrograph ×112,000 courtesy of J. D. Berlin;[15] drawing from Mollenhauer and Morré.[11])*

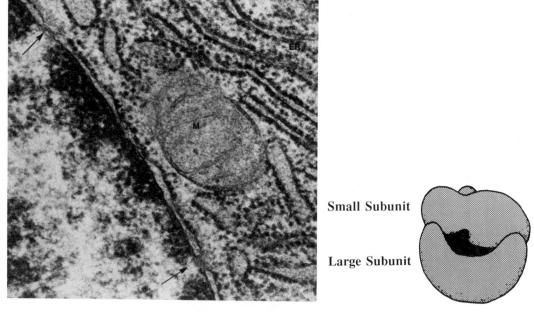

Small Subunit

Large Subunit

Figure 2-33. Arrangement of ribosomes on the rough endoplasmic reticulum *(electron micrograph ×59,000 courtesy of Wolfe[21])*. The diagram shows the large and small subunits of a ribosome.

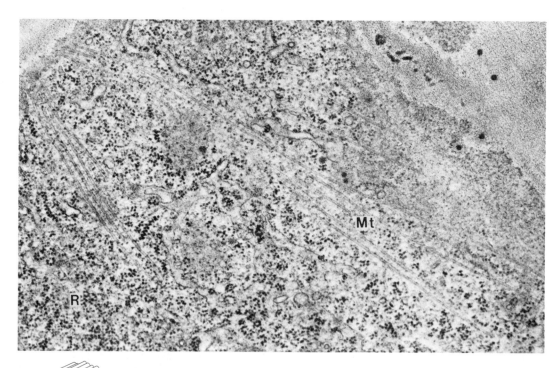

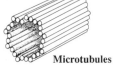

Microtubules

Figure 2-34. Electron micrograph of microtubules (Mt) in a young plant cell. The drawing illustrates the arrangement of the fibril making up the elongated tubular nature of the microtubule. *(Electron micrograph courtesy of W. W. Thompson; drawing after Ting.[18])*

gration. Colchicine, a chemical used widely by plant breeders to double the number of chromosomes per cell, binds to tubulin, preventing chromosome migration during anaphase. Thus the cell retains both sets of chromosomes.

Microtubules also appear to be involved in the coordination of the development of the cell wall. They are found in greatest numbers in the peripheral cytoplasm, adjacent to regions with growing cell walls. Disruption of the microtubules with colchicine also appears to alter the orderly arrangement of the new wall. Microtubules also appear to impart a portion of the structure of the cytoplasm and may also function in intracellular movements.

REFERENCES

1. Baker, E. A. 1982. Chemistry and morphology of plant epicuticular waxes. In *The Plant Cuticle,* D. F. Cutler, K. A. Alvin, and C. E. Price (eds.). Academic Press, London, pp. 139–165.
2. Bidwell, R. G. S. 1979. *Plant Physiology.* Macmillan, New York, 643p.
3. Bonner, J., and R. W. Galston. 1958. *Principles of Plant Physiology.* Freeman, San Francisco, 499p.
4. Coombe, B. G. 1976. The development of fleshy fruits. *Ann. Rev. Plant Physiol.* **27:**207–228.
5. Esau, K. 1977. *Anatomy of Seed Plants.* John Wiley, New York, 550p.
6. Fahn, A. 1967. *Plant Anatomy.* Pergamon Press, Oxford, England, 534p.
7. Gilbert, L. E. 1971. Butterfly-plant coevolution: Has *Passiflora adenopoda* won the selectional race with Heliconine butterflies? *Science* **172:**585–586.
8. Juniper, B. E., and C. E. Jeffree. 1983. *Plant Surfaces.* Edward Arnold, London, 93p.
9. Kramer, P. J., and T. T. Kozlowski. 1979. *Physiology of Woody Plants.* Academic Press, New York, 811p.
10. Ledbetter, M. C., and K. R. Porter. 1970. *Introduction to the Fine Structure of Plant Cells.* Springer-Verlag, Berlin, 188p.
11. Mollenhauer, H. H., and D. J. Morré. 1966. Golgi apparatus and plant secretion. *Ann. Rev. Plant Physiol.* **17:**27–46.
12. Onwueme, I. C. 1978. *The Tropical Tuber Crops.* John Wiley and Sons, New York, 234p.
13. Priestley, D. A., and A. C. Leopold. 1983. Lipid changes during natural aging of soybean seed. *Plant Physiol.* **59:**467–470.
14. Rachmilevitz, T., and A. Fahn. 1975. The floral nectary of *Tropaeolum majus* L.—The nature of the secretory cells and the manner of nectar secretion. *Ann. Bot.* **39:**721–728.
15. Ramsey, J. C., and J. D. Berlin. 1976. Ultrastructural aspects of early stages in cotton fiber elongation. *Am. J. Bot.* **63:**868–876.
16. Rosso, S. W. 1968. The ultrastructure of chromoplast development in red tomatoes. *J. Ultrastructure Res.* **25:**307–322.
17. Singer, R. 1961. *Mushrooms and Truffles: Botany, Cultivation, and Utilization.* Interscience, New York, 272p.
18. Ting, I. P. 1982. *Plant Physiology.* Addison-Wesley, Reading, Mass., 642p.
19. Wilson, C. L., and W. E. Loomis. 1967. *Botany,* 4th ed. Holt, Rinehart and Winston, New York, 626p.
20. Wilson, D. O., Jr., and M. B. McDonald. 1986. The lipid peroxidation model of seed aging. *Seed Sci. Technol.* **14:**269–300.
21. Wolfe, S. L. 1985. *Cell Ultrastructure.* Wadsworth, Belmont, Calif., 144p.

ADDITIONAL READINGS

Barnet, J. R. (ed). 1981. *Xylem Cell Development*. Castle House, Turnbridge Wells, Kent, England, 307p.

Brett, C. T., and J. R. Hillman (eds.). 1985. *Biochemistry of Plant Cell Walls*. Cambridge University Press, Cambridge, Mass., 312p.

Bryant, J. A., and V. L. Dunham (eds). 1988. *DNA Replication in Plants*. CRC Press, Boca Raton, Fla., 193p.

Cronshaw, J. 1980. Phloem structure and function. *Ann. Rev. Plant Physiol.* **32**:465–484.

Cronshaw, J., W. J. Lucas, and R. T. Giaquinta (eds). 1985. *Phloem Transport: Proceedings of an International Conference on Phloem Transport*. A. R. Liss, New York, 650p.

Dawidowicz, E. A. 1987. Dynamics of membrane lipid metabolism and turnover. *Ann. Rev. Biochem.* **56**:43–62.

Esau, K. 1953. *Plant Anatomy*. John Wiley, New York, 735p.

Fahn, A. 1969. *Plant Anatomy*. Pergamon Press, Oxford, England, 534p.

Giaquinta, R. T. 1983. Phloem loading of sucrose. *Ann. Rev. Plant Physiol.* **34**:347–387.

Glass, A. D. M. 1983. Regulation of ion transport. *Ann. Rev. Plant Physiol.* **34**:311–326.

Gunning, B. E. S., and A. R. Hardham. 1982. Microtubles. *Ann. Rev. Plant Physiol.* **33**:651–698.

Harris, N. 1986. Organization of the endomembrane system. *Ann. Rev. Plant Physiol.* **37**:73–92.

Heber, U., and H. W. Heldt. 1981. The chloroplast envelope: Structure, function, and role in leaf metabolism. *Ann. Rev. Plant Physiol.* **32**:139–168.

Heidecker, G., and J. Messing. 1986. Structural analysis of plant genes. *Ann. Rev. Plant Physiol.* **37**:439–466.

Hoober, J. K. 1984. *Chloroplasts*. Plenum Press, New York, 280p.

Jacobs, S., and P. Cuatrecassas. 1981. *Membrane Receptors. Methods for Purification and Characterization*. Chapman and Hall, New York, 240p.

Jarvis, P. G., and T. A. Mansfield (eds.). 1981. *Stomatal Physiology*. Cambridge University Press, Cambridge, Mass., 295p.

Kirk, J. T. O., and R. A. E. Tilney-Bassett. 1978. *The Plastids. Their Chemistry, Structure, Growth and Inheritance*. Elsevier/North-Holland Biomedical Press, Amsterdam, 960p.

Kuhlemeier, C., P. J. Green, and N. Chua. 1987. Regulation of gene expression in higher plants. *Ann. Rev. Plant Physiol.* **38**:221–257.

Marin, B. (ed). 1986. *Plant Vacuoles. Their Importance in Solute Compartmentation in Cells and Their Applications in Plant Biotechnology*. Plenum, New York, 562p.

Marme, D., E. Marre, and R. Hertal (eds.). 1981. *Plasmalemma and Tonoplast: Their Functions in the Plant Cell*. Elsevier Biomedical, Amsterdam, 446p.

McNeil, M., A. G. Darvill, S. C. Fry, and P. Albersheim. 1984. Structure and function of the primary cell walls of plants. *Ann. Rev. Biochem.* **53**:625–663.

Moorby, J. 1981. *Transport Systems in Plants*. Longman, London, 169p.

Newport, J. W., and D. J. Forbes. 1987. The nucleus: Structure, function, and dynamics. *Ann. Rev. Biochem.* **56**:535–566.

Noller, H. F. 1984. Structure of ribosomal RNA. *Ann. Rev. Biochem.* **53**:119–162.

Possingham, J. V. 1980. Plastid replication and development in the life cycle of higher plants. *Ann. Rev. Plant Physiol.* **31**:113–129.

Reinhold, L., and A. Kaplan. 1984. Membrane transport of sugars and amino acids. *Ann. Rev. Plant Physiol.* **35**:45–83.

Robards, A. W. (ed.). 1974. *Dynamic Aspects of Plant Ultrastructure.* McGraw-Hill, New York, 546p.

Robinson, D. G. 1985. *Plant Membranes. Endo- and Plasma Membranes of Plant Cells.* John Wiley and Sons, New York, 331p.

Sancar, A., and G. B. Sancar. 1988. DNA repair enzymes. *Ann. Rev. Biochem.* **57:**29–67.

Schiff, J. A. (ed). 1982. *On the Origins of Chloroplasts.* Elsevier/North Holland, Amsterdam, 336p.

Serrano, R. 1985. *Plasma Membrane ATPase of Plants and Fungi.* CRC Press, Boca Raton, Fla., 174p.

Spiker, S. 1985. Plant chromatin structure. *Ann. Rev. Plant Physiol.* **36:**235–253.

Taiz, L. 1984. Plant cell expansion: Regulation of cell wall mechanical properties. *Ann. Rev. Plant Physiol.* **35:**585–657.

Thomson, W. W., and J. M. Whatley. 1980. Development of nongreen plastids. *Ann. Rev. Plant Physiol.* **31:**375–394.

Trelease, R. N. 1984. Biogenesis of glyoxysomes. *Ann. Rev. Plant Physiol.* **35:**321–347.

Tzagoloff, A. 1982. *Mitochondria.* Plenum, New York, 342p.

Walker, G. C. 1985. Inducible DNA repair systems. *Ann. Rev. Biochem.* **45:**425–457.

Willmer, C. M. 1983. *Stomata.* Longman, London, 166p.

Zeiger, E. 1983. The biology of stomatal guard cells. *Ann. Rev. Plant Physiol.* **34:**441–475.

Zeiger, E., G. D. Farguhar, and I. R. Cowan (eds.). 1987. *Stomatal Function.* Stanford University Press, Stanford, Calif., 503p.

Zimmermann, M. H. 1983. *Xylem Structure and the Ascent of Sap.* Springer-Verlag, Berlin, 143p.

3

METABOLIC PROCESSES IN HARVESTED PRODUCTS

Metabolism represents the entirety of the many chemical activities that occur within cells. Many specific components of metabolism, expecially those that are beneficial or detrimental to the quality of harvested products, are of major interest to postharvest physiologists. The acquisition and storage of energy and the utilization of this stored energy are two of the central processes in the control of the overall metabolism of plants. The acquisition of energy through photosynthesis and its recycling via the respiratory pathways are compared in table 3-1. They are often viewed in a very general way as opposing forces. As we will see later in this chapter, photosynthesis does not occur in all postharvest products and in fact is not possible in products devoid of chlorophyll.

From the standpoint of carbon acquisition, allocation, and storage, there is within intact plants a high degree of specialization of function in the various organs. Leaves, for example, photosynthesize but seldom act as long-term storage sites for photosynthates. Petioles and stems transport fixed carbon but typically have only a limited photosynthetic potential, and when utilized for storage, often act only as temporary sinks (e.g., the stems of the Jerusalem artichoke (*Helianthus tuberosus,* L.)). Floral, root, tuber, and other organs or tissues likewise have relatively specific roles with regard to the overall acquisition of carbon. While attached to the plant, many of these organs derive the energy required to carry out their specific functions from photosynthesizing leaves. There is, therefore, in intact plants an interdependence between these different organs with divergent primary functions. Severing these organs from the plant at harvest disrupts this interdependence and can, therefore, influence postharvest behavior. For example, the detaching of leaves, whose primary function is to fix carbon dioxide rather than to store fixed carbon, markedly restricts or terminates photosynthesis (i.e., the ability to trap for subsequent utilization an external source of energy), leaving them with extremely low reserves that can be used to maintain the organ. Storage organs, on the other hand, if sufficiently mature, have substantial stored carbon that can be recycled for utilization in maintenance and synthetic reactions.

Table 3-1. General Comparison of Photosynthesis and Respiration in Plants

	Photosynthesis	*Respiration*
Function	Energy acquisition	Energy utilization and formation of carbon skeletons
Location in cells	Chloroplasts	Mitochondria and cytoplasm
Role of light	Essential	Not involved
Substrates	CO_2, H_2O, light	Stored carbon, O_2
End products	O_2, stored carbon	CO_2, H_2O, energy
Overall effect	Increases the weight	Decreases the weight of the plant or plant part
General reaction	$6CO_2 + 6H_2O \xrightarrow[\text{chloroplasts}]{\text{energy}}$ $C_6H_{12}O_6 + 6O_2$	$C_6H_{12}O_6 + 6O_2 \xrightarrow{\text{mitochondria}}$ $6CO_2 + 6H_2O + \text{energy}$

In contrast to this high degree of specialization between organs in the acquisition of energy, respiration occurs in all living cells and is essential for the maintenance of life in products after harvest. The following section discusses the factors affecting energy acquisition (photosynthesis) and energy utilization (respiration). These processes are affected by both internal (commodity) and external (environmental) factors that often interact. Important commodity factors include species, cultivar, type of plant part, stage of development, surface-to-volume ratio, surface coating, previous cultural and handling conditions, and chemical composition. Among the major external factors influencing respiratory rate are temperature, gas composition, moisture conditions, light, and other factors that can induce stress conditions within the harvested product.

RESPIRATION

Respiration is a central process in living cells that mediates the releases of energy through the breakdown of carbon compounds and the formation of carbon skeletons necessary for maintenance and synthetic reactions after harvest. From a postharvest point of view, rate of respiration is important because of these main effects; however, the rate of respiration also gives an indication of the overall rate of metabolism of the plant or plant part. All metabolic changes occurring after harvest are important, especially those that have a direct bearing on the quality of the product. The central position of respiration in the overall metabolism of a plant or plant part and its relative ease of measurement allow respiration to be used as a general measure of metabolic rate. This relationship between respiration and metabolism, however, is very general since specific metabolic changes may occur without measurable changes in net respiration. This situation is illustrated by comparing changes in a number of the physical and chemical properties of pineapple (*Ananas comosus*, (L.) Merrill) fruit during development, maturation, and senescence (fig. 3-1). Neither changes in the concentration of chlorophyll nor

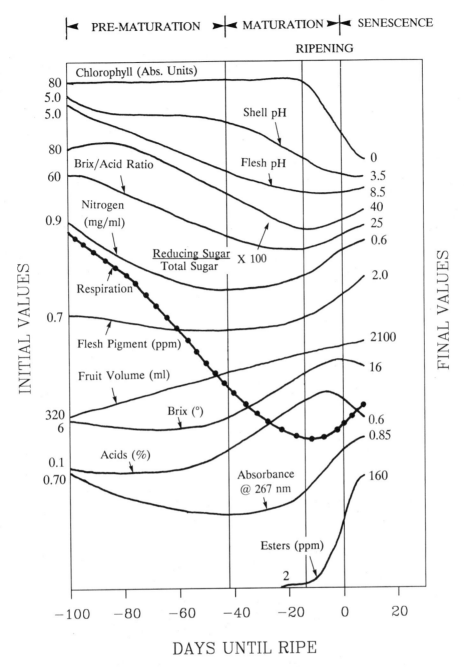

Figure 3-1. Chemical and physical changes in the pineapple (*Ananas comosus*, (L.) Merr.) fruit during prematuration, maturation, ripening, and senescence (*after Gortner et al.*[37]). While respiratory rate can give a general indication of the overall rate of metabolism, it often does not correlate with specific changes occurring, for example, changes in the percent of acids.

reducing sugars, acidity, carotenoids, or esters correlate well with changes in respiratory rate. Therefore, it is important that respiration is viewed as it fits into the overall process of metabolism in harvested products rather than as an end in itself.

There are two general types of respiratory processes in plants: those that occur at all times regardless of the presence or absence of light (dark respiration) and those that occur only in the light (light or photorespiration).

Dark Respiration Pathways

The living cells of harvested plant products respire continuously, utilizing stored reserves and oxygen from the surrounding environment and releasing carbon dioxide. This ability to respire is an essential component of the metabolic processes that occur in live harvested products. The absence of respiration is the major distinction between processed plant products and living products. Respiration is the term used to represent a series of oxidation-reduction reactions where a variety of substrates found within the cells are oxidized to carbon dioxide. At the same time, oxygen absorbed from the atmosphere is reduced to form water. In its simplest form, the complete oxidation of glucose can be written as:

$$C_6H_{12}O_6 + 6O_2 \rightarrow 6CO_2 + 6H_2O + 686 \text{ kcal}$$

glucose   oxygen   carbon   water   energy
                                  dioxide

The products of this reaction are carbon dioxide, water, and most importantly, energy, which is required for essential processes within the cells. Much of the energy generated in respiration of harvested produce is lost as heat; however, significant amounts are retained by the cells in chemical forms that may be used for these essential processes.

Respiration is in fact much more complex than the generalized reaction just presented. The glycolytic, tricarboxylic acid, pentose phosphate, and cytochrome system pathways are involved in the breakdown of many of the common substrates utilized by the cell. Often during the oxidation of a substrate the conversion to carbon dioxide is not complete and the intermediates formed are utilized by the cells for synthetic reactions such as the formation of amino acids, nucleotides, pigments, lipids, and flavor compounds. Hence the respiratory pathways provide precursors, often called carbon skeletons, that are required for the formation of a large number of plant products.

During the preharvest growth of a plant or plant product, a major portion of the carbon trapped during photosynthesis is diverted into synthetic reactions. Hence, during rapid growth (synthetic reactions) common substrates such as hexose sugars often are not completely oxidized to carbon dioxide but only proceed partway through the respiratory system pathways, yielding carbon skeletons. It is through the respiratory pathways that the carbon from photosynthesis begins its transformation into the majority of the other compounds in the plant. Since the synthesis of these compounds also requires energy, derived from the respiratory pathways, a portion of the photosynthetic carbon fixed is utilized for this purpose. Therefore, a balance is reached between respiratory substrate availability and the demands for energy production and carbon skeletons. Since neither availability nor demands are static, the system is continually changing this balance during the day and over the developmental cycle of the plant or plant part.

At harvest the relationship between carbon acquisition and utilization is radically changed when the plant product is severed from its readily replenishable supply of carbon provided by photosynthesis. Hence, a new balance must be reached: Energy and carbon skeletons must now come from sources already existing within the severed product. Mature plant parts that function as carbon storage organs (i.e., seeds, roots, bulbs, tubers, etc.) have substantial stores of carbon that can be utilized via the respiratory pathways for an extended period. Organs such as leaves or flowers that do not function as carbon storage sites have very little reserves and as a consequence the balance between supply and demand moves from a dynamic equilibrium.

This general scenario is true even for harvested intact plants in that after production an attempt is made to maintain the product as close to its harvested condition as possible, that is, growth is normally considered undesirable. Postharvest conditions for these products often result in photosynthesis being extensively reduced or totally eliminated, and thus a reliance on existing respiratory substrate reserves. The respiratory pathways that are operative after harvest in both intact plants and severed plant parts are the same as those prior to harvest. The major change is the now finite supply of respiratory substrate to be used in the pathways and the maintenance of a balance between the supply of this substrate and the tissue demand for it.

The series of steps in the respiratory oxidation of sugar or starch involves three major interacting pathways. The initial pathway is that of glycolysis where sugar is broken down into pyruvic acid, a three-carbon compound. This process takes place in the cytoplasm and does not require oxygen. The second pathway is the tricarboxylic acid (TCA) or Krebs cycle. Here pyruvic acid is oxidized to carbon dioxide in the mitochondria. Oxygen, although not reacting directly in these steps, is required for the pathway to proceed, as are several organic acids. The third pathway, the electron transport system, transfers hydrogen atoms that have been removed from organic acids in the tricarboxylic acid cycle and from 3-phosphoglyceraldehyde during glycolysis. They are moved through a series of oxidation-reduction steps that terminate upon uniting with oxygen, forming water. Energy is trapped chemically in the form of adenosine triphosphate (ATP)* that can be utilized to drive various energy-requiring reactions within the cell. A fourth respiratory pathway, the pentose phosphate system, while not essential for the complete oxidation of sugars, functions by providing carbon skeletons, reduced NADP (nicotinamide adenine dinucleotide phosphate) required for certain synthetic reactions, and ribose-5-phosphate for nucleic acid synthesis. It appears to be operative in varying degrees in all respiring cells.

While oxygen is not required for the operation of the glycolytic pathway, it is essential for the tricarboxylic acid cycle, the pentose phosphate pathway,

---

* Other phosphorylated nucleotides, such as uridine triphosphate (UTP), guanosine triphosphate (GTP), and cytidine triphosphate (CTP), can also play an important role in cellular energy transfer.

and the electron transport system. Glycolysis can proceed therefore under anaerobic conditions, that is, in the absence of oxygen. The occurrence of anaerobic conditions poses a serious problem in the postharvest handling of plant products. When the oxygen concentration within the tissue falls below a threshold level, pyruvic acid can no longer proceed through the tricarboxylic acid cycle and instead forms ethanol, which may accumulate in toxic levels. Prolonged exposure to anaerobic conditions, therefore, results in death of the cells and loss of the harvested product. Exposure for short periods often results in the formation of off-flavors in edible products, which may or may not, depending on the tissue and length of exposure, be eliminated upon returning to aerobic conditions.

Glycolysis

The glycolytic pathway is also known as the Embden-Meyerhof-Parnas (EMP) pathway after the three German scientists, G. Embden, O. Meyerhof, and J. K. Parnas, whose work in the early 1900s led to its elucidation. The glycolytic pathway breaks down glucose from starch or sucrose in a sequence of steps to form pyruvic acid. In the initial step glucose must have phosphorus added (phosphorylated) (fig. 3-2). If the starting compound is free glucose, the reaction is catalyzed by the enzyme hexokinase and forms glucose-6-phosphate. If, as found in many postharvest products, the glucose is found as part of a starch molecule, phosphorus is added by the enzyme starch phosphorylase, forming glucose-1-phosphate, which is subsequently converted to glucose-6-phosphate. The phosphorylation of free glucose requires energy, in the form of an ATP molecule, while the phosphorylation of glucose when it is part of a starch molecule does not.

The 6-carbon glucose molecule progresses through fructose-1-phosphate to fructose-1,6-bisphosphate before being split by the enzyme aldolase into two 3-carbon compounds: dihydroxyacetone phosphate and 3-phosphoglyceraldehyde. The 3-phosphoglyceraldehyde molecule is the first compound to lose electrons in the respiratory pathway, forming 1,3-diphosphoglyceric acid, when 2 hydrogen atoms are removed and are accepted by either NAD (nicotinamide adenine dinucleotide) or NADP (nicotinamide adenine dinucleotide phosphate). 1,3-Diphosphoglyceric acid then progresses through three additional enzymatic steps, resulting in the formation of pyruvic acid.* None of the reactions in the preceding steps from glucose or starch requires oxygen so the glycolytic pathway can proceed normally under anaerobic conditions.

If anaerobic conditions occur in the harvested tissue due to restricted entry of oxygen or an insufficient supply in the atmosphere surrounding the com-

---

* The glycolytic pathway can proceed in both directions, that is, toward the formation of pyruvic acid (glycolytic direction) or in the opposite (gluconeogenic) direction with the formation of sugars. Carbon often flows in both directions at the same time within a product. For example, when banana fruit ripen, starch located in plastids is broken down into three carbon compounds (triose phosphates) that enter glycolysis midway in the pathway. Some of this carbon is used for energy (glycolytic direction), but most is converted to sugars, giving the banana its characteristic sweetness when ripe.[5]

modity, pyruvic acid cannot enter the tricarboxylic acid cycle and be oxidized. When this situation occurs, pyruvic acid accumulates and is usually decarboxylated to form carbon dioxide and acetaldehyde, which is subsequently reduced to ethanol. The reaction requires energy that is provided by reduced NAD or NADP formed during the oxidation of 3-phosphoglyceraldehyde previously in the pathway. The overall reaction in simplified form is:

$$\text{glucose} + 2\text{ATP} + 2\text{Pi} + 2\text{ADP} \rightarrow 2 \text{ ethanol} + 2 \text{ CO}_2 + 4\text{ATP}$$

When ethanol is produced, two ATP molecules are required but four are formed from each free glucose molecule, giving a net yield of two ATPs. This amount represents only one third the energy yield that would be derived from the glycolytic pathway when sufficient oxygen is present and is only one eighteenth that derived when glucose is fully oxidized (glycolysis and the tricarboxylic acid cycle). In addition, alcohol accumulates within the tissue. As the cells switch their flow of carbon toward alcohol, the production of carbon dioxide increases (fig. 3-2). This increase is due to the reduced energy yield under anaerobic conditions; therefore, much more glucose must be oxidized to meet the energy requirements of the cell. The complete oxidation of 1 glucose molecule under aerobic conditions yields thirty-six ATP equivalents whereas under anaerobic conditions only two ATPs are formed from a glucose molecule. Although anaerobiosis has disastrous consequences for live tissue in terms of loss of stored reserves and accumulation of undesirable compounds, it is also the basis of a very important processing technique: fermentation. Potential energy remains stored in the form of alcohol.

Tricarboxylic Acid Cycle

Pyruvic acid produced by the glycolytic pathway is further broken down in the Krebs cycle, named after the English scientist, H. A. Krebs, who first proposed a cyclic series of reactions. The pathway is also known as the TCA cycle and occasionally as the citric acid cycle. Tricarboxylic acid refers to the three carboxyl groups that are present on some of the acids in the cycle and citric acid is an important early intermediate in the sequence of reactions.

The reactions of the TCA cycle occur in the mitochondria, probably on the surface of the inner membranes. Pyruvic acid, therefore, must move from the cytoplasm, where glycolysis occurs, into the mitochondria for further oxidation to proceed.

In the initial step, pyruvic acid loses CO_2 and combines with Coenzyme A, forming the two-carbon compound acetyl CoA (fig. 3-3). Acetyl CoA then combines with the four-carbon molecule of oxaloacetic acid, yielding citric acid that moves through a series of reactions ending with the formation of oxaloacetic acid, allowing the cycle to begin again. Energy is produced as reduced NAD (NADH + H$^+$) upon the conversion of isocitric acid to α-ketoglutaric acid, α-ketoglutaric acid to succinyl CoA, and malic acid to oxaloacetic acid. A single ATP is yielded on the transfer of succinyl CoA to succinic acid and $FADH_2$ in the conversion of succinic acid to fumaric acid. Carbon dioxide is liberated upon removal from pyruvic, isocitric, and α-ketoglutaric acids.

Each cycle of a pyruvate molecule through the tricarboxylic acid cycle

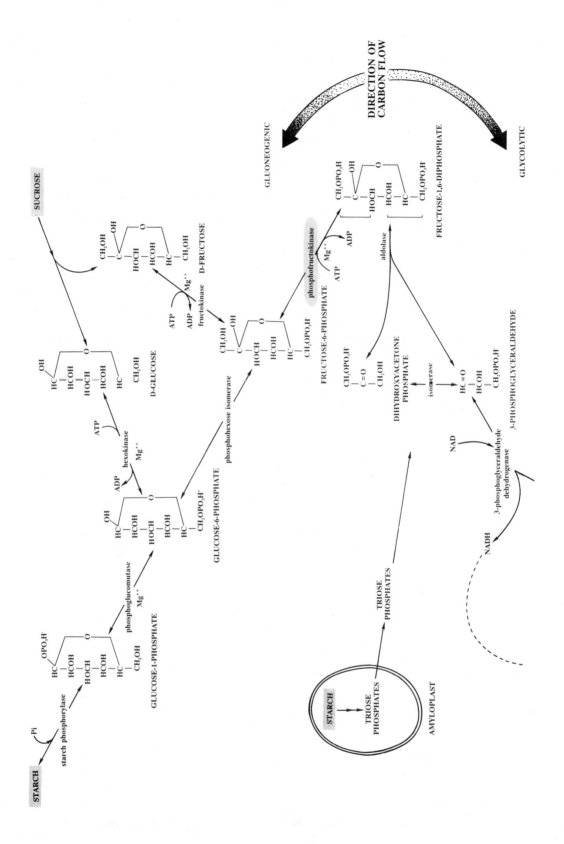

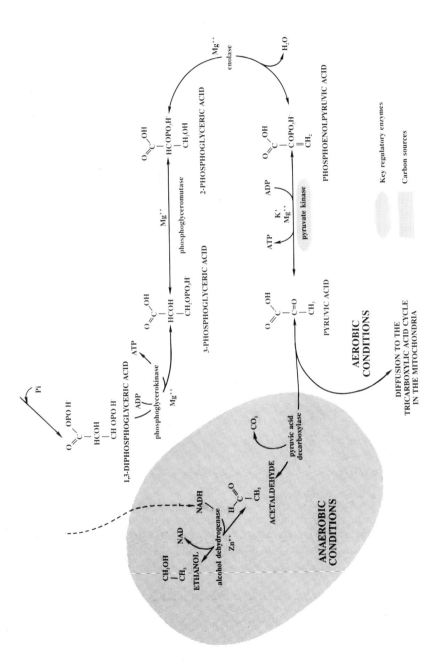

Figure 3-2. The glycolytic pathway for the aerobic oxidation of glucose or glucose-1-phosphate to pyruvate. In most instances, carbon moves in the pathway in a glycolytic direction; however, it can also flow in the opposite direction. For example, during the ripening of banana (*Musa* spp.) there is a pronounced flow in the gluconeogenic direction forming sugars. (Key regulatory enzymes and several possible carbon sources are shaded.) Under anaerobic conditions the movement of pyruvate into the tricarboxylic acid cycle is inhibited and NADH formed in the oxidation of 3-phosphoglyceraldehyde is utilized to reduce acetaldehyde to ethanol by way of the shaded portion of the pathway.

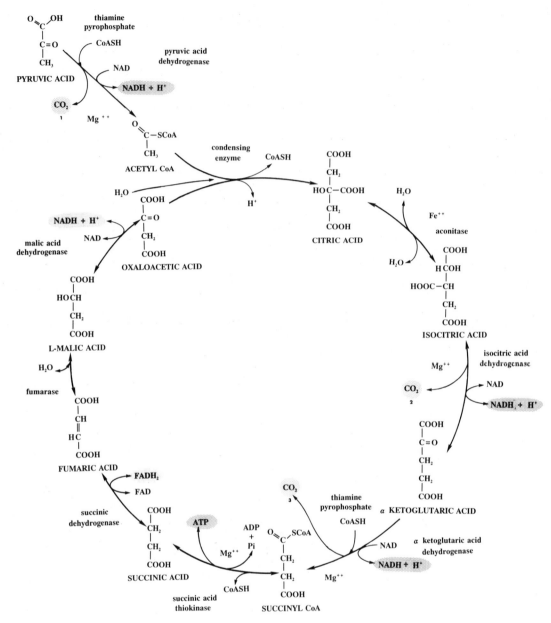

Figure 3-3. The tricarboxylic acid cycle results in the complete oxidation of one pyruvate molecule, with each complete sequence through the cycle forming CO_2, ATP, NADH + H, and $FADH_2$. NADH + H and $FADH_2$ are then oxidized in the electron transport system (fig. 3-4).

releases three CO_2 molecules and produces four electron pairs in the form of four NADH + H molecules and one electron pair as one $FADH_2$ molecule (X2). Combining this release with the two electron pairs (2NADH + H) from the glycolytic pathway gives a total of twelve electron pairs formed with the complete oxidation of a single glucose molecule. Only twelve of these electrons (H) are from glucose; the remaining twelve are from H_2O that is added at various steps in the cycle.

Electron Transport or Cytochrome System

NAD that has been reduced to NADH in the Krebs cycle, during glycolysis, and by other reactions in the cell is recycled by the removal of the electrons. NADH cannot, however, combine directly with oxygen to form water. The electrons are removed through a series of reactions, forming a positive potential gradient, from compounds of low-reduction potentials to higher-reduction potentials (i.e., from lower to greater tendency to accept electrons), culminating in a reaction with oxygen that has the greatest tendency to accept electrons. During this process some of the free energy liberated is conserved in a biologically usable form as ATP. This energy can be used to drive reactions, especially those of a synthetic type that require energy inputs. In actively metabolizing cells the efficiency of this energy trapping in the electron transport system is only around 54%. A mole of glucose has a calorie potential of approximately 686,000 gcal or 686 kcal. Only a small amount of energy is lost in the initial transfer of energy as electron pairs to NAD, FAD, and ADP in glycolysis and the tricarboxylic acid cycle, with most of it being retained.

However, during the transfer of this energy to ATP in the electron transport system, the energy potential drops to approximately 263,000 gcal/mole. The remaining energy escapes as respiratory ("vital") heat, a normally detrimental factor that must be dealt with in the postharvest handling and storage of plant products. Therefore, the overall function of the electron transport system is to trap energy in a biologically usable form (ATP) and recycle NAD and FAD required for the functioning of the major pathways. While the electron transport system is not completely elucidated, figure 3-4 shows the possible sequence of steps involved. Each enzyme is specific and can only accept electrons from the previous enzyme in the chain. NADH and $FADH_2$, being different in energy potential, enter the chain at different points. The formation of ATP from ADP and Pi in the electron transport system is called oxidative phosphorylation and occurs at three points in the pathway.

Both carbon monoxide (CO) and hydrogen cyanide (HCN) are potent inhibitors of electron transport in that they combine with the iron atom in cytochrome oxidase, the final enzyme of the chain. In certain plants or plant parts, HCN does not inhibit the terminal cytochrome oxidase but in fact stimulates the rate of respiration. In these tissues there is a second pathway called the alternate or cyanide resistant electron transport system. It follows the normal electron path until reaching a branch point at ubiquinone. Here the electrons are transferred to a flavoprotein of intermediate potential, then to an alternate oxidase, and subsequently to oxygen, forming water. It is important to note that only one ATP is generated in contrast to three ATPs in the normal pathway; the remainder of the energy is lost as heat. Therefore, the alternate pathway represents an inefficient energy-conserving system that bypasses the normal pathway, substantially increasing the respiratory heat load of the product.

The alternate electron transport pathway is found in a number of plants and plant parts that are harvested and stored (e.g., potato (*Solanum tuberosum*, L.) tubers, parsnip (*Pastinaca sativa*, L.) and carrot (*Daucus carota*, L.) roots, avocado (*Persea americana*, Mill.) and banana (*Musa* spp.) fruits). In nature, it appears to function in the skunk cabbage (*Symplocarpus foetidus*

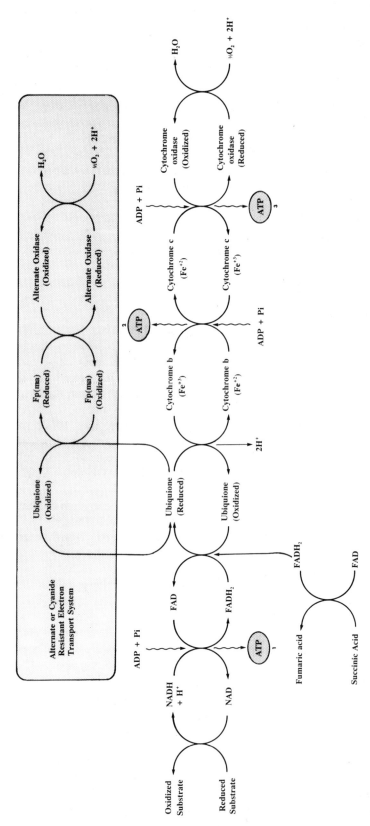

Figure 3-4. Tentative scheme of the electron transport system, located on the mitochondrial membranes, controlling the transfer of electrons obtained from the oxidation of organic acids (TCA cycle), down an energy gradient to the terminal acceptor, oxygen. The decrease in energy at three steps is sufficient to phosphorylate ADP to ATP. The alternative or cyanide-resistant pathway (shaded) is thought to branch from the normal pathway at ubiquinone. (Whether the alternate pathway branches at ubiquinone or in fact represents an entirely separate pathway remains a subject of debate.) When electrons are moved through the alternative pathway, only one ATP is formed. ($F_{p(ma)}$ = flavoprotein of medium potential)

(L.) Nutt.) as a means of frost protection. In some flowers or inflorescences of some species belonging to the families Annonaceae, Araceae, Aristolochiaceae, Cyclanthaceae, and Nymphaeaceae, the elevated temperature (e.g., up to 15°C[68]) caused by the pathway in the flower results in the volatilization of an array of insect attractants that help facilitate pollination. The alternative pathway is activated on the day of flowering and remains active for only a few hours. Although the existence of the alternative pathway has been known for more than 60 years,[34] the physiological significance in most situations (e.g., storage tissues) has not yet been ascertained. It is possible that it functions under certain conditions when high levels of intermediates are required for synthetic reactions within the cells. If the rate of production of these intermediates is limited by the rate at which the normal electron transport system proceeds, the cyanide resistant system may provide an unrestricted alternative that permits accelerated respiration and production of these intermediates.

Oxidative Pentose Phosphate Pathway

In addition to glycolysis and the TCA cycle, the pentose phosphate pathway can be used to break down sugars to CO_2. The name is derived from the fact that many of the intermediates in the pathway are five carbon (penta)phosphorylated sugars. The pentose phosphate system is found in the cytoplasm and its main function does not appear to be energy production via the formation of ATP in the electron transport system but rather it is a source of ribose-5-phosphate for nucleic acid production, a source of reduced NADP for synthetic reactions, and a means of interconversion of sugars to provide 3, 4, 5, 6, and 7 carbon skeletons for biosynthetic reactions. One example is the formation of erythrose-4-phosphate that can lead to shikimic acid and aromatic amino acids. In addition, NADPH is required for the production of fatty acids and sterols from acetyl CoA. A new pentose phosphate pathway in which two 8-carbon sugars are formed has been isolated from mammalian liver tissue. This new pathway is known as the L scheme and its presence or absence in plant tissue is yet to be established.

A major difference between the pentose phosphate pathway and the TCA-glycolysis systems is that NADP rather than NAD accepts electrons from the sugar molecule. NADPH cannot enter directly into the electron transport system but is used primarily for biosynthetic reactions in the cytoplasm. However, in some circumstances, NADPH can pass its reducing power to the electron transport system indirectly by being used for the synthesis of malic acid from oxaloacetic acid. Upon entering the mitochondria, malic acid is then converted back to oxaloacetic acid in the TCA cycle, yielding NADH (fig. 3-5).

Initial reactions in the pentose phosphate pathway include the irreversible oxidation of glucose-6-phosphate from glycolysis to a 6-phosphogluconic acid yielding a reduced NADP (fig. 3-6). Subsequently, 6-phosphogluconic acid is converted through the removal of CO_2 and a hydrogen to a 5-carbon sugar, ribulose-5-phosphate, which upon isomerization forms ribose-5-phosphate that is essential for nucleic acid synthesis. The conversion of phosphogluconic acid to ribulose-5-phosphate is also not reversible and re-

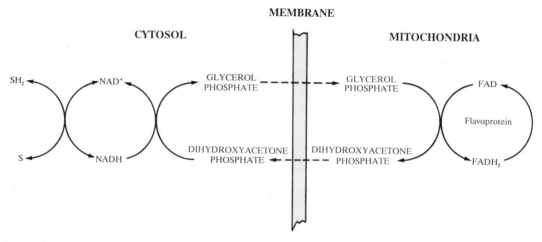

Figure 3-5. While reduced NADs produced in the glycolytic pathway can not transverse the mitochondrial membranes, their energy can be transported from the cytosol into the mitochondria (site of electron transport pathway) via the glycerol phosphate shunt. NADH reduces dihydroxyacetone phosphate to glycerol phosphate, which can transverse the mitochondrial membrane. Once on the interior, glycerol phosphate is oxidized back to dihydroxyacetone phosphate, producing one reduced FAD. The result is a loss of one ATP equivalent for each reduced NAD, hence a net yield of 36 ATP equivalents for each glucose molecule that is completely oxidized instead of 38, since 2 reduced NADs are formed in the glycolytic pathway per glucose molecule.

duced NADP is formed. The two initial reactions are the only oxidative (removal of hydrogen) steps in the pathway and the second is the only point in the entire pathway in which CO_2 is removed. Subsequent steps are reversible and can recycle back to glucose-6-phosphate, the initial substrate.

Since the pentose pathway is an alternative means of oxidizing sugars, it is of interest to know which system is operative in a harvested tissue. Existing evidence indicates that both the tricarboxylic acid and the pentose phosphate pathways are operative to some extent in all tissue; however, it is difficult to accurately measure the precise contribution of each pathway. In tomato fruits, the pentose phosphate pathway accounts for only around 16% of the total carbohydrates oxidized, a level probably fairly common in many tissues. However, in some tissues such as storage roots the pentose pathway is thought to be responsible for 25% to 50% of the oxidation of sugars.

Photorespiration

The acquisition of carbon via photosynthesis and the loss of carbon through respiration can be seen as opposing processes in chlorophyll-containing plant tissues. Growth is achieved when the gain in carbon exceeds losses. This occurs when the CO_2 concentration is above the carbon dioxide compensation point. In most species, it is known that the respiratory rate of chlorophyllous tissue as measured by the loss of CO_2 from the tissue proceeds at a

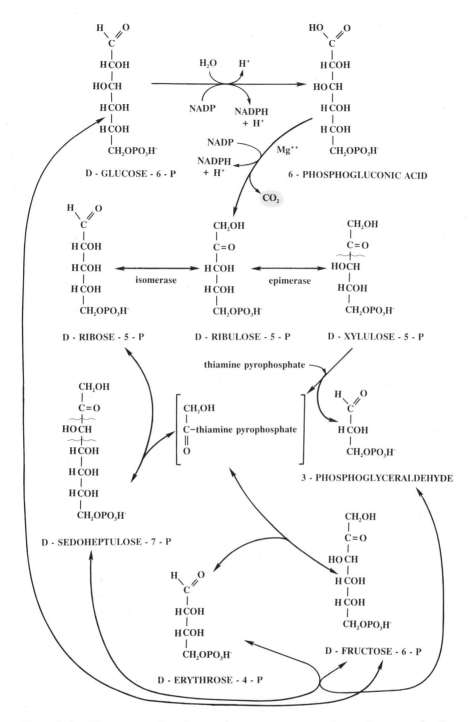

Figure 3-6. The pentose phosphate pathway represents an alternative means for the oxidation of sugars and provides a ready mechanism of the formation of 3, 4, 5, 6, and 7 carbon skeletons for synthetic reactions. The pathway also provides NADPH + H$^+$ and ribose-5-phosphate, which are needed for nucleic acid production.

higher rate in the light than in the dark. This light-stimulated loss of carbon, termed photorespiration, is a process that is in addition to or superimposed on the normal dark respiratory processes in the plant discussed previously.

If using a strict interpretation, photorespiration should not be considered a respiratory process since there is no net gain in the energy trapped during the transfer of energy between molecules, which is a classical requirement for respiration. Rather, it represents a form of oxidative photosynthesis. A significant portion of the carbon that is fixed into sugars in many species actually moves through this pathway. Since it is generally viewed as a respiratory process, for continuity this conventional approach is used here.

In contrast to photorespiration, dark respiration (glycolysis, tricarboxylic acid, pentose phosphate pathway, and electron transport system) proceeds at essentially the same rate whether in the dark or in the light. It has been estimated that 30–50% of the photosynthetically assimilated carbon in the leaves of some C_3 plants may be lost through the process of photorespiration.[95]

The relative importance of photorespiration, and for that matter photobiology in general, during the postharvest period, has not been studied to any appreciable extent. As a consequence, the degree of concern with detrimental effects of light and the potential usefulness of light during this time frame remains to be ascertained. Because photorespiration occurs in chlorophyllous tissues that are actively photosynthesizing, it is assumed to be of greater importance in intact plants (e.g., bedding plants, woody ornamentals, transplants) than in detached plant parts. Since photorespiration decreases with both decreasing light intensity and oxygen concentration—both conditions common in postharvest handling—its rate could be readily altered.

The primary objective during the postharvest period is to maintain the product as close to the preharvest condition as possible (i.e., no significant growth in intact plants). As a consequence, the balance between photosynthesis and respiratory losses may be more critical than the actual rates of each process. Since photorespiration remains to date a largely unknown quantity in postharvest biology, this section contains only a brief overview of the process rather than an in-depth description.

Of the three primary photosynthetic carbon fixation pathways operative in higher plants, approximately 500 species utilize the C_4 pathway, 250 species utilize the CAM pathway and the remaining 300,000 are generally thought to utilize the C_3 pathway. In comparing the two primary groups, C_3 and C_4, there are a number of important characteristics that distinguish them. For example, plants having the C_3 photosynthetic pathway for carbon fixation have distinctly higher levels of photorespiration and CO_2 compensation points than do C_4 species (table 3-2). C_3 species, which comprise the majority of the woody and herbaceous ornamentals and transplants in postharvest handling and marketing, also differ in a number of other important characteristics. Photosynthesis in C_3 species is significantly inhibited by ambient oxygen levels (21%) and as a consequence, net photosynthesis is elevated and photorespiration is depressed with low oxygen conditions. In addition, photosynthesis in many C_3 species also tends to saturate at lower light intensities than in C_4 species and the optimum temperature for photosynthesis is significantly lower (table 3-2).

Table 3-2. Several Characteristics That Distinguish C₃ and C₄ Species

Characteristics	C_3 Plants	C_4 Plants
Leaf anatomy	No significant differentiation between mesophyll and bundle sheath cells	Bundle sheath cells containing large numbers of chloroplasts and other organelles
Major pathway of CO_2 fixation in light	Red. Pentose-P cycle (i.e., Calvin Benson cycle)	C_4 pathway plus Red. Pentose-P cycle
Photorespiration (glycolate oxidation)	High	Low
Inhibitory effect of O_2 on photosynthesis and growth	Yes	No
CO_2 compensation concentration in photosynthesis (ppm CO_2)	30–70	0–10
Net photosynthesis vs. light intensity	Saturation at ca. 1000–4000 ftc	No saturation
Maximum rate of net photosynthesis (mg CO_2/dm² leaf area/hr)	15–35	40–80
Optimum temperature for net photosynthesis (°C)	15–25	30–45
Transpiration rate (gH_2O/g dry wt)	450–950	250–350

Source: After Kanai and Black.[52]

During photosynthesis in C_3 species, a relatively high level of glycolic acid is synthesized; however, this molecule cannot be metabolized in the chloroplasts. Upon movement out of the chloroplast and into peroxisomes, glycolic acid is oxidized to glyoxylic acid, which is subsequently converted to glycine (fig. 3-7). Glycine then moves into adjacent mitochondria where two molecules of glycine react to produce one molecule of serine and carbon dioxide. Since the oxidation step is not linked to ATP formation, photorespiration results in both a loss of energy and photosynthetic carbon from the plant.

The inhibition of photosynthesis by oxygen was first observed by O. Warburg in 1929 and has subsequently been known as the Warburg effect, in the same manner as the inhibition of sugar breakdown by oxygen was named the Pasteur effect after L. Pasteur. This inhibition of photosynthesis by oxygen involves the competition between molecules of carbon dioxide and oxygen for ribulose bisphosphate carboxylase,* the primary photosynthetic carboxylating enzyme. The higher the oxygen level the more favored the oxygenation reaction and the greater the production of glycolic acid, which is the substrate for photorespiration.

---

* Also called ribulose diphosphate carboxylase.

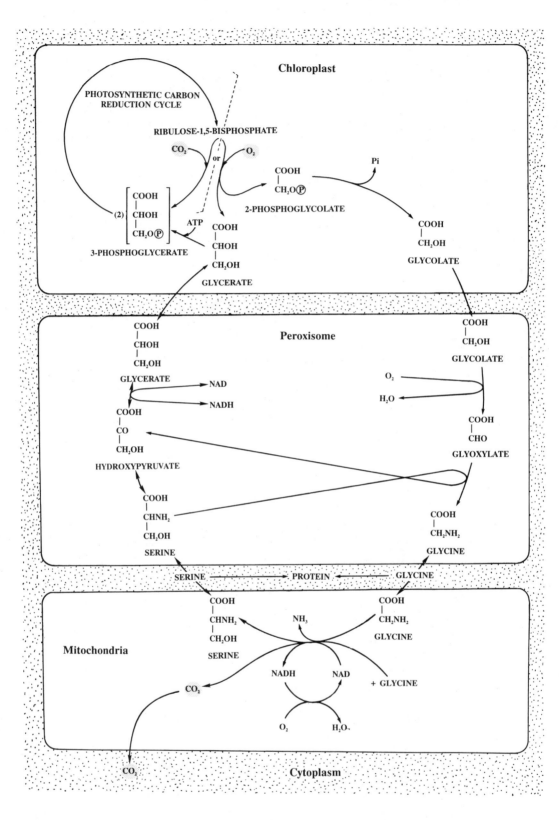

Figure 3-7 *(at left).* The pathway for carbon metabolism during photorespiration. Glycolate produced in the chloroplasts is transported to peroxisomes where it is oxidized to glycine. Glycine is then converted in the mitochondria, forming serine and liberating carbon dioxide. Serine can then be cycled back through the peroxisomes and converted to glycerate, which reenters the C_3 cycle, giving a net loss of one molecule of carbon dioxide per molecule of glycolate formed.

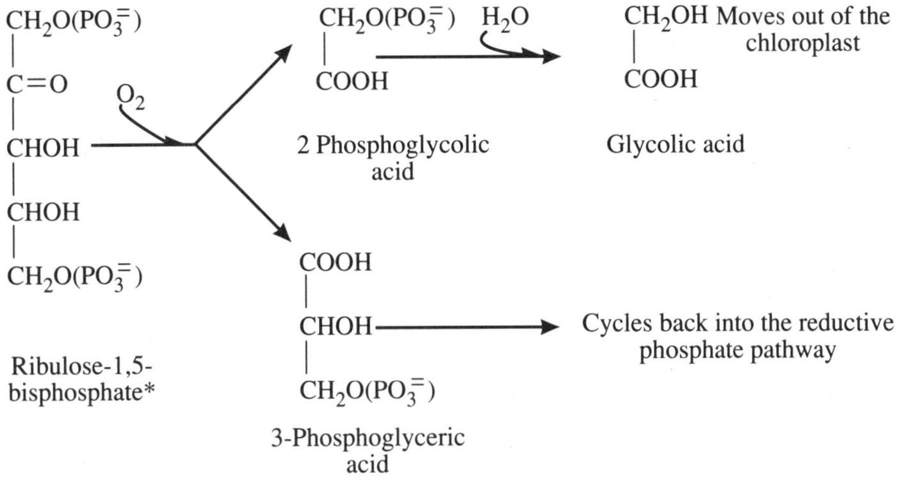

$CH_2O(PO_3^=)$
|
$C=O$ O_2
|
$CHOH$ ———→
|
$CHOH$
|
$CH_2O(PO_3^=)$

Ribulose-1,5-
bisphosphate*

$CH_2O(PO_3^=)$ H_2O
|
$COOH$ ————————→

2 Phosphoglycolic
acid

CH_2OH Moves out of the
| chloroplast
$COOH$

Glycolic acid

$COOH$
|
$CHOH$ ————————→ Cycles back into the reductive
| phosphate pathway
$CH_2O(PO_3^=)$

3-Phosphoglyceric
acid

When the oxygen concentration is lowered, the carboxylation reaction is increasingly favored.

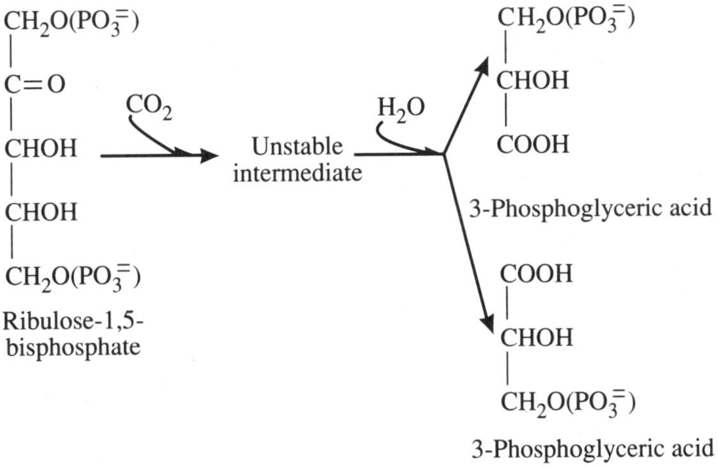

$CH_2O(PO_3^=)$
|
$C=O$ CO_2
|
$CHOH$ ———→   Unstable
| intermediate
$CHOH$
|
$CH_2O(PO_3^=)$

Ribulose-1,5-
bisphosphate

H_2O

$CH_2O(PO_3^=)$
|
$CHOH$
|
$COOH$

3-Phosphoglyceric acid

$COOH$
|
$CHOH$
|
$CH_2O(PO_3^=)$

3-Phosphoglyceric acid

The rate of photorespiration is difficult to measure precisely in an illuminated leaf since a portion of the carbon dioxide respired is photosynthetically refixed before it escapes from the leaf. The carbon dioxide compensation point, the concentration of carbon dioxide in the atmosphere where carbon dioxide fixed equals that respired, is often used as an index of photorespiration. Species that have high compensation points (30–70 ppm carbon dioxide) have high rates of photorespiration and, conversely, those with low

---

* Also referred to as ribulose-1,5-diphosphate.

compensation points (0–10 ppm carbon dioxide) have low photorespiration rates. For C_3 species, the difference in the rate of photosynthesis at 21% oxygen and 2% oxygen is also used as a measure of photorespiration since photorespiration is almost totally blocked (the oxidation of ribulose-1,5-bisphosphate) by low oxygen.

PHOTOSYNTHESIS

Photosynthesis, the process by which green plants capture light energy and convert it into chemical energy, is not commonly considered a significant postharvest metabolic process. This is no doubt due to the fact that many of the products traditionally associated with postharvest physiology contain few chloroplasts and are usually stored in the dark. A significant number of products, however, have the potential to photosynthesize and many, although not all, may derive benefit from this process upon removal from the production area. These products fall into two major groups: (1) intact plants such as ornamentals, leafy cuttings, and tissue cultures; and (2) chlorophyll-containing detached plant parts such as green apples or pepper fruits, petioles, shoots, and leaves. Therefore, a distinct group of postharvest products are, at least theoretically, not totally severed from an external source of energy that may be used for the maintenance of homeostasis. In some cases, this situation may substantially reduce or eliminate the products dependence on stored reserves.

With intact plants, there are two general options for handling the product. Conditions can be created or selected that will maintain the plants' photosynthetic environment. The requirements are light, an appropriate CO_2 concentration and temperature, and sufficient water to maintain an adequate moisture balance within the plant. In contrast to the site of production, the postharvest environment is maintained at a lower level of these requisites, a level that will ensure maintenance of the product rather than enhanced growth and development. In many postharvest environments for intact plants, appropriate plant moisture status is the parameter that is perhaps most commonly handled improperly.

A postharvest environment may also be selected for intact plants that will minimize the metabolic rate of the product. Therefore, in contrast to an environment that is conducive to photosynthesis and the acquisition of sufficient external energy to maintain the plant, an environment can be selected that will simply minimize the utilization of stored energy reserves. This option is the primary one selected for the handling of detached plant parts, and in both intact plants and detached plant parts it is accomplished largely by product temperature management.

Products that were photosynthetic organs prior to being severed from the plant at harvest (i.e., lettuce (*Lactuca sativa* L.), amaranths (*Amaranthus* spp.), spinach (*Spinacea oleracea* L.)) might be considered as logical candidates to derive a benefit from light during storage. This, however, is rarely the case. One reason is that the light energy trapping efficiency of plants even under optimum conditions is very low (usually under 5%), the remaining energy being dissipated primarily as heat. This situation elevates the leaf temperature, which results in a counterproductive acceleration of the use of stored energy reserves via the respiratory pathways. In intact plants, leaf temperature is decreased through the cooling effect of evapotranspiration. One gram of water removes 540 calories of heat upon being transformed from a liquid to a gas. Severed plant parts, however, do not have a readily replenishable source of water that can be used for cooling via evapotranspiration or in photosynthetic reactions and turgor maintenance. As a consequence the product temperature increases.

An additional problem with utilizing photosynthesis to help maintain harvested chlorophyll-containing plant parts is that the temperatures at which the products are normally stored are substantially below those required for optimum photosynthesis. These lower temperatures are essential, however, for successful storage since they decrease the metabolic rate of the product and the subsequent utilization of stored energy reserves.

In products that benefit from photosynthesis after harvest, the amount of external energy needed prior to harvest differs from that required after harvest. This difference is based on a distinction made between the primary goals of the product before and after harvest. Prior to harvest, growth is a primary goal; therefore, carbon and energy acquisition must be greater than respiratory utilization. After harvest, during the postharvest handling period, growth is seldom desirable. Rather, the objective is to maintain the product in as close to its preharvest condition as possible (i.e., minimize change). Therefore, photosynthesis after harvest is seen as a way of maintaining the energy balance within the plant rather than as a means of providing excess energy for the purpose of carbon accumulation.

Light Reactions

Photosynthesis occurs within specialized plastids, called chloroplasts, found primarily in leaves. The most important pigment in these plastids is chlorophyll, but other pigments such as carotenoids and phycobilins also participate in photosynthesis. The simplified overall reaction occurring in photosynthesis can be written as:

$$6CO_2 + 6H_2O + \text{light (h}\nu) \xrightarrow{\text{green plant}} C_6H_{12}O_6 + 6O_2$$

where CO_2 is fixed and oxygen from water is released. This process can be divided into three major steps, two that require light (light reactions) and one that does not (dark reactions). The light reactions involve the splitting of water with the release of oxygen:

$$\text{light (h}\nu) + H_2O + NADP \longrightarrow \tfrac{1}{2}O_2 + NADPH + H$$

and the light-driven formation of ATP from ADP and Pi (photophosphorylation). The reactions trap light energy (photons) and transport it in the form of electrons from water through a series of intermediates to NADP where it can be stored (fig. 3-8). Two separate light reactions act cooperatively in elevating the electrons to the energy level required for their transfer to NADP. In this process, the electrons are transported via an electron transport chain, which while operating on the same alternating oxidation-reduction principle as the respiratory electron transport system, is distinctly different.

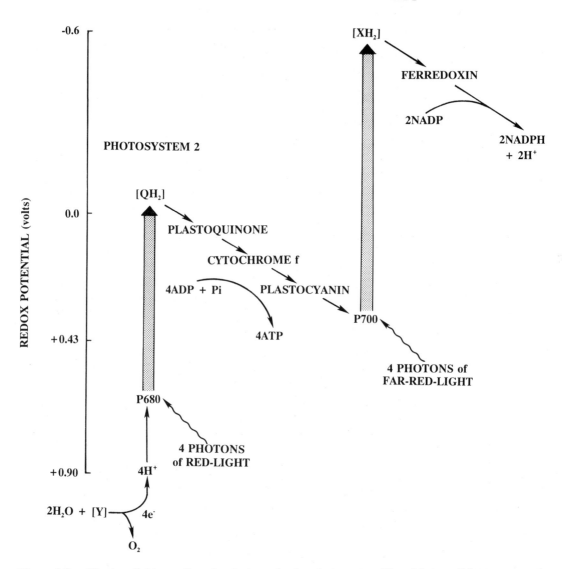

Figure 3-8. The two light reactions in photosynthesis, photosystem II and I, trap light energy and convert it to ATP and NADPH + H$^+$; oxygen is also liberated in the process. Energy is absorbed by photosystem II, which results in the splitting of water (photolysis) releasing O$_2$ and H$^+$ and the excitation of electrons to a high-energy level where they can be accepted by the first carrier in a series that transfers the electrons to the chlorophyll of photosystem I. Additional light energy absorbed by the chlorophyll molecule increases the energy level of the electrons that are trapped by an electron acceptor and subsequently transferred to ferredoxin. NADP is reduced, utilizing the H$^+$ formed in the photolysis of water, yielding NADPH + H$^+$.

Dark Reactions

The energy trapped in the light reactions as NADPH + H and ATP can be used in a number of reactions within the plants; however, its primary role is in the fixation of carbon from atmospheric CO$_2$ (dark reactions). In plants commonly encountered in postharvest conditions, there are three primary means of fixation of CO$_2$: the C$_3$, C$_4$, and CAM pathways.

Reductive Pentose Phosphate Pathway

The C_3 or reductive pentose phosphate cycle is operative within the majority of plant species. Here CO_2 is fixed by reacting with ribulose-1,5-bisphosphate (5-carbon sugar), forming two 3-carbon phosphoglyceric acid (PGA) molecules.

Ribulose-1,5-bisphosphate   (2) Phosphoglyceraldehydes

Light provides the energy required to convert PGA back into ribulose-1,5-bisphosphate (fig. 3-9) for the continuation of the process. A total of nine ATP equivalents are required for the fixation of one molecule of CO_2: three as ATP and two NADPH molecules. The dark fixation pathway is also referred to as the Calvin-Benson cycle after M. Calvin and A. A. Benson, who were largely responsible for elucidating the series of reactions.

C_4 Pathway

In some species of plants, CO_2 reacts with phosphoenolpyruvic acid, forming the 4-carbon compound, oxaloacetic acid. Hence, the name C_4 pathway. Oxaloacetic acid is then converted to malic acid (fig. 3-10) from which CO_2 is removed (decarboxylated), forming pyruvic acid and allowing the resynthesis of phosphoenolpyruvic acid, completing the cycle. The CO_2 removed is not lost but is fixed via the reductive pentose phosphate cycle in adjacent cells. Therefore, in C_4 plants, the enzymes required for both the C_4 and C_3 pathways are present.

Crassulacean Acid Metabolism

A third means of fixing carbon is found in crassulacean acid metabolism (CAM) plants. Here CO_2 is fixed at night when the stomates of these plants are open through the action of the enzyme phosphoenolpyruvate carboxylase forming oxaloacetic acid from phosphoenolpyruvic acid (fig. 3-11). During

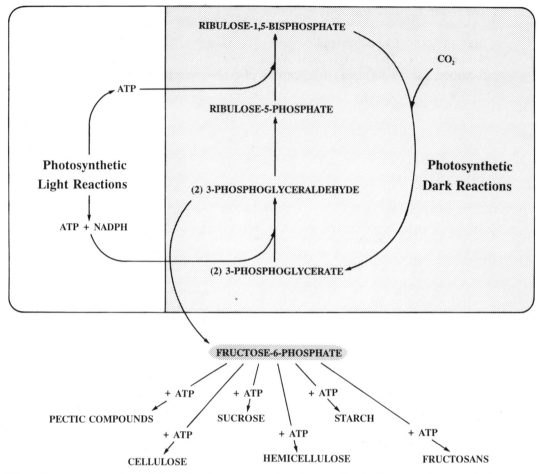

Figure 3-9. An overview of the reductive pentose phosphate or Calvin-Benson photosynthetic cycle. NADPH and ATP formed in the light reactions (fig. 3-8) are used to convert 3-phosphoglyceraldehyde back to ribulose-1,5-bisphosphate to complete the cycle.

the day, malic acid, which is formed from oxaloacetic acid, has CO_2 removed (decarboxylated) and refixed via the reductive pentose phosphate cycle. In CAM plants, both the C_3 and C_4 cycles are operative and found within the same cells as the CAM cycle.

METABOLIC CONSIDERATIONS IN HARVESTED PRODUCTS

Dark Respiration

As we have noted in the general equation for the oxidation of a hexose sugar (page 76), substrate and oxygen are converted into carbon dioxide, water, and energy. These chemical changes have a number of significant effects on both the stored product and the environment surrounding the product.

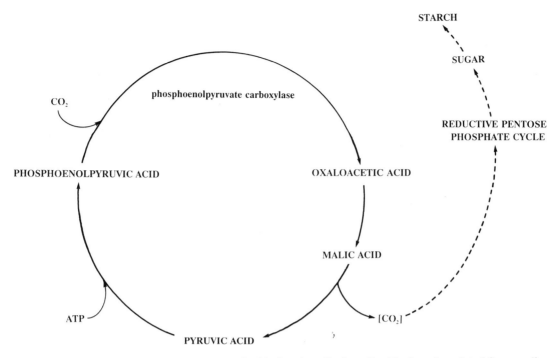

Figure 3-10. The C$_4$ pathway for carbon dioxide fixation. Carbon dioxide decarboxylated from malic acid is refixed via the reductive pentose phosphate cycle and is therefore not lost from the plant.

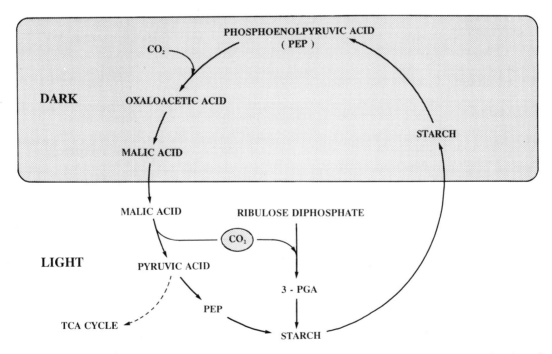

Figure 3-11. The CAM or crassulacean acid cycle found in some plants involves the dark fixation of carbon dioxide and the formation of malic acid recycled in a series of reactions eventually forming phosphoenolpyruvic acid.

Effects of Respiration

Two of the primary functions of respiration are the release of energy stored chemically as sugars, lipids, and other substrates and the formation of carbon skeletons that can be used in various synthetic and maintenance reactions. These processes are described in more detail in the section on dark respiration pathways in this chapter. Secondary effects of respiration include the utilization of substrate in the stored product, the consumption of oxygen from the surrounding environment, and the production of carbon dioxide, water, and heat. Several of these effects substantially alter the methods that must be employed in handling and/or storing many products and as a consequence are of considerable importance at the commercial level.

The loss of substrate from stored plant products results in a decrease in energy reserves within the tissue, which in turn decreases the length of time the product can effectively maintain its existing condition. Loss of energy reserves, which eventually results in tissue starvation and accelerated senescence, is especially critical in products such as leaves, flowers, and other structures that do not function as carbon storage sites. Respiration also decreases the total food value (i.e., energy content). Finally, in a marketing system based on weight, respiratory losses of carbon represent weight losses in the product; hence, a decreased value. The rate of respiration can in fact be used to predict the loss of dry weight from stored products (fig. 3-12).

Another extremely important effect of respiration is the removal of oxygen from the storage environment. If the ambient oxygen concentration is allowed to be excessively depleted, anaerobic conditions occur that can rapidly spoil many plant products. As a consequence, the rate of respiration is important for determining the amount of ventilation required in the storage area. It is also critical in determining the type and design of packaging material that can be used as well as the use of artificial surface coatings on the product [e.g., waxes on citrus or rutabagas (*Brassica napus*, L., Napobrassica group)]. The respiratory reduction in oxygen concentration in the storage environment can also be used as a tool to extend the storage life of a product. Since oxygen concentration has a pronounced effect on the rate at which respiration proceeds, a respiration-mediated decrease in the ambient oxygen concentration can be used to decrease respiration, creating a modified environment. This principle, used since Roman times, represents the basis for present-day storage practices for several highly perishable products.

Elevated ambient levels of carbon dioxide resulting from the respiratory process can be used to decrease respiration since its accumulation impedes the rate at which the process proceeds. The effectiveness of the inhibition of respiration by carbon dioxide and the sensitivity of the tissue to carbon dioxide, however, vary widely between products. Carbon dioxide produced during the respiratory process, if allowed to accumulate, can be harmful to many stored products. For example, lettuce,[63] mature green tomatoes (*Lycopersicon esculentum*, Mill.), bell peppers (*Capsicum annuum*, L.),[72] and other products are damaged by high levels of carbon dioxide. As a consequence, it is essential that the carbon dioxide concentration be maintained at a safe level through adequate ventilation or absorption of the molecule.

Water is produced during the respiratory process (termed metabolic water) and becomes part of all of the water present within the tissue. Metabolic

Muskmelons (100 kg) which are 90% moisture are stored at 5°C and have a respiratory rate of 9 mg CO_2/kg · hr and a fresh weight loss of 3%/day.

RATE OF DRY WEIGHT LOSS

For every 180 g of sugar oxidized, 264 g of CO_2 is produced by the product. Therefore, the rate of dry matter loss in grams of glucose/kg fwt of fruit/day is equal to:

The muskmelons lose: $\left[\dfrac{9 \text{ mg } CO_2/kg \cdot hr}{1000}\right][.68][24] = 0.147$ g/kg fwt/day or with 100 kg of fruit

= 14.7 g/load/day.

RESPIRATORY OR VITAL HEAT PRODUCED

One mole of glucose yields 686 kcal, therefore, for every 6 moles of CO_2 given off, 686/6 kcal has been produced. Then 1 mole of CO_2 represents 114 kcal or 114,000 cal/44 g (weight of 1 mole of CO_2) = 2.591 cal/mg CO_2. One Btu = 252 cal. Then the number of Btu's produced by 1 ton when 1 mg of CO_2/kg · hr is given off can be calculated by:

$$\left[\frac{1 \text{ mg}}{kg \cdot hr}\right]\left[\frac{2.591 \text{ cal}}{mg}\right]\left[\frac{1000 \text{ kg}}{metric \text{ ton}}\right]\left[\frac{1 \text{ Btu}}{252 \text{ cal/Btu}}\right]\left[\frac{24 \text{ hr}}{day}\right] = 247 \text{ Btu/metric ton} \cdot day \text{ or } 224 \text{ Btu/British ton} \cdot day.$$

Therefore, the 100 kg of melons will produce the following number of Btu's/day:

$$\left[\frac{9 \text{ mg } CO_2/kg \cdot hr}{kg \cdot hr}\right]\left[\frac{247 \text{ Btu}}{247 \text{ Btu}}\right][100 \text{ kg fruit weight}] = 222 \text{ Btu/day}$$

METABOLIC WATER PRODUCED

The ratio of the weight of CO_2 to water produced = 264/108. Therefore, the melons produce the following metabolic water: $\dfrac{264 \text{ g } CO_2}{108 \text{ g } H_2O} \times 9 \text{ mg } CO_2/kg \cdot hr = 3.68 \text{ mg } H_2O$/kg · hr or x mg H_2O/kg · hr

.00000368 kg H_2O/kg · hr.

The total amount of water produced by all of the melons per week
= $\left[\dfrac{.00000368 \text{ kg } H_2O}{kg \cdot hr}\right]\left[\dfrac{24 \text{ hr}}{day}\right]\left[\dfrac{7 \text{ days}}{week}\right][100 \text{ kg fruit}] = .0618$ kg H_2O/week. The percent of the total water

that is metabolic water in one week = $\dfrac{.0618 \text{ kg } H_2O/100 \text{ kg} \cdot week}{90 \text{ g } H_2O \text{ in } 100 \text{ kg fruit}] = .0687\%$ of the total water in the fruit.

Therefore, the 3% weight loss/week represents: 3 kg fresh weight - .103 kg CO_2/100 kg · week + .0618 kg metabolic water/100 kg · week = 2.96 kg fresh weight lost due to evapotranspiration.

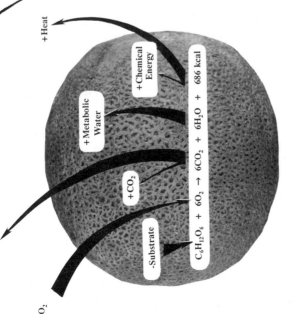

-O₂

+CO₂

+Heat

+Chemical Energy

+Metabolic Water

+CO₂

-Substrate

$$C_6H_{12}O_6 + 6O_2 \rightarrow 6CO_2 + 6H_2O + 686 \text{ kcal}$$

	GLUCOSE	OXYGEN	CARBON DIOXIDE	WATER	ENERGY
moles	1	6	6	6	
weight	180 g +	192 g =	264 g +	108 g	

Figure 3-12. Respiration results in the utilization of substrate (e.g., glucose) and oxygen and the formation of carbon dioxide, water, and energy. Knowing the rate of respiration (mg CO_2 produced per kg of product each hour) the rate of dry weight loss and the amount of heat and metabolic water produced can be calculated. Examples of these calculations for muskmelons (*Cucumis melo*, L., Reticulatis group) are illustrated in the figure.

water, however, represents only a very minor addition to the total volume of water within the tissue and as a consequence is probably of only minimal significance (fig. 3-12).

Energy, the final product in the respiratory equation, has a significant influence both on the maintenance of the product and the storage environment in which the product is held. The complete oxidation of one mole of a six carbon sugar such as glucose results in the formation of 686 kcal (686,000 gcal) of energy. In actively growing tissues, a significant portion of this energy is utilized in chemical forms by the cell for synthetic and maintenance reactions. A substantial amount of energy, however, is lost as vital heat.* In actively metabolizing tissues, estimates of around 62% of the total respiratory energy (423 kcal/mole of glucose) is lost as heat. This amount, however, varies between different types of plants and organs and the general condition of the tissue. In general, the amount of vital heat produced by the product can be calculated fairly accurately (i.e., within 10%[38]) directly from the respiratory rate of the product (fig. 3-12). Knowledge of the amount of vital heat produced is important in determining the amount of cooling required for a product and the size of the refrigeration system needed to maintain the desired temperature of the storage room. It also influences the amount of air movement required around the product in storage and consequently the package design and stacking method.

Respiratory Substrates

Respiration depends on the presence of a substrate. In many tissues, this substrate is a storage form of carbohydrate such as starch in the sweetpotato (*Ipomoea batatas,* (L.) Lam.) root or inulin in Jerusalem artichoke (*Helianthus tuberosus,* L.) tubers. These more complex molecules are broken down into simple sugars that can enter the respiratory pathways, providing energy for the plant. In some species, carbon may be stored for subsequent use as lipids. For example, the avocado fruit is comprised of approximately 25% lipid on a fresh-weight basis and pecan (*Carya illinoensis,* (Wang.) C. Koch.) kernels are comprised of approximately 74% lipid. Organic acids, proteins, and other molecules may also be utilized as respiratory substrates in plants, although in most cases they are not normally produced for this purpose. Under conditions where the tissue is depleted or "starved" of carbohydrate or lipid reserves, these secondary respiratory substrates are utilized. This situation is more likely to occur in postharvest products such as leaves or flowers that do not represent storage organs and therefore have relatively little reserve substrate. Proteins may be hydrolyzed into their component amino acids, which can then be catabolized in the glycolytic pathway and tricarboxylic acid cycle.

These various substrates utilize, when completely oxidized, different amounts of oxygen in relation to the amount of carbon dioxide given off by the plant. Thus, an analysis of the ratio of the number of molecules of carbon

---

* Energy transfers between molecules within the plant are not 100% efficient. Much of this lost energy is given off as heat, collectively referred to as vital heat to distinguish it from other sources of heat (e.g., solar radiation).

dioxide given off to the number of molecules of oxygen absorbed, called the respiratory quotient (RQ), can provide a general indication of what particular substrate is being used as the major source of respiratory energy. For example, the oxidation of a common carbohydrate, lipid, and organic acid gives the following respiratory quotients:

Type of Substrate	Substrate	Reaction	Respiratory Quotient (CO_2/O_2)
Carbohydrate	Glucose	$C_6H_{12}O_6 + 6O_2 \rightarrow 6CO_2 + 6H_2O$	1.00
Lipid	Palmitic acid	$C_{16}H_{32}O_2 + 11O_2 \rightarrow C_{12}H_{22}O_{11} + 4CO_2 + 5H_2O$	0.36
Organic acid	Malic acid	$C_4H_6O_5 + 3O_2 \rightarrow 4CO_2 + 3H_2O$	1.33

The respiratory quotient was of much greater interest in the earlier part of the twentieth century when the repertoire of analytical techniques for studying plant biochemistry was much more limited. Even then there was a great diversity of opinion as to the actual value of the measure. Along with tissue type, temperature, and tissue age, a number of other factors significantly alter the respiratory quotient. In addition, substrates are not always completely oxidized and/or several types of substrates may be used simultaneously by the cells, each of which greatly complicates the interpretation of the ratios obtained.

Control Points in the Respiratory Pathway

Changes in the internal environment of the cells of a storage product, whether mediated by external (e.g., temperature) or internal (e.g., substrate availability) factors, often result in significant changes in respiration. These changes may be due to shifts in the points at which the regulatory pathways are controlled or may be due to changing priorities in the level of operation of specific pathways. Control of respiration in plant cells can be regulated at various points in the respiratory pathways and by a number of means. Substrate supply may exert a control on the rate of respiration by regulating the availability of the substrate for a particular reaction. For example, if glucose-6-phosphate levels are high, the reaction catalyzed by phosphoglucoisomerase shifts carbon movement toward the formation of fructose-6-phosphate in order to maintain an equilibrium. Substrate control is probably more important when demand for intermediates in the tricarboxylic acid cycle is high.

Enzymatic control, both through the activity of an enzyme and to a lesser extent by the concentration of the enzyme, represents a major form of control in respiring cells. Enzyme activities may be modulated by substrate and product concentration, cofactors such as metal ions, activation or inhibition of the enzyme, and concentration of the enzyme. Enzyme concentration for rate-limited reactions is thought to represent a means of course control for the reaction or series of reactions. In contrast, enzyme activation is considered a means of fine control.

The availability of phosphate acceptors (ADP) represents an extremely important means of respiratory control. Restricting the rate of flow of electrons through the electron transport chain and, hence, the rate of oxidation of NADH, limits the rate of a number of reactions. However, if NADH is reoxidized by an alternative reaction, oxidative phosphorylation is diminished in its regulatory role. High levels of ATP also directly inhibit certain enzyme reactions, for example, phosphofructokinase and pyruvate kinase. Therefore, the levels of ADP, NAD, and NADP and their reduced products represent important modulators of the respiratory pathways.

The tricarboxylic acid cycle appears to be largely regulated by the latter means: mitochondrial energy status. However, low oxygen and high carbon dioxide are known to have a pronounced effect on the rates of specific enzymes in the cycle. High carbon dioxide is known to inhibit the conversion of succinate to malate and malate to pyruvate in apple fruit tissue.[59] Key enzymes controlling the rate of the glycolytic pathway, on the other hand, are phosphofructokinase and pyruvate kinase, while in the pentose phosphate pathway the activity of glucose-6-phosphate dehydrogenase is controlled by the NADPH/NADP ratio.

Factors Affecting the Rate of Dark Respiration

The postharvest respiratory responses of plant products can be strongly influenced by a number of commodity and environmental factors. Since elevated respiratory rates are in many products closely correlated with shortened product life expectancy, proper management of the factors that affect respiration is imperative for maximizing storage life.

TEMPERATURE

Temperature has a pronounced effect on the metabolic rate of a harvested product. As the product temperature increases, reaction rates increase, although not all reactions within a tissue will have the same relative rates of change (e.g., the optimum temperature for photosynthesis is usually lower than the optimum temperature for respiration). These changes in the rates of reactions due to temperature are commonly characterized using a measure called the Q_{10}, which is the ratio of the rate of a specific reaction at one temperature (T_1) versus the rate at that temperature + 10°C (rate at T_1 + 10°C/rate at T_1). Q_{10}'s are often quoted for respiration as respiration gives a very general estimate of the effect of temperature on the overall metabolic rate of the tissue. There are, however, many metabolic exceptions; for example, the respiratory rate of potato tubers decreases with decreasing temperature and the formation of sugars from starch increases dramatically below 10°C (fig. 3-13). For many products the Q_{10} for respiration is between 2.0 and 2.5 for the 5°C to 25°C temperature range. Therefore, for every 10°C increase in temperature, the respiratory rate increases 2.0 to 2.5 times.

As a consequence, if the objective is to maintain a product as close to the condition it was in at harvest, the use of low temperature to depress changes due to metabolism of the product is essential. As the temperature increases from 25°C into the 30°C to 35°C range, the Q_{10} for most products declines, and at very high temperatures reaction rates are actually depressed, probably due

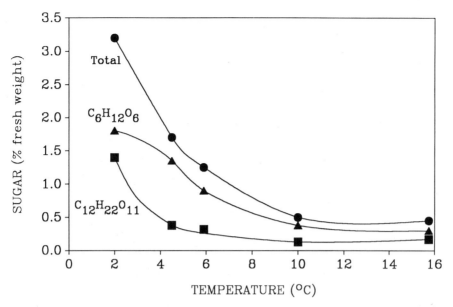

Figure 3-13. The effect of temperature on the formation of sugars from starch in potato tubers (*Solanum tuberosum*, L. cv. 'Majestic') *(after Burton[22])*.

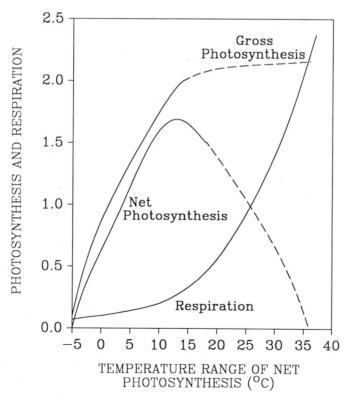

Figure 3-14. The effect of temperature on photosynthesis, respiration, and net or apparent photosynthesis of Swiss stone pine (*Pinus cembra*, L.) seedlings (broken lines represent estimates) *(after Tranquillini[88])*.

to the denaturing of enzymes. The actual range in temperature, over which there is a linear increase in Q_{10}, and the maximum and minimum temperature for a particular metabolic process varies substantially between species and the type of tissue monitored (fig. 3-14). For example, respiration in *Populus tremuloides,* Michx. stems can be measured at $-11°C,$[32] a temperature that would freeze an apple (*Malus sylvestris,* Mill.) fruit solid, terminating respiration.

It is important to note that while the ambient temperature at which the product is stored is of critical importance in determining the metabolic rate, the product temperature is typically slightly higher than the ambient temperature due to the heat liberated from the respiratory process. This often slight difference in temperature is quite important due to its effect on maintaining the moisture balance of the harvested product (see chap. 7).

GAS COMPOSITION

The composition of the gaseous atmosphere to which postharvest products are exposed can markedly influence both the respiratory and general metabolic rate of the commodity. Oxygen, carbon dioxide, and ethylene are the most important in influencing respiration; however, gases such as sulfur dioxide, ozone, and propylene, which are generally considered pollutants, can have a significant effect if their concentration becomes sufficiently high.

During the normal growth and development of most plant products there are seldom large or long-term alterations in the composition of the gas atmosphere surrounding the tissue. The organ maintains a dynamic equilibrium with the changing environment in which it is growing. Upon harvest, however, the plant products are normally bulked tightly together and placed in containers and storage areas that restrict airflow (fig. 3-15). This situation creates additional resistances to normal gas movement into and out of the product that can significantly alter the concentration of gases within the tissue. With most products, equilibrium is shifted toward a decrease in internal oxygen and an increase in carbon dioxide; however, the opposite may be true for some crops such as submerged aquatics (Chinese water chestnut, *Eleocharis dulcis,* L. or lotus root, *Nelumbo nucifera,* Gaertn.) and root and tuber crops that grow normally under conditions of high diffusion resistance. Therefore, postharvest conditions commonly result in significant alterations in the gas environment to which the product has been accustomed and these changes can have a significant influence on the metabolic activity of the tissue.

The effect of oxygen concentration on harvested plant products has been known since around the beginning of the nineteenth century (see review by Hill[47]). Berard in 1821[8] noted that fruits held in an environment devoid of oxygen would not ripen, and if they were not kept too long under those conditions, they would ripen normally upon return to air.

Oxygen is closely tied to the rate of respiration of harvested products. As the internal concentration decreases, respiration decreases (fig. 3-16) until the oxygen concentration reaches what is called the extinction point or critical concentration. Here, the rate of respiration begins to increase with additional decreases in oxygen concentration and represents the point at

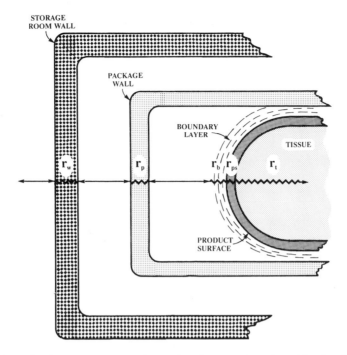

Figure 3-15. Resistances to gas exchange in a harvested product. Total resistance is r_w (storage walls) + r_p (package wall) + r_b (boundary layer) + r_{ps} (product surface) + r_t (tissue). The resistances for r_w, r_p, and r_{ps} are often manipulated for certain postharvest products to extend storage life.

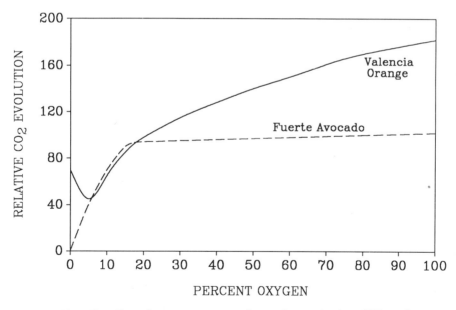

Figure 3-16. The effect of oxygen concentration on the respiration of 'Fuerte' avocados (*Persea americana*, Mill.) and 'Valencia' oranges (*Citrus sinensis*, (L.) Burm. f.). Note the "Pasteur effect" (stimulated respiration) on oranges at very low oxygen levels *(after Biale[9,11])*.

which aerobic respiration via the tricarboxylic acid cycle is blocked and anaerobic fermentation begins. This increase in respiration is known as the Pasteur effect after Louis Pasteur, who first studied the phenomenon in microorganisms. Although found widely in plants, it does not occur, at least to a significant degree, in all harvested products. For example, there is not an increase in respiration in the avocado fruit even at 0% oxygen (fig. 3-16).

The precise reason why the glycolytic pathway is stimulated by very low oxygen concentrations in most products is not known, although a number of possible explanations have been suggested. For example, the enzyme phosphofructokinase, which catalyzes the conversion of fructose-6-phosphate to fructose-1,6-bisphosphate in the glycolytic pathway, is inhibited by ATP and citric acid, both of which are formed in the oxygen-dependent TCA cycle and may represent the modulating factors. It is also possible that oxygen inhibits one or more of the glycolytic enzymes.

In contrast to the stimulation of respiration at very low oxygen, the low oxygen concentration in the range normally utilized in controlled atmosphere storage decreases the respiratory rate of the tissue (i.e., 1–3% O_2). This effect is pronounced and extremely important from a postharvest context. For example, the respiratory rate of apple fruit can be depressed and the onset of ripening can be delayed using low oxygen. Respiration is decreased through an effect on several of the enzymes in the TCA cycle. In green banana fruit, low oxygen limits the operation of the enzymatic steps between either oxaloacetate or pyruvate and citrate and between 2-oxoglutarate and succinate.[68]

The respiratory rate of most stored products can be decreased by lowering the oxygen concentration to a concentration that is not below the extinction point for that product (table 3-3). The actual critical concentration of oxygen appears to vary with the wide range of products handled. In addition, the external concentration of oxygen, which will give the appropriate internal concentration, varies with the rate of utilization of oxygen by the tissue, the diffusive resistance of the tissue, and the differential in partial pressures for the gas between the interior and the exterior. Therefore, at higher temperatures, the minimum external concentration of oxygen will increase due to the more rapid utilization of the molecule within the tissue.

In general, substantial decreases in the respiratory rate for most stored products do not occur until the external oxygen concentration is reduced below 10%. The optimum external concentration for a number of products held in cold storage is in the 1% to 3% range; however, there are many exceptions. For example, sweetpotatoes shift to anaerobic conditions at external oxygen concentrations below 5% to 7%.[23] While species variation contributes to these differences, the external oxygen concentration is only one of a number of factors that controls the internal level of oxygen, which is the critical parameter. Much of the variation in optimum external oxygen concentrations between types of products, and even between cultivars, can be accounted for by factors other than the external oxygen concentration that governs the internal concentration (e.g., environmental and physiological conditions that affect diffusion resistance and rate of oxygen use).

The use of low oxygen for stored products has the potential to decrease overall metabolic rate of the product and a diverse array of specific biochemical changes associated with it. These specific effects include such factors as rate of use of respiratory substrate and the production of energy. At the

Table 3-3. Effect of Temperature and Oxygen Concentration on the Respiratory Rate of Various Commodities

| | Carbon Dioxide Production ($mg \cdot kg^{-1} \cdot hr^{-1}$) | | | | | |
| | In Air | | | In 3% O_2 | | |
Temperature (°C)	0	10	20	0	10	20
Asparagus	28	63	127	25	45	75
Beans						
broad	35	87	145	40	55	80
runner	21	36	90	15	25	46
Beetroot						
storing	4	11	19	6	7	10
bunching with leaves	11	22	40	7	14	32
Blackberries, Bedford Giant	22	62	155	15	50	125
Blackcurrants, Baldwin	16	39	130	12	30	74
Brussels sprouts	17	50	90	14	35	70
Cabbage						
Primo	11	30	40	8	15	30
January King	6	26	57	6	18	28
Deccma	3	8	20	2	6	12
Carrots						
storing	13	19	33	7	11	25
bunching with leaves	35	74	121	28	54	85
Calabrese	42	105	240	—	70	120
Cauliflower, April Glory	20	45	126	14	45	60
Celery, white	7	12	33	5	9	22
Cucumber	6	13	15	5	8	10
Gooseberries, Leveller	10	23	58	7	16	26
Leeks, Musselburgh	20	50	110	10	30	57
Lettuce						
Unrivalled	18	26	85	15	20	55
Kordaat	9	17	37	7	12	25
Klock	16	31	80	15	25	45
Onion, Bedfordshire Champion	3	7	8	2	4	4
Parsnip, Hollow Crown	7	26	49	6	12	30
Potato						
maincrop (King Edward)	6	4	6	5	3	4
"new" (immature)	10	20	40	10	18	30
Peas (in pod)						
early (Kelvedon Wonder)	40	130	255	29	84	160
main crop (Dark Green Perfection)	47	120	250	45	60	160
Peppers, green	8	20	35	9	14	17
Raspberries, Malling Jewel	24	92	200	22	56	130
Rhubarb (forced)	14	35	54	11	20	42
Spinach, Prickly True	50	80	150	51	87	137
Sprouting broccoli	77	170	425	65	115	215
Strawberries, Cambridge Favourite	15	52	127	12	45	86
Sweetcorn	31	90	210	27	60	120
Tomato, Eurocross BB	6	15	30	4	6	12
Turnip, bunching with leaves	15	30	52	10	19	39
Watercress	18	80	207	19	72	168

Note: Data from Robinson et al.[81]

product level, the net effect may be seen as retarded ripening, aging, or the development of several storage disorders in certain postharvest commodities. Thus, it may appear surprising that low oxygen environments are not used more extensively during the postharvest period. However, the potential benefit of using low oxygen to restrict the overall metabolic rate of a product must be weighed against the cost of utilizing it. Often the very short time span between harvest and retail sale of many products precludes its use. In addition, for most products, the costs are substantially greater than the benefits. There are, however, notable exceptions. More than 2.6 billion pounds of apples are stored each year in the United States utilizing low oxygen conditions. These conditions extend the availability of the crop 4 to 6 months over conventional storage practices, greatly increasing the net worth of the industry.

Elevation of the ambient concentration of carbon dioxide in a number of postharvest tissues impedes the forward movement of the respiratory pathway, resulting in a net and often quite significant decrease in respiration. This effect of carbon dioxide, although not universal, has been shown in seedlings, intact plants, and detached plant parts and has been found under both aerobic and anaerobic respiratory conditions. The degree to which respiration is impeded increases in relation to the concentration of carbon dioxide in the atmosphere. For example, in pea (*Pisum sativum,* L.) seedlings the inhibitory action of carbon dioxide at concentrations of up to 50% increased approximately with the square root of the concentration.[54] Carbon dioxide, therefore, appears to retard the rate of respiration but does not totally block it.

Under aerobic conditions, the effect of high carbon dioxide is closely tied to the level of respiratory energy reserves within the tissue. Respiration is impeded by high carbon dioxide when sufficient respiratory reserves are present, a common condition with most postharvest products. However, under conditions where the tissue is depleted of a ready source of stored carbon for respiratory processes, respiration is no longer decreased by carbon dioxide.

The precise mechanism(s) of action of high carbon dioxide resulting in a decrease in the respiratory process has not been adequately explored. It is known that the inhibitory effect is not due to permanent injury to the tissue, since upon removal of the carbon dioxide, respiration returns to normal. High carbon dioxide concentration under aerobic conditions has been shown to affect the tricarboxylic acid cycle in apple fruits at two points: the conversion of succinate to malate and malate to pyruvate.[59] Succinate dehydrogenase appears to be the enzyme most significantly affected, with the influence on other TCA cycle enzymes being negligible. High concentrations of carbon dioxide (e.g., 15%) have been shown to result in toxic levels of succinate accumulating in apples, causing damage to the tissue.[49]

Increasing the carbon dioxide concentration during storage does not depress respiration in all tissues. In some cases, respiration may be unaffected or significantly increased by carbon dioxide. The respiratory rate of potato tubers, onion (*Allium cepa,* L.) and tulip bulbs (*Tulipa* spp.), and beet roots (*Beta vulgaris,* L.) has been shown to be substantially increased, in some instances up to 200%, upon exposure to extremely high levels of carbon dioxide (30–70%).[87] This increase has also been shown in lemon (*Citrus limon,* (L.) Burm.) fruits at 10% carbon dioxide. Carrot roots, on the other hand, were not affected.

The mechanisms resulting in the stimulation of respiration by high carbon dioxide may be in part due to fixation of the molecule by malic enzyme and/or phosphoenolpyruvate carboxylase (fig. 3-17). The initial products formed in lemon fruits after brief exposure to $^{14}CO_2$ are malic, citric, and aspartic acids.[94] High concentrations of carbon dioxide may, therefore, facilitate the formation of tricarboxylic acid cycle intermediates, which stimulate respiration. The stimulation in respiration may also be related to secondary effects of the carbon dioxide molecule on the cytoplasm such as pH alteration. The effect of elevated carbon dioxide on cellular pH is complex. The molecule is readily soluble in the cytoplasm and vacuole, existing as bicarbonate and hydrogen ions with the dissociation of carbonic acid. The magnitude of pH changes varies with the tissue in question. Lettuce exposed to 15% carbon dioxide had a reduction in pH (i.e., 0.4 units in the cytoplasm, 0.1 in the vacuole) but when moved back into air the pH elevated to near pretreatment levels.[83] The change in pH of freshly harvested green peas exposed to elevated carbon dioxide was compensated for by a decrease in malic acid concentration, giving essentially no net change.[91] Short transitory changes in pH caused by returning the tissue to ambient carbon dioxide conditions may activate the carboxylation of phosphoenolpyruvate to oxalacetate and subsequently malate.[92] At present the effects of even small changes in pH on respiration are not known.

The respiratory rate of a number of postharvest products can be significantly stimulated by the hormone ethylene. This was first illustrated by the work of Denny[27,28] on citrus fruit and later on bananas.[45] This effect of ethylene is of considerable interest in postharvest biology in that harvested products produce ethylene that can alter the respiratory rate of the product if allowed to accumulate. In most cases, however, an increase in respiratory rate per se represents only a minor concern in relation to other major biochemical changes in quality that may also be induced by ethylene (e.g., accelerated floral senescence, loss of chlorophyll, abscission).

A relatively wide range of vegetative and reproductive tissues respond to ethylene with an increase in respiratory rate (table 3-4) and the increase is dependent on the continued presence of the gas.[79,85] However, respiration is not stimulated in all tissues by ethylene. For example, the respiration of strawberry (*Fragaria* × *Ananassa,* Duchesne) fruits, *Pisum sativum,* L.,[36] and *Triticum dicoccum* Schubler[64] seedlings and peanut (*Arachis hypogaea,* L.) leaves[53] is not stimulated. Flowers typically undergo a decline in respiration after harvest followed by an increase as they begin to senesce. Enhanced respiration by ethylene in the case of flowers may be largely associated with an indirect effect, through an acceleration of the senescence process, rather than a direct effect on respiration. Many of the tissues in which respiration is accelerated by ethylene have significant storage energy reserves. It has been suggested that the ability of a tissue to respond to ethylene with an increase in respiration is closely correlated with the presence of the alternate or cyanide-resistant electron transport pathway (table 3-4). In these tissues, both HCN and ethylene stimulate respiration although through different mechanisms. The respiratory increases in both cases do not necessarily involve the induction of ripening or stimulated ethylene synthesis since they can be found in potato tuber tissue, which neither ripens nor has autocatalytic synthesis of ethylene. With both ethylene

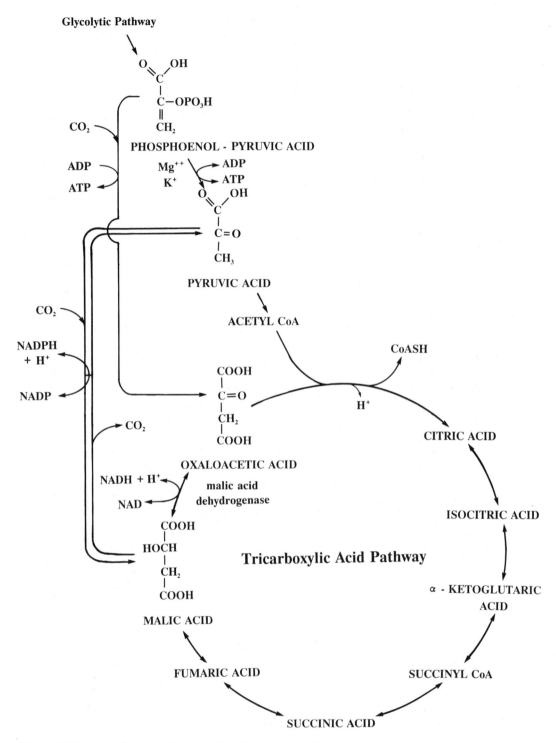

Figure 3-17. Dark fixation of carbon dioxide in plant cells can proceed via at least two mechanisms. Pyruvic acid can be converted to malic acid by malic enzyme. Likewise, phosphoenolpyruvate carboxylase is capable of catalyzing phosphoenolpyruvate to oxaloacetic acid in the dark.

Table 3-4. Effect of Ethylene and Cyanide on the Respiratory Rate as Measured by Oxygen Uptake of Various Types of Plant Tissue

| Tissue | $\mu l\ O_2 \cdot g^{-1}$ fresh wt $\cdot$ hour^{-1} | | |
	Control	Ethylene	Cyanide
Fruit			
Apple, *Malus sylvestris*	6	16	18
Avocado, *Persea americana*	35	150	150
Cherimoya, *Annuva cherimola*	35	160	152
Lemon, *Citrus limon*	7	16	21
Grapefruit, *Citrus grandis*	11	30	40
Stem			
Irish Potato, *Solanum tuberosum*	3	14	14
Rutabaga, *Brassicanapus* Napobrassica group	9	18	23
Root			
Beet, *Beta vulgaris*	11	22	24
Carrot, *Daucus carota*	12	20	30
Sweetpotato, *Ipomoea batatas*	18	22	24

Note: Data from Solomos and Biale.[85]

and HCN, the respiratory increases involve a stimulation of the glycolytic pathway and an increase in ATP formation.

MOISTURE CONTENT OF THE TISSUE

The moisture content of postharvest products can have a pronounced effect on their rate of respiration. In general, respiration and metabolic processes decrease with decreasing moisture content of the tissue. There are, however, many interacting factors such as species, tissue type, and physiological condition that can significantly alter the plants' or plant parts' response to a particular moisture condition. This change in respiratory rate can be due to moisture changes after harvest or to preharvest differences in moisture content. For example, the respiratory rate of the storage roots of sweetpotato cultivars is closely related to the cultivars' percent moisture.

The effect of postharvest changes in the product's moisture status on respiration is dramatically illustrated by seeds (fig. 3-18) where respiratory increases quite closely parallel seed moisture content as the seeds imbibe or lose water. Respiration is also decreased in the leaves of intact plants by decreasing water content; however, under severe dehydration it may temporarily increase.[19] Leaf water stress conditions can be induced by either soil moisture deficits or excesses, both of which result in leaf desiccation.[24] In fleshy fruits, respiratory responses to postharvest changes in moisture status normally occur only after a substantial change in the internal water concentration has occurred. With most fleshly or succulent postharvest products, decreasing the internal moisture content is not a viable method for controlling the rate of respiration of the tissue. In these tissues, moisture content is often closely tied to product quality and decreases in moisture are counterproductive. In many other products, however, longevity can be greatly extended through the repression of respiratory and metabolic activity with reduced product moisture. For example, in the storage of grains, seeds, dates, and

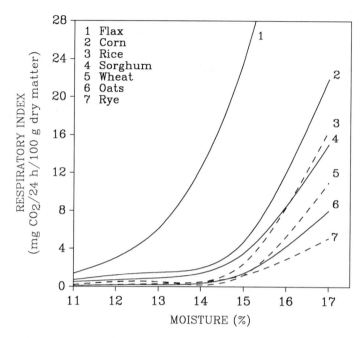

Figure 3-18. Effect of seed moisture level on respiration rates *(after Bailey[4]).*

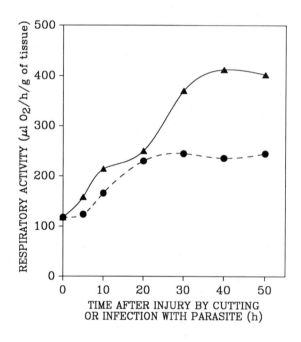

Figure 3-19. The effect of mechanical wounding (●---●) or infection by fungus *Ceratocystis fimbriata* (◄——►) on the respiratory rate of sweetpotato storage root tissue *(after Uritani and Asahi[90]).*

most nuts, moisture status alteration represents an excellent means of extending the useful life of the product.

WOUNDING

Wounding of plant tissue stimulates the respiratory rate of the affected cells, a response that has been known for nearly a century. Boehm[17] demonstrated that cutting potato tubers resulted in an abnormal rate of carbon dioxide production. Cutting the roots of trees[43,44] and handling the leaves[3,35] also results in a stimulation of respiration.

Respiratory increases due to wounds to plant tissue are often grouped into two general classes: (1) those caused by mechanical damage, that is, wound respiration; and (2) those caused by infection by another organism such as fungi or viruses, that is, infection-induced respiration.[90] This classification is not, however, definitive in that wounds induced by other means, although normally less frequent, are found (e.g., chemical sprays, light, pollutants). In addition, several types of wounds may be found concurrently in the same tissue. Mechanically induced wounds include those caused by harvesting, handling, wind, rain, hail, insects, animals, and others. It is advantageous to separate these wounds into subclasses based on the presence or absence of surface punctures, cuts, or lesions. Injuries that facilitate the diffusion of gasses result in a substantial but transient increase in apparent respiration due to the escape of carbon dioxide that has accumulated in the intercellular spaces of the tissue. As a consequence, it is often difficult to make a clear distinction between facilitated diffusion and wound effects on respiration when carbon dioxide production is used as the sole means of measuring respiration.

Uritani and Asahi[90] have characterized the differences in respiratory response between mechanically wounded and infected tissue (fig. 3-19), illustrating two distinct patterns. In both cases, increases in respiration coincided with increases in storage carbohydrate catabolism with an increase in soluble sugars in some tissues. Both the glycolytic[51] and pentose phosphate pathways are stimulated in response to increased demand for both primary and secondary plant products needed for wound healing and/or reaction. These respiratory increases in response to mechanical wounds are related to the healing process. Healing includes the formation of lignin, suberin, and in some cases, callus. Wound respiration, therefore, facilitates the supply of precursors and cofactors required for the biosynthesis of these wound-healing layers (see Bloch[15,16]).

Infection-induced respiratory increases are related to primary and secondary defense reactions by the cells. Plants have evolved various techniques to combat the invasion of microorganisms. For example, rapid cell death resulting in necrotic areas confines the mycelia, limiting the number of cells infected. In addition, secondary products such as phytoalexins may be formed to minimize invasion. These processes, like those of mechanical wounding, require respiratory energy and products resulting in the observed increases in respiration of the tissue.

TYPE OF PLANT AND PLANT PART

Extremely large differences in respiratory rates can be found between different plant parts. Benoy[7] monitored the rate of respiration of harvested vegetables representing a range of plant parts and found that on a dry-weight basis they could be ranked in the following order: asparagus (*Asparagus officinalis,* L.), lettuce, green bean (*Phaseolus vulgaris,* L.), okra [*Abelmoschus esculentus,* (L.) Moench.], green onion (*Allium fistulosum,* L.), carrot, tomato, beetroot, green mango (*Mangifera indica,* L.), and red pimento (*Capsicum annuum,* L.). While these differences include the variation between species and cultivar, work by Burlakow[21] illustrated the wide potential range between plant parts of the same plant. The respiratory activity of the embryo of wheat seeds was shown to be twenty times that of the endosperm.

The intensity of respiration of harvested products can vary widely between similar plant parts from different species. For example, seeds of flax (*Linum usitatissimum,* L.) have a respiratory rate that is fourteen times that of barley (*Hordeum vulgare,* L.) at the same temperature and moisture content (11%).[4] Fruits of avocado have a maximum rate of respiration at their climacteric peak of nearly eight times that of apple.[10]

CULTIVAR

While a significant potential range in respiratory rates between species would be anticipated, differences at the cultivar level may in some cases also be substantial. For example, cut flowers of the *Chrysanthemum × morifolium,* Ramat. cultivar 'Indianapolis White' had a respiratory rate of 1.6 times that of the cultivar 'Indianapolis Pink' based on flower fresh weight and the rate was 4.3 times greater expressed on a per flower basis.[62] Likewise 'McIntosh' apple fruits have been shown to have preharvest respiratory rates double that of fruits of the cultivar 'Delicious'.[39,40] Although all cultivars do not exhibit this degree of variation, it is common, and depending on the postharvest conditions, it may be a factor of consideration.

STAGE OF DEVELOPMENT

The stage of development of a plant or plant part can also have a pronounced effect on the respiratory and metabolic rate of the tissue after harvest. In general, young, actively growing cells tend to have higher respiratory rates than older, more mature cells. A number of factors, however, affect this relationship between maturity and respiratory rate; for example, species, plant part considered, and the actual range in stages of maturity are often critical. Therefore, generalizations on the effect of maturity should be kept within fairly strict commodity-species bounds and in some cases even cultivar or exceptions become more numerous than the rule.

The effect of maturity and its variation within a general commodity type is well illustrated in cut flowers. Carnations (*Dianthus caryophyllus,* L. cv. 'White Sims') harvested at varying stages of maturity between the tight bud stage and when fully opened display significant differences in their respiratory rate expressed on either a weight or per blossom basis.[62] When the flowers were harvested at a more mature stage of development, the respiratory rate rose. In addition, these differences tended to be maintained during the postharvest period. In contrast, however, chrysanthemums (cv. 'Indianapolis White') displayed the opposite trend, with the respiratory rate decreasing with maturity (fig. 3-20). Even more pronounced effects of maturation on respiratory rate can be seen in climacteric fruits when the range in maturities

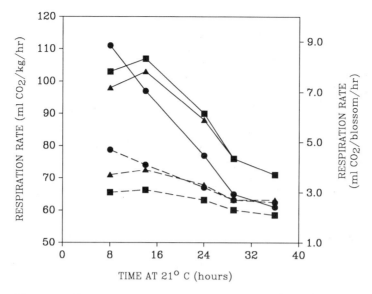

Figure 3-20. The effect of chrysanthemum (*Chrysanthemum × morifolium*, Ramat., cv. 'Indianapolis White') flower maturity (■, intermediate; ▲, mature; ●, fully mature) on flower respiration (---, ml CO_2/kg/hr; —, ml CO_2/blossom/hr) at harvest (0 time) and during storage (*after Kuc and Workman*[62]).

tested includes preclimacteric, climacteric, and postclimacteric stages.[30] Young apple leaves and stems have from three to seven times the rate of respiration of corresponding fully developed organs from the same plant.[75]

SURFACE AREA TO VOLUME RATIO

The surface-to-volume ratio may also influence the respiratory rate of some products due to its effect on gas exchange. As an object increases in size, assuming the shape is not altered, there is a progressive decrease in surface area relative to its volume. This is because volume increases as the cube of length (length × length × length) while surface area increases only as the square of length (length × length). This fact is illustrated in figure 3-21 with the comparison of two sizes of spherical fruits placed in a cubic box (20 cm × 20 cm × 20 cm). One sphere 20 cm in diameter will fit into the container or eight spheres 10 cm in diameter will fit. The composite surface area of the smaller spheres is double that of the larger sphere while the total volumes are equal. This difference provides in a harvested product a larger surface area for gas exchange for the underlying cells, shifting the uptake-utilization (O_2) and production-emanation (CO_2) equilibria.

The shape of the majority of harvested products (leaves, flowers, nuts, etc.) deviates substantially from spherical and in some cases these products also have rough or uneven surfaces. This circumstance increases the area-to-volume ratio, facilitating diffusion. In products where the surface represents a significant barrier to diffusion, this increased surface area may be quite important. Although the actual surface-to-volume ratio can be substantially altered by environmental conditions during growth (e.g., the effect of thin-

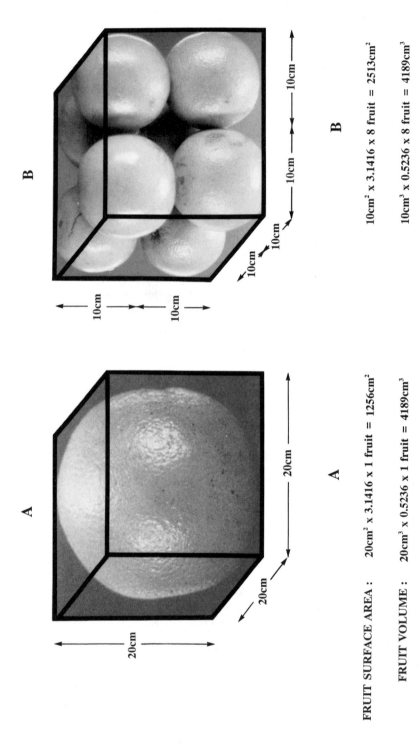

A

FRUIT SURFACE AREA : $20cm^2$ x 3.1416 x 1 fruit = $1256cm^2$

FRUIT VOLUME : $20cm^3$ x 0.5236 x 1 fruit = $4189cm^3$

B

$10cm^2$ x 3.1416 x 8 fruit = $2513cm^2$

$10cm^3$ x 0.5236 x 8 fruit = $4189cm^3$

VOLUME B / VOLUME A = 1

SURFACE AREA B / SURFACE AREA A ≈ 2

Figure 3-21.   The relationship between product size and the surface area available for the diffusion of gases into and out of tissue.

ning on fruit size), little control can be exerted over it during the postharvest period. However, the postharvest environment can be adjusted to compensate for the surface/volume conditions of a specific product, thus preventing undesirable internal conditions from developing within the product.

NATURE OF SURFACE OF THE HARVESTED PRODUCT

The composition of the gas atmosphere within most harvested products can have a pronounced effect on their respiratory rate. Both high carbon dioxide and low oxygen have been shown to decrease the respiratory rate of cells. The internal composition of carbon dioxide and oxygen is controlled by the rate of use of oxygen and production of carbon dioxide by the tissue, differences in the partial pressures of these gases between the interior and exterior environment, and the gas permeability of the tissue. The nature of the surface of the harvested product, therefore, has a pronounced influence on the latter factor through its effect on the diffusion resistance to gases. High diffusion resistance results in a greater difference between the internal and external gas atmosphere. If the differential between internal and external oxygen and carbon dioxide concentration is sufficiently large, the respiratory rate of the internal cells is altered.

Normally surface resistances are much larger than internal diffusion resistances due to the significant volume of intercellular air space within the tissue, the usually much tighter surface cell arrangement, and the presence of surface compounds that resist gas movement (e.g., cutin, waxes). Therefore, the nature of the surface of harvested products and postharvest practices that alter these surface characteristics (e.g., application of waxes) can exert a considerable influence over respiratory and metabolic rates. Lenticels, stomates, surface cuts or abrasions, stem or peduncle scars on fruits, and other openings provide localized areas that have lower diffusion resistances than the majority of the surface of the product. Natural surface coatings of epicuticular waxes and cutin tend to increase the diffusive resistance to oxygen, carbon dioxide, and water movement into and out of the tissue.[60]

In some cases, surface waxes are not extensive enough to have a pronounced effect on diffusion; however, in many cases diffusion is highly restricted. Thus, when gas diffusion is sufficiently restricted, the internal concentration of carbon dioxide increases and oxygen decreases significantly. For example, apples of the 'Granny Smith' cultivar held in air had an internal oxygen concentration of 17% at 7°C and 2% at 29°C while the respective internal carbon dioxide concentrations were 2% and 17%.[89] The surface of the tomato fruit restricts all but around 5% of the total gas exchange between the interior and the exterior, the primary path of exchange being via the stem scar.

In some postharvest products, it is advantageous to apply an artificial coating of wax on the surface. Citrus fruits, apples, cucumbers (*Cucumis sativus*, L.), rutabagas, and dormant rose plants are commonly waxed. Waxing not only alters the internal gas concentration of the product[29] but also has the additional advantage of decreasing water loss and, in many cases, enhancing the appearance of the product by imparting a shiny gloss to the surface.

PREHARVEST CULTURAL AND POSTHARVEST HANDLING CONDITIONS

The respiratory behavior of harvested products can be significantly influenced by a diverse array of prestorage, and in many cases, preharvest factors. The nutrient composition of the harvested product is strongly influenced by the nutrition of the parent plant. Preharvest nutrition alters not only the elemental composition of the product but also the relative amount of many organic compounds.[73] Plant tissues low in potassium and calcium often have substantially higher respiratory rates.[1,2,31,41] This is illustrated by the correlation between the calcium content of the peel of apples and their subsequent postharvest respiratory rate (fig. 3-22). Apples with a low calcium concentration had higher respiratory rates at the preclimacteric stage, at the climacteric peak, and during the postclimacteric period than apples with a higher calcium content.[18] High tissue nitrogen concentration (apples, strawberries) is also correlated with elevated respiration; in the case of apple fruits the effect of high nitrogen is pronounced only when the fruit calcium concentration is low.[31,77]

Other factors such as preharvest sprays,[84] rough handling,[65] orchard temperature,[70] production year,[18] and acclimatization can also significantly influence postharvest respiratory responses.

Methods of Measuring Respiration

The respiratory rate of a stored product can be used as an indicator for adjusting the storage conditions to maximize the longevity of the commodity.

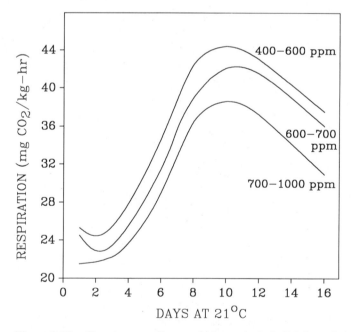

Figure 3-22. Respiratory climacteric in apple fruit (*Malus sylvestris,* Mill. cv. 'Baldwin') with varying peel calcium contents (*after Bramlage et al.[18]*).

As a consequence, it is often desirable to measure respiration in commercial storage houses. These measures can also be used in many cases as a general index of the potential storage life of the product. In addition, the rate of respiration can be used to calculate the loss of dry matter from the product during storage and the rate of oxygen removal from storage room air.

As discussed previously, respiration consumes oxygen from the surrounding environment and substrate from the commodity. Carbon dioxide, water, and energy (both chemical and heat) are produced:

$$\text{substrate} + O_2 \rightarrow CO_2 + H_2O + \text{energy (chemical and heat)}$$

Theoretically, changes in any of these reactants or products could be used as a measure of respiration. In general practice, however, measurements of oxygen utilization or carbon dioxide production are used due to their relative ease of measurement and accuracy. Since the respiratory reactions take place in an aqueous medium, the small quantity of water produced in relation to the total volume of water present in the tissue cannot be accurately measured. Similarly, relatively large rates of respiration over a short measurement period result in only small changes in total substrate or dry matter. Energy production, whether chemically trapped or liberated as heat, is also difficult to measure precisely. As a consequence, the utilization of oxygen or the production of carbon dioxide is almost invariably used to monitor respiration.

Several techniques may be used for collecting gas samples from a product in which the respiration is to be monitored. The product may be placed in a closed (gas-tight) container and the decrease in oxygen or increase in carbon dioxide can be measured over a known period of time by withdrawing small samples of the enclosed atmosphere and measuring either or both gases (fig. 3-22). By measuring the change (Δ) in concentration as a function of time (i.e., concentration of oxygen at t_1 minus the concentration of oxygen at t_2 divided by t_2 minus t_1 gives ΔO_2/unit of time), the volume of free space in the container, and the weight of the enclosed product, the respiratory rate expressed as weight of gas per weight of product per unit of time can be readily calculated (see fig. 3-23). With a closed system, care must be taken not to leave the product enclosed too long since the decreasing oxygen and increasing carbon dioxide concentrations will subsequently affect the rate of respiration of the product. For many products, it is not desirable to let the carbon dioxide concentration increase to much above 0.5%.

A second technique employs a continuous flow of air or gas of known composition through the container holding the product (fig. 3-23). The difference (Δ) between the concentration of oxygen and/or carbon dioxide going into the container and that leaving the container is used to calculate the respiratory rate. In addition to the difference in gas concentration, the rate of airflow through the container and the weight of the product must be known. Care must be taken to adjust the airflow rate through the container to an appropriate level. An excessively high flow rate results in extremely low differences between incoming and exiting gases, making measurement with an acceptable level of accuracy difficult. Airflow rates that are too slow result in the same problem that can be encountered with a closed system, the buildup of carbon dioxide or depletion of oxygen, which affects the rate of respiration. Care should also be taken to allow the system sufficient time to develop a steady-state equilibrium before measurements are made.

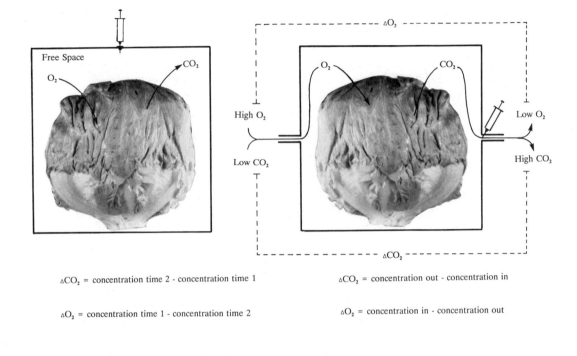

CLOSED SYSTEM   CONTINUOUS FLOW SYSTEM

ΔCO_2 = concentration time 2 - concentration time 1 ΔCO_2 = concentration out - concentration in

ΔO_2 = concentration time 1 - concentration time 2 ΔO_2 = concentration in - concentration out

$$\frac{(\Delta \% \times 10)(\text{free space volume of container in liters})}{(\text{product fwt in kg})(\text{time container is closed in hours})} = \text{ml kg}^{-1} \text{hr}^{-1}$$ $$\frac{(\Delta \% \times 10)(\text{air flow rate in ml/min} \times 60)/1000}{(\text{product fwt in kg})} = \text{ml kg}^{-1} \text{hr}^{-1}$$

*Milliliters of gas are normally converted to milligrams to remove the effect of temperature on the volume of gas so that direct comparisons can be made. To do this, a temperature correction must be used.

One mole of gas is equal to 22.4 L at 0°C at 1 atmosphere, therefore, its volume (V_1) at the temperature of the product can be calculated with the following equation:

$$V_1 = 22.4 \left(1 + \frac{\text{Temperature of product in °C}}{273°\text{Kelvin}} \right)$$

For example, the volume of 1 mole of CO_2 at 25°C = 24.45 L. The volume of gas per gram is calculated by dividing the correct volume by the molecular weight of the gas (CO_2 = 44, O_2 = 32), i.e. 24.45 L/44 = .556 L/g or 556 ml/1000 mg (the volume of CO_2 at 25°C divided by its molecular weight). Then the weight of gas from the respiration sample can be calculated by:

$$\frac{556 \text{ ml}}{1000 \text{ mg}} = \frac{\text{measured ml from sample}}{x}$$

Corrections for common temperatures are:
        0°C = 509 ml CO_2/1000 mg or 700 ml O_2/1000 mg
        5°C = 518 ml CO_2/1000 mg or 712 ml O_2/1000 mg
        10°C = 528 ml CO_2/1000 mg or 726 ml O_2/1000 mg
        15°C = 537 ml CO_2/1000 mg or 738 ml O_2/1000 mg
        20°C = 546 ml CO_2/1000 mg or 751 ml O_2/1000 mg
        25°C = 556 ml CO_2/1000 mg or 764 ml O_2/1000 mg
        30°C = 565 ml CO_2/1000 mg or 777 ml O_2/1000 mg

Figure 3-23. Techniques for collecting respiratory gas samples and calculating respiratory rates of harvested products.

The oxygen and/or carbon dioxide concentrations in gas samples from either system can be measured utilizing any of a number of different methods. A general comparison of several of the methods of analysis is presented in table 3-5.

GAS CHROMATOGRAPHY

Both carbon dioxide and oxygen can be measured with a gas chromatograph equipped with a thermal conductivity detector and dual columns. Gas chromatographic analysis is used widely due to its high potential level of accuracy and since quite small gas samples (0.1–5 ml) can be tested.

INFRARED GAS ANALYZER

This instrument is used to measure carbon dioxide in a continuous flow of gas and has the advantage over many techniques in that it is extremely accurate in the very low carbon dioxide concentration range. Molecules of carbon dioxide in the sample absorb infrared radiation at a specific wavelength and this absorption is used as a measure of carbon dioxide concentration in the airstream.

PARAMAGNETIC OXYGEN ANALYZERS

Oxygen is strongly paramagnetic, and since no other gases commonly present in the air exert a magnetic influence,* this characteristic can be monitored and used as a measure of the oxygen concentration in a continuous stream of air.

TITRATION/COLORIMETRY

A stream of air is passed through a sodium hydroxide or calcium hydroxide solution that absorbs the carbon dioxide:

$$NaOH + CO_2 \quad \rightarrow \quad NaHCO_3 \quad \rightarrow \quad Na_2CO_3 + H_2O$$

sodium hydroxide   sodium bicarbonate   sodium carbonate

$$Ca(OH)_2 + CO_2 \quad \rightarrow \quad CaCO_3 + H_2O$$

calcium hydroxide   calcium carbonate

The change in alkalinity (decrease in pH) is used to determine the quantity of carbon dioxide absorbed[25] and can be used for measuring carbon dioxide concentrations of up to 1.0%. The pH change is measured either by titration or colorimetrically with the addition of bromthymol blue[78] and monitored with a spectrophotometer. As with other continuous flow systems, the flow rate of air must be known to calculate the final respiratory rate.† This technique has the advantage of being relatively inexpensive in that only a limited amount of equipment is needed.

---

* Oxides of nitrogen are also paramagnetic; however, their concentrations are normally sufficiently low so as not to cause a problem.
† If all of the carbon dioxide is collected and compared with the carbon dioxide collected from an air control with an equal flow rate, then the flow rate is not needed in the calculation; only the length of time and the weight of the produce are needed.

Table 3-5. Comparison of Techniques Available for Measuring Respiration of Plants

Technique	Gas Measured	Sample Type	Sample Size	Initial Expense	Recurring Expense	Requirement for Electricity
Gas chromatography	CO_2 and/or O_2	Discrete	0.2–5.0 ml	Very high	Medium	Yes
Infrared	CO_2	Continuous flow		Very high	Low	Yes
Paramagnetic	O_2	Continuous flow		High	Very low	Yes
Titration/colorimetry	CO_2	Continuous flow		Low	Low	No
Kitagawa	CO_2 or O_2	Discrete		Very low	Very low	No
Pressure/volume changes	CO_2 or O_2	Discrete	100 ml	Medium	Low	No
Polarography	O_2	Continuous flow		Medium	Low	Yes

KITAGAWA DETECTORS

Carbon dioxide, oxygen, and a number of other gases can be measured quickly, relatively accurately, and without significant expense. Gas is pulled into a reaction tube (specific for each gas monitored) where it is absorbed and reacts with a specific chemical reagent. The color change produced is used as a measure of the concentration of the gas. Carbon dioxide can be accurately measured between .01% and 2.6% and oxygen between 2% and 30%.

PRESSURE/VOLUME CHANGES

Samples of gas are placed in sealed containers of known volume and either carbon dioxide or oxygen is absorbed by a suitable reactant. The change in the internal pressure of the container or in the volume of gas within the container is used as a measure of the concentration of the respective gas. The absorption principle is similar to that used in the titration/colorimetry method; however, instead of measuring changes in the absorbing material (e.g., pH), pressure or volume changes within the chamber are determined.

In one of several commercially available analyzers that utilize this principle, a calibrated buret containing the air sample is connected to an absorption tube of potassium hydroxide that absorbs carbon dioxide and a second tube of either alkaline pyrogallol or ammoniacal cuprous chloride that absorbs oxygen. Relatively large gas samples (e.g., 100 ml) are required; however, analyses are accurate to approximately 0.5%.

POLAROGRAPHY

The oxygen concentration in a gas sample can be measured with a polarographic oxygen electrode. The differential in electrical potential across a pair of electrodes is measured.

Respiratory Patterns

As previously noted (pages 104–120), a number of external (environmental) and internal (commodity) factors have a pronounced influence on the rate of respiration of plant tissues. While environmental factors such as temperature are routinely studied with each postharvest product, considerable research interest has been directed toward understanding the more elusive commodity factors that influence respiration. The marked changes in the rate of respiration of climacteric fruits have long fascinated postharvest physiologists. Although these fruits represent an extremely small percentage of the plant products handled in agriculture, their dramatic shift in respiration during ripening has stimulated research into the function and control of respiratory patterns in plant products. By monitoring the changes in respiratory rate during growth, development, and senescence of a plant or plant part under standard conditions, distinctive patterns are evident and these can often be related to other functional processes occurring concurrently.

One of the early studies on respiratory patterns in plants was conducted in the 1920s by Kidd and West on sunflower plants and component parts during an entire growing season (fig. 3-24). In general, they found that respiration was closely correlated with the rate of growth of the plant, that is, young,

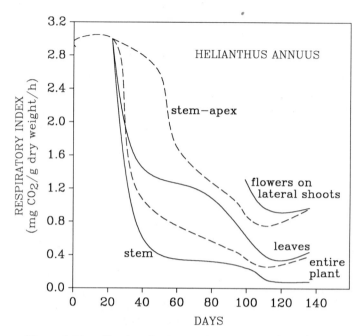

Figure 3-24. Changes in the respiratory rate of sunflower plants and selected plant parts during development (data from Kidd et al.[57]).

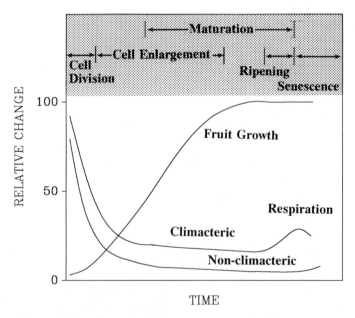

Figure 3-25. General respiratory pattern of climacteric and nonclimacteric fruits during development, maturation, ripening, and senescence *(after Biale[12])*.

rapidly metabolizing cells have the highest respiratory rates. The high demand for energy and carbon compounds in actively growing cells results in a stimulation of respiration. As the age of the plant or individual organ (stem, leaves, flowers) increased, the respiratory rate decreased. This decline in respiration of the whole plant could not be attributed simply to an increased percentage of nonrespiring structural material in the plant since the initial respiratory rate of successive new leaves also decreased with the age of the plant. Hence, internal factors have a pronounced influence on respiration.

Distinct changes in respiratory rate also occur in detached plant parts. Fruits are typically classed into one of two groups based on these trends in respiratory behavior during the final stages of ontogeny of the organ (fig. 3-25). Fruits classified as having a respiratory climacteric exhibit a marked upsurge in respiratory activity at the end of the maturation phase. The respiratory climacteric represents a transition between maturation and senescence. Nonclimacteric fruits (fig. 3-25) do not exhibit an upsurge in respiration but rather a progressive, slow decline during senescence until microbial or fungal invasion.

The climacteric rise in respiration was described as early as 1908[74] in apple and pear (*Pyrus communis*, L.) fruits. Later, Kidd and West[56] detailed the relationship between changes in respiratory rate and changes in quality attributes occurring during the climacteric period. The respiration of an unripe fruit declines after harvest to what is termed the preclimacteric minimum, occurring just before the climacteric rise in respiration (fig. 3-25). Subsequently, respiration increases dramatically, often to levels that are two to four times that of the preclimacteric minimum. A similar trend occurs if the fruits are allowed to ripen on the tree, although the respiratory pattern is modified somewhat in both the rate at which it proceeds (slower) and the peak values (higher). An exception is with some cultivars of avocado and mango in which the respiratory upsurge is inhibited while the fruit is attached to the tree.

The respiratory climacteric is substantially altered by temperature. At both low and high temperatures, the climacteric can be suppressed. As storage temperature decreases from around 25°C, the duration of the climacteric rise is prolonged and the rate of respiration at the climacteric peak is depressed. In addition, the ambient oxygen and carbon dioxide concentration can markedly alter the respiratory climacteric. Low oxygen and high carbon dioxide (up to approximately 10%) can prolong the length of time to the climacteric peak in a number of fruits, thus extending their storage life. Respiration can also be depressed in many nonclimacteric tissues by low oxygen and high carbon dioxide. There are exceptions, however; for example, high carbon dioxide tends to stimulate the respiration of lemon fruits,[13] probably through the fixation and incorporation of the molecule into organic acids utilized in the tricarboxylic acid cycle.

The unsaturated hydrocarbon gas, ethylene, stimulates the respiration of a wide range of plant tissues. The response of fruits, however, differs between climacteric and nonclimacteric types (fig. 3-26). Exposure of climacteric fruits to relatively low levels of ethylene decreases the length of time of the preclimacteric period without a substantial effect on the rate of respiration at the climacteric peak. The concentration required for maximum acceleration varies with different fruits (e.g., 1 $\mu l \cdot l^{-1}$ for banana, 10 $\mu l \cdot l^{-1}$ for avocado)

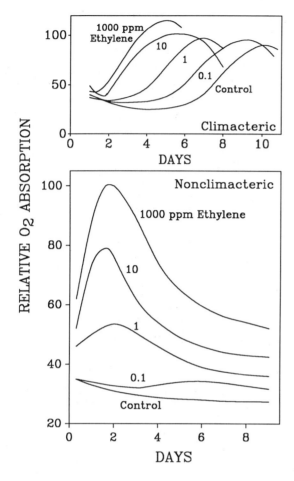

Figure 3-26. Oxygen uptake by climacteric and nonclimacteric fruits exposed to varying levels of ethylene *(after Biale[12])*.

and the shortening of the preclimacteric period is approximately proportional to the logarithm of the concentration applied.[20] Once ripening is initiated, removal of the external ethylene has no effect on the subsequent respiratory rate or pattern. With nonclimacteric fruits, respiration is likewise stimulated by ethylene; however, upon removal of ethylene the respiratory rate returns to the base value found prior to treatment (fig. 3-26).

Ethylene is known to be a natural product of ripening fruits.[33] The synthesis and the internal concentration increase in most climacteric fruits around the same time as the upsurge in respiration. Ethylene is thought to act in some fruits as a natural triggering mechanism for the induction of the respiratory climacteric. Exposure to an external source of ethylene results in an increased synthesis of ethylene by the tissue. This stimulated synthesis is the result of the ethylene effect on the de novo synthesis of ACC synthase, a critical enzyme in the ethylene synthesis pathway. Exogenous ethylene, therefore, can be substituted for the ethylene normally produced by the plant to trigger the respiratory climacteric and ripening response, allowing earlier ripening of the fruit.

Theories as to the precise cause of the respiratory climacteric have been numerous. Blackman and Parija in 1928[14] proposed that the increase in respiration was due to the loss of organizational resistance between enzymes and substrates. Several of the other theories that have enjoyed popularity at various stages of the evolution of postharvest research are (1) presence of "active" substrate,[55] (2) availability of phosphate acceptors, (3) availability of cofactors, (4) uncouplers of oxidation and phosphorylation,[71] (5) shifts in metabolic pathways,[50] and (6) increases in mitochondria content and/or activity.[46] Suffice it to say that after more than 50 years of research the precise cause of the respiratory climacteric in fruits has yet to be elucidated, although understanding of the physiological, chemical, and enzymatic changes occurring has increased tremendously.

Leaves undergo distinct changes in their respiratory behavior at certain stages of their ontogeny (see review by Thimann[86]). Generally, there is an increase in respiration during the early stages of senescence (the period of chlorophyll degradation) followed by a steady decline in the later stages. This respiratory increase, although not universal, occurs in a wide range of species in both attached[42] and detached leaves. Severing the leaf from the plant enhances the rate of senescence[82]; however, the timing of the respiratory increase relative to other biochemical and physical changes occurs at essentially the same stage in the senescence process. Low light (100–200 lux) delays senescence in detached leaves (*Avena sativa,* L.) and appears to alter the respiratory strategy utilized.[85] In leaves held in the dark, approximately 25% of the respiratory increase could be accounted for by increases in free amino acids and sugars due to catabolic events occurring. The remaining respiration (75%) appeared to be due to a partial uncoupling of respiration from phosphorylation. In leaves exposed to low light, however, respiration increases were due to increases in available respiratory substrates.

Many flowers also undergo marked changes in respiration, the pattern and control of which have many parallels with the respiratory changes in climacteric fruits. In fact, the term *respiratory climacteric* is increasingly used in studies on flower storage and senescence. Respiration in many species of cut flowers declines after harvest and then increases as the flowers begin to senesce.[76] This trend, however, is not universal. For example, cut roses progressively decline in respiration following harvest (fig. 3-27). In flowers that exhibit a postharvest respiratory rise, the increase in respiration, like that in many climacteric fruits, appears to be closely tied to the endogenous synthesis of ethylene by the flower (fig. 3-28). There is an autocatalytic synthesis of ethylene in flowers like the carnation and the increase in ethylene precedes changes in membrane permeability and other senescence phenomena.[67] In addition, chemicals such as rhizobitoxine and Ag^{2+} ions, which inhibit ethylene synthesis or action, delay senescence in a wide range of flowers.

As the flower proceeds toward the final stages of senescence, there is a gradual decline in respiration that may reflect a decline in respiratory substrate availability. Carbohydrates are known to be transported from the petals into the ovary during this period, the movement of which is stimulated by ethylene.

It can be seen from this section that many plant products undergo substantial changes in their respiratory pattern after harvest. These changes often reflect significant alterations in metabolism and concurrent physical and

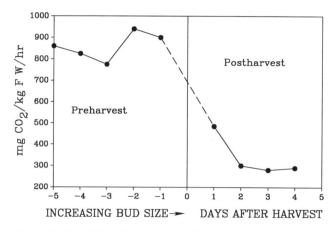

Figure 3-27. The effect of time of harvest on the respiratory rate of 'Velvet Times' roses held at 21°C *(after Coorts et al.[26]).* Zero time corresponds to the stage of development that the cultivar would be commercially harvested.

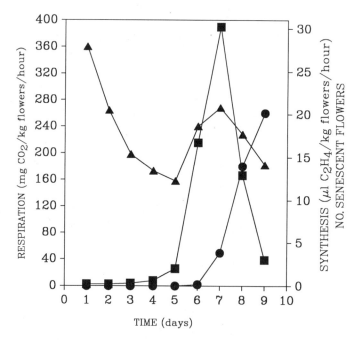

Figure 3-28. Changes in the rate of respiration (▲) and ethylene synthesis (■) with the development of senescence (●) after harvest of 'Improved White Sin' carnation flowers at 20°C *(after Maxie et al.[66]).*

chemical alterations within the tissues. Changes in respiration are of interest from an applied point of view in that specific handling strategies may be required. Likewise, they in part reflect the physiological state of the commodity that can help in predicting the product's future storage potential and life expectancy.

Photosynthesis

The potential for photosynthesis and the rate at which it proceeds in harvested products varies widely and is influenced by both internal and external factors. Many plant products are devoid of chlorophyll or are held after harvest under environmental conditions not conducive to photosynthesis. The importance of photosynthesis in detached chlorophyll-containing tissues is probably minimal. However, it is a subject that has not been adequately studied by postharvest physiologists.

A significant number of postharvest products not only have the potential for photosynthesis but need to photosynthesize to maintain the product's existing level of quality. Intact plants such as actively growing herbaceous and woody ornamentals, vegetable transplants, and rooted cuttings are typical examples of types of postharvest products that normally photosynthesize during the handling and marketing period. The rate at which these plants photosynthesize after removal from the production area is governed by a number of internal and external factors. While the internal factors, such as

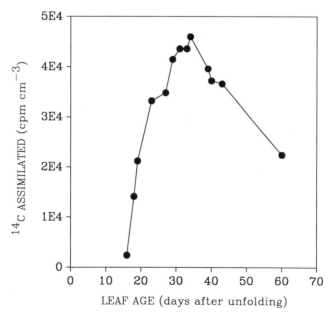

Figure 3-29. The effect of leaf age on the rate of photosynthesis ($^{14}CO_2$ assimilated) of grape (*Vitis vinifera,* L.) leaves *(after Kriedemann et al.[61]).*

stomatal number and photosynthetic pathway, influence the rate of photosynthesis, it is not possible to influence these factors to any significant extent. However, a number of external factors can be altered such as light, temperature, moisture, carbon dioxide, and exogenous chemicals. Manipulation of these external factors therefore provides the potential to exert a significant level of control over the rate of photosynthesis in many harvested products.

Tissue Type and Condition

The ability to photosynthesize and the rate at which the process proceeds varies considerably between species of plants as well as with the types of tissue within a plant. Chloroplasts are found primarily in leaf tissue; however, they are also commonly found in petioles, stems, and specialized floral parts in many plants and some fruit. Roots, tubers, and other structures normally devoid of chlorophyll are capable of synthesizing the molecule when exposed to sufficient light. The preharvest or postharvest formation of chlorophyll in some products (e.g., potatoes, Jerusalem artichokes) is detrimental to the quality of the product and should be avoided. The contribution of chloroplasts in organs other than leaves to the total assimilation of carbon is generally small due in part to the low number or absence of stomates, minimizing the availability of carbon dioxide. In some instances, however, such as the corticular tissue of "dormant" dogwoods (*Cornus florida*, L.), photosynthesis by nonleaf structures may offset a significant portion of respiratory loss of carbon.

Leaf age also affects the rate at which photosynthesis proceeds (fig. 3-29). Photosynthetic rates are commonly highest when leaves are first mature and shortly thereafter but tend to decline gradually with age.[61] This general phenomenon is found in annuals and perennials, including evergreen species.

Light

Whether from the sun or from an artificial source, light provides the energy required by plants to fix carbon from CO_2, allowing them to offset respiratory losses incurred. Light is one of the most important postharvest external variables affecting photosynthesis. During the postharvest handling of plants, especially intact plants, light intensity, quality, and duration are important and should be given attention.

Individual leaves, when exposed to increasing light intensity, exhibit a typical light response pattern (fig. 3-30). As the intensity of the light increases initially, the light compensation point is reached. Here CO_2 trapped is equal to CO_2 lost from the tissue due to respiratory processes. Additional increases in intensity result in a proportional elevation of photosynthetic rate. Progressive increases in light eventually reach a point at which photosynthesis becomes light saturated. At this point, additional increases in light intensity have only a slight effect on elevating the rate of carbon fixation. Light saturation of individual leaves of full sun plants is often only one fourth to one half that of full sunlight; however, with entire plants saturation is seldom reached. This situation is due to mutual shading of the leaves within the

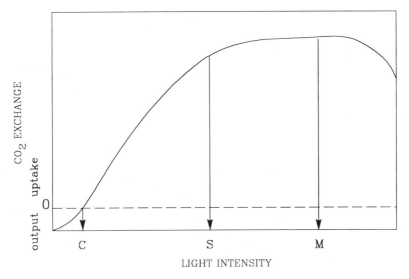

Figure 3-30. The relationship between light intensity and the rate of carbon uptake by photosynthesizing leaves (C, light compensation point; S, light saturation intensity; M, maximum photosynthetic rate) *(after Rabinowitch[79])*. At the light compensation point (C), the amount of carbon fixed by the leaf is equal to the amount lost through respiration. At very high light intensities, photosynthesis is inhibited.

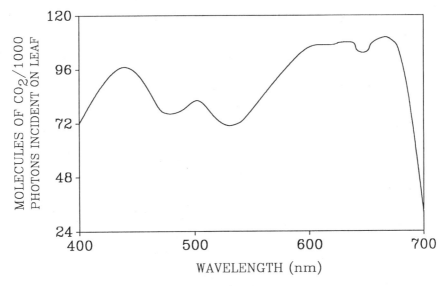

Figure 3-31. The change in photosynthetic activity with changes in wavelength of the light striking the leaf *(after Balegh and Biddulph[5])*.

canopy. With further increases in light, the point of maximum photosynthesis is reached and additional increases result in a decrease in carbon fixation (fig. 3-30).

Light is necessary for the formation of chlorophyll in plants and there is a continuous turnover (synthesis and degradation) of the molecule under normal conditions. Insufficient light, therefore, can result in the net loss of leaf chlorophyll. With prolonged exposure, an indirect loss occurs through abscission of leaves, decreasing the plant's surface area of photosynthetic tissue. Normally, this loss progresses from the oldest leaves to the youngest, with leaves that are shaded due to their position at the bottom or interior of the canopy being more susceptible to abscission.

Plant species vary widely in their tolerance to light and excess light may also present serious postharvest problems. For example, prolonged exposure of full shade plants such as the African violet (*Saintpaulia ionantha,* Wendl.) to full sunlight may result in chlorophyll degradation and leaf burning, which decreases net photosynthesis. The effect of excess light can also be seen in heliotropic responses where the plant's growth pattern is altered in the direction of light that may potentially be in an undesirable orientation.

In addition to light intensity, the duration of exposure is also important. Since it is normally desirable to maintain the product in the same qualitative condition that existed at harvest (removal from the production zone for intact plants), the net carbon balance (acquisition vs. utilization) is important. The precise postharvest photosynthetic requirements for individual species at various stages in their life cycle are not currently known. It is probable, however, that the requirements for photosynthetic carbon input needed to maintain the existing condition of the plant are somewhat above the plant's gross respiratory utilization. This is due to nonrespiratory uses of carbon in maintenance reactions. Photosynthetic acquisition of carbon can be maintained at or above this critical maintenance point with a range of light intensity-duration combinations. In some species, however, if exposed to prolonged periods at higher light intensities, the chloroplasts are unable to properly store the additional starch formed. As a consequence, photosynthesis is inhibited.

Of the total spectrum of radiant energy, plants only utilize light from a region between 400 nm and 700 nm for photosynthesis (fig. 3-31). Peak photosynthesis and absorption of light are in the red and blue portions of the spectrum. In natural canopies, light entering the upper leaves is selectively absorbed in the red and blue regions of the spectrum with less photosynthetically active light being transmitted to the lower and inner leaves. As the spectral distribution of the light shifts toward a greater percentage of green light, the rate of photosynthesis per photon of photosynthetically active light declines.

Changes in the amount of energy the plant receives at particular wavelengths (light quality) is often also dramatically altered after harvest. These changes occur primarily from the use of artificial light, which does not have the same spectral quality as sunlight, and from the use of shading material that selectively absorbs light from certain regions of the spectrum. While photomorphogenic changes in the plant due to light quality are normally of greater postharvest importance than changes in the net photosynthetic rate, prolonged exposure of plants to light that is not spectrally suited for photosynthesis will compromise the maintenance of product quality.

Temperature

The rate of photosynthesis is highly dependent on temperature. Within the plant kingdom, this dependence varies widely, with some species capable of photosynthesizing at temperatures near 0°C whereas others require substantially higher temperatures. Generally, net photosynthesis (total photosynthesis minus respiration) increases with temperature until reaching a maximum after which it declines (fig. 3-14). This decline is probably mediated by several factors, one of which is the elevation in respiration that occurs with increasing temperature. With progressive increases in temperature, respiratory losses will eventually be greater than carbon fixation through photosynthesis, giving a net loss of carbon. During the postharvest handling period of photosynthetically active products, both excessively low and high temperatures present potential problems in maintaining an adequate carbon input-output balance. At very low temperatures, photosynthesis is insufficient, while at high temperatures, respiratory losses are greater than CO_2 fixation.

Moisture Stress

The availability of water is an important factor governing the rate of photosynthesis in plants (fig. 3-32). Upon removal of plants from the production zone, the potential for moisture stress increases substantially due in part to the altered environmental conditions to which the plants are exposed. In addition, the control over water management often becomes less organized and structured. Lack of precise control over the postharvest product's moisture balance can result in either water deficit or excess, each of which can substantially impede photosynthesis in that both decrease leaf water content. This situation has a direct effect on water's chemical role in photosynthesis and indirect effect on hydration of the protoplasm and closure of stomates. Since the total amount of water within the leaf that is participating directly in the biochemical reactions of photosynthesis is extremely small, indirect effects on stomatal aperture appear to present the greater hindrance to photosynthesis under conditions of low moisture. The degree of inhibition of photosynthesis due to water deficit depends to a large extent on both the level of stress imposed and the species involved. Wilted leaves of *Helianthus annuus*, L. are as much as ten times less efficient in photosynthesis than turgid leaves. As a consequence, prolonged closure of stomates can result in a serious deficit in carbohydrates within the plant. However, preconditioning* and postharvest environmental parameters can have a pronounced influence on the maintenance of quality after harvest.

---

* Preconditioning involves altering the physiological condition of the postharvest product through the manipulation of environmental factors (e.g., exposure to reduced temperature or moisture conditions) to enhance its potential to withstand stress and ensure the maintenance of quality.

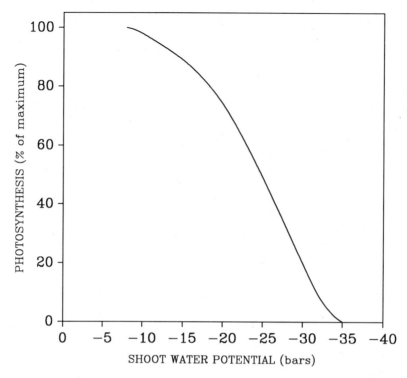

Figure 3-32. The effect of shoot water potential on the photosynthetic rate of Douglas fir trees *(after Brix[19])*.

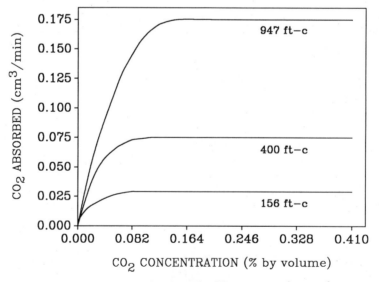

Figure 3-33. The effect of carbon dioxide concentration on photosynthetic rate (CO_2 absorbed) at varying light intensities *(after Hoover et al.[48])*.

Carbon Dioxide

Photosynthesis in plants exposed to sufficient light is limited primarily by the low CO_2 concentration in the ambient atmosphere (approximately 330 $\mu l \cdot l^{-1}$). As CO_2 concentration increases up to 1,000 to 1,500 $\mu l \cdot l^{-1}$, photosynthesis also increases in many species. At higher light intensities, this effect of additional CO_2 is even greater (fig. 3-33).

Elevated CO_2 concentration has been used successfully to increase the growth rate of a number of agricultural crops grown in controlled environment conditions (e.g., greenhouses).[93] To date, the beneficial effect of elevated CO_2 concentration on the photosynthetic rate of plants has been used only in the production zone for these crops and not during the postharvest handling and sales period.

Chemicals

A number of chemical compounds that are either applied directly to plants (pesticides, residue from water) or are found in the ambient atmosphere (air pollutants) can depress photosynthesis. This response may be due to a direct effect on the biochemical reactions of photosynthesis or through indirect effects such as increasing the diffusive resistance of CO_2 into the leaf, changes in optical properties of the leaf, changes in the leaf's thermal balance, or the loss of photosynthetic surface area. Common air pollutants that can reduce photosynthesis are sulfur dioxide, ozone, fluorides, ethylene, and particulate matter such as dusts. Brief exposure to sulfur dioxide (4.1 mg·m^{-3} for 2 hours) has been shown to inhibit photosynthesis by as much as 80%. Although the effect of chemicals on postharvest photosynthesis may be pronounced, more direct effects such as the formation of lesions, chlorosis, discoloration, and leaf and flower drop are normally economically much more significant due to reduced product quality than impaired photosynthesis.

REFERENCES

1. Alban, E. K., H. W. Ford, and F. S. Nowlett. 1940. A preliminary report on the effect of various cultural practices with greenhouse tomatoes on the respiration rate of the harvested fruit. *Proc. Am. Soc. Hort. Sci.* **52**:385–390.
2. Amberger, A. 1953. Zur Rolle des Kaliums bei Atmung svorgängen. *Biochem. Z.* **323**:437–438.
3. Audus, L. J. 1935. Mechanical stimulation and respiration rate in the cherry laurel. *New Phytol.* **34**:386–402.
4. Bailey, C. H. 1940. Respiration of cereal grains and flaxseed. *Plant Physiol.* **15**:257–274.
5. Balegh, S. E., and O. Biddulph. 1970. The photosynthetic action spectrum of the bean plant. *Plant Physiol.* **46**:1–5.
6. Beaudry, R. M., R. F. Severson, C. C. Black, and S. J. Kays. 1989. Banana ripening: Implications of changes in glycolytic intermediate concentrations, glycolytic and gluconeogenic carbon flux, and fructose 2,6-bisphosphate concentration. *Plant Physiol.* **91**:1436–1444.
7. Benoy, M. P. 1929. The respiration factor in the deterioration of fresh vegetables at room temperature. *J. Agric. Res.* **39**:75–80.

8. Berard, J. E. 1821. Memoire sur la maturation des fruits. *Annal. Chim. Phys.* **16:**152–183, 225–251.

9. Biale, J. B. 1946. Effect of oxygen concentration on respiration of the Fuerte avocado fruit. *Am. J. Bot.* **33:**363–373.

10. Biale, J. B. 1950. Postharvest physiology and biochemistry of fruits. *Ann. Rev. Plant Physiol.* **1:**183–206.

11. Biale, J. B. 1954. Physiological requirements of citrus fruits. *Citrus Leaves* **34:**6–7, 31–33.

12. Biale, J. B. 1960. Respiration of fruits. In *Handbuch der Pfanzenphysiologie,* vol. XII/2, W. Rukland (ed.). Springer-Verlag, Berlin, pp. 536–592.

13. Biale, J. B., and R. E. Young. 1962. The biochemistry of fruit maturation. *Endeavor* **21:**164–174.

14. Blackman, F. F., and P. Parija. 1928. Analytic studies in plant respiration. II. The respiration of a population of senescent ripening apples. *Roy. Soc. London Proc.,* ser. B, **103:**422–445.

15. Bloch, R. 1941. Wound healing in higher plants. *Bot. Rev.* **7:**110–146.

16. Bloch, R. 1964. Wound healing in higher plants. *Bot. Rev.* **18:**655–679.

17. Boehm, J. A. 1887. Ueber die respiration der kartoffel. *Bot. Ztg.* **45:**671–675, 680–691.

18. Bramlage, W. J., M. Drake, and J. H. Baker. 1974. Relationships of calcium content to respiration and postharvest conditions of apples. *J. Am. Soc. Hort. Sci.* **99:**376–378.

19. Brix, H. 1962. The effect of water stress on the rates of photosynthesis and respiration on tomato plants and loblolly pine seedlings. *Physiol. Plant.* **15:**10–20.

20. Burg, S. P., and E. A. Burg. 1962. Role of ethylene in fruit ripening. *Plant Physiol.* **37:**179–189.

21. Burlakow, G. 1898. Ueber athmung des keimes des weizens, *Triticum vulgare. Bot. Cent.* **74:**323–324.

22. Burton, W. G. 1965. The sugar balance in some British potato varieties during storage. I. Preliminary observations. *Eur. Potato J.* **8:**80–91.

23. Chang, L. A., and S. J. Kays. 1981. Effect of low oxygen on sweet potato roots during storage. *J. Am. Soc. Hort. Sci.* **106:**481–483.

24. Childers, N. F., and D. G. White. 1942. Influence of submersion of the roots on transpiration, apparent photosynthesis, and respiration of young apple trees. *Plant Physiol.* **17:**603–618.

25. Claypool, L. L., and R. M. Keefer. 1942. A colorimetric method for CO_2 determination in respiration studies. *Proc. Am. Soc. Hort. Sci.* **40:**177–186.

26. Coorts, G. D., J. B. Gartner, and J. P. McCollum. 1965. Effect of senescence and preservative on respiration in cut flowers of *Rosa hybrida,* 'Velvet Times'. *Proc. Am. Soc. Hort. Sci.* **86:**779–790.

27. Denny, F. E. 1924. Hastening the coloration of lemons. *J. Agric. Res.* **27:**757–771.

28. Denny, F. E. 1924. Effect of ethylene upon respiration of lemons. *Bot. Gaz.* (*Chicago*) **7:**327–329.

29. Eaks, I. L., and W. A. Ludi, 1960. Effects of temperature, washing, and waxing on the composition of the internal atmosphere of orange fruits. *Proc. Am. Soc. Hort. Sci.* **76:**220–228.

30. Emmert, F. H., and F. W. Southwick. 1954. The effect of maturity, apple emanations, waxing and growth regulators on the respiration and red color development of tomato fruit. *Proc. Am. Soc. Hort. Sci.* **63:**393–401.

31. Faust, M., and C. B. Shear. 1972. The effects of calcium on respiration of apples. *J. Am. Soc. Hort. Sci.* **97:**437–439.

32. Foote, K. C., and M. Schaedle. 1976. Diurnal and seasonal patterns of photosynthesis and respiration by stems of *Populus tremuloides* Michx. *Plant Physiol.* **58:**651–655.

33. Gane, R. 1934. Production of ethylene by some ripening fruits. *Nature* **7**:1465–1470.

34. Genevois, M. L. 1929. Sur la fermentation et sur la respiration chez les végétaux chlorophylliens. *Rev. Gen. Bot.* **41**:252–271.

35. Godwin, H. 1935. The effect of handling on the respiration of cherry laurel leaves. *New Phytol.* **34**:403–406.

36. Goeschl, J. D., L. Rappaport, and H. K. Pratt. 1966. Ethylene as a factor regulating the growth of pea epicotyls subjected to physical stress. *Plant Physiol.* **42**:877–884.

37. Gortner, W. A., G. G. Dull, and B. Krauss. 1967. Fruit development, maturation, ripening, and senescence: A biochemical basis for horticultural terminology. *HortScience* **2**:141.

38. Green, W. P., W. V. Hukill, and D. H. Rose. 1941. Colorimetric measurements of the heat of respiration of fruits and vegetables. *USDA Tech. Bull.* **771**:1–21.

39. Greene, D. W., W. J. Lord, and W. J. Bramlage. 1977. Mid-summer applications of ethephon and daminozide on apples. I. Effect on 'McIntosh'. *J. Am. Soc. Hort. Sci.* **102**:491–494.

40. Greene, D. W., W. J. Lord, and W. J. Bramlage. 1977. Mid-summer applications of ethephon and daminozide on apples. II. Effect on 'Delicious'. *J. Am. Soc. Hort. Sci.* **102**:494–497.

41. Gregory, F. G., and F. J. Richards. 1929. Physiological studies in plant nutrition. I. The effect of manurial deficiency on the respiration and assimilation rate of barley. *Ann. Bot.* **43**:119–161.

42. Hardwick, K., M. Wood, and H. W. Woolhouse. 1968. Photosynthesis and respiration in relation to leaf age in *Perilla frutescens* (L.) Britt. *New Phytol.* **67**:79–86.

43. Harris, G. H. 1929. Studies on tree root activities. I. An apparatus for studying root respiration and factors which influence it. *Sci. Agric.* **9**:553–565.

44. Harris, G. H. 1930. Studies on tree root activities. II. Some factors which influence tree root respiration. *Sci. Agric.* **10**:564–585.

45. Harvey, R. B. 1928. Artificial ripening of fruits and vegetables. *Minn. Agric. Exp. Sta. Bull. 247,* 36p.

46. Hatch, M. D., J. A. Pearson, A. Millerd, and R. N. Robertson. 1959. Oxidation of Krebs cycle acids by tissue slices and cytoplasmic particles from apple fruit. *Aust. J. Biol. Sci.* **12**:167–174.

47. Hill, G. R. 1913. Respiration of fruits and growing plant tissues in certain gases with reference to ventilation and fruit storage. *Cornell Agric. Exp. Sta. Bull.* **330**:374–408.

48. Hoover, W. H., E. S. Johnston, and F. S. Brackett. 1933. Carbon dioxide assimilation in a higher plant. *Smithsonian Inst. Misc. Collect.* **87**:1–19.

49. Hulme, A. C. 1956. Carbon dioxide injury and the presence of succinic acid in apples. *Nature* **178**:218–219.

50. Hulme, A. C., J. D. Jones, and L. S. C. Wooltorton. 1963. The respiratory climacteric in apple fruits. *Roy. Soc. London Proc.,* ser. B., **158**:514–535.

51. Kahl, G. 1974. Metabolism in plant storage tissue slices. *Bot. Rev.* **40**:263–314.

52. Kanai, R., and C. C. Black, Jr. 1972. Biochemical basis for net CO_2 assimilation in C_4-plants. In *Net Carbon Dioxide Assimilation in Higher Plants.* C. C. Black, Jr. (ed.). Symposium sponsored by Southern Section, American Society of Plant Physiologists and Cotton Inc., Mobile, Ala.

53. Kays, S. J., and J. E. Pallas, Jr. 1980. Inhibition of photosynthesis by ethylene. *Nature* **285**:51–52.

54. Kidd, F. 1917. The controlling influence of carbon dioxide. Part III. The retarding effect of carbon dioxide on respiration. *Royal Soc. London Proc.* **89B**:136–156.

55. Kidd, F. 1934. The respiration of fruits. Royal Institute of Great Britain, weekly evening meeting 1934 (as cited by Biale[12]).

56. Kidd, F., and C. West. 1925. The course of respiratory activity throughout the life of an apple. *Great Br. Dept. Sci. Ind. Res., Food Invest. Board Rep. 1924,* pp. 27–33.

57. Kidd, F., C. West, and G. E. A. Briggs. 1921. A quantitative analysis of the growth of *Helianthus annuus.* Part I. The respiration of the plant and of its parts throughout the life cycle. *Roy. Soc. London Proc.* **B92:**368–384.

58. Kirk, J. T., and R. A. Tilney-Bassett. 1967. *The Plastids; Their Chemistry, Structure, Growth, and Inheritance.* W. H. Freeman, San Francisco.

59. Knee, M. 1973. Effects of controlled atmosphere storage on respiratory metabolism of apple fruit tissue. *J. Sci. Food Agric.* **24:**1289–1298.

60. Kolattukudy, P. E. 1980. Cutin, suberin and waxes. In *The Biochemistry of Plants,* vol. 4, P. K. Stumpf and E. E. Conn (eds.). Academic Press, New York, pp. 571–645.

61. Kriedemann, P. E., W. M. Kleiwer, and J. M. Harris. 1970. Leaf age and photosynthesis in *Vitis vinifera* L. *Vitis* **9:**97–104.

62. Kuc, R., and M. Workman. 1964. The relationship of maturity to the respiration and keeping quality of cut carnations and chrysanthemums. *Proc. Am. Soc. Hort. Sci.* **84:**575–581.

63. Lipton, W. J. 1977. Toward an explanation of disorders of vegetables induced by high CO_2 and low O_2. *2nd Nat. Controlled Atmos. Res. Conf. Proc., Mich. State Univ. Hort. Rep.* **28:**137–141.

64. Mack, W. B., and B. E. Livingstone. 1933. Relation of oxygen pressure and temperature to the influence of ethylene or carbon dioxide production and shoot elongation in very young wheat seedlings. *Bot. Gaz.* **94:**625–687.

65. Massey, L. M., Jr., B. R. Chase, and M. S. Starr. 1982. Effect of rough handling on CO_2 evolutions from 'Howes' cranberries. *HortScience* **17:**57–58.

66. Maxie, E. C., D. S. Farnham, F. G. Mitchell, N. F. Sommer, R. A. Parsons, R. G. Snyder, and H. L. Rae. 1973. Temperature and ethylene effects on cut flowers of carnations (*Dianthus carophyllus* L.). *J. Am. Soc. Hort. Sci.* **98:**568–572.

67. Mayak, S., and A. H. Halevy. 1980. Flower senescence. In *Senescence in Plants,* K. V. Thimann (ed.). CRC Press, Boca Raton, Fla., pp. 131–156.

68. McGlasson, W. B., and R. B. H. Wills. 1972. Effects of oxygen and carbon dioxide on respiration, storage life and organic acids of green bananas. *Aust. J. Biol. Sci.* **25:**35–42.

69. Meeuse, B. J. D. 1975. Thermogenic respiration in Aroids. *Ann. Rev. Plant Physiol.* **26:**117–126.

70. Mellenthin, W. M., and C. Y. Wang. 1976. Preharvest temperatures in relation to postharvest quality of 'd'Anjou' pears. *J. Am. Soc. Hort. Sci.* **101:**302–305.

71. Millerd, A., J. Bonner, and J. B. Biale. 1953. The climacteric rise in plant respiration as controlled by phosphorylative coupling. *Plant Physiol.* **28:**521–531.

72. Morris, L. L., and A. A. Kader. 1977. Physiological disorders of certain vegetables in relation to modified atmospheres. *2nd Nat. Controlled Atmos. Res. Conf. Proc., Mich. State Univ. Hort. Rep.* **28:**142–148.

73. Mulder, E. G. 1955. Effect of mineral nutrition of potato plants on respiration of the tubers. *Acta Bot. Neerlandica* **4:**429–451.

74. Müller-Thurgan, H., and O. Schneider-Orelli. 1908. Reifevorgänge bei Kernobst früchten. *Landwirtsch. J. B. Schwei.* **22:**760–774.

75. Nicholas, G. 1918. Contribution á l'etude des variatins de la respiration des végétaux avec l'age. *Rev. Gen. Bot.* **30:**214–225.

76. Nichols, R. 1968. The response of carnations (*Dianthus caryohyllus*) to ethylene. *J. Hort. Soc.* **43:**335–349.

77. Overholser, E. L., and L. L. Claypool. 1931. The relation of fertilizers to respiration and certain physical properties of strawberries. *Proc. Am. Soc. Hort. Sci.* **28:**220–224.

78. Pratt, H. K., and D. B. Merrdoza, Jr. 1979. Colorimetric determination of carbon dioxide for respiration studies. *HortScience* **14:**175–176.

79. Rabinowitch, E. I. 1951. *Photosynthesis and related processes. Vol. II Pt. I. Spectroscopy and Fluorescence of Photosynthetic Pigments: Kinetics of Photosynthesis.* Interscience, New York, 1,208p.

80. Rhodes, M. J. C., and L. S. C. Wooltorton. 1971. The effect of ethylene on the respiration and on the activity of phenylalanine ammonia lyase in swede and parsnip root tissue. *Phytochemistry* **10:**1989–1997.

81. Robinson, J. E., K. M. Browne, and W. G. Burton. 1975. Storage characteristics of some vegetables and soft fruits. *Ann. Appl. Biol.* **81:**399–408.

82. Simon, E. W. 1967. Types of leaf senescence. *Soc. Exp. Biol. Symp.* **21:**215–230.

83. Siripanich, J., and A. A. Kader. 1986. Changes in cytoplasmic and vacuolar pH in harvested lettuce tissue as influenced by CO_2. *J. Am. Soc. Hort. Sci.* **111:**73–77.

84. Smock, R. M., L. J. Edgerton, and M. B. Hoffman. 1954. Some effects of stop drop auxins and respiratory inhibitors on the maturity of apples. *Proc. Am. Soc. Hort. Sci.* **63:**211–219.

85. Solomos, T., and J. G. Biale. 1975. Respiration and fruit ripening. *Colloq. Int. CNRS* **238:**221–228.

86. Thimann, K. V. 1980. The senescence of leaves. In *Senescence in Plants,* K. V. Thimann (ed.). CRC Press, Boca Raton, Fla., pp. 85–115.

87. Thornton, N. C. 1933. Carbon dioxide storage. III. The influence of carbon dioxide on oxygen uptake by fruits and vegetables. *Boyce Thompson Inst. Contrib.* **5:**371–402.

88. Tranquillini, W. 1955. Die bedeutung des lichtes und der temperatur fur die kohlensaureassimilation von *Pinus cembra* jungwuchs an einem hochalpinen standort. *Planta* **46:**154–178.

89. Trout, S. A., E. G. Hall, R. N. Robertson, M. V. Hackney, and S. M. Sykes. 1942. Studies in the metabolism of apples. *Aust. J. Exp. Biol. Med. Sci.* **20:**219–231.

90. Uritani, I., and T. Asahi. 1980. Respiration and related metabolic activity in wounded and infected tissue. In *The Biochemistry of Plants,* vol. 2, D. D. Davis (ed.). Academic Press, New York, pp. 463–485.

91. Wager, H. G. 1974. The effect of subjecting peas to air enriched with carbon dioxide. I. The path of gaseous diffusion, the content of CO_2 and the buffering of the tissue. *J. Exp. Bot.* **25:**330–337.

92. Wager, H. G. 1974. The effect of subjecting peas to air enriched with carbon dioxide. II. Respiration and the metabolism of the major acids. *J. Exp. Bot.* **25:**338–351.

93. Wittwer, S. H., and W. Robb. 1964. Carbon dioxide enrichment of greenhouse atmospheres for food crop production. *Econ. Bot.* **18:**34–56.

94. Young, R. E., and J. B. Biale. 1956. Carbon dioxide fixation by lemons in a CO_2 enriched atmosphere. *Plant Physiol.* **32**(suppl.):23.

95. Zelitch, I. 1979. Photorespiration: Studies with whole tissues. In *Encyclopedia of Plant Physiology,* vol. 6, *Photosynthesis II,* M. Gibbs and E. Latzko (eds.). Springer-Verlag, Berlin, pp. 353–367.

ADDITIONAL READINGS

Anderson, J. M. 1986. Photoregulation of the composition, function, and structure of thylakoid membranes. *Ann. Rev. Plant Physiol.* **37:**93–136.

Barber, J. (ed.). 1987. *The Light Reactions.* Elsevier, Amsterdam, 565p.

Beevers, H. 1961. *Respiratory Metabolism in Plant.* Row, Peterson, Evanston, Ill., 232p.

Blackman, F. F. 1954. *Analytic Studies in Plant Respiration.* Cambridge University Press, Cambridge, England, 231p.

Buchanan, B. B. 1980. Role of light in the regulation of chloroplast enzymes. *Ann. Rev. Plant Physiol.* **31**:341–374.

Cottrell, H. J. (ed.). 1987. *Pesticides on Plant Surfaces.* Wiley, New York, 86p.

Dennis, D. T. 1987. *The Biochemistry of Energy Utilization in Plants.* Blackie, Glasgow, 145p.

Douce, R., and M. Neuburger. 1989. The uniqueness of plant mitochondria. *Ann. Rev. Plant Physiol.* **40**:371–414.

Farquhar, G. D., and T. D. Sharkey. 1982. Stomatal conductance and photosynthesis. *Ann. Rev. Plant Physiol.* **33**:317–345.

Goodwin, T. W., and E. I. Mercer (eds.). 1983. *Introduction to Plant Biochemistry.* Pergamon, Oxford, England, 677p.

Haehnel, W. 1984. Photosynthetic electron transport in higher plants. *Ann. Rev. Plant Physiol.* **35**:659–693.

Hall, D. O., and K. K. Rao. 1987. *Photosynthesis.* E. E. Arnold, Baltimore, Md., 122p.

Hart, J. W. 1988. *Light and Plant Growth.* Allen and Unwin, Boston, 204p.

Hipkins, M. F., and N. R. Baker (eds.). 1986. *Photosynthesis: Energy Transduction, A Practical Approach.* IRL Press, Oxford, England, 199p.

Kowallik, W. 1982. Blue light effects on respiration. *Ann. Rev. Plant Physiol.* **33**:51–72.

Kyle, D. J., C. B. Osmond, and C. J. Arntzen (eds.). 1987. *Photoinhibition.* Elsevier, Amsterdam, 315p.

Laties, G. G. 1982. The cyanide-resistant, alternative path in higher plant respiration. *Ann. Rev. Plant Physiol.* **33**:519–555.

Lorimer, G. H. 1981. Events in photosynthesis and photorespiration. *Ann. Rev. Plant Physiol.* **32**:349–383.

Ogren, W. L. 1984. Photorespiration: Pathways, regulation, and modification. *Ann. Rev. Plant Physiol.* **35**:415–442.

Palmer, J. M. (ed.). 1984. *The Physiology and Biochemistry of Plant Respiration.* Cambridge University Press, Cambridge, England, 195p.

Sironval, C., and M. Bruwers (eds.). 1984. *Protochlophyllide Reduction and Greening.* Kluwer Academic Pub., Hingham, Mass., 408p.

Ting, I. P. 1985. Crassulacean acid metabolism. *Ann. Rev. Plant Physiol.* **36**:595–622.

Ting, I. P., and M. Gibbs (eds.). 1982. Crassulacean acid metabolism. *5th Ann. Symp. Bot. Proc.,* University of California, Riverside, 316p.

van der Veen, R., and G. Meijer. 1959. *Light and Plant Growth.* Phillips Technical Library, Eindhoven, Holland, 161p.

Vanngard, T., and L. Andréasson. 1988. Photosynthetic electron transport in higher plants. *Ann. Rev. Plant Physiol.* **39**:379–411.

Velthuys, B. R. 1980. Mechanisms of electron flow in photosystem II and toward photosystem I. *Ann. Rev. Plant Physiol.* **31**:545–567.

Whatley, J. M., and F. R. Green. 1980. *Light and Plant Life.* E. Arnold, London, 91p.

Woodrow, I. E., and J. A. Berry. 1988. Enzymatic regulation of photosynthetic CO_2 fixation in C_3 plants. *Ann. Rev. Plant Physiol.* **39**:533–594.

4

SECONDARY METABOLIC PROCESSES AND PRODUCTS

The synthesis and degradation of carbohydrates, organic acids, proteins, lipids, pigments, aromatic compounds, phenolics, vitamins, and phytohormones are classified as secondary processes (i.e., secondary to respiration and photosynthesis), but the distinction is somewhat arbitrary. The metabolism of most of these products is absolutely essential in both the preharvest and postharvest life of a product. During the postharvest period, there is continued synthesis of many compounds [e.g., the volatile flavor components of apples (*Malus sylvestris,* Mill.)] and a degradation of other compounds to provide energy and precursors for synthetic reactions. Many of these changes occurring after harvest, however, are not desirable. As a consequence, we strive to store products in a manner that minimizes the development of undesirable changes.

Excluding the synthesis of carbohydrates and proteins, there are three primary pathways that lead to the diverse array of chemical compounds found in plants. These include (1) the shikimic acid pathway that leads to the formation of lignin, coumarins, tannins, phenols, and various aromatics; (2) the acetate-malonate pathway that forms the precursors of fatty acids, phospholipids, glycerides, waxes, glycolipids, and polyketides; and (3) the acetate-mevalonate pathway that results in various terpenoids (gibberellins, carotenoids, abscisic acid) and steroids (fig. 4-1).

This chapter covers the major classes of plant constituents in postharvest products, looks at some of the quantitative and qualitative changes that occur after harvest, and outlines a number of the important environmental factors that accelerate these postharvest alterations.

CARBOHYDRATES

Carbohydrates are the most abundant biochemical constituent in plants, representing 50–80% of the total dry weight. They function as forms of stored energy reserves and make up much of the structural framework of the cells. In addition, simple carbohydrates such as the sugars sucrose and fructose

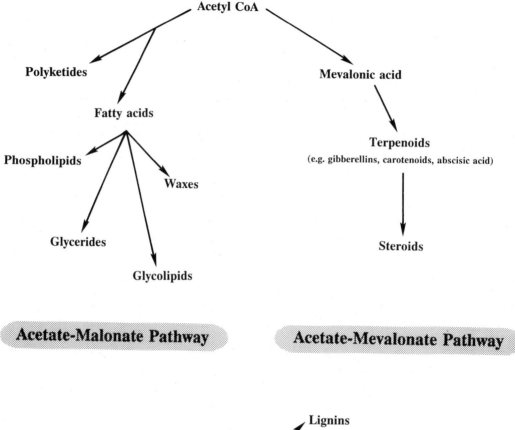

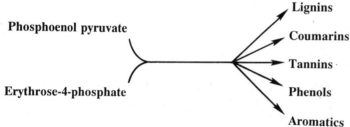

Figure 4-1. The three dominant pathways responsible for the synthesis of secondary plant products.

impart important quality attributes to many harvested products. The concentration of sugars alone can range from slight, as in the lime, to as much as 61% of the fresh weight of the product, as in the date (*Phoenix dactylifera,* L.) (table 4-1).

Carbohydrates are molecules comprised of carbon, hydrogen, and oxygen; however, many may also contain other elements such as nitrogen and phosphorus. As a group, they are defined as polyhydroxy aldehydes or ketones or substances that yield either of these compounds upon hydrolysis.

Table 4-1. General Range of Total Sugars and Relative Amounts of Specific Sugars Found in a Cross-Section of Fruits

| | Total Sugars[b] (% Fresh Weight) | Specific sugars[a] | | |
		Glucose	Fructose	Sucrose (% Fresh Weight) of Edible Portion
Apple	11.6	1.7	6.1	3.6
Avocado	0.4			
Currant, red	5.1	2.3	1.9	0.2
Date	61.0	32.0	23.7	8.2
Grape	14.8	8.2	7.3	
Lime	0.7			
Pineapple	12.3	2.3	1.4	7.9
Pear	10.0	2.4	7.0	1.0
Tomato	2.8	1.6	1.2	

[a] Data from a separate study by Widdowson and McCance.[98]
[b] Data from Biale[6]; Money[65]; Money and Christian[66]; and Swisher and Higby.[89]

Glucose and fructose, structural isomers of each other (both are $C_6H_{12}O_6$), illustrate the differences between the two types of sugars, aldoses and ketoses.

$$
\begin{array}{cc}
\overset{\displaystyle H \quad O}{\underset{\displaystyle C}{\diagdown \,/\!/}} & \overset{\displaystyle H}{\underset{\displaystyle HCOH}{|}} \\
HCOH & C=O \\
HOCH & HOCH \\
HCOH & HCOH \\
HCOH & HCOH \\
CH_2OH & CH_2OH \\
\text{glucose} & \text{fructose} \\
\text{(aldose)} & \text{(ketose)}
\end{array}
$$

In addition, sugars that have a free or potentially free aldehyde group are classified as reducing sugars based on their ability to act as a reducing agent (accept electrons) in an alkaline solution. Most of the common sugars in plants are reducing sugars (e.g., glucose, fructose, galactose, mannose, ribose, and xylose). Sucrose and raffinose are the most common nonreducing sugars. The level of reducing sugars is important in several postharvest products [e.g., white potato, *Solanum tuberosum*, L., to be used for chips (crisps)] in that when the reducing sugar concentration is high there is a much greater incidence of undesirable browning reactions during processing. The handling and storage conditions prior to processing can significantly alter the level of free reducing sugars in the product, leading to a lower-quality processed product.

Carbohydrates can be further classified, based on their degree of polymerization, into monosaccharides, oligosaccharides, and polysaccharides. Simple sugars or monosaccharides represent the most fundamental group and cannot be further broken down into smaller sugar units. These basic units of carbohydrate chemistry are subclassed based on the number of carbon atoms they contain (fig. 4-2). This number ranges from three in the triose sugars to seven in the heptose sugars, although occasionally octuloses are found. Both the glycolytic and pentose pathways are important in the synthesis of these molecules that form the building blocks for more complex carbohydrates. Monosaccharides may also be modified to form several types of compounds that are essential metabolic components of the cells. For example, amino and deoxy sugars and sugar acids and alcohols, although seldom high in concentration in harvested products, are common (fig. 4-2). Less common are branched sugars such as apiose.

Monosaccharides

In many products the monosaccharides comprise a major portion of the total sugars present (table 4-1). Glucose and fructose are the predominant simple sugars found, especially in fruits; however, mannose, galactose, arabinose, xylose, and various others are found in a number of harvest products.

Oligosaccharides

Oligosaccharides are more complex sugars that yield two to six molecules of simple sugars upon hydrolysis. For example, a disaccharide such as sucrose yields two monosaccharides (glucose and fructose) upon hydrolysis and a pentasaccharide yields five monosaccharide molecules. Both the monosaccharides and oligosaccharides are water soluble and together they comprise the total sugars in a product. The most abundant oligosaccharide is sucrose, which is the primary transport form of carbohydrate in most plants. While its synthesis is not thoroughly understood, sucrose phosphate synthetase appears to be the prevalent in vivo enzyme catalyzing the reversible reaction:

$$UDP\text{-glucose} + \text{fructose-6-P} \rightarrow \text{sucrose-P} + UDP$$
$$\text{sucrose-P} \rightarrow \text{sucrose} + Pi$$

Sucrose can also be synthesized by sucrose synthase. This reaction is also reversible; however, the opposite direction (i.e., the hydrolysis of sucrose) is generally favored. In addition to sucrose synthase, sucrose may also be hydrolyzed by the enzyme invertase, yielding glucose and fructose. Other common oligosaccharides are the disaccharide maltose found in germinating seeds and the trisaccharide raffinose and tetrasaccharide stachyose that act as translocatable sugars in several species (fig. 4-3).

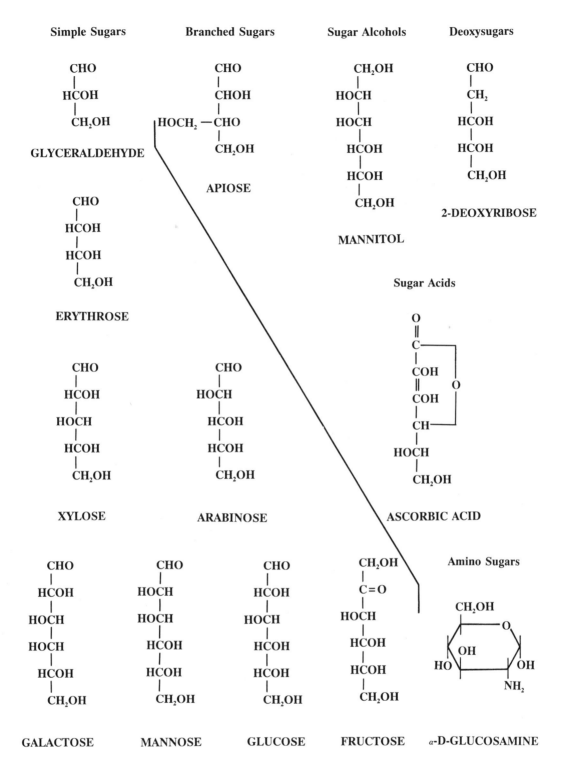

Figure 4-2. Common sugars and sugar derivatives.

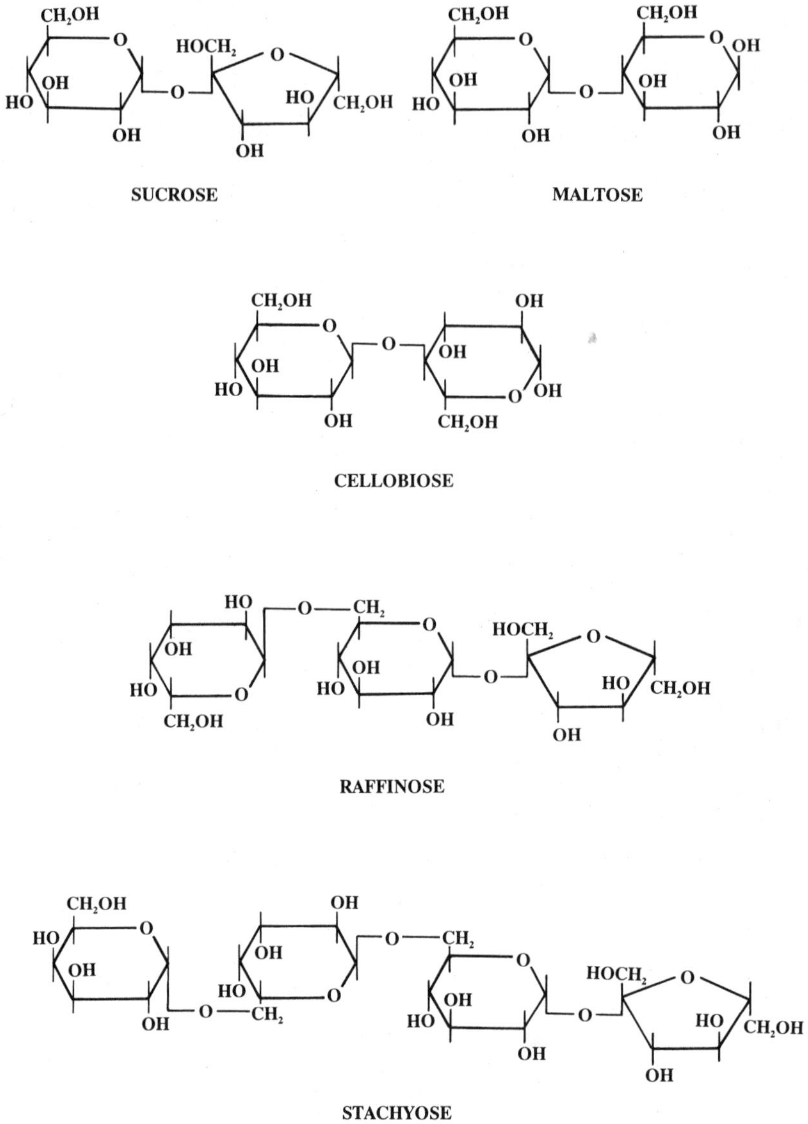

Figure 4-3. Molecular structure of some common oligosaccharides found in plants.

Polysaccharides

Cellulose

Cellulose is a straight-chain polymer of glucose and represents one of the most abundant compounds in intact plants. Cellulose is not, however, present in large quantities in many storage organs where storage carbohydrates or lipids often abound. For example, the cellulose content of dates is only 0.8%[12] whereas cotton (*Gossypium hirsutum*, L.) fibers are 98% cellulose. Individual molecules can be extremely long, 1,000–10,000 glucose sub-units, with molecular weights of 200,000–2,000,000. The attachment between neighboring glucose molecules in the chain is a β-linkage between carbon 1 of one glucose molecule and the number 4 carbon of the next (fig. 4-4). Cellulose is found largely in the primary and, when present, secondary cell walls of the tissue.

The synthesis of cellulose is not entirely understood. The basic subunit for insertion into the chain is cellobiose (two glucoses) rather than individual glucose molecules. UDP-glucose and lipid-pyrophosphate sugars appear to be essential participants in the synthesis scheme.[13] In most detached products where there is little, if any, growth, cellulose synthesis is usually quite limited.

Cellulose molecules are extremely stable and can be broken down only with strong acids or by enzymes, called cellulases, capable of hydrolyzing the molecule. Cellulases are found in low amounts in harvested products; however, they do not appear to be important in softening. There is often little change in the cellulose structure in fruits during ripening [e.g., peach (*Prunus persica,* Batsch.)[87]] and the level of activity of the enzyme does not correlate with softening changes that occur during ripening.[40] Cellulases are known to function during the abscission of leaves or other organs from the parent plant; however, these appear to be isozymes that differ from the general cellulase found in most cells.

Starch

Starch, composed of a mixture of branched and straight-chained glucose polymers, represents the major storage carbohydrate in most postharvest products. It is stored in the form of starch grains found in specialized storage plastids (amyloplasts) and in chloroplasts in leaves. Starch is comprised of two compounds: a straight-chained molecule, amylose, that contains 200–1,000 glucose subunits and amylopectin, a branched-chain molecule that is substantially larger, 2,000–200,000 subunits. Amylose has individual glucose molecules linked by α-(1-4) glucosidic bonds (fig. 4-4), the bonding angle of which imparts a helical structure to the molecule. Amylopectin has similar α-(1-4) bonds between glucose subunits; however, every 20–25 glucose mole-cules along the chain there is a branch formed via an α-(1-6) glucosidic linkage (fig. 4-4). The ratio of amylose to amylopectin in a particular species is genetically controlled with amylopectin comprising the dominant form, for example, 60% to greater than 95%.

Biosynthetic or hydrolytic changes in the starch concentration are ex-tremely important during the postharvest period for many commodities. For example, in banana (*Musa* spp.) and many other climacteric fruits, the conversion of starch to sugars in the fruit is an important component of the ripening process, giving the fruit its distinctive sweet flavor as well as pre-cursors for many of the aromatic flavor compounds. On the other hand, in some products [e.g., some of the older cultivars of sweet corn (*Zea mays* var. *rugosa,* Bonaf.)] free sugars are readily converted to starch after harvest, decreasing the quality of the product. Because of these differing postharvest scenarios, an understanding of the metabolism of starch is advantageous.

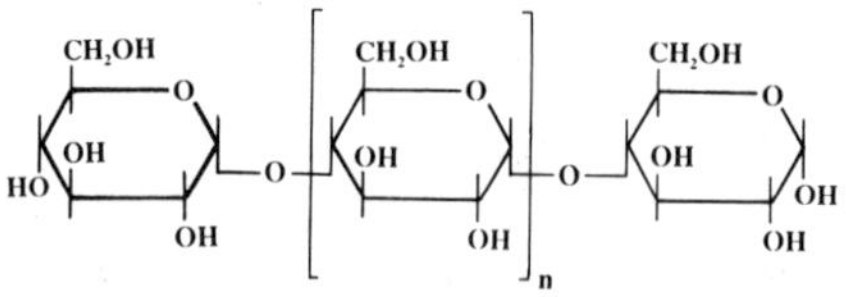

Figure 4-4. The structure of several common polysaccharides found in plants.

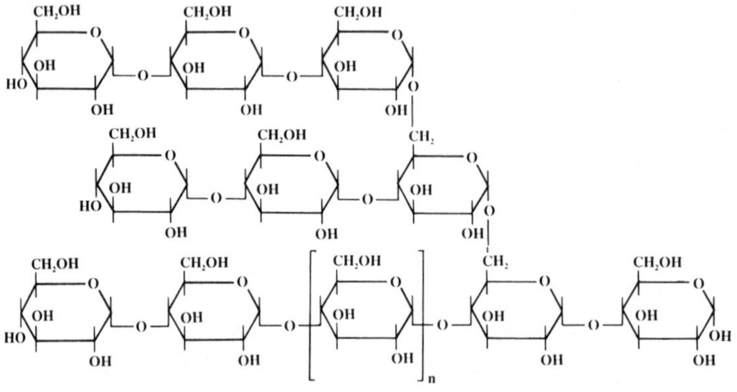

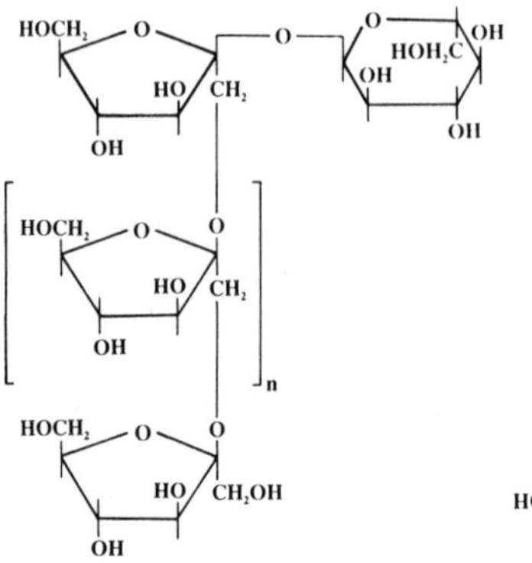

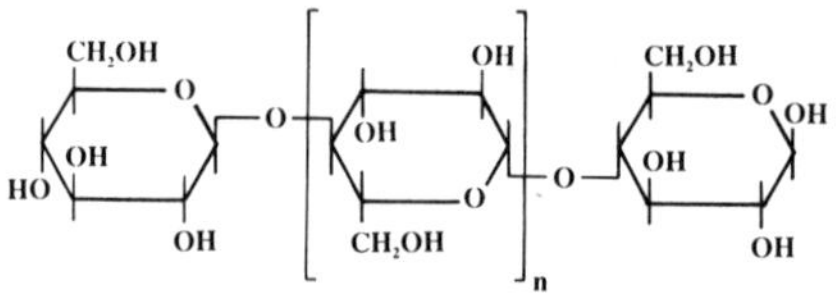

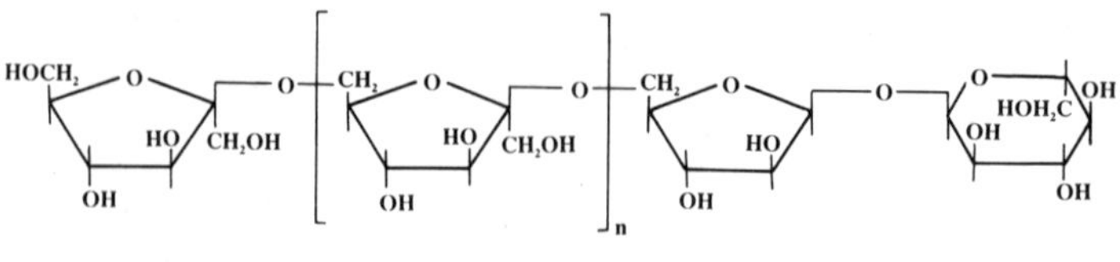

STARCH SYNTHESIS

The synthesis of starch is not thoroughly understood. Synthesis probably begins with sucrose, the primary transport carbon source, which is converted to nucleotide diphosphate glucose. Subsequent polymerization is carried out by the enzyme starch synthetase, which requires an existing α-(1-4) glucan primer molecule of at least two glucose residues (e.g., maltose). There are several forms of starch synthetase (e.g., ADP-glucose and UDP-glucose transglucosylase) that catalyze the addition of glucose from either ADP or UDP-glucose nucleotides. Sucrose can be converted to ADP-glucose either directly by sucrose synthase or indirectly through the action of invertase.

a. ADP-glucose formation from sucrose

$$\text{sucrose} + H_2O \xrightarrow[\text{invertase}]{} \text{glucose} + \text{fructose}$$

$$\text{glucose} + ATP \longrightarrow \text{glucose-6-P} \longrightarrow \text{glucose-1-P}$$
$$\text{glucose-1-P} + ATP \longrightarrow \text{ADP-glucose} + PPi$$

b. Addition of glucose to an existing glucan chain

$$n \text{ ADP-glucose} + \alpha\text{-(1-4) glucan primer} \longrightarrow \text{starch} + n\text{ADP}$$

The formation of amylopectin, requiring the addition of α-(1-6) branches, proceeds similarly; however, a branching enzyme (Q enzyme or amylo-(1,4-1,6)-transglycosylase) catalyzes the branch addition every 20–25 glucose residues.

Several mutants of sweet corn have been found that give both higher sugar levels in the endosperm and decreased incorporation of the sugars into starch after harvest.[28] Preliminary evidence indicates that in the case of *ae du* mutant both the branching enzyme and starch synthase levels are considerably reduced.[73]

STARCH BREAKDOWN

Starch can be converted back to glucose by at least three different enzymes: α-amylase, β-amylase, and starch phosphorylase. The amylases hydrolyze starch into two glucose segments (maltose) that are then further hydrolyzed by the enzyme maltase:

$$\text{starch} + n\ H_2O \xrightarrow[\text{amylase}]{} n\text{ maltose}$$

$$\text{maltose} + H_2O \xrightarrow[\text{maltase}]{} 2\text{ glucose}$$

The α-amylases rapidly hydrolyze the α-(1-4) linkages of amylose at random points along the chain, forming fragments of approximately 10 glucose subunits called maltodextrins. These are more slowly hydrolyzed to maltose by the enzyme. Alpha-amylase also attacks the α-(1-4) linkages of amylopectin; however, in the regions of the α-(1-6) branch points it is inactive, leaving what are called limit dextrins (>3 glucosyl units).[54] Beta-amylase removes maltose units starting from the nonreducing end of the starch chain and hydrolyzes up to a α-(1-6) branching point. This reaction yields maltose and limit dextrins.

Starch phosphorylase also attacks α-(1-4) linkages, but forms glucose-1-phosphate. Unlike the hydrolysis reactions of the amylases where a single water molecule is used in each bond cleavage, the phosphorylase enzyme incorporates phosphate:

$$\text{starch} + \text{n Pi} \xrightarrow{\text{starch phosphorylase}} \text{n glucose-1-P}$$

Neither starch phosphorylase nor the amylases will attack the α-(1-6) branch points of amylopectin, so complete breakdown by these enzymes is not possible. Several debranching enzymes have been isolated from plants; however, neither their importance nor action are well understood at present.

Pectic Substances

The bulk of the primary cell walls in plants is comprised of dense gel-like noncellulosic polysaccharides called pectic substances. Pectin is found extensively in the middle lamella where it functions as the binding agent between neighboring walls. There are three primary forms of pectic substances: pectic acids, pectins, and protopectins. Each is comprised largely of α-(1-4) linked D-galacturonic acid subunits, although a number of other monosaccharides may be present (i.e., arabian, galactan, rhamnose, xylose) (table 4-2). The structure of these pectic substances varies widely with source. For example, Jackfruit, *Artocarpus heterophyllus,* Lam.,[4] is comprised of only galacturonic acid residues; however, most species have rhamnose interspersed between segments of galacturonic acid residues.

Pectic acids, found in the middle lamella and the primary cell wall, are the smallest of the three polymers and are usually around 100 galacturonic acid subunits in size. Pectic acids are soluble in water but may become insoluble if many of the carboxyl groups combine with Ca^{++} or Mg^{++} to form salts (see the Ca bridge in fig. 4-5).

Pectins are usually larger than pectic acids (e.g., 200 subunits) and have many of their carboxyl groups esterified by the addition of methyl groups. They are also found in the middle lamella, in the primary cell wall, and in some cells as constituents of the cytoplasm.

Table 4-2. Gross Composition of the Cell Wall for Apple and Strawberry: Monomers Yielded Upon Hydrolysis of Wall Polymers

Component Monomers	% Total Accounted for	
	Apple	Strawberry
Rhamnose	0.4	1.1
Fucose	0.7	ND
Arabinose	19.5	6.5
Xylose	5.9	1.9
Mannose	1.9	0.7
Galactose	5.8	7.6
Glucose	47.5	31.1
Galacturonic acid	16.6	40.3
α-Amino acid	1.7	10.7
Hydroxyproline	0.04	0.1

Source: After Knee.[50]
ND = not detected.

Protopectins (fig. 4-5) are larger in molecular weight than pectins and intermediate in the degree of methylation between pectic acids and pectins. They are insoluble in hot water and are found primarily in cell walls.

SYNTHESIS OF PECTIC SUBSTANCES

Pectic substances are synthesized from UDP-galacturonic acid on other UDP-sugars, although several other nucleotides may, in some cases, function in place of uridine diphosphate. Methylation appears to occur after the subunit is placed in the chain, rather than before. The methyl donor is S-adenosylmethionine and the reaction is catalyzed by methyl transferase (fig. 4-5).

BREAKDOWN OF PECTIC SUBSTANCES

The activity of pectic enzymes has been shown to be closely correlated with increases in softening in many ripening fruits and the concurrent increase in soluble pectins. For example, the soluble pectin content of apples increases more than threefold during a 1.4-kg decrease in fruit firmness.[5] This increase is due to hydrolytic cleavage of the long pectic chains increasing their solubility. In some cases, however, softening is not associated with solubilization and textural changes may be through the action of pectic enzymes cleaving the linkages between the molecule and other cell wall components. The principal pectic enzymes are pectinesterase, endopolygalacturonase, and exopolygalacturonase.

Pectinesterase or pectinmethylesterase catalyzes the hydrolysis of methyl esters along the pectic chain, producing free carboxyl groups (fig. 4-5). The enzyme deesterifies in a linear manner, moving down the chain and producing segments with free carboxyl groups. The precise role of deesterification is not entirely clear. An increase in free carboxyl groups would lead to an increase in interaction with Ca^{++}, which would decrease solubility. It appears possible that the deesterification by pectinesterase must precede degradation by polygalacturonases.

The polygalacturonases represent a class of pectolytic enzymes that degrade deesterified pectin chains into smaller molecular weight polymers and component monosaccharides. Often two polygalacturonases are found in fruit tissues. Exopolygalacturonase cleaves single galacturonic acid subunits from the nonreducing end of the protopectin molecule (fig. 4-5), whereas endopolygalacturonase attacks the chain randomly. Cleavage within the chain by endopolygalacturonase has a much more pronounced effect on the degree of solubilization of the pectic molecule. Hence, attack by endopolygalacturonase results in much more rapid softening of the tissue. Both enzymes have been found in a number of fruits and increases in their activity tend to parallel the formation of water-soluble pectins and softening during ripening.[74]

Hemicelluloses

The hemicelluloses represent a heterogenous group of polysaccharide compounds that are closely associated with cellulose; hence the name (*half*) cellulose. They are one of the major components of cell walls and are quite

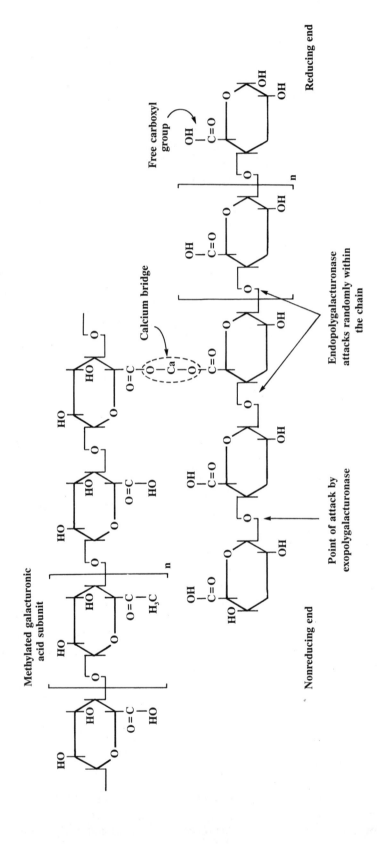

Methylated galacturonic acid subunit

Calcium bridge

Free carboxyl group

Reducing end

Endopolygalacturonase attacks randomly within the chain

Point of attack by exopolygalacturonase

Nonreducing end

A. ADDITION OF METHYL GROUPS DURING SYNTHESIS

OH
|
C=O
|
−G−G−G−G−   + S-adenosyl methionine   —————————→   −G−G−G−G−   + S-adenosyl homocysteine
                  (methyl donor)   Methyltransferase

Deesterified galacturonic
acid subunit

CH₃
|
C=O
|
Methylated
subunit

B. REMOVAL OF METHYL GROUPS DURING DEGRADATION

(free carboxyl group)

CH₃
|
C=O
|
−G−G−G−G−   —————————→   −G−G−G−G−
              Pectinesterase

(methylated subunit)

O
|
C=O
|

Figure 4-5. The structure of protopectin and a deesterified protopectin molecule cross-linked by a Ca^{++} salt bridge. *A*: The addition of methyl groups during synthesis. *B*: Their removal during degradation.

stable once formed. Typically, with perhaps the exception of some of the mannans, hemicelluloses do not represent carbohydrate reserves that may be recycled as an energy source for the cell. Hemicelluloses are composed largely of glucose, galactose, mannose, xylose, and arabinose molecules linked in various combinations and with varying degrees of branching. The total number of subunits may range from 40 to 200. In monocots, the major hemicellulosic components of the cell walls are arabinoxylans and in dicots they are xyloglucans. Hemicellulose appears to be synthesized from nucleotide sugars (e.g., UDP-xylose, UDP-arabinose, UDP-glucuronic acid) in reactions catalyzed by transferase enzymes.

Fructosans

Some plant species store polymers of fructose as carbohydrate reserves rather than glucose. The polymers include the inulins and levans and are particularly common in the Asteraceae (Compositae), Campanulaceae, and Poaceae (Gramineae) families. Inulin is composed of a chain of 25 to 35 fructose subunits joined by β-linkages through C-1 of one molecule and C-2 of the adjacent [β-(2-1)] and is terminated with a sucrose molecule. Inulin is a straight-chained polymer; however, some fructosans are branched. They are substantially smaller than starch molecules and more soluble in water. Inulins are more commonly found stored in roots and tubers rather than in aboveground plant parts. The tubers of Jerusalem artichokes (*Helianthus tuberosus*, L.) and Dahlia (*Dahlia pinnata*, Cav.), the bulbs of *Iris* spp., and the roots of dandelion (*Taraxacum officinale*, Weker) and chicory (*Chicorium intyvus* L.) are high in inulin.

The synthesis of inulin begins with the addition of fructose to a terminal sucrose molecule forming a trisaccharide.[15] The complete synthesis of the polymer requires several enzymes:

- Sucrose-sucrose 1-fructosyltransferase (SST), which catalyzes the formation of the trisaccharide from two sucrose molecules:

$$\text{Glu}\sim\text{Fru} + \text{Glu}\sim\text{Fru} \rightarrow \text{Glu}\sim\text{Fru}-\text{Fru} + \text{Glu}$$

- Glucose is subsequently converted via several steps into sucrose;
- β-(2-1) fructan 1- fructosyltransferase (FFT), which catalyzes the transfer of a fructose subunit from a donor to an acceptor, both of which are trisaccharides or greater in size:

$$\underset{\text{donor}}{\text{Glu}\sim\text{Fru}-\text{Fru}_N} + \underset{\text{acceptor}}{\text{Glu}\sim\text{Fru}-\text{Fru}_M} \rightarrow \underset{\text{donor}}{\text{Glu}\sim\text{Fru}-\text{Fru}_{N-1}} + \underset{\text{acceptor}}{\text{Glu}\sim\text{Fru}-\text{Fru}_{M+1}}$$

This reaction is reversible and also functions during depolymerization.

The degradation or depolymerization of inulin can follow one of two possible pathways. During cold storage, inulin is broken down into shorter chain-length oligomers. This breakdown involves the action of hydrolases, β-(2-1') fructan 1-fructosyl-transferase (FFT), and the enzymes involved in sucrose synthesis. During sprouting, inulin is degraded completely to fructose by hydrolases that are then converted to sucrose for export to the growing apexes. Several yeasts have been isolated that have the enzymes required to hydrolyze the inulin polymer and subsequently convert the subunits to alcohol, thus enhancing the attractiveness of using inulin as a carbon substrate for alcohol production.[31]

Levans, another type of fructose polymer, are formed through a β-linkage between C-2 and C-6 of two adjacent fructose subunits. As with inulin, levans are also terminated with a sucrose molecule. Levans are found primarily in the Poaceae (Gramineae) family; for example, a levan called phlein is found stored in the roots of timothy (*Phleum pratense*, L.).

Gums and Mucilages

Gums and mucilages are composed of a wide cross-section of sugar subunits and as a consequence generalizations about their individual composition cannot easily be made. Hydrolysis of gum from plum (*Prunus domestica*, L.) fruits yields a mixture: D-galactose, D-mannose, L-arabinose, D-xylose, L-rhamnose, D-glucuronic acid, and traces of 4-0-methyl glucuronic acid.

These polymers may be found free in the cytoplasm or in some cases sequestered in specialized cells. As a group they are generally hydrophilic in nature and their function in the plant is not well understood. Gums are thought to be involved in sealing mechanical or pathogenic wounds to the plant while mucilages may function by modulating water uptake, as in seeds, or water loss from some succulent species.

ORGANIC ACIDS

A number of harvested plant products contain significant concentrations of organic acids, many of which play a central role in metabolism. In addition, the levels of organic acids present often represent an important quality parameter; this is especially so in many fruits (table 4-3).

Organic acids are small mono-, di-, and tricarboxylic acids that exhibit acidic properties due to the presence of their carboxyl group(s) (COOH), which can give up a hydrogen. They exist as free acids or anions, or are combined as salts, esters, glycosides, or other compounds. Organic acids are found in active pools that are utilized in the cytoplasm for metabolism and to a greater extent as storage pools in the vacuole. For example, only about 30% of the malate is in the mitochondria, with the remainder thought to be largely sequestered in the vacuole. In some plant cells, certain organic acids may, to a large extent, be in the form of insoluble salts, for example, calcium oxalates in rhubarb or potassium bitartrate in grapes. When in the ionized anion form ($-COO^-$) the name of the acid ends in "ate" (e.g., malate); while in the protonated state ($-COOH$), the ending is "ic" (e.g., malic acid).

Organic acids can be classified or grouped in a number of ways. For example, they may be grouped based on the number of carbon atoms present (typically two to six) or on their specific function(s) within the cell. Another means of separation is based on the number of carboxylic groups present. This nomenclature gives a greater indication of how the acid will act chemically. A number of the common organic acids found in postharvest products are illustrated in figure 4-6. These compounds include monocarboxylic acids, monocarboxylic acids with alcohol, ketone or aldehyde

Table 4-3. Organic Acids in Apple, Pear, Grape, Banana, and Strawberry

	Apple	*Pear*	*Grape*	*Banana*	*Strawberry*
Glycolic	+	+	+	+	tr
Lactic	+	+	+	+	
Glyceric	+	+	+	+	tr
Pyruvic	+		+	+	
Glyoxylic	+		+	+	
Oxalic	+		+	+	
Succinic	+	+	+	+	+
Fumaric	+		+		
Malic	++	++	++	++	+
Tartaric			++		
Citramalic	+	+		+	
Citric	+	+	+	+	+++
Isocitric	+		+		
Cis-aconitic			+		
Oxaloacetic	+		+	+	
α-Oxoglutaric	+	+	+	+	
Galacturonic	+	+	+		
Glucuronic	+		+		
Caffeic	+		+		
Chlorogenic	+	+	+		
p-Coumarylquinic	+				
Quinic	+	+	+	+	+
Shikimic	+	+	+	+	tr

Source: After Ulrich.[94]
Notes: Data derived from Hane[35]; Hulme and Wooltorton[44]; Kliewer[48]; Kollas[53]; Steward et al.[88]; and Wyman and Palmer.[101]
Increasing number of + signs denotes increased concentration; tr = trace.

groups, monocarboxylic carbocyclic aromatic and alicyclic acids, dicarboxylic acids, and tricarboxylic acids. The type of acid present and the absolute and relative concentration of each varies widely between different postharvest products.

Table 4-3 illustrates a comparison of the acids present in the apple, pear (*Pyrus communis,* L.), grape (*Vitis vinifera,* L.), banana, and strawberry (*Fragaria chiloensis,* Duchesne.) with an indication of the relative concentration of each. Most organic acids are found in only trace amounts; however, several, such as malic, citric, and tartaric, tend to be found in abundance in some tissues. The concentrations of these abundant acids vary widely between products, for example, lemon (*Citrus limon,* Burm.) fruits contain 70–75 meq of citric acid/100 g fresh weight whereas the banana fruit has only 4 meq of malic acid/100 g fresh weight. In addition, in the first case (lemon), high acidity is a desirable flavor attribute whereas in the second it is not.

Organic acids play a central role in the general metabolism of postharvest products. A number of organic acids are essential components of the respiratory tricarboxylic acid cycle and phosphoglyceric acid plays an essential role in photosynthesis. Many organic acids have multiple functions in the plant. In some tissues where they are found in high concentrations, organic acids represent a readily available source of stored energy that can be utilized after

1. MONOCARBOXYLIC ACIDS

3. ALIPHATIC TRICARBOXYLIC ACIDS

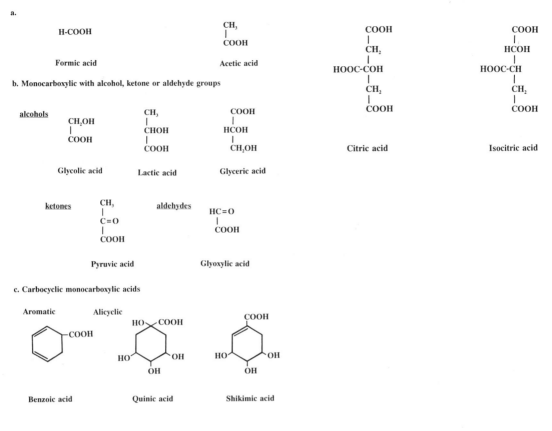

2. ALIPHATIC DICARBOXYLIC ACIDS

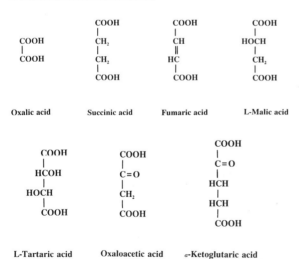

Figure 4-6. Structures of a number of the common organic acids found in plants.

the product is severed from the plant. In addition, in food products organic acids may impart a significant portion of the characteristic flavor, both taste and odor. Aromatic compounds, such as esters of organic acids given off by various plant products, are diverse in type (table 4-4), and in some cases, such as isopentyl acetate in banana fruits, they represent character impact compounds that impart the major portion of the characteristic aroma.

Organic acids are synthesized primarily through oxidations, decarboxylations, and in some cases, carboxylations in the respiratory tricarboxylic acid pathway. Some, however, are formed from sugars during the early stages of the photosynthetic dark reactions. Therefore, in most cases, the immediate precursors of organic acids are either other organic acids or sugars.

After harvest and during storage, the concentration of total organic acids tends to decline. Postharvest changes vary with the specific acid in question, the type of tissue, handling and storage conditions, cultivar, year, and a number of other parameters. For example, the concentration of citric and malic acids in the juice of 'Shamouti' oranges (*Citrus sinensis*, (L.) Osbeck) declined with storage time while malonic, succinic, and adipic acids increased (fig. 4-7). Changes in albedo (white interior portion of the peel) and flavedo (pigmented exterior portion of the peel) organic acids, however, did not parallel changes in the juice.[81] Likewise, both cultivar and year of production can have a pronounced effect on the concentration of specific acids and total titratable acidity (table 4-5).

Controlled atmosphere storage has been shown to alter the changes in organic acids occurring after harvest. The juice of Valencia oranges stored at 3% O_2 and 5% CO_2 (3.5°C) lost less acid than oranges held in air storage (0°C).

Table 4-4. Organic Acids Evolved as Esters by Apples (cv. 'Cabville Blanc')

Methyl formate	tr	Methyl butyrate	tr
Propyl formate	tr	Ethyl butyrate	
Hexyl formate	tr	Propyl butyrate	tr
Isobutyl formate	tr	Butyl butyrate	
Methyl acetate	tr	Amyl butyrate	
Ethyl acetate		Hexyl butyrate	tr
Propyl acetate		Isopropyl butyrate	tr
Butyl acetate		Isobutyl butyrate	tr
Amyl acetate		Isoamyl butyrate	tr
Hexyl acetate		Ethyl isobutyrate	tr
Isobutyl acetate		Ethyl valerianate	tr
sec-Butyl	tr	Butyl valerianate	tr
Isoamyl acetate		Methyl caproate	tr
Methyl propionate	tr	Ethyl caproate	
Ethyl propionate	tr	Butyl caproate	
Propyl propionate	tr	Ethyl octanoate	tr
Butyl propionate			
Amyl propionate	tr		
Isoamyl propionate			

Notes: Data from Paillard[69]; tr = trace.

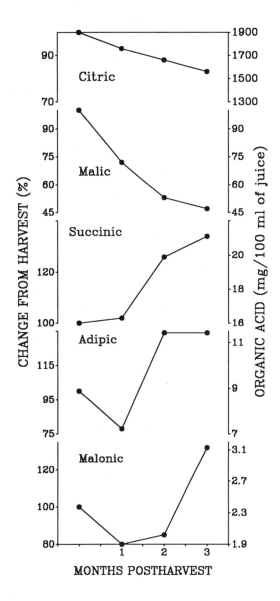

Figure 4-7. Changes in the organic acid composition of the juice of 'Shamouti' oranges (*Citrus sinensis,* (L.) Osbeck.) during storage at 17°C. Changes are expressed as the percent change from harvest and as mg/100 ml of juice (*after Sasson and Monselise[81]*).

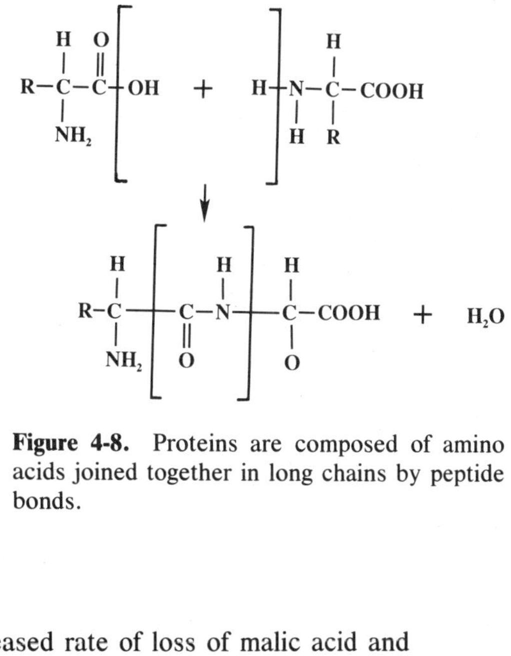

Figure 4-8. Proteins are composed of amino acids joined together in long chains by peptide bonds.

Principal differences were in the decreased rate of loss of malic acid and increases in quinic and shikimic acids in controlled-atmosphere stored fruit.[53] These effects may in part be due to the dark fixation of CO_2 by the fruit (see page 112).

PROTEINS AND AMINO ACIDS

Proteins are extremely important components of living cells in that they regulate metabolism, act as structural molecules, and in some products, represent storage forms of carbon and nitrogen. Proteins are composed of chains of amino acids each joined to the next in the sequence by a peptide

Table 4-5. Differences in the Organic Acid Concentration and Titratable Acidity for 12 Muscadine Grape Cultivars (*Vitis rotundifolia*, Michx) and Differences Between Seasons for the Cultivar Roanoke

	Malate (%)	Tartrate (%)	Citrate (%)	Titratable Acidity (%)
Albemarle	0.49[a]	0.25	0.05	0.75
Carlos	0.54	0.19	0.04	0.85
Chowan	0.50	0.23	0.05	0.82
Dearing	0.45	0.23	0.04	0.71
Hunt	0.54	0.32	0.04	0.93
Magnolia	0.33	0.23	0.05	0.64
Magoon	0.43	0.41	0.02	0.94
Pamlico	0.62	0.24	0.04	1.05
Roanoke	0.70	0.29	0.04	0.10
Scuppernong	0.52	0.28	0.04	0.87
Thomas	0.51	0.27	0.05	0.84
Topsail	0.37	0.19	0.04	0.51
Roanoke 1965	0.26	0.23	0.05	0.52
1966	0.67	0.27	0.06	1.22
1967	1.16	0.38	0.02	1.55

Note: Data from Carroll et al.[10]
[a] Mean of three seasons.

bond (fig. 4-8). Thus, they are often referred to as polypeptides.* Many of the properties of a protein or polypeptide are a function of the amino acids of which it is composed and their particular sequence in the molecule. In plants there are approximately 22 amino acids that are commonly found; however, over 100 nonprotein amino acids have been identified.

Each of the amino acids that makes up a protein has distinct properties that, in turn, influence the properties of the polypeptide in which it is found. Amino acids are small in size and soluble in water. Each contains both a carboxyl group ($-COOH$) and an amino group ($-NH$) and some may also have hydroxyl groups ($-OH$), sulfhydryl groups ($-SH$), or amide groups ($-CONH$):

---

* The term *polypeptide* typically refers to amino acid chains from 10 to 100 amino acids in length. Chains greater than 100 amino acids are called proteins; those less than 10 are peptides.

Typically, amino acids are classified into one of six types based on the properties of their R group (fig. 4-9):

1. Neutral amino acids, where R is a hydrogen or an aliphatic, or hydroxyl group (glycine, alanine, valine, leucine, and isoleucine).
2. Basic amino acids (arginine and lysine).
3. Acidic amino acids (aspartic acid, glutamic acid), and their amides (asparagine and glutamine).
4. Hydroxylated amino acids (serine and threonine).
5. Aromatic and heterocyclic amino acids (phenylalanine, tyrosine, proline, hydroxyproline, tryptophan, and histidine*).
6. Sulfur-containing amino acids (cysteine, cystine, and methionine).

Protein Synthesis

During the postharvest period, the metabolic processes within plant cells continue, requiring specific proteins in appropriate quantities and at precise times. Synthesis and degradation are the two primary means of modulating the level of a specific protein. As a consequence, protein synthesis, especially the synthesis of specific proteins after harvest, is of interest to postharvest physiologists. The sequence for each protein, as illustrated by one of the component groups of the storage protein zein (fig. 4-10), is found coded in a section of the cell's DNA (fig. 4-11, page 169). This coded sequence is transcribed by the formation of a special type of ribonucleic acid, the messenger RNAs (mRNA), thus transferring the required amino acid sequence to a molecule that can move from the nucleus into the cytoplasm where the actual synthesis of the protein molecule will occur. The messenger RNA is subsequently attached to a ribosome that translates the code from the mRNA and assembles the polypeptide. Several ribosomes are often agglomerated together on an mRNA strand to form a polysome. One protein molecule is assembled per ribosome at a time. In some cases, there are modifications in the protein after assembly of the amino acids; this process is called posttranslational modification.

A very general overview of the steps in protein synthesis is given in figure 4-12 (page 168). As may be anticipated, the precise method in which DNA is transcribed to RNA and RNA is translated to form the polypeptide is much more complex than indicated by this brief overview. Several references listed in this chapter give a more detailed account.[37,55]

Protein Structure

The proteins formed have a three-dimensional structure that is a function of the kind, number, and sequence of the amino acids present in the peptide chain and the type of nonprotein (prosthetic) groups attached. Protein structure is typically broken down into four levels of organization: primary, secondary, tertiary, and quaternary. The primary structure is due to the kind,

---

* Histidine is a heterocyclic basic amino acid.

Neutral Amino Acids

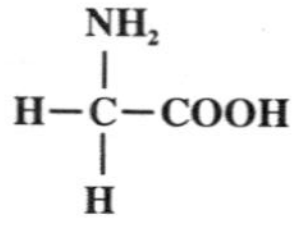

$$NH_2$$
$$H-C-COOH$$
$$H$$

L-GLYCINE

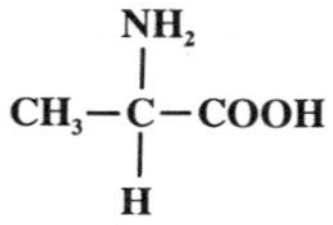

$$NH_2$$
$$CH_3-C-COOH$$
$$H$$

L-ALANINE

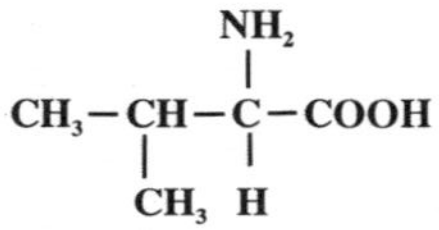

$$NH_2$$
$$CH_3-CH-C-COOH$$
$$CH_3\ \ H$$

L-VALINE

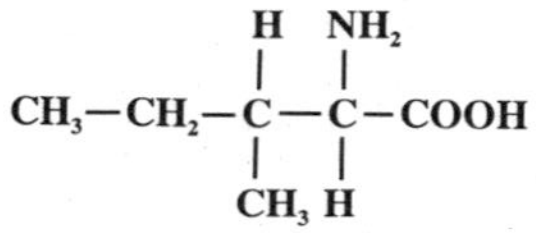

$$H\ \ NH_2$$
$$CH_3-CH_2-C-C-COOH$$
$$CH_3\ H$$

L-ISOLEUCINE

$$NH_2$$
$$CH_3-CH-CH_2-C-COOH$$
$$CH_3\ \ \ \ \ \ \ \ \ \ H$$

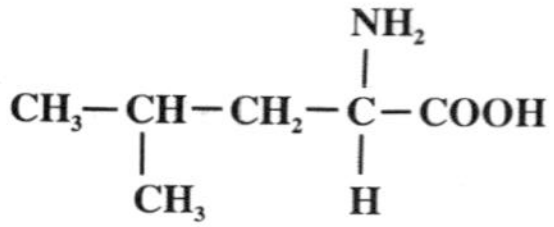

L-LEUCINE

Basic Amino Acids

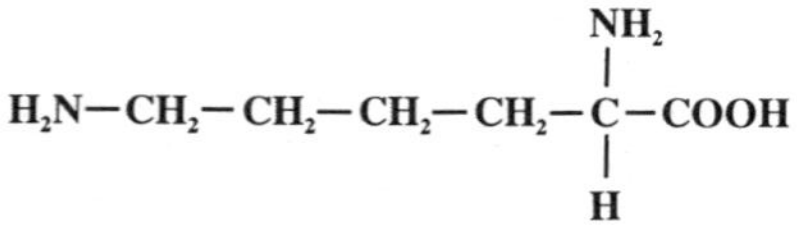

$$NH_2$$
$$H_2N-CH_2-CH_2-CH_2-CH_2-C-COOH$$
$$H$$

L-LYSINE

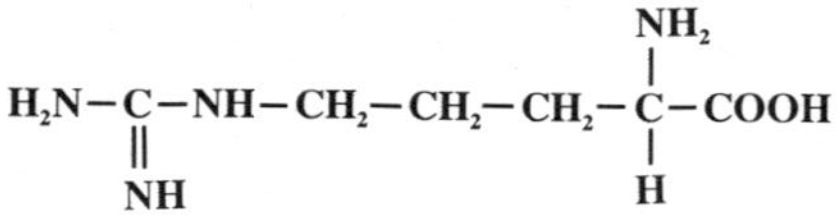

$$NH_2$$
$$H_2N-C-NH-CH_2-CH_2-CH_2-C-COOH$$
$$NH \qquad\qquad\qquad\qquad H$$

L-ARGININE

$$NH_2$$
$$HC=C-CH_2-C-COOH$$
$$N\ \ NH \qquad\quad H$$
$$\diagdown C \diagup$$
$$H$$

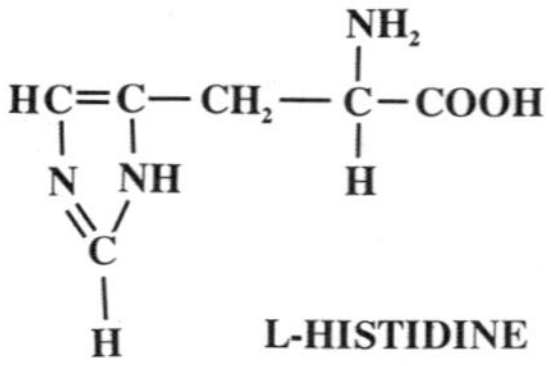

L-HISTIDINE

Acidic Amino Acids

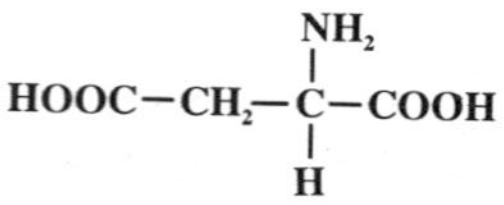

$$NH_2$$
$$HOOC-CH_2-C-COOH$$
$$H$$

L-ASPARTIC

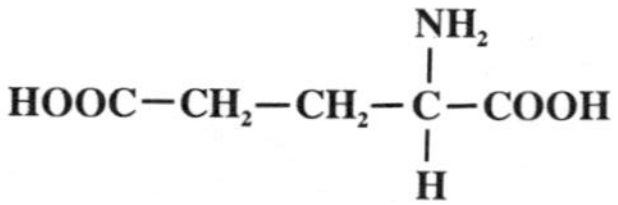

$$NH_2$$
$$HOOC-CH_2-CH_2-C-COOH$$
$$H$$

L-GLUTAMIC

Amides

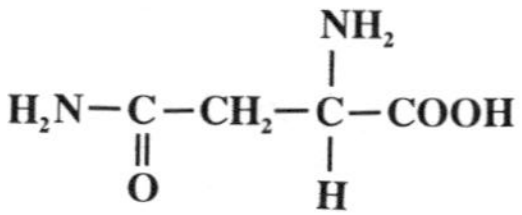

$$NH_2$$
$$H_2N-C-CH_2-C-COOH$$
$$O \qquad\quad H$$

L-ASPARAGINE

$$NH_2$$
$$H_2N-C-CH_2-CH_2-C-COOH$$
$$O \qquad\qquad\qquad H$$

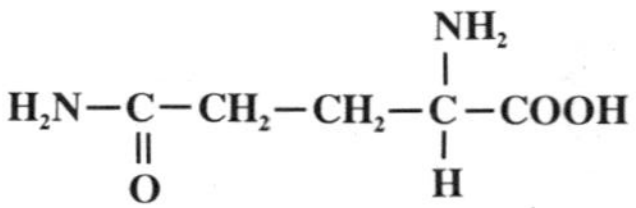

L-GLUTAMINE

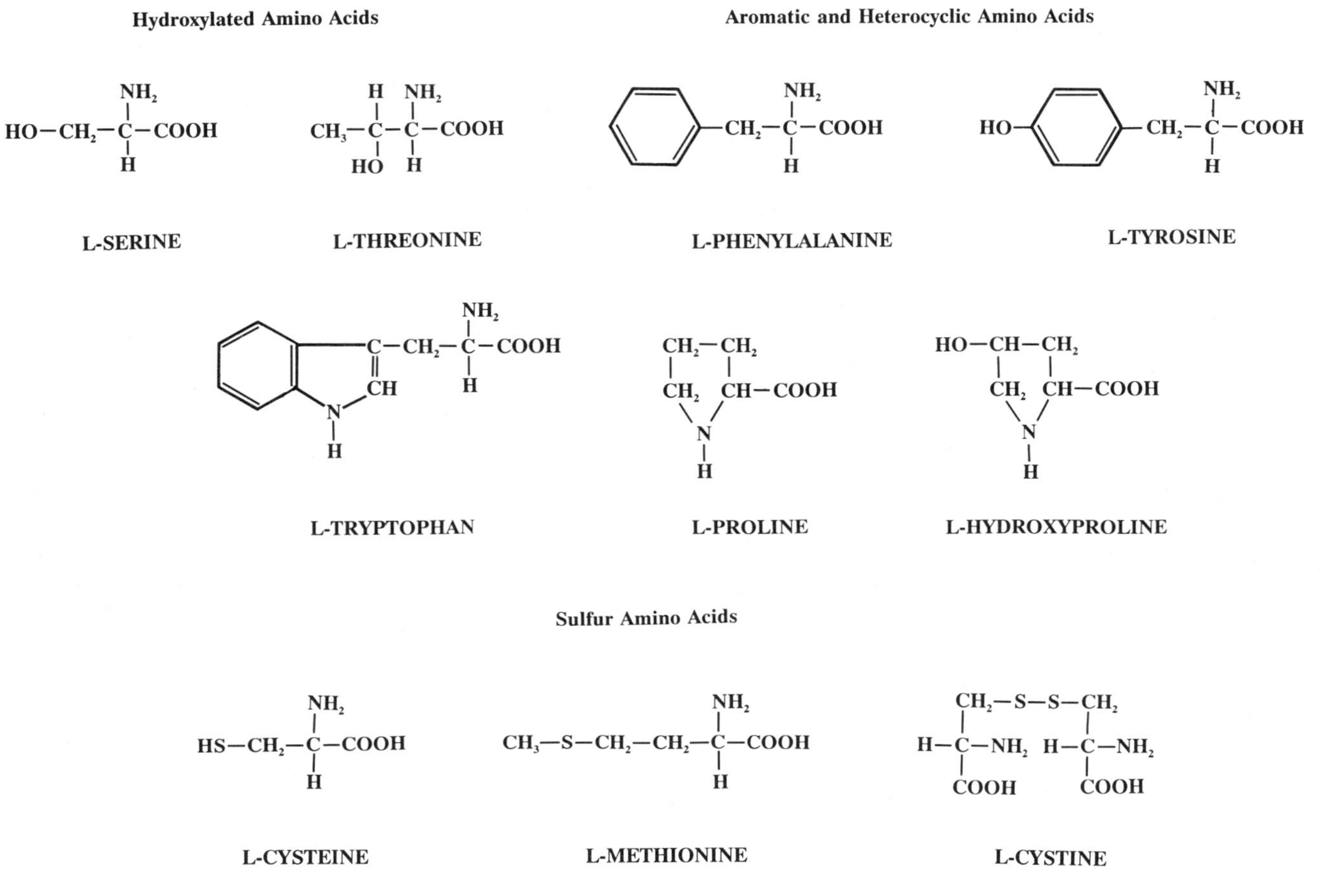

Hydroxylated Amino Acids

$$HO-CH_2-\overset{\overset{\displaystyle NH_2}{|}}{\underset{\underset{\displaystyle H}{|}}{C}}-COOH$$

L-SERINE

$$CH_3-\overset{\overset{\displaystyle H}{|}}{\underset{\underset{\displaystyle HO}{|}}{C}}-\overset{\overset{\displaystyle NH_2}{|}}{\underset{\underset{\displaystyle H}{|}}{C}}-COOH$$

L-THREONINE

Aromatic and Heterocyclic Amino Acids

$$\text{HO} - \underset{\text{(benzene ring)}}{\bigcirc} - CH_2-\overset{\overset{\displaystyle NH_2}{|}}{\underset{\underset{\displaystyle H}{|}}{C}}-COOH$$

L-TYROSINE

$$\underset{\text{(benzene ring)}}{\bigcirc} - CH_2-\overset{\overset{\displaystyle NH_2}{|}}{\underset{\underset{\displaystyle H}{|}}{C}}-COOH$$

L-PHENYLALANINE

L-PROLINE

$$\begin{array}{c} CH_2-CH_2 \\ | \qquad | \\ CH_2 \quad CH-COOH \\ \diagdown \ N \ \diagup \\ | \\ H \end{array}$$

L-HYDROXYPROLINE

$$\begin{array}{c} HO-CH-CH_2 \\ | \qquad | \\ CH_2 \quad CH-COOH \\ \diagdown \ N \ \diagup \\ | \\ H \end{array}$$

L-TRYPTOPHAN

$$\begin{array}{c} NH_2 \\ | \\ C-CH_2-C-COOH \\ \| \qquad | \\ CH \qquad H \\ \diagdown N \diagdown \\ | \\ H \end{array}$$

Sulfur Amino Acids

L-CYSTEINE

$$HS-CH_2-\overset{\overset{\displaystyle NH_2}{|}}{\underset{\underset{\displaystyle H}{|}}{C}}-COOH$$

L-METHIONINE

$$CH_3-S-CH_2-CH_2-\overset{\overset{\displaystyle NH_2}{|}}{\underset{\underset{\displaystyle H}{|}}{C}}-COOH$$

L-CYSTINE

$$\begin{array}{c} CH_2-S-S-CH_2 \\ | \qquad\qquad | \\ H-C-NH_2 \ \ H-C-NH_2 \\ | \qquad\qquad | \\ COOH \qquad\quad COOH \end{array}$$

Figure 4-9. The classes and structures of the amino acids found in plant proteins. Plants also contain a relatively diverse cross-section of nonprotein amino acids.

Met-Ala-Thr-Lys-Ile-Leu-Ala-Leu-Leu-Ala-Leu-Leu-Ala-Leu-Leu-Val-Ser-Ala-
Thr-Asn-Ala-Phe-Ile-Ile-Pro-Gln-Cys-Ser-Leu-Ala-Pro-Ser-Ala-Ser-Ile-Pro-Gln-
Phe-Leu-Pro-Pro-Val-Thr-Ser-Met-Gly-Phe-Glu-His-Pro-Ala-Val-Gln-Ala-Tyr-
Arg-Leu-Gln-Leu-Ala-Leu-Ala-Ala-Ser-Ala-Leu-Gln-Gln-Pro-Ile-Ala-Gln-Leu-
Gln-Gln-Gln-Ser-Leu-Ala-His-Leu-Thr-Leu-Gln-Thr-Ile-Ala-Thr-Gln-Gln-Gln-
Gln-Gln-Gln-Phe-Leu-Pro-Ser-Leu-Ser-His-Leu-Ala-Met-Val-Asn-Pro-Val-Thr-
Tyr-Leu-Gln-Gln-Gln-Leu-Leu-Ala-Ser-Asn-Pro-Leu-Ala-Leu-Ala-Asn-Val-Ala-
Ala-Tyr-Gln-Gln-Gln-Gln-Gln-Leu-Gln-Gln-Phe-Met-Pro-Val-Leu-Ser-Gln-Leu-
Ala-Met-Val-Asn-Pro-Ala-Val-Tyr-Leu-Gln-Leu-Leu-Ser-Ser-Ser-Pro-Leu-Ala-
Val-Gly-Asn-Ala-Pro-Thr-Tyr-Leu-Gln-Gln-Gln-Leu-Leu-Gln-Gln-Ile-Val-Pro-
Ala-Leu-Thr-Gln-Leu-Ala-Val-Ala-Asn-Pro-Ala-Ala-Tyr-Leu-Gln-Gln-Leu-Leu-
Pro-Phe-Asn-Gln-Leu-Ala-Val-Ser-Asn-Ser-Ala-Ala-Tyr-Leu-Gln-Gln-Arg-Gln-
Gln-Leu-Leu-Asn-Pro-Leu-Ala-Val-Ala-Asn-Pro-Leu-Val-Ala-Thr-Phe-Leu-Gln-
Gln-Gln-Gln-Gln-Leu-Leu-Pro-Tyr-Asn-Gln-Phe-Ser-Leu-Met-Asn-Pro-Ala-Leu-
Gln-Gln-Pro-Ile-Val-Gly-Gly-Ala-Ile-Phe

Figure 4-10. The amino sequence for one of four distinct groups forming the storage protein zein (approximate molecular weight = 22,000), as determined by sequencing its mRNA (*after Marks and Larkins*[56]).

number, and sequence of the amino acids present in the chain while the secondary structure is the conformation of the chain of amino acids due to hydrogen bonding between the oxygen of a carboxyl group and a nitrogen of a neighboring amino group. Tertiary structure, which causes the folding or bending of the chain, is due to interactions between side chains on certain amino acids and adjacent portions within the chain. These interactions may be held by hydrogen bonds, ionic bonds, hydrophobic bonds, or disulfide covalent bonds. Some proteins also form a quaternary structure where individual subunits aggregate, often into globular or fibrous forms. The quaternary structure is maintained by hydrophobic and electrostatic forces. An example of the tertiary and quaternary structure of the storage protein zein is illustrated in figure 4-13.

Protein Classification

Proteins can be classified based on (1) their physical and/or chemical properties, (2) the type of molecules that may be joined to the protein, or (3) the function of the protein within the cell. The first type of classification typically separates proteins into groups based on their size, structure, solubility, or degree of basicity (kind and number of basic amino acids, e.g., lysine, arginine, and histidine). The second means of classification separates proteins into classes such as lipoproteins (lipid prosthetic group), nucleoproteins (nucleic acid prosthetic group), chromoproteins (pigment prosthetic group), metaloproteins (metal prosthetic group), and glycoproteins (carbohydrate prosthetic group). Classifications based on function are more nebulous. While proteins can be grouped into three general classes based on function—structural proteins (membrane and cell wall proteins), storage proteins, and enzymes—these classes often overlap. For example, many enzymes are also components of membranes and may, therefore, have a multiple role within the cell.

Storage proteins, found in abundance in seeds, serve as a source of nitrogen and amino acids that are utilized during germination. Cereal grains contain an average 10% protein, which is largely storage protein, and legumes have 20–30%. Together these make up the major portion of the protein consumed by man, about 70%. Most other edible crops are lower in protein (table 4-6).

Enzymatic proteins are extremely important in that they regulate virtually all of the biochemical reactions within the cells of harvested plant products. Primarily through enzyme synthesis, activation, and degradation, control is exerted over the rate of specific processes, thus allowing the plant product to adjust its metabolism to changes in the environment in which it is held and to genetically controlled metabolic shifts (e.g., ripening).

Enzymes are grouped based on their type of catalytic function:

Oxidoreductases catalyze oxidation-reduction reactions; an example would be malate dehydrogenase.

Transferases catalyze the transfer of a specific group from one molecule to another, for example, methionine transferase in the ethylene synthesis pathway.

Hydrolases catalyze hydrolysis by the addition of water; examples would be the amylases.

Lyases catalyze the addition or removal of groups without the involvement of water; an example would be phosphoenolpyruvate carboxylase.

Isomerases catalyze isomerizations, for example, the conversion of glucose-6-phosphate to fructose-6-phosphate by phosphoglucoisomerase.

Ligases catalyze condensing reactions; an example would be pyruvate carboxylase.

Protein Degradation

Many of the proteins within cells are in a continuous state of synthesis and degradation. After synthesis, they begin to progress toward eventual degradation and recycling of their component parts. The length of their life expectancy varies widely between different proteins; it may be from as short as a few minutes to as long as years. In comparison to protein synthesis, we presently know very little about the degradation of proteins and the relationship of this process to the control of metabolism within cells. The greatest interest, thus far, has been directed toward studying the degradation of storage proteins in seeds during germination. The turnover of proteins in other plant parts such as leaves, fruits, roots, and tubers is poorly understood.

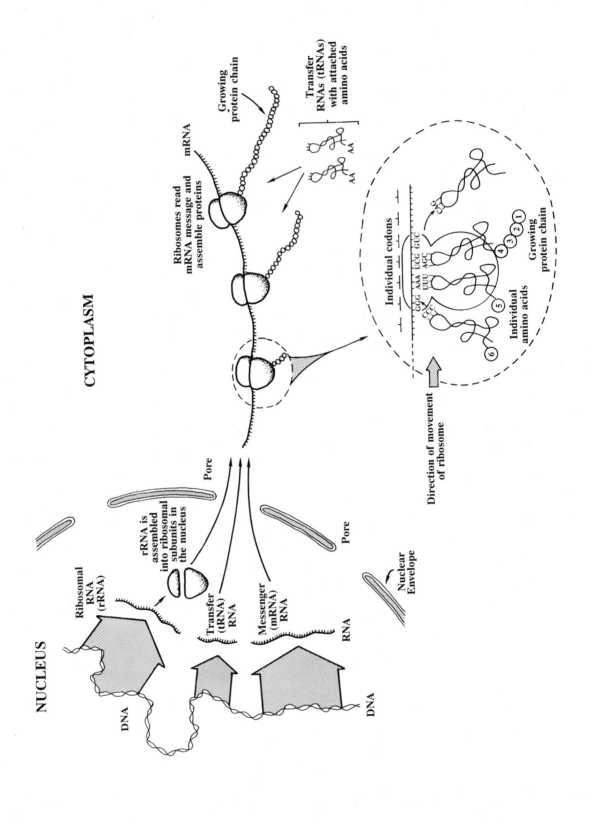

Figure 4-11 *(at left).* Synthesis of a protein. The amino acid sequence for an individual protein is transcribed with the synthesis of a specific messenger RNA (mRNA) for the protein. Ribosomal RNA (rRNA) and transfer RNA (tRNA), also derived from the DNA molecule, are likewise present. Once in the cytoplasm, several ribosomes attach to an mRNA molecule, and moving down the molecule, each translate the code. Insertion of the appropriate amino acid, carried to the ribosome by its tRNA, allows assembly of the protein in the appropriate sequence, one amino acid at a time. The inset illustrates how the three nucleotide codons on the mRNA assure the correct amino acid is added. Many proteins undergo some posttranslational modifications once the initial amino acid sequence is assembled.

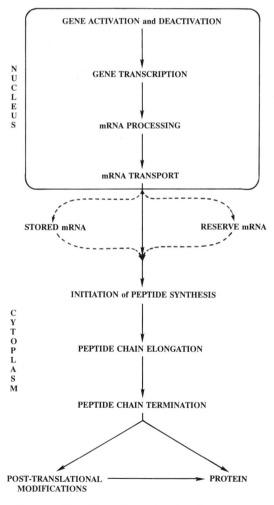

Figure 4-12. Flow chart showing the major sequence of steps in protein synthesis and their location in the cell.

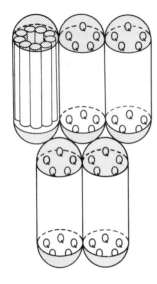

Figure 4-13. The proposed tertiary (individual cylinders, top-left) and quaternary structure (composite of cylinders into a capsulelike structure) of zein proteins within a plane and the stacking of molecular planes. Individual cylinders (top-left) represent the tertiary structure of a single protein molecule. Glutamine residues (Q) at each end would allow for hydrogen bond interactions among molecules in neighboring planes (adapted from Argos et al.[2]).

Enzymes that cleave the peptide bonds in protein chains (peptidases or proteases) are classified as either endopeptidases or exopeptidases. Endopeptidases cleave peptide bonds within the chain whereas exopeptidases cleave single amino acids from either the carbon or nitrogen end of the molecule. These two classes are further broken down into subgroups based on the enzymes' mechanism of action. There are currently four types of endopeptidases and two types of exopeptidases. The site of degradation may

Table 4-6. Amino Acid Composition of Several Types of Postharvest Products

Food Group	Plant Part Used	Common Name	Protein Content $(g \cdot 100g^{-1})$	Isoleucine	Leucine	Lysine	Methionine	Cystine	Phenylalanine	Tyrosine	Threonine
				Amino Acids (mg/g of nitrogen)							
Cereal	Seed	Rice	7.1	306	563	219	225	100	350	300	288
	Seed	Wheat	12.2	204	417	179	94	159	282	187	183
Pulses	Seed	Chickpea	20.1	277	468	428	65	74	358	183	235
	Seed	Bean	22.1	262	476	450	66	53	326	158	248
Roots and	Tuber	White potato	2.0	236	377	299	81	37	251	171	235
Tubers	Root	Sweetpotato	1.3	230	340	214	106	69	241	146	236
Vegetables	Leaves	Lettuce	1.3	238	394	238	112	—	319	169	256
	Stems	Celery	1.1	244	425	150	138	—	281	—	213
	Flowers	Cauliflower	2.8	302	436	356	99	—	225	—	264
	Immature seed	Pea	6.6	260	435	456	58	60	275	194	235
	Immature seed	Green bean	2.4	234	432	344	81	53	266	209	241
Fruits	Fruit	Apple	0.4	220	390	370	49	84	160	94	230
	Fruit	Banana	1.2	181	294	256	125	169	244	163	213
Nuts	Nut	Coconut	6.6	244	419	220	120	76	283	167	212
	Nut	Brazil nut	14.8	175	431	175	363	131	244	169	163
Milk	Cow's milk		3.5	295	596	487	157	51	336	297	278
Eggs	Hen egg		12.4	393	551	436	210	152	358	260	320

Note: Data from FAO.[18]

be in the cytoplasm, the vacuole, or exterior to the plasma membrane, and individual steps at a specific locale may be sequestered in separate compartments within the cell.

Changes in Amino Acids and Proteins After Harvest

The relative change in the protein and amino acid content and composition in harvested plant parts is greatest in those products that undergo significant changes in homeostasis during the postharvest period. Hence, products such as seeds, which are relatively stable when properly stored, do not undergo substantial changes in their protein composition. On the other hand, two general phenomena that precipitate large changes in protein and amino acid content and composition are the onset of senescence and fruit ripening.

Leaf tissue has been used widely as a model system for studying senescence, since the changes in the tissue are substantial and occur quite rapidly. Senescence in leaves is characterized by a decline in photosynthesis and the loss of protein and chlorophyll (fig. 4-14). Proteolysis, the breakdown of protein, begins fairly rapidly after harvest, especially if the individual leaf is detached from the parent plant.[57,58] Peptidases (proteases) that cleave proteins are always present within the leaves; however, their concentrations increase substantially during the onset of senescence (fig. 4-15). While most enzymes are declining, certain specific enzymes appear to increase in activity and/or concentration during senescence. For example, glutamate dehydrogenase activity increases by as much as 400% in spinach leaves held in the dark[46] and appears to stimulate deamination of amino acids.

					Amino Acids (mg/g of nitrogen)						
Tryptophane	Valine	Arginine	Histidine	Alanine	Aspartic Acid	Glutamic Acid	Glycine	Proline	Serine	Total Essential Amino Acids	Total Amino Acids
73	463	644	188	—	275	731	381	313	363	2887	—
—	276	288	143	226	308	1866	245	621	287	2049	6033
—	284	588	165	271	726	991	251	263	318	2426	5998
—	287	355	177	262	748	924	237	223	347	2389	5662
—	292	311	94	278	775	639	237	235	259	2082	4910
—	283	307	84	298	825	541	234	219	255	1972	4735
—	338	281	100	269	719	638	256	325	206	—	—
81	300	289	120	—	—	—	—	—	—	—	—
86	347	250	94	—	—	—	—	—	—	—	—
—	296	548	133	281	620	910	246	240	281	2332	5591
—	306	266	147	275	750	669	238	238	334	2253	5170
58	250	170	120	280	1300	700	240	200	270	1905	5205
—	250	469	469	275	656	575	263	256	244	1968	5175
—	339	822	128	279	553	1171	281	233	303	2148	5918
119	269	831	144	219	463	1163	275	300	269	2239	5903
—	362	205	167	217	481	1390	123	571	362	2947	6463
—	428	381	152	370	601	796	207	260	478	3201	6446

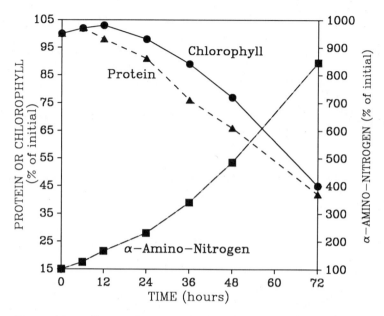

Figure 4-14. Changes in concentration of chlorophyll, protein, and α-amino nitrogen in detached *Avena* leaves held for varying intervals in the dark *(after Martin and Thimann[57])*.

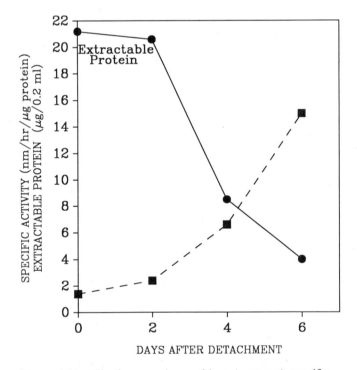

Figure 4-15. The increase in peptidase (protease) specific activity and concurrent decrease in protein concentration with time in detached *Avena* leaves held in the dark *(after Martin and Thimann[57])*.

Therefore, while proteins are being broken down and the component amino acids are recycled, a small but extremely important number of specific proteins are also being synthesized. Their importance in the development of proteolysis and senescence can be inferred from the fact that inhibitors of either RNA or protein synthesis strongly decrease the rate at which senescence proceeds. The amino acids formed are largely transported, although often after conversion to glutamine, to other parts of the plant and this transport is greatest to areas that represent strong sinks (high demand) such as reproductive organs.[11] In leaves detached at harvest, transport out of the organ is not possible; thus, the decomposition products tend to accumulate.

During the onset of ripening of several climacteric fruits, it has been shown that the actual concentration of protein increases (fig. 4-16). In apples, avocados (*Persea americana,* Mill.), and several other climacteric fruits, enhanced synthesis of both RNA and protein occurs.[43,77,78] The net effect is an enhanced activity of certain enzymes during ripening (table 4-7). As with protein synthesis during the onset of leaf senescence, these new proteins appear to be essential since ripening is inhibited if protein synthesis is inhibited. The increase in synthesis of specific enzymes has been monitored using labeled amino acids and then identifying the enzymes that are radioactive. Malic enzyme, which catalyzes the decarboxylation of malic acid, the primary organic acid in apples and certain pear cultivars, is an example of one enzyme that increases markedly during the climacteric. The increase in malic

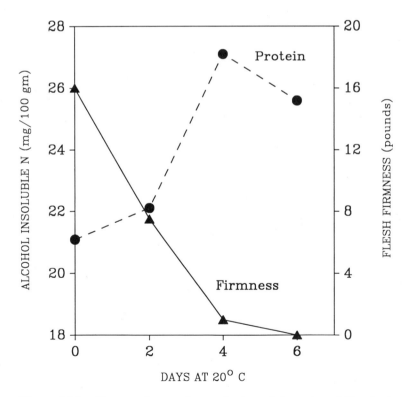

Figure 4-16. Changes in protein synthesis and firmness of 'Bartlett' pears (*Pyrus communis,* L.) during ripening *(after Frenkel et al.[24]).*

Table 4-7. Changes in the Activity of Various Enzymes and Isoenzymes During the Development and Ripening of a Tomato Fruit

Enzyme	Stage with Peak Activity	Stage with Maximum Number of Bands	Activity Trend During Development	Number of Bands at		
				SG	MG	OR
Tyrosinase	LG	MG	Decrease	1	4	2
Peroxidase	OR	MG-OR	Increase	1	4	4
Esterase	MG	MG	Peak at MG	11	13	8
Acid phosphatase	SG	SG	Decrease	8	6	3
Glycerophosphatase	SG	red	Decrease	1	3	2
ATPase	SG	SG	Decrease	6	5	5
NADH$_2$-diaphorase	MG	MG	Peak at MG	10	12	11
Fumarase	MG	MG	Peak at MG	1	4	2
Malate dehydrogenase	LG red	all	Maximum LG-OR	4	4	4
NADP$^+$ malic enzyme	MG	MG	Peak at MG	3	4	2
Iso-citrate dehydrogenase	SG	MG-OR	Decrease	1	2	2
Glutamate dehydrogenase	MG	SG-MG	Peak at MG	6	6	1
Phosphofructokinase	MG?	MG	Peak at MG?	2	4	3
6-Phosphogluconate dehydrogenase	MG	MG	Peak at MG	2	6	2
Phosphoglucomutase	MG	MG	Peak at MG	3	6	1
Phosphohexose isomerase	SG-MG	MG	SG-MG then a decrease	6	10	6
Glucose-6-phosphate dehydrogenase	SG-MG	MG	Peak at MG	3	4	3
Glutamate-oxaloacetate transaminase	SG-MG	all	Decrease	4	4	4
Leucine aminopeptidase	red?	all	Peak at red?	3	3	3

Source: After Hobson.[41]
Notes: SG, small green; LG, large green (preclimacteric); MG, mature green (close to the beginning of the climacteric rise); red, nearly fully ripe; and OR, overripe (postclimacteric).

enzyme activity increases the concentration of pyruvic acid, the product of the reaction, which can then enter the respiratory tricarboxylic acid cycle.

LIPIDS

Plant lipids represent a very broad group of compounds with diverse roles in the physiology and metabolism of harvested products. In addition, the absolute concentration of these compounds varies widely between different species and plant parts. Most postharvest products, however, are relatively low in total lipids. In contrast, avocados, olives (*Olea europaea,* L.), and many seeds are very high in lipids (table 4-8). A major portion of the lipids present is in the form of storage compounds, which in the case of seeds can be used as an energy source during germination. Plant lipids, in addition to representing a storage form of carbon, also function as components of cellular membranes, as cuticular waxes forming a protective surface on many products, and in

Table 4-8. Lipid Content of Several Types of Harvested Products

		Lipid Content of Edible Portion	
	Botanical Name	*% Dry Weight*	*% Fresh Weight*
Fruits			
avocado	*Persea americana*, Mill.	63.0	16.4
banana	*Musa* spp.	0.8	0.2
olive	*Olea europea*, L. cv. 'Ascolano'	69.0	13.8
Seeds			
peanut	*Arachis hypogaea*, L.	50.3	47.5
rice	*Oryza sativa*, L.	0.5	0.4
walnut	*Juglans regia*, L.	61.2	59.3
Leaves			
amaranth	*Amaranthus* spp.	3.8	0.5
cabbage	*Brassica oleracea*, L. Capitata group	2.6	0.2
lettuce	*Lactuca sativa*, L.	2.2	0.1
Roots and tubers			
parsnip	*Pastinaca sativa*, L.	2.4	0.5
potato	*Solanum tuberosum*, L.	0.4	0.1
radish	*Raphanus sativus*, L.	1.8	0.1

Note: Data from Watt and Merrill.[96]

some cases as vitamins, pigments, sterols, and secondary products such as rubber.

Biochemically, lipids are normally grouped into neutral lipids, waxes, phospholipids, glycolipids, and terpenoids. Neutral lipids are comprised of fats and oils and represent primarily carbon storage compounds. Waxes are typically long-chain fatty acids or esters of fatty acids and long-chain alcohols, although numerous other compounds may be found.[51] These compounds form the thin, waxy layer on the surface of leaves, fruits, and other plant parts. Phospholipids and glycolipids are components of cellular membranes, and terpenoids are primarily water-insoluble acyclic and cyclic compounds such as steroids, essential oils, and rubber.

Fatty Acids

A substantial portion of the physical and chemical properties of lipids is due to the long chains of component fatty acids present. These fatty acids may be saturated (no double bonds present) or unsaturated in varying degrees. The most common fatty acids in plants range from 4 to 26 carbons in size (table 4-9), with oleic and linoleic being the most prevalent in nature. Their structure has a zigzag configuration (fig. 4-17) and double bonds, as seen in the example of oleic and linoleic acids, tend to result in curvature of the molecule.

Several methods are utilized to designate fatty acids. They may be referred to by their common name, systematic name, or quite commonly by the use of an abbreviation. The short or abbreviated form simply denotes the number of carbon atoms in the molecule and the number of double bonds present. For

Table 4-9. Common Fatty Acids in Harvested Plant Products

Abbreviation	Systematic Name	Common Name	Formula
Saturated Fatty Acids			
4:0	Butanoic	Butyric acid	$CH_3(CH_2)_2COOH$
6:0	Hexanoic	Caproic acid	$CH_3(CH_2)_4COOH$
8:0	Octanoic	Caprylic acid	$CH_3(CH_2)_6COOH$
10:0	Decanoic	Capric acid	$CH_3(CH_2)_8COOH$
12:0	Dodecanoic	Lauric acid	$CH_3(CH_2)_{10}COOH$
14:0	Tetradecanoic	Myristic acid	$CH_3(CH_2)_{12}COOH$
16:0	Hexadecanoic	Palmitic acid	$CH_3(CH_2)_{14}COOH$
18:0	Octadecanoic	Stearic acid	$CH_3(CH_2)_{16}COOH$
20:0	Eicosanoic	Arachidic acid	$CH_3(CH_2)_{18}COOH$
22:0	Docosanoic	Behenic acid	$CH_3(CH_2)_{20}COOH$
24:0	Tetracosanoic	Lignoceric acid	$CH_3(CH_2)_{22}COOH$
Unsaturated Fatty Acids			
16:1	9-Hexadecenoic[a]	Palmitoleic acid	$CH_3(CH_2)_5CH = CH(CH_2)_7COOH$
18:1	9-Octadecenoic	Oleic acid	$CH_3(CH_2)_7CH = CH(CH_2)_7COOH$
18:2	9,12-Octadecadienoic	Linoleic acid	$CH_3(CH_2)_3(CH_2CH = CH)_2(CH_2)_7COOH$
18:3	9,12,15-Octadecatrienoic	Linolenic acid	$CH_3(CH_2CH = CH)_3(CH_2)_7COOH$

[a] The double bond is between carbons 9 and 10.

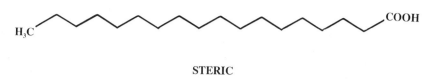

STERIC

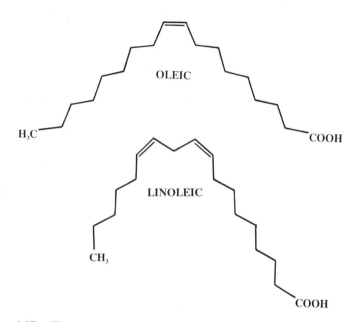

OLEIC

LINOLEIC

Figure 4-17. The structure of three fatty acids. Note the change in geometry with the addition of each double bond.

example, linoleic acid (18:2) has 18 carbons with 2 double bonds. The position of the double bond is designated at the beginning of the systematic name or, in some cases, at the end of the common name of the unsaturated fatty acids (table 4-9).

Most fatty acids have an even number of carbons, although trace amounts of straight-chain, odd-numbered carbon compounds from C_7 to C_{35} have been detected. For example, in the pecan kernel (*Carya illinoensis,* (Wang) C. Koch.) (table 4-10), $C_{15:0}$, $C_{15:1}$, $C_{15:2}$, $C_{17:0}$, $C_{17:1}$, $C_{17:2}$, and $C_{21:0}$ fatty acids have been found but their concentration represents only approximately 2.1% of the total fatty acids present.[83,84]

Triacylglycerols

Triacylglycerols (previously called triglycerides) are comprised of three fatty acids linked through ester bonds to a glycerol molecule. Monoacylglycerols and diacylglycerols may also be present with the fatty acid-free position(s) on

Table 4-10. Fatty Acid Composition of Pecans.

Abbreviation	Systematic Name	Common Name	Concentration (g/100 g nut meat)
Saturated Fatty Acids			
10:0	Decanoic	Capric	<0.20
12:0	Dodecanoic	Lauric	<0.20
14:0	Tetradecanoic	Myristic	1.20
15:0	Pentadecanoic		<0.20
16:0	Hexadecanoic	Palmitic	4.09
17:0	Heptadecanoic	Margaric	0.27
18:0	Octadecanoic	Stearic	1.51
20:0	Eicosanoic	Arachidic	0.44
21:0	Heneicosanoic		<0.20
Unsaturated Fatty Acids			
12:1	Dodecenoic		<0.20
14:1	Tetradecenoic		<0.20
14:2	Tetradecadienoic		<0.20
15:1	Pentadecenoic		<0.20
15:2	Pentadecadienoic		<0.20
16:1	Hexadecenoic	Palmitoleic	0.42
16:2	Hexadecadienoic		<0.20
17:1	Heptadecenoic		0.26
17:2	Heptadecadienoic		<0.20
18:1	Octadecenoic	Oleic	37.90
18:2	Octadecadienoic	Linoleic	22.53
18:3	Octadecatrienoic	α-Linolenic	1.53
20:1	Eicosenoic		0.54
20:2	Eicosadienoic		<0.20

Notes: Data from Senter and Horvat.[83,84]
Data for the major fatty acids represent means of six cultivars.

the glycerol molecule being bonded to other compounds, thus yielding a wide range of other types of lipids.

CH$_2$OH	CH$_2$OCOR*	CH$_2$OCOR$_1$	CH$_2$OCOR$_1$	CH$_2$OCOR$_1$
CHOH	CHOH	R$_2$COOCH	HOCH	R$_2$COOCH
CH$_2$OH	CH$_2$OH	CH$_2$OH	CH$_2$OCOR$_2$	CH$_2$OCOR$_3$
glycerol	1-monoacyl-glycerol	1,2-diacyl-glycerol	1,3 diacyl-glycerol	triacyl-glycerol

*R = alkyl group

Triacylglycerols that are present in nature are normally mixtures and these mixtures are often very complex due to the different fatty acids that can be esterified to each of the three hydroxyl positions. For example, the combination of only two fatty acids at all possible positions on the molecule yields six potential triacylglycerol compounds.

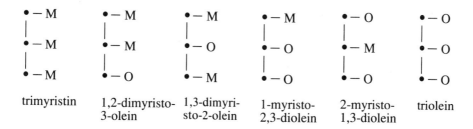

| trimyristin | 1,2-dimyristo-3-olein | 1,3-dimyristo-2-olein | 1-myristo-2,3-diolein | 2-myristo-1,3-diolein | triolein |

Neutral Lipids

The neutral lipids are largely triacylglycerols that make up the fats and oils found in plants. Waxes are also grouped here; however, due to their distinctly different physiological role in harvested products, they will be treated separately. As a group, neutral lipids do not have charged functional groups. The distinction between fats and oils is based simply on their physical form at room temperature. Oils are liquid and tend to contain a larger percentage of unsaturated fatty acids (e.g., oleic, linoleic, and linolenic) than fats, which are solid at room temperature. Fats have saturated fatty acids as their major component.

The role of these storage lipids has been studied most extensively in oil-containing seed crops. Here they represent a source of energy and carbon skeletons during the germination period. In the avocado, which has high levels of lipids in the mesocarp tissue, existing evidence suggests that it does not represent a source of respiratory substrate during the ripening process.

Waxes, Cutin, and Suberin

The outer surface of plants is protected by three general types of lipid compounds. Waxes, typically esters of a larger molecular weight fatty acid and a higher aliphatic alcohol, and cutin, hydroxy fatty acid polymers, act as a protective coating on much of the aboveground parts of the plant (fig. 2-4). Likewise, suberin, a lipid-derived polymeric material, is found on underground plant parts and on healed surfaces of wounds. Like cutin, suberin is often embedded with waxes.

Plant waxes are extremely important during the postharvest storage and marketing of plant products in that they function by limiting the water loss from the tissue and impeding invasion of pathogens. As a group, waxes are chemically very heterogenous. In addition to being esters of a higher fatty acid and a higher aliphatic alcohol, waxes contain alkanes, primary alcohols, long-chain free fatty acids, and other groups. Alkanes may represent more than 90% of the hydrocarbons in waxes of apple fruits and *Brassica oleracea*, L.[52,62] Some plant products (e.g., rutabagas (*Brassica napus*, L. Napobrassica group), oranges) may benefit from the application of supplemental wax after harvest since there may be insufficient natural waxes on the product or surface waxes during washing operations may have been inadvertently removed.

The precise chemistry and structure of both cutin and suberin are not yet fully elucidated, although substantially more is known about the former. Both represent very complex heterogenous compounds. Cutin, a polyester, is composed primarily of C_{16} and/or C_{18} monomers. General differences in the composition of cutin and suberin are given in table 4-11 with tentative models in figure 4-18.

Phospholipids and Glycolipids

Phospholipids and glycolipids are important components of cellular membranes. Phospholipids are often diacylglycerols that yield inorganic phosphate upon hydrolysis. They are characteristically high in linoleic acid in the fatty acid positions of the molecule. The fatty acid portion forms the hydrophobic tail, which strongly influences the orientation of the molecule. Common examples of phospholipids in plants are phosphatidyl serine, phosphatidyl ethanolamine, phosphatidyl choline, phosphatidyl glycerol, and phosphatidyl inositol. They represent important components of both the cytoplasmic and mitochondrial membranes.

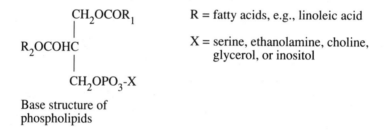

$$CH_2OCOR_1$$
$$R_2OCOHC$$
$$CH_2OPO_3\text{-}X$$

R = fatty acids, e.g., linoleic acid

X = serine, ethanolamine, choline, glycerol, or inositol

Base structure of phospholipids

Glycolipids, in contrast, have carbohydrate substitutions without phosphate and are important components of the chloroplast membranes. Common examples would be monogalactosyl diglyceride and digalactosyl diglyceride. Approximately 70% of the fatty acid component in these photosynthetic membranes is linolenic acid.

Table 4-11. Differences in the Monomer Composition of Cutin and Suberin

	Cutin	*Suberin*
Dicarboxylic acids	Minor	Major
In-chain substituted acids	Major	Minor; in some cases, substantial
Phenolics	Low	High
Very long-chain (C_{20}–C_{26}) acids	Rare and minor	Common and substantial
Very long-chain alcohols	Rare and minor	Common and substantial

Source: After Kolattukudy.[51]

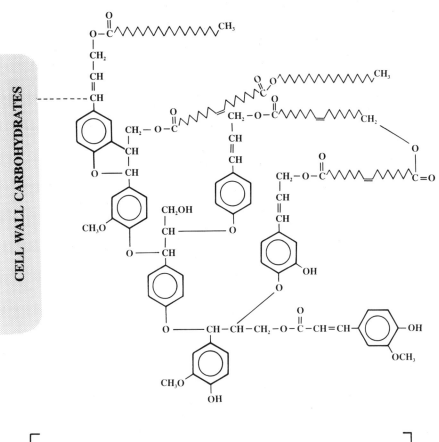

Figure 4-18. Models proposed for the structure of suberin (*top*) and cutin (*bottom*) (*after Kolattukudy*[51]).

Fatty Acid and Lipid Synthesis

Fatty acids are synthesized within the cytosol of the cell and in many cases within certain plastids (e.g., chloroplasts, chromoplasts). There are two distinct pathways operative, one forming saturated fatty acids and one forming unsaturated fatty acids. With saturated fatty acids, acetyl subunits are condensed via a series of enzymatically controlled steps, forming fatty acids of up to 16 carbons in length. Longer fatty acids require a separate chain-lengthening sequence, usually building from a palmitic acid base unit.

With the synthesis of unsaturated fatty acids, the introduction of the first double bond forming oleic acid is fairly well established. It can occur via one of two options: an anaerobic system or an aerobic one. Introduction of the second and third double bonds forming linoleic and α-linolenic acids, however, is not well understood at present.

Triacylglycerols are synthesized in plants either directly from carbohydrates or through the modification of existing glycerides. The glycerol backbone is derived from glycerol-3-phosphate from either the glycolytic or pentose phosphate pathway, or by direct phosphorylation of glycerol. The esterification of positions 1 and 2 (addition of a fatty acid) is by specific acyltransferases. In some cases, the monoacyl (position 1) is formed from dihydroxyacetone phosphate in a separate two-step sequence; however, the number 2 position is esterified by the 2-specific acyltransferase. The phosphate moiety is then cleaved from the molecule by phosphatidate phosphohydrolase to yield diacylglycerol, which then serves as a precursor for triacylglycerols, phosphoglycerides, and glycosylglycerides. The final step in the synthesis of a triacylglycerol is the esterification of the 3-hydroxy position by 3-specific transferase.

Lipid Degradation

Each class of lipids found in plants undergoes varying degrees of degradation during the postharvest period as the product approaches senescence, or in the case of seeds, as they begin to germinate. Of the constituent lipids, the storage lipids are known to undergo marked changes in many products. Storage lipids are composed primarily of triacylglycerols that represent the most common lipids found in the plant kingdom, although they are not the predominant form in all plants. In that their degradation has been fairly well elucidated, the discussion will focus on this class of lipids.

The first step in the recycling of carbon stored as triacylglycerols is the removal (hydrolysis) of the three acyl (fatty acid) units from the glycerol molecule. This step is accomplished through the action of the enzyme lipase (acyl hydrolase) or more specifically triacylglycerol acyl hydrolase (see review by Galliard[26]). The sequence involves the removal of the fatty acid from the number 3 position, yielding a 1,2-diacylglycerol, followed by the removal of a second fatty acid, yielding either 1- or 2-monoacylglycerol. The final acyl group is then hydrolyzed, leaving glycerol. The free acids can then be converted to acetate, a starting point for many synthetic reactions and a respiratory substrate. In addition, acetate can also be converted to sucrose, the primary transport form of carbon, an essential step during the germination of many seeds.

Free fatty acids can be metabolized by several possible mechanisms in the plant. The most prevalent mechanism for the degradation of fatty acids is β-oxidation (fig. 4-19). It results in the formation of acetyl-CoA, in which a major portion of the stored energy remains trapped in the thioester bond. This trapped energy can either be converted to ATP by the movement of acetyl-CoA through the tricarboxylic acid cycle or acetyl-CoA can move through the glyoxylate cycle that provides carbon skeletons for synthetic reactions. This second option (i.e., via the glyoxylate cycle) does not appear to be operative in most harvested products; however, it is extremely important in germinating seeds, which are high in lipids.

β-Oxidation occurs in the cytosol and in the glyoxysomes in many oil-containing seeds. In this scheme the two terminal carbons of the fatty acid are cleaved sequentially, moving down the chain. With each acetyl-CoA produced, five ATP equivalents (one FADH₂ and one NADH) are produced. Note that only one ATP is required to activate the fatty acid for complete degradation regardless of the number of carbon atoms in the chain. In the

β-OXIDATION

$$R-CH_2-CH_2-CO_2H + CoA + ATP \xrightarrow[\text{thiolase}]{} R-CH_2-CH_2-\overset{O}{\overset{\|}{C}}-SCoA + ADP + Pi$$

$$R-CH_2-CH_2-\overset{O}{\overset{\|}{C}}-SCoA + FAD \xrightarrow[\text{acyl-CoA dehydrogenase}]{} R-CH=CH-\overset{O}{\overset{\|}{C}}-SCoA + FADH_2$$

$$R-CH=CH-\overset{O}{\overset{\|}{C}}-SCoA + H_2O \xrightarrow[\text{enoyl-CoA hydrase}]{} R-CHOH-CH_2-\overset{O}{\overset{\|}{C}}-SCoA$$

$$R-CHOH-CH_2-\overset{O}{\overset{\|}{C}}-SCoA + NAD \xrightarrow[\substack{\text{β-hydroxy acyl-CoA} \\ \text{dehydrogenase}}]{} R-\overset{O}{\overset{\|}{C}}-CH_2-\overset{O}{\overset{\|}{C}}-SCoA + NADH$$

$$R-\overset{O}{\overset{\|}{C}}-CH_2-\overset{O}{\overset{\|}{C}}-SCoA + CoASH \xrightarrow[\text{β-ketoacyl-CoA thiolase}]{} R-\overset{O}{\overset{\|}{C}}-SCoA + CH_3-\overset{O}{\overset{\|}{C}}-SCoA$$

α-OXIDATION

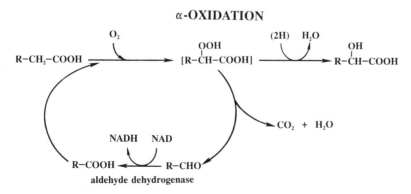

Figure 4-19. A comparison of ß-oxidation and α-oxidation of free fatty acids *(after Shine and Stumpf[86]).*

initial step (fig. 4-19) of the reaction, catalyzed by the enzyme thiolase, the free fatty acid combines with coenzyme A, a step requiring ATP to form acyl-CoA. An oxidative reaction, catalyzed by acyl-CoA dehydrogenase, then produces a double bond between the number 2 and 3 carbons and results in the formation of $FADH_2$. This process is followed by the addition of water to the double bond (carbon 3) by enoyl-CoA hydrase and subsequent oxidation of the hydroxyl of the number 3 carbon, with the production of NADH. In the final step, acetyl-CoA (the two terminal carbons) is cleaved from the fatty acid molecule. This series of steps is then repeated, removing additional acetyl-CoA's.

Free fatty acids may also be degraded, yielding CO_2, H_2O, and energy by the α-oxidation pathway; however, its role appears to be only minor, at least from the standpoint of energy production from stored lipids. Unlike β-oxidation where the reactions involve an acyl thioester, α-oxidation acts directly on the free fatty acids. The proposed scheme for α-oxidation is presented in figure 4-19.

In most cases, direct oxidation of fatty acids to CO_2 does not appear to be the primary physiological role of α-oxidation in plants. Under extreme conditions, for example, wounding of tissue slices, the initial rise in respiration does appear to be largely due to α-oxidation. With normally metabolizing cells, however, the α-oxidation mechanism is thought to function by creating odd-numbered fatty acids by the removal of a terminal carbon. It may also be used as an adjunct to β-oxidation for the removal of a carbon when the number 3 carbon of the fatty acid has a side group preventing the β-oxidation process.

Lipid Peroxidation

The oxidation of lipids in harvested plant products may occur either in biologically mediated reactions catalyzed by lipoxygenases or through direct chemical or photochemical reactions. In enzymatically controlled oxygenation reactions, polyunsaturated fatty acids are attacked, producing hydroperoxides that can be further degraded, often forming characteristic tastes and odors, both desirable and undesirable (e.g., rancidity). For example, linolenic acid (18:3) can be oxidized, forming a number of hydroperoxides that decompose into aldehydes (table 4-12). In cucumber (*Cucumis sativus,* L.) fruit, linoleic acid is attacked by lipoxygenase, forming both 9-hydroperoxide and 13-hydroperoxide, which are cleaved to form the volatile flavor components *cis*-3-nonenal and hexenal, respectively.[27] While these enzymes occur widely in higher plants, the level of activity does not appear to follow any set botanical or morphological pattern. Very high activities have been recorded for bean (*Phaseolus vulgaris,* L.) seed, potato (*Solanum tuberosum,* L.) tubers, eggplant (*Solanum melongena,* L.) fruit, and immature artichoke (*Cynara scolymus,* L.) flowers.[72]

Peroxidation reactions also include autocatalytic oxidation, the direct reaction with oxygen, and a nonautocatalytic process mediated by light. As with enzymatically controlled reactions, hydroperoxides are formed. The sequence of autooxidation is presented in figure 4-20. In general, the rate of lipid oxidation is largely dependent on the degree of unsaturation of the

Table 4-12. Hydroperoxides and Aldehydes (with Single Oxygen Function) Possibly Formed in Autoxidation of Some Unsaturated Fatty Acids.

Fatty acid	Methylene Group Involved	Isomeric Hydroperoxides Formed From Structures Contributing to Intermediate Free Radical Resonance Hydrid	Aldehydes Formed by Decomposition of Hydroperoxides
Oleic	11	11-Hydroperoxy-9-ene	Octanal
		9-Hydroperoxy-10-ene	2-Decenal
	8	8-Hydroperoxy-9-ene	2-Undecenal
		10-Hydroperoxy-8-ene	Nonanal
Linoleic	11	13-Hydroperoxy-9,11-diene	Hexanal
		11-Hydroperoxy-9,12-diene	2-Octenal
		9-Hydroperoxy-10,12-diene	2,4-Decadienal
Linolenic	14	16-Hydroperoxy-9,12,14-triene	Propanal
		14-Hydroperoxy-9,12,15-triene	2-Pentenal
		12-Hydroperoxy-9,13,15-triene	2,4-Heptadienal
	11	13-Hydroperoxy-9,11,15-triene	3-Hexenal
		11-Hydroperoxy-9,12,15-triene	2,5-Octadienal
		9-Hydroperoxy-10,12,15-triene	2,4,7-Decatrienal
Arachidonic	13	15-Hydroperoxy-5,8,11,13-tetraene	Hexanal
		13-Hydroperoxy-5,8,11,14-tetraene	2-Octenal
		11-Hydroperoxy-5,8,12,14-tetraene	2,4-Decadienal
	10	12-Hydroperoxy-5,8,10,14-tetraene	3-Nonenal
		10-Hydroperoxy-5,8,11,14-tetraene	2,5-Undecadienal
		8-Hydroperoxy-5,9,11,14-tetraene	2,4,7-Tridecatrienal
	7	9-Hydroperoxy-5,7,11,14-tetraene	3,6-Dodecadienal
		7-Hydroperoxy-5,8,11,14-tetraene	2,5,8-Tetradecatrienal
		5-Hydroperoxy-6,8,11,14-tetraene	2,4,7,10-Hexadecatetraenal

Source: From Badings[3].
Note: Only the most active methylene groups in each acid are considered.

component fatty acids. Numerous other factors, both internal (e.g., antioxidants, prooxidants) and external (e.g., oxygen concentration, temperature and light intensity), exert a pronounced influence.

Galactolipases and Phospholipases

Plant cells contain enzymes capable of breaking down both glycolipids and phospholipids. They function in the normal turnover of these molecules in cells of harvested products; however, they also may have specific but at present poorly defined additional roles. For example, galactolipases may be important in the breakdown of prolamellar bodies during the greening of etioplasts.[91] In addition, both types of enzyme appear to function during the onset of senescence of harvested products. During this period there is general breakdown of lipids with subsequent disorganization of the integrity of the cellular membranes.

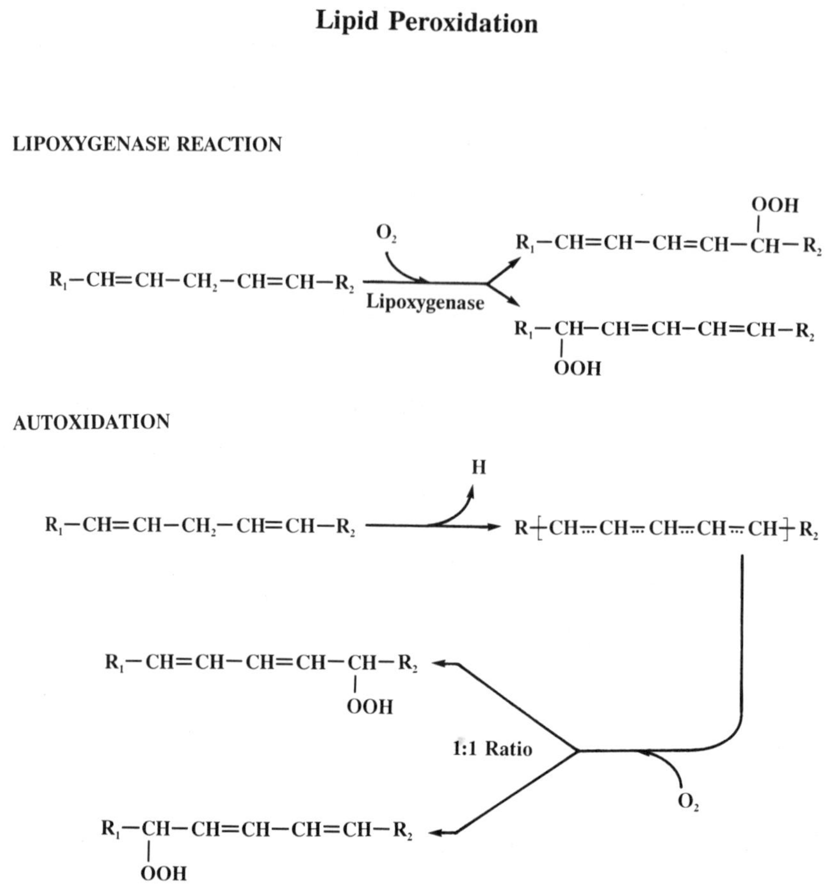

Lipid Peroxidation

LIPOXYGENASE REACTION

AUTOXIDATION

Figure 4-20. Enzymatic and nonenzymatic lipid peroxidation reactions occurring within the cell.

Enzymes capable of attacking galactolipids, the predominant form of plant glycolipids, are found both in the cytosol and in the chloroplasts. Lipolytic acyl hydrolase catalyzes the hydrolysis of the ester bonds, yielding free fatty acids and the corresponding galactosylglycerols. The galactosylglycerols are subsequently attacked by α- and β-galactosidases, giving galactose and glycerol.

Phospholipids are broken down by four specific types of phospholipases, each of which attacks the parent molecule in a distinct manner. For example, the C type hydrolyzes the ester bond of carbon 3 of glycerol and phosphoric acid and in the case of phosphatidyl choline yields a 1,2-diacylglycerol and phosphorylcholine.

Postharvest Alterations

Changes in the lipid fraction of horticultural crops during the postharvest period have not been studied thoroughly. In general, in oleaginous plant products such as pecan kernels (approximately 74% lipid), changes in lipids are largely qualitative rather than quantitative. Approximately 98% of the lipid fraction is triacylglycerols of which 90% are unsaturated.[83,84] Oxidation represents the primary qualitative alteration in lipids during the storage and marketing period. Large quantitative and qualitative changes in oil seed crops do occur during germination when stored lipids are recycled; however, they are not considered postharvest alterations.

In the fruit of avocado, the composition of the oil does not change during maturation and storage. While there is a large increase in fruit respiration during the ripening climacteric, mesocarp lipids do not appear to represent the source of carbon utilized.[7]

Substantial changes in cellular lipid in nonoleaginous tissues occur during senescence. Here there are significant alterations in both glycolipids and phospholipids. In cucumber cotyledons—a model system used for studying lipid changes during the onset and development of senescence—phosphatidyl choline, the major phospholipid present, begins to disappear once the cotyledons reach their maximum fresh weight. As senescence progresses, rapid desiccation of the tissue begins. By this time, 56% of the phosphatidyl choline has been broken down and phosphatidyl ethanolamine begins to be metabolized (fig. 4-21). The glycolipids begin to be lost concurrently with chlorophyll, approximately two weeks prior to initial weight loss of the tissue (fig. 4-22).[20] These changes in lipid composition may mediate alterations in the structure of the membranes, resulting in abnormal permeability and decreased activity of membrane sequestered enzymes, thus accelerating senescence.

PLANT PIGMENTS

Our lives are surrounded and in many ways dominated by plant colors. These colors are due to the presence of pigments within the plant and their interaction with light striking them. Sunlight is composed of a number of different wavelengths, the composite of which is called the spectrum (fig. 4-23). When light strikes a plant, part of the wavelengths are absorbed by the component pigments while others are either reflected or transmitted through the tissue. What is seen as a specific plant color, such as the blue of grape-hyacinth (*Hyacinthus orientalis*, L.) flowers, is due to the absorption by pigments of all of the other wavelengths in the visible spectrum, except the blue region, which is reflected from the flower tissue.

In biological tissues there are two types of reflected light: surface (also called regular) and body reflectance. With surface reflectance, the light striking the product is reflected from the surface without penetrating the tissue. It represents only about 4% of the light striking most biological samples. Much more important is body reflectance where the light actually penetrates into the tissue, becomes diffused (spread out in all directions) upon interacting with internal surfaces and molecules, and is eventually either absorbed or reaches the surface and escapes from the tissue. Part of this light is absorbed by the component pigments and the remainder moves back out of the tissue, becoming the color that is perceived.

Plant pigments can be separated into four primary classes based on their chemistry: the chlorophylls, carotenoids, flavonoids, and betalains, the latter

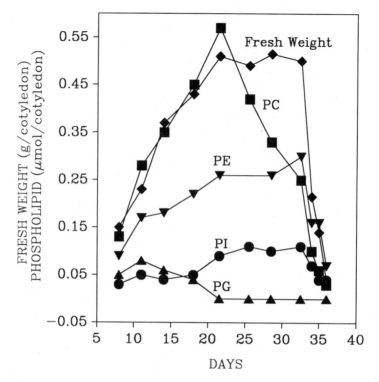

Figure 4-21. Changes in membrane phospholipids in cotyledons during the onset and development of senescence (PC, phosphatidyl choline; PE, phosphatidyl ethanolamine; PG, phosphatidyl glycerol; PI, phosphatidyl inositol) *(after Ferguson and Simon[20])*.

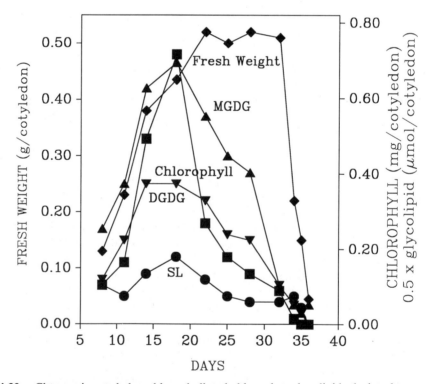

Figure 4-22. Changes in cotyledon chlorophyll and chloroplast glycolipids during the onset and development of senescence (MGDG, monogalactosyl diglyceride; DGDG, digalactosyl diglyceride; SL, sulpholipid) *(after Ferguson and Simon[20])*.

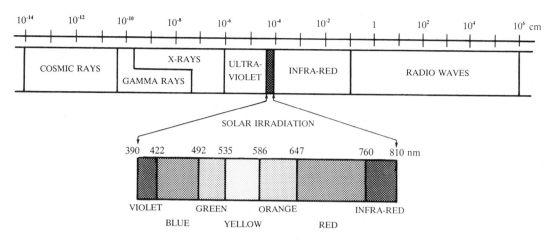

Figure 4-23. Visible light is a part of the spectrum of electromagnetic radiation from the sun. Radiation travels in waves; the length of each wave ranges from very short in gamma or cosmic rays (10^{-8} to 10^{-14} cm) to extremely long in radio waves (1 to 10^6 cm). What is perceived as the color of a harvested product is due to the absorption of part of the visible spectrum by pigments in the surface cells and reflection (body reflectance) of the remainder, part of which is detected by the human eye.

being fairly limited in distribution (table 4-13). There are, of course, additional pigments (e.g., quinones, phenalones, phyrones); however, these typically make only a minor contribution to the color of plants.

In nature, pigments serve a number of functions. The chlorophylls and carotenoids trap light energy in photosynthesis. The carotenoids also act as protecting agents for chlorophyll molecules in photosynthetic tissues to prevent photooxidation. Pigments are of paramount importance in many plant species for their role in facilitating pollination.[75] Flower colors are known to attract certain insects, birds, and in some cases, bats, which pollinate the flower. In the wild, the color of some fruits and seeds also plays an important role in dispersal. Dispersal helps to minimize the potential for competition between the parent and its progeny, enhancing the potential for colonization of new areas.

For man, color, form, and freedom from defects are three of the primary parameters used to ascertain quality in postharvest plant products. Pigmenta-

Table 4-13. Plant Pigments

	Color	Cellular Localization	Solubility
Chlorophyll	Blue-green, yellow-green	Chloroplasts	Insoluble in water, soluble in acetone, ether, alcohols
Carotenoids	Yellow, orange-red	Chloroplasts, chromoplasts	Insoluble in water, soluble in acetone, ether, alcohols
Flavonoids	Yellow, orange, red, blue	Vacuole	Water soluble
Betalains	Yellow, orange, red, violet	Vacuole, cytosol	Water soluble

tion provides quality information, such as the degree of ripeness of fruits (e.g., banana) or the mineral nutrition of ornamentals (e.g., chlorosis). It is especially important with ornamentals since visual impressions are almost exclusively relied on for quality judgments.

Classes of Pigment Compounds

Chlorophylls

The plant world is dominated by the color green, which is the result of the presence of the chlorophyll pigments. The chlorophylls are the primary light-accepting pigments found in plants that carry out photosynthesis through the fixation of carbon dioxide and the release of oxygen. In nature, there are two predominant forms, chlorophyll a and chlorophyll b, differing only slightly in structure (fig. 4-24). Both are normally found concurrently within the same plant and usually at a 2.5–3.5:1 ratio between a and b. Two other chlorophylls, c and d, are found in a relatively limited number of species. For example, chlorophyll c is found in several marine plants.

Each form of chlorophyll is a magnesium-containing porphyrin formed from four pyrrole rings (fig. 4-24). Attached to the propionyl group of a pyrrole ring of chlorophylls a, b, and d is a 20-carbon phytol chain ($C_{20}H_{39}OH$). Chlorophylls a and b differ structurally only in the replacement of a methyl group on chlorophyll a with an aldehyde ($-CHO$). Unlike the flavonoid pigments, the chlorophylls are hydrophobic, and as a consequence, they are not soluble in water. The primary function of chlorophyll is to absorb light energy and convert it to chemical energy. This process occurs in the chloroplasts.

Carotenoids

The carotenoids are a large group of pigments associated with chlorophyll in the chloroplasts and are also found in the chromoplasts. Their colors range from red, orange, and yellow to brown and are responsible for much of the autumn leaf pigmentation.

Chemically carotenoids are terpenoids comprised of eight isoprenoid units (fig. 4-25). Nearly all carotenoids are composed of 40 carbon atoms. They are divided into two subgroups: the carotenes and their oxygenated derivatives, the xanthophylls. Both are insoluble in water, although the xanthophylls tend to be less hydrophobic than the carotenes.

The nomenclature of the carotenoids is based on their nine carbon end groups of which there are seven primary types (fig. 4-25). These can be arranged in various combinations on the methylated straight-chain portion of the molecule, for example, α-carotene is β,ϵ-carotene while β-carotene is β,β-carotene. Leutin, a common xanthophyll, is structurally like α-carotene, differing only in the presence of a hydroxyl group on carbon 3 of both the β and ϵ end groups. Thus, a tremendous range in potential structural variations is possible and in some tissues a quite large assortment of specific compounds is found. For example, in the juice of the 'Shamouti' orange, 32 carotenoids have been identified (table 4-14).

In photosynthetic tissue, carotenoids function both in the photosynthesis process per se and as protectants, preventing the chlorophyll molecules from being oxidized (photooxidation) in the presence of light and oxygen. In flowers and fruits, carotenoids appear to act as attractants that aid in securing pollination or dispersal; however, in underground structures such as roots and tubers their role is not understood.

Flavonoids

While green is the dominant color in plants, other colors have a tremendous attraction both for man and other animals. Many of the intense colors of flowers, fruits, and some vegetables are the result of flavonoid pigments and closely related compounds. These represent a large class of water-soluble compounds with a diverse range of colors. For example, there are yellows, reds, blues, and oranges. Numerous variations in color are derived both from structural differences between compounds and the relative concentration of specific pigments within the cells. The flavonoids are found both in the cytosol and in vacuoles.

The basic structure of flavonoid pigments is present in figure 4-26 (page 196). It consists of two benzene rings (A and B) joined by a 3-carbon link that forms a γ-pyrone ring through oxygen. Various classes of flavonoids differ only in the state of oxidation of the 3-carbon link, while individual compounds within these classes differ mainly by the number and orientation of the hydroxy, methoxy, and other groups substituted on the two benzene rings. Individual classes of flavonoids include anthocyanidins, flavones, catechins, flavonols, flavanones, dihydroflavonols, and the flavan-3,4-diols or proanthocyanidins (fig. 4-26). Most flavonoid pigments exist in live plant tissue as glycosides where one or more of their hydroxyl groups is joined to a sugar. In some anthocyanins, an organic acid may be esterified to one of the hydroxyls on the sugar, giving an acylated compound. This is the case with grapes where p-coumaric acid, a derivative of cinnamic acid, can be found attached to both the mono- and diglucoside anthocyanin pigments (fig. 4-27, page 198).

Closely related to flavonoid compounds are the chalcones, dihydrochalcones, isoflavones, neoflavones, and aurones (fig. 4-26). These do not have the 2-phenylchroman base structure (fig. 4-26) but are closely akin both chemically and in their biosynthesis.

Betalains

The betalains represent a fourth but substantially restricted group of plant pigments. They are found in the flowers, fruits, and in some cases in other plant parts, giving colors of yellow, orange, red, and violet. Perhaps the best example is the red-violet pigment from the root of the beet, *Beta vulgaris,* L., the first betalain isolated in crystalline form; hence the derivative name betalain.

As a group they are characterized by being water-soluble nitrogenous pigments found in the cytosol and in vacuoles. Chemically betalains are subdivided into two groups: the red-violet betacyanins illustrated by the

Chlorophylls

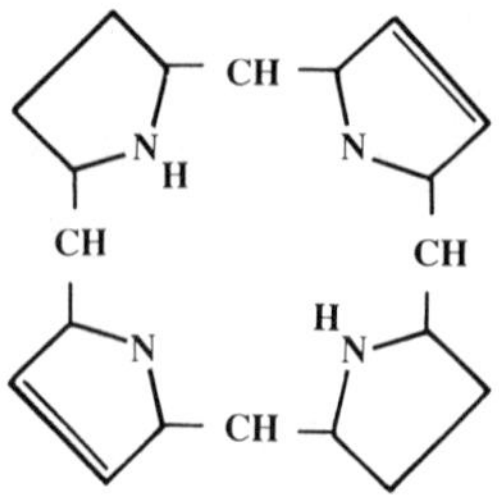

PORPHYRIN BASE STRUCTURE

PYRROLE RING

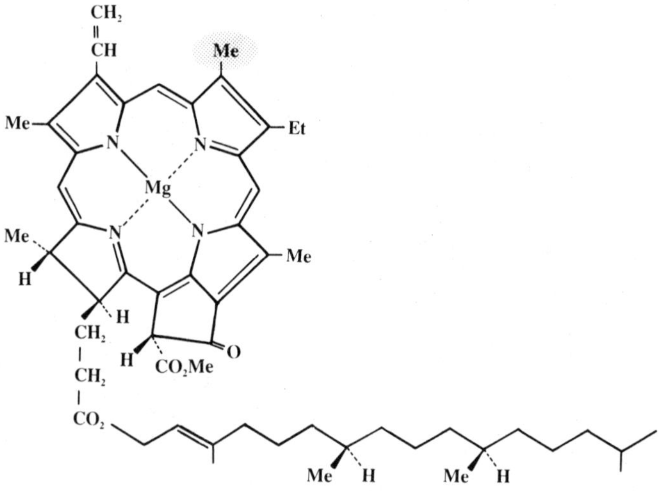

CHLOROPHYLL A

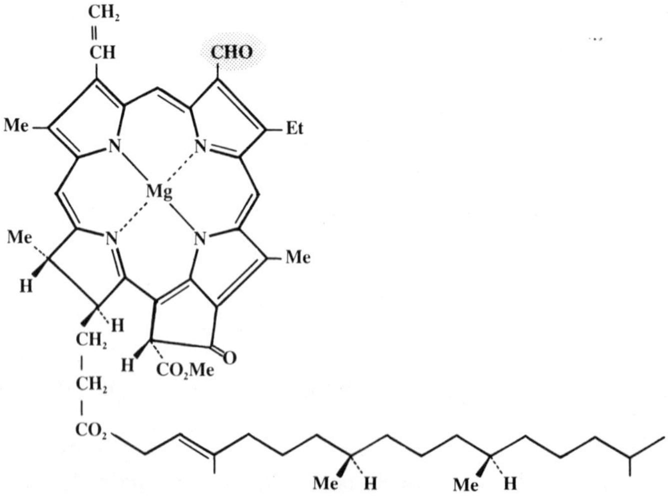

CHLOROPHYLL B

Figure 4-24 *(at left)*. The chlorophylls are magnesium-containing porphyrins derived from four pyrrole rings. The two prevalent chlorophylls, a and b, differ only in a single side group (shaded). Chlorophyll c, however, is structurally quite distinct from a, b, and d in that the phytol tail is not present.

End Group Designation

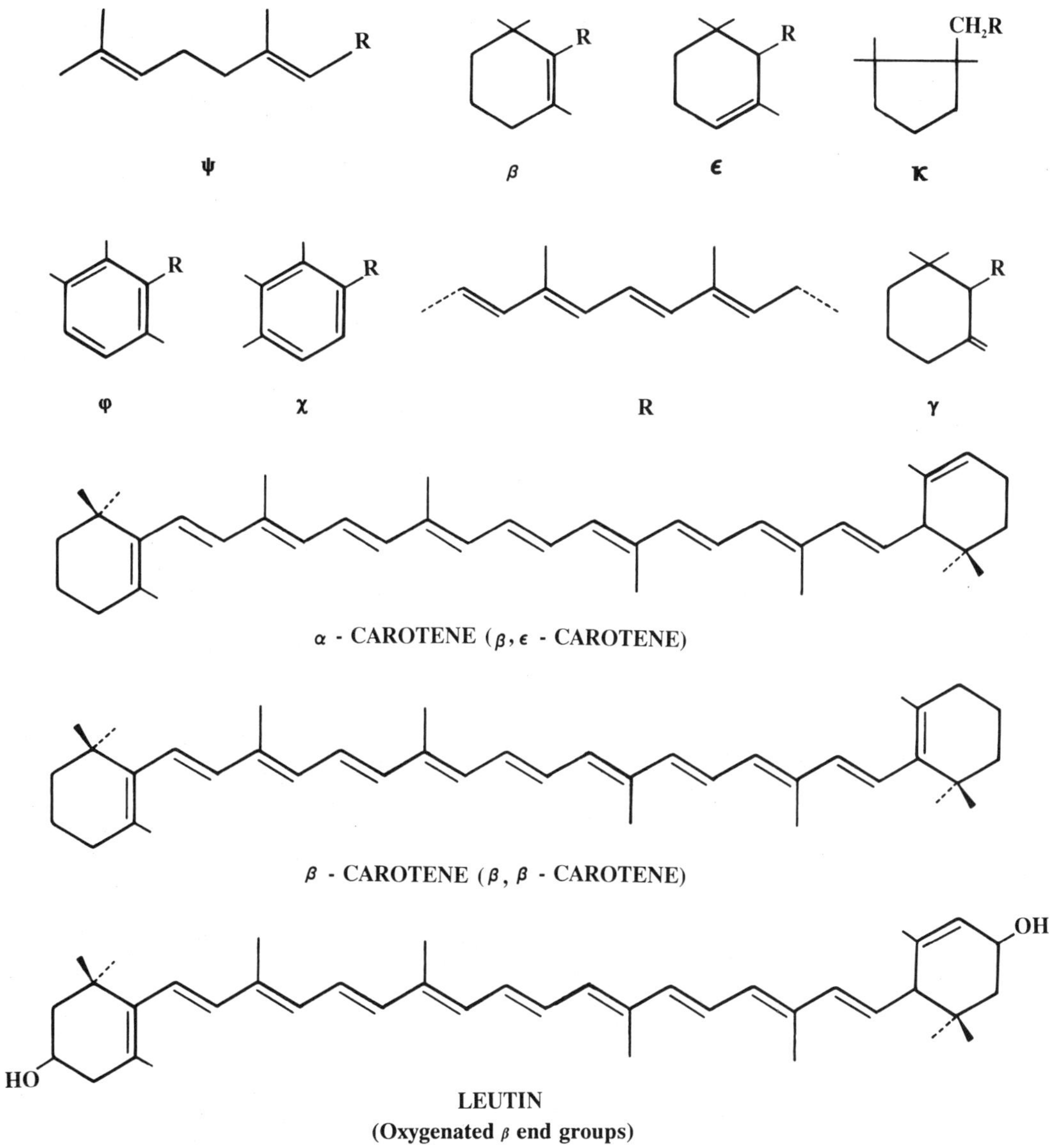

α - CAROTENE (β,ε - CAROTENE)

β - CAROTENE (β, β - CAROTENE)

LEUTIN
(Oxygenated β end groups)

Figure 4-25. The base structures of various carotenoid end groups and the representative structures of two carotenes (α-carotene and ß-carotene). The figure shows the end group designation in naming. Xanthophylls, carotenes oxygenated on one or both end groups, are represented by the pigment leutin.

Table 4-14. Quantitative Composition of Carotenoids in the Juice of 'Shamouti' Oranges

Carotenoid	% Total Carotenoids
Mutatoxanthin	15.13
Cryptoxanthin	12.88
Trollixanthin	9.64
Luteoxanthin-like	6.92
Antheraxanthin	6.83
Zeaxanthin	6.20
Phytoene	5.70
Lutein	5.20
Isolutein	4.00
Luteoxanthin	3.98
Neoxanthin	3.97
Violaxanthin	3.00
OH-Sintaxanthin	2.48
Phytofluene	2.40
cis-Cryptoxanthin	2.00
Trollichrome	1.46
ζ-Carotene	1.40
β-Carotene	1.24
poly-cis-Cryptoxanthin	1.05
Carbonyl 422	0.78
Cryptoflavin	0.70
Auroxanthin	0.58
OH-α-Carotene	0.50
Pigment 426	0.39
α-Carotene	0.36
Citraurin	0.11
Cryptoxanthin diepoxide	0.10
Chrysanthemaxanthin	0.07
Sintaxanthin	0.07
Rubixanthin	0.05
β-Apo-10'-carotenal	0.03
Mutatochrome	0.02

Note: Data from Gross et al.[30]

structure of betanidin and betanin (fig. 4-28) and the yellow betaxanthins, characterized by vulgaxanthin I and II. A number of the naturally occurring betaxanthins have the tail portion of the molecule either partially or totally closed into a ring structure as in dopaxanthin from the flowers of *Glottiphyllum longum,* (Haw.) N.E. Br.[45]

While the precise function of the betalains is not known, it is possible that they may function like the anthocyanins in flowers and fruits, enhancing insect or bird pollination and seed dispersal. No role has been presently ascribed for their presence in plant parts such as roots, leaves, and stems.

Pigment Biosynthesis and Degradation

Chlorophylls

Chlorophyll synthesis is modulated by a number of external influences, two of the most important being light and mineral nutrition. The initial pyrrole ring, porphobilinogen, is formed from two molecules of δ-amino levulinic acid derived from glycine and succinate (fig. 4-29, page 200). Four molecules of porphobilinogen are polymerized, producing a ring structure, uroporphyrinogen, which has acetyl and propinoyl groups attached to each of the component pyrroles. After a series of decarboxylation reactions, protoporphyrin is formed and in subsequent steps magnesium is inserted followed by the addition of the phytol tail. Chlorophyll a, which is blue-green in color, differs from chlorophyll b (yellow-green) only in the presence of a single methyl group instead of a formyl group.

Decomposition of chlorophyll may, in many cases, be quite rapid and dramatic in effect as in the autumn coloration of deciduous trees in the northern temperate zones or the ripening of bitter melon, *Momordica charantia,* L. In many tissues this loss of chlorophyll is part of a transition of the chloroplasts into chromoplasts containing yellow and red carotenoid pigments. The loss of chlorophyll can be mediated through several processes such as the action of the enzyme chlorophyllase, enzymatic oxidation, or photodegradation. While the precise sequence of biochemical steps is not known, the initial reactions appear to be much like the reverse of the final steps in the synthesis pathway. Phytol may be removed to yield chlorophyllide or both magnesium and phytol to give pheophoride. In subsequent degradative steps, the low molecular weight products that are formed are colorless.

Carotenoids

Carotenoids, typically 40 carbon compounds, are built up from 5 carbon isoprene subunits, the most important of which is isopentenyl pyrophosphate. These initial subunits are formed in a series of steps from acetyl-CoA and acetoacetyl-CoA in the terpenoid pathway. Isoprene subunits are sequentially added, build up to a 20-carbon intermediate, geranylgeranyl pyrophosphate, two of which condense to give phytoene with the typical carotenoid skeleton (fig. 4-30, pages 202–203). Subsequent steps involve ring closure and, in the case of the xanthophylls, the addition of one or more oxygens. Some carotenoids (e.g., carotenols) may be esterified to long-chain fatty acids (e.g., oleate or palmitate) or other compounds, which occurs with leaf coloration in the fall[92] and in the peel of apple fruit during ripening. While carotenol esters were once thought to represent breakdown products formed during senescence (i.e., esterified from fatty acids formed during senescence (i.e., esterified from fatty acids formed during membrane degradation), current evidence points toward a controlled synthesis of the pigments.[49]

With xanthophyll synthesis, oxygens are added to the cyclic portions of the molecule, from carbon 1 to carbon 6, and often as hydroxyls. For example, the most common structural feature of xanthophylls is the presence of a hydroxyl at the C-3 and C-3′ positions, giving lutein (β,ε-carotene-3,3′-diol)

Flavonoids

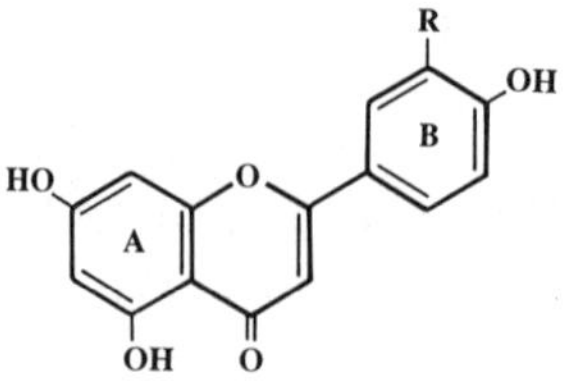

FLAVONES

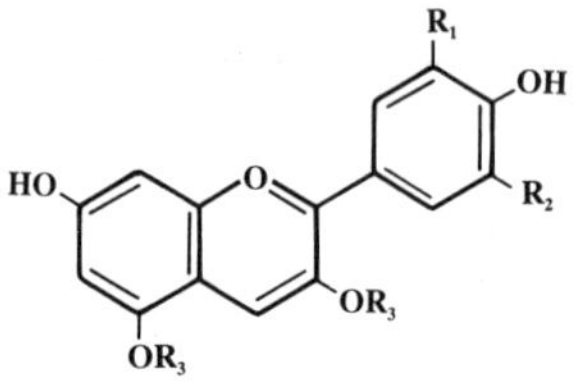

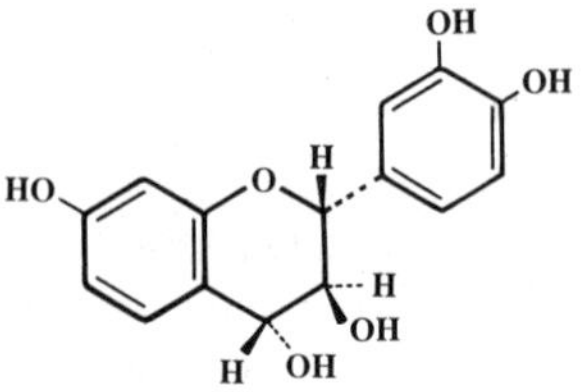

FLAVAN - 3,4 - DIOLS

Closely Related Pigments

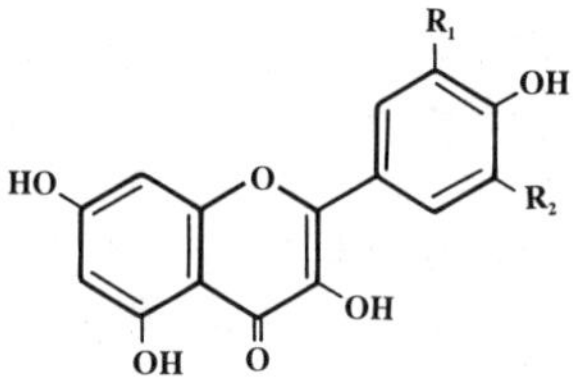

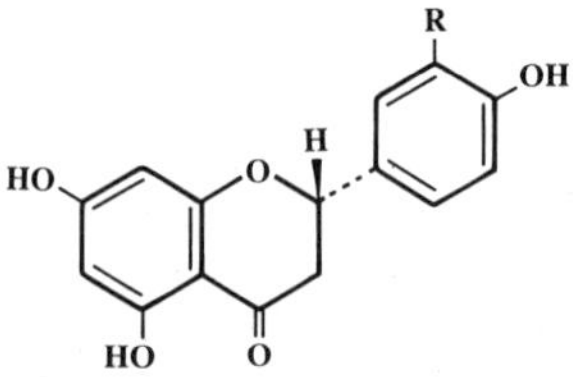

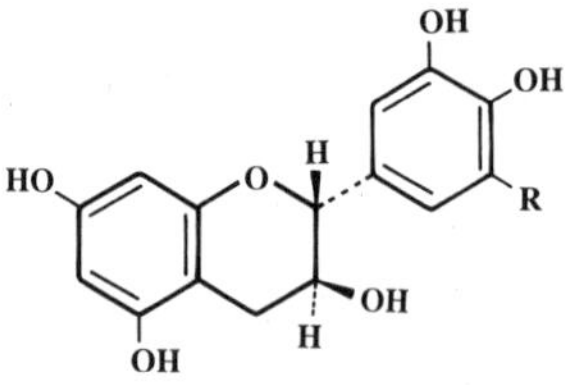

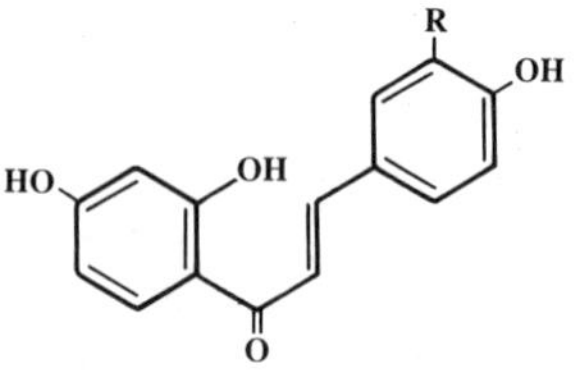

ANTHOCYANIDINS

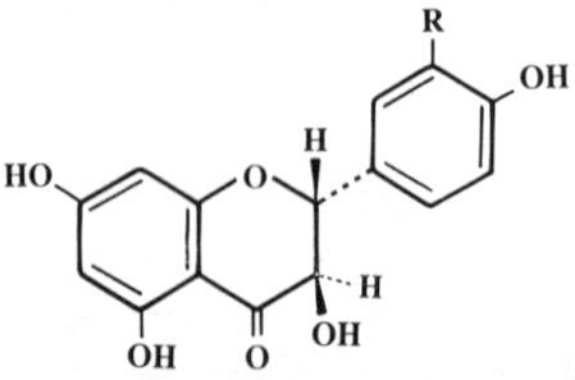

CHALCONES

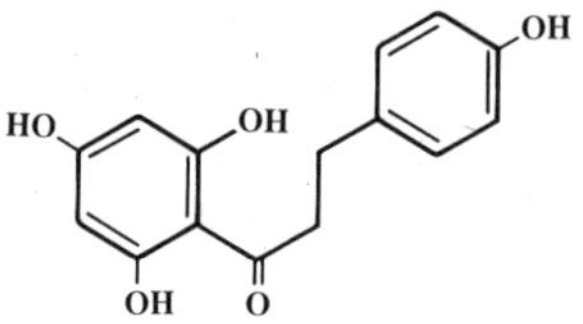

FLAVONOLS

DIHYDROCHALCONES

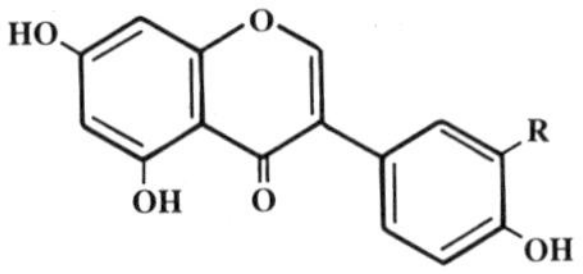

FLAVANONES

ISOFLAVONES

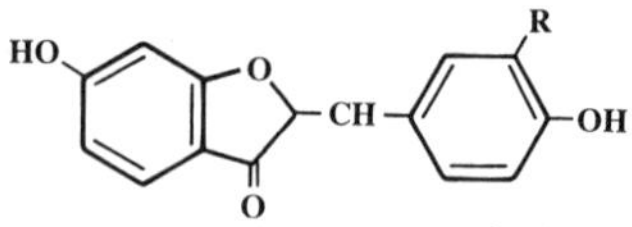

CATECHINS

AURONES

DIHYDROFLAVONOLS

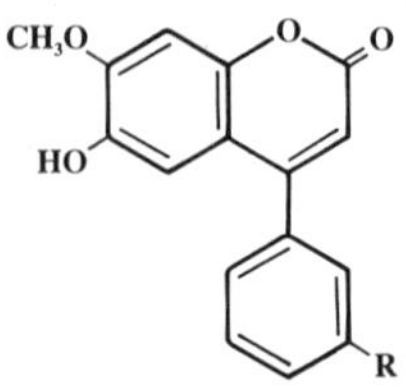

NEOFLAVONES

and zeaxanthin (β,β-carotene-3,3'-diol). Because of the large number of potential combinations for the placement of one or more oxygens, the xanthophylls are numerous. Of the more than 300 naturally occurring carotenoids that have been identified, approximately 87% are xanthophylls. For additional details on carotenoid synthesis see Britton.[9]

The stability of carotenoids is highly variable. In some cases, such as in narcissus flowers, degradation occurs in only a few days,[8] whereas with stored corn over 50% of the carotenoids may be present after 3 years of storage.[76] A number of factors affect the rate of loss of carotenoids. These include the specific type of pigment, storage temperature, product moisture level, type of product, and prestorage treatments (e.g., drying of corn). In edible products, the breakdown of β-carotene is of special concern due to its role as a precursor of vitamin A.

In the breakdown of carotenoids, the initial degradative steps involve oxygenation of the molecule. This process differs from oxygenation reactions that occur largely on the ring structures and result in the formation of xanthophylls. Carotenoid-degrading enzyme systems have been found in chloroplasts and mitochondria. The double bonds within the linear portion of the molecules are subject to attack by lipoxygenase[39] and activity is accelerated by the availability of oxygen, light, and certain metals. A variety of shorter-chain-length terpenoids are formed, a number of which are volatile and in some cases represent distinct odors.[97]

As a group, xanthophylls are characteristically more stable than the carotenes. In leaf tissue in the autumn, the xanthophylls are released into the cytoplasm upon the disruption of the chloroplasts. These molecules are subsequently esterified, which appears to substantially enhance their stability, especially in contrast to the carotenes.

Flavonoids

Flavonoid biosynthesis begins with the formation of the basic $C_6C_3C_6$ skeleton through the combination of three malonyl-CoA molecules with one cinnamyl-CoA (fig. 4-31, page 204). A chalcone is yielded and with ring closure it is converted to a flavanone. In the second series of steps, the flavanone may be converted to each of the different classes of flavonoids, for example, flavone, flavonol, anthocyanidin. In the final stage these are converted to individual compounds such as cyanidin, myricetin, and apigenin. The anthocyanins range from reds to blues. As the degree of methylation increases the individual compounds become increasingly red while hydroxylation results in deeper blues. In addition, blue coloration can result from complexes formed by the chelation of Al^{+3} and Fe^{+3} to the hydroxyls of the A ring.

Figure 4-26 *(at left).* The basic structure of flavonoid pigments. Each consists of two benzene rings (A and B) joined by a 3-carbon link. The various classes of flavonoids differ only in the state of oxidation of the 3-carbon link. Within each class is a wide range of individual pigments varying in the number and position of groups (e.g., OH, CH_3) attached to the two rings. Also illustrated are several closely related pigments (e.g., chalcones, dihydrochalcones, isoflavones, aurones, and neoflavones).

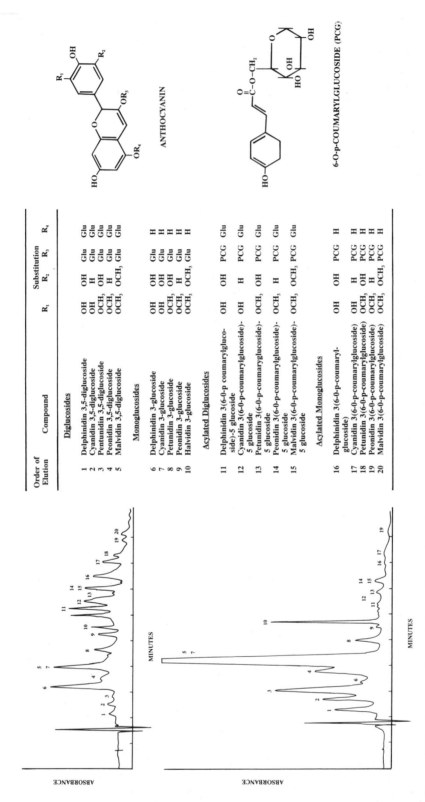

Order of Elution	Compound	R₁	R₂	R₃	R₄
	Diglucosides				
1	Delphinidin 3,5-diglucoside	OH	OH	Glu	Glu
2	Cyanidin 3,5-diglucoside	OH	H	Glu	Glu
3	Pentunidin 3,5-diglucoside	OCH₃	OH	Glu	Glu
4	Peonidin 3,5-diglucoside	OCH₃	H	Glu	Glu
5	Malvidin 3,5-diglucoside	OCH₃	OCH₃	Glu	Glu
	Monoglucosides				
6	Delphinidin 3-glucoside	OH	OH	Glu	H
7	Cyanidin 3-glucoside	OH	OH	Glu	H
8	Petunidin 3-glucoside	OCH₃	OH	Glu	H
9	Peonidin 3-glucoside	OCH₃	H	Glu	H
10	Halvidin 3-glucoside	OCH₃	OCH₃	Glu	H
	Acylated Diglucosides				
11	Delphinidin 3(6-0-p coumarylglucoside)-5 glucoside	OH	OH	PCG	Glu
12	Cyanidin 3(6-0-p-coumarylglucoside)-5 glucoside	OH	H	PCG	Glu
13	Petunidin 3(6-0-p-coumarylglucoside)-5 glucoside	OCH₃	OH	PCG	Glu
14	Peonidin 3(6-0-p-coumarylglucoside)-5 glucoside	OCH₃	H	PCG	Glu
15	Malvidin 3(6-0-p-coumarylglucoside)-5 glucoside	OCH₃	OCH₃	PCG	Glu
	Acylated Monoglucosides				
16	Delphinidin 3(6-0-p-coumaryl-glucoside)	OH	OH	PCG	H
17	Cyanidin 3(6-0-p-coumarylglucoside)	OH	H	PCG	H
18	Petunidin 3(6-0-p-coumarylglucoside)	OCH₃	OH	PCG	H
19	Peonidin 3(6-0-p-coumarylglucoside)	OCH₃	H	PCG	H
20	Malvidin 3(6-0-p-coumarylglucoside)	OCH₃	OCH₃	PCG	H

Figure 4-27. A comparison of the anthocyanins of two grape cultivars (*top*, 'Concord,' *bottom*, 'DeChaunac') separated by high-pressure liquid chromatography. Substitutions in the basic structure of the anthocyanin molecule occur at any of the four positions, denoted as R₁ – R₄. These substitutions are given for twenty of the anthocyanins found in the two grape cultivars. Part of the anthocyanins have an organic acid esterified to the hydroxyl group of their attached sugar (termed acylated). In this case, p-coumaric (PCG), a derivative of cinnamic acid, is present (*after Williams and Hrazdina*[99]).

Betacyanins

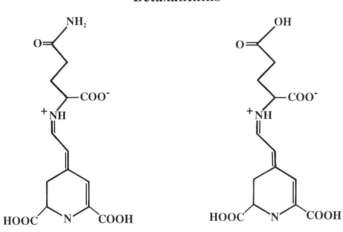

BETANIDIN   BETANIN

Betaxanthins

VULGAXANTHIN I   VULGAXANTHIN II

Figure 4-28. Examples of the two primary structural groups of betalains: the betacyanins and the betaxanthins *(after Piattelli et al.[70,71]).* Betanidin, vulgaxanthin I, and vulgaxanthin II are found in the root of the common garden beet, *Beta vulgaris,* L. Epimers of betacyanins are formed by alterations in the configuration; an example would be the epimer of betanin, isobetanin.

Of the flavonoids, the decomposition of anthocyanin has been studied in greatest detail. The vulnerability of individual pigments tends to vary; substitutions at specific positions on the molecule can significantly affect its stability. For example, a hydroxyl at the 3′ position enhances the pigment's propensity for degradation.

Enzymes that have the potential to degrade anthocyanins have been iso-

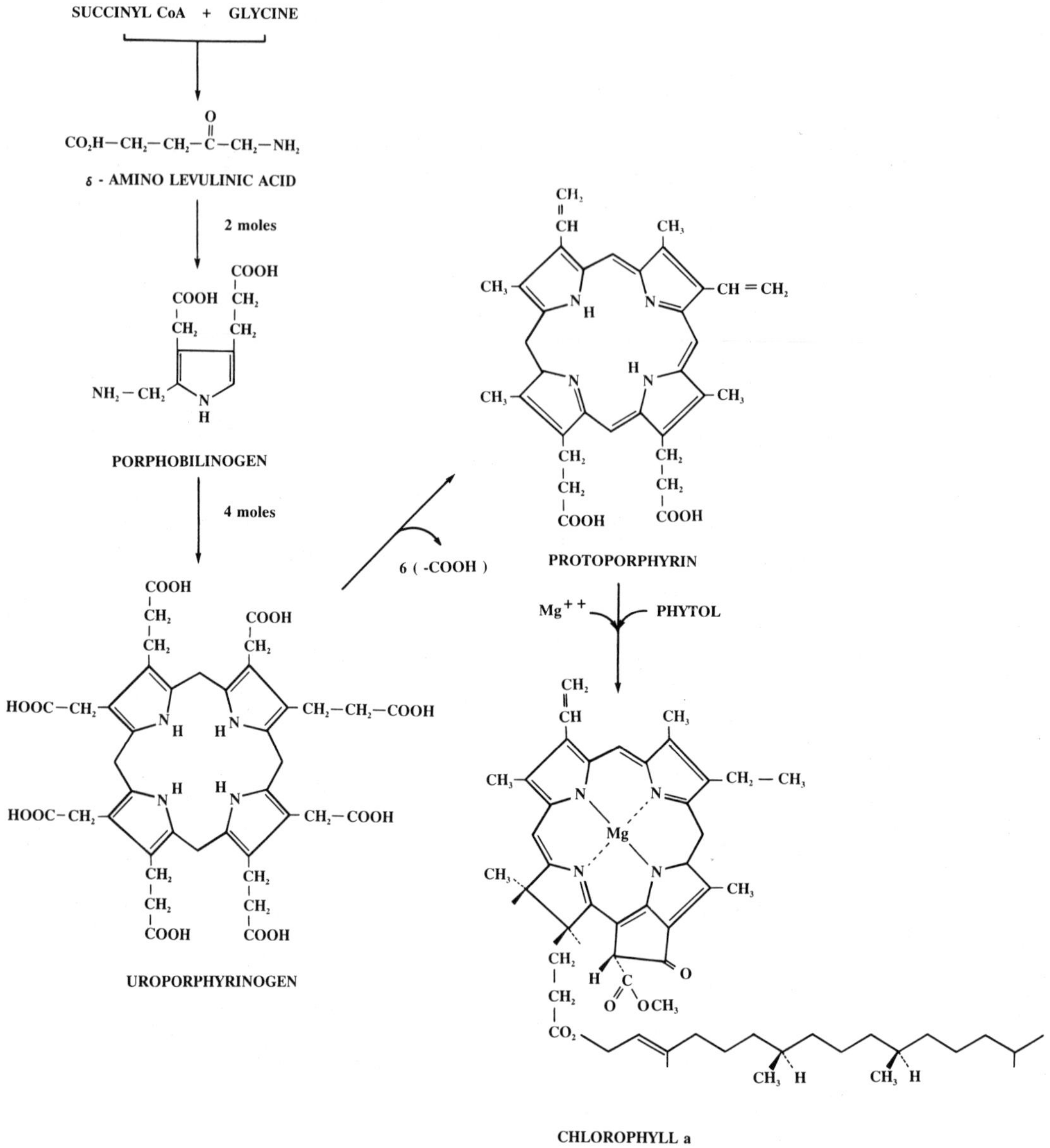

Figure 4-29. The general biosynthetic pathway for chlorophyll. Insertion of Mg^{++} and the addition of the phytol tail occur in the last series of steps.

lated from a number of different tissues (e.g., flowers, fruits, and others). These tend to fall into two classes: glucosidases and polyphenol oxidases. Both have the ability to produce colorless products. Other possible mechanisms of pigment alteration and breakdown include pH alterations, which accompany ripening in some fruits, and attack of the charged portion of the molecule by naturally occurring nucleophiles (e.g., ascorbic acid).

Betalains

The betalains appear to be derived from 3,4-dihydroxyphenylalanine (L-DOPA). The betacyanins are formed from two of these molecules, one of which has an oxidative opening of the aromatic ring followed by subsequent closure to yield betalamic acid (fig. 4-32). The second forms cyclodopa and upon condensation with betalamic acid yields betanidin, the base molecule for the formation of the various betacyanins. The betaxanthins are formed through the condensation of betalamic acid with an amino or imino group (other than cyclodopa).

Research on the breakdown of the betalains has centered to a large extent on color alterations occurring in beet roots, although these pigments may also be found in members of the Aizoaceae, Amaranthaceae, Basellaceae, and Cactaceae families, which are of postharvest interest. This research was, in part, prompted by the elimination of the use of red dye number 2 in foods. Betalains were briefly considered as possible replacements.Nearly all of the studies to date have been done on either extracted pigments or processed tissue, and as a consequence, the extent to which these reactions can be implicated in color changes in stored fresh beet roots (*Beta vulgaris,* L.) is not clear.

Beet root tissue exposed to low pH (3.5–5.5) retains its color relatively well whereas at higher pH (7.5–8.5) discoloration occurs.[33] Action by β-glucosidase results in the removal of the sugar side group, converting betanin and isobetanin to their aglucones, betanidin and isobetanidin. In addition, exposure to air and/or light results in the degradation of betalains, often causing a browning discoloration.

Postharvest Alterations in Pigmentation

During both the preharvest and postharvest period many products undergo significant changes in their pigment composition (figs. 4-33 and 4-34, page 206). These changes include both the degradation of existing pigments and the synthesis of new pigments; in many cases, both processes may occur concurrently. Pigmentation changes are of paramount importance in many products in that they are used as a primary criterion for assessing quality.

The degradation of pigments can be subdivided into two general classes: pigment losses that are beneficial to quality and those that are detrimental. Many of the beneficial losses center around the degradation of chlorophyll with concurrent synthesis of other pigments or the unmasking of preexisting pigments within the tissue. Examples would be the degreening of oranges, during which time carotenoids are being synthesized, and the loss of chlorophyll in banana, allowing the expression of the pigments that are already present. Detrimental losses of pigments after harvest can be seen in the color fading of flowers and in chlorophyll losses in broccoli (*Brassica oleracea,* L. Italica group) florets or leaf crops. As with the degradation of pigments, the synthesis of pigments after harvest can be either beneficial or undesirable. The development of red coloration in the fruit of the tomato (*Lycopersicon esculentum,* Mill.) after harvest is highly desirable while the formation of

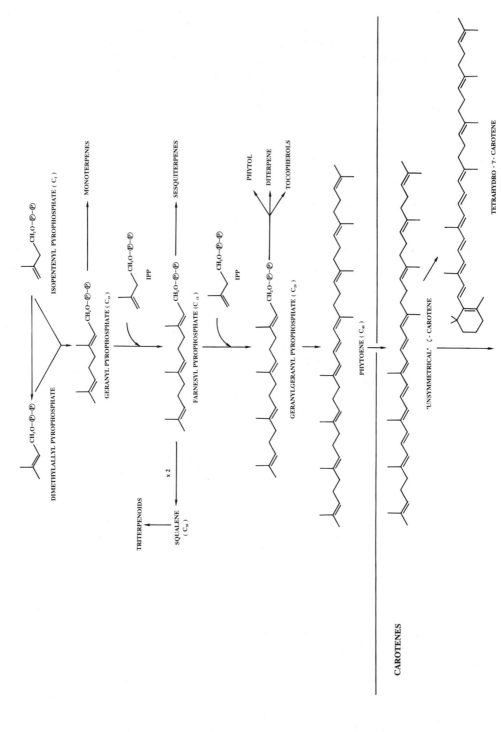

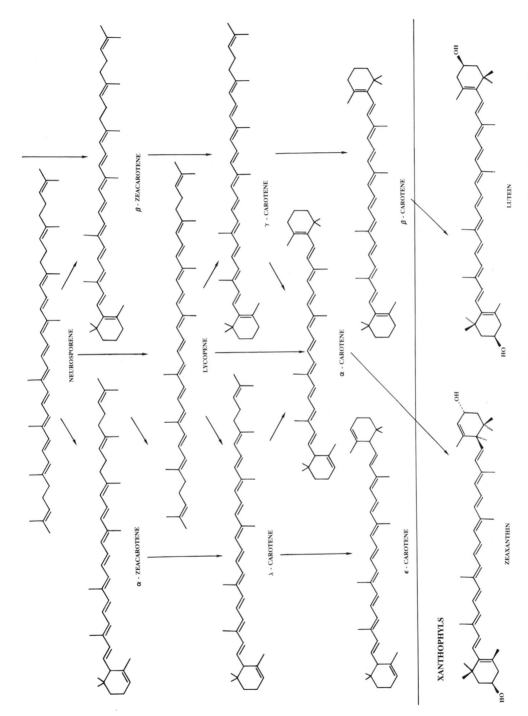

Figure 4-30. Generalized pathway for the biosynthesis of carotenes and their oxygenation to form xanthophylls.

NEUROSPORENE

β - ZEACAROTENE

α - ZEACAROTENE

LYCOPENE

γ - CAROTENE

λ - CAROTENE

α - CAROTENE

β - CAROTENE

ε - CAROTENE

LUTEIN

XANTHOPHYLS

ZEAXANTHIN

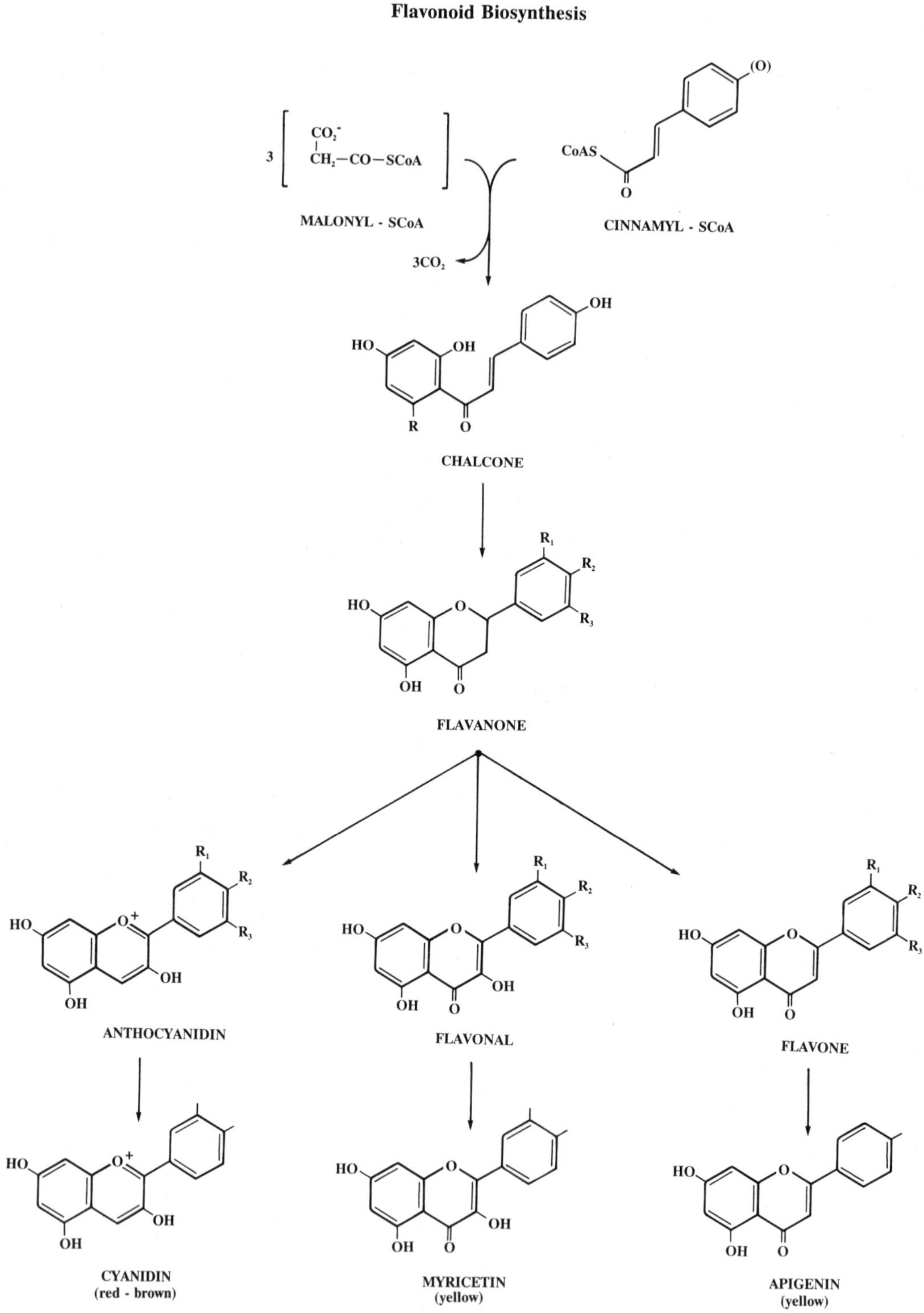

Flavonoid Biosynthesis

Figure 4-31. Flavonoid biosynthesis occurs in three stages starting with the combination of three malonyl-SCoA molecules with one cinnamyl-SCoA molecule forming a chalcone. It is followed by ring closure forming a flavanone and conversion to the various classes of flavonoids. Final steps involve the formation of individual compounds with various additions to the two rings.

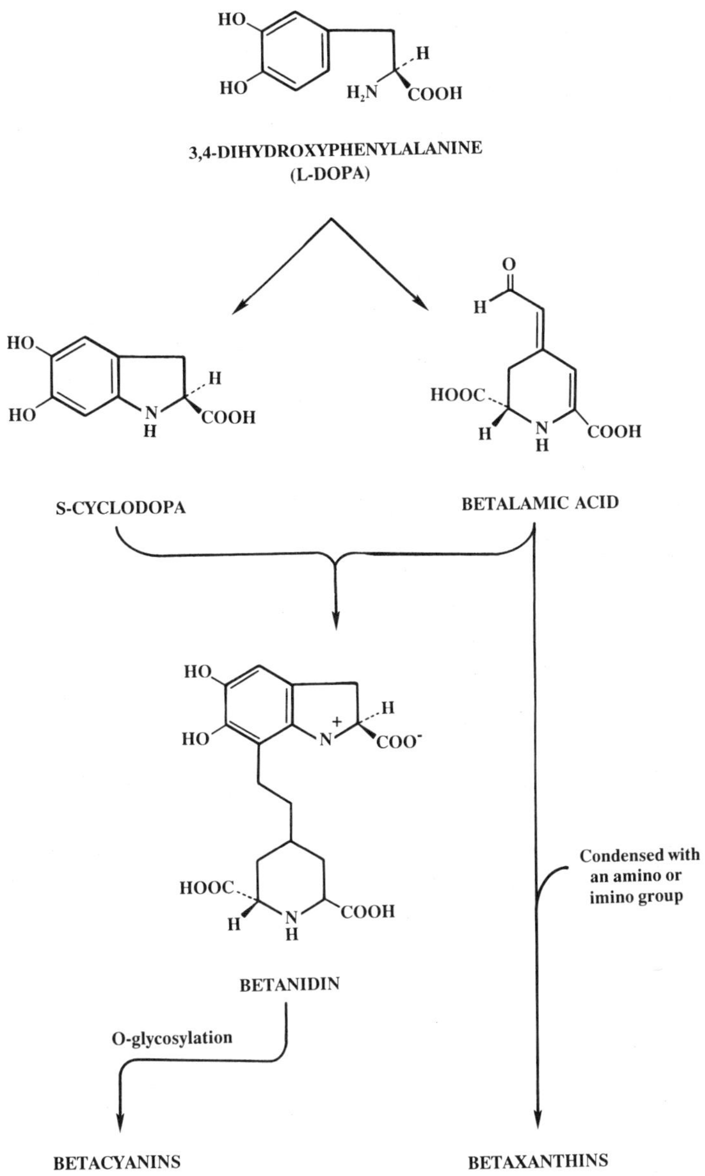

Figure 4-32. The proposed biosynthetic pathways for the betalains.

chlorophyll in harvested potatoes or the synthesis of carotenoids in the bitter melon is undesirable.

Many postharvest factors affect the degree of change in pigmentation after harvest, the most important of which are light and temperature. Light is essential for the synthesis of chlorophyll and its presence delays the loss of these pigments in detached leaves. It also appears to be important in stimulating the synthesis of anthocyanins and lycopene in some products; however, not β-carotene in tomato fruits. Changes in the pigmentation of many tissues

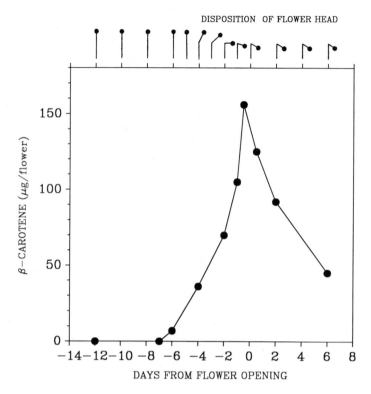

Figure 4-33. Time course of the synthesis and degradation of ß-carotene in the corona of *Narcissus majalis,* Curtis flowers. Day 0 represents anthesis *(after Booth[8]).*

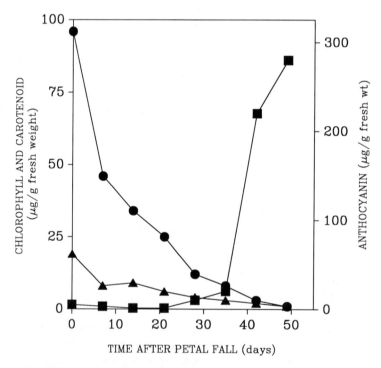

Figure 4-34. Changes in the pigment concentration in developing strawberry fruits (● chlorophyll; ▲ carotenoid; ■ anthocyanin *(after Woodward[100]).*

are temperature dependent. The effect of temperature, however, varies with the specific pigment, the tissue of interest, and whether synthetic or degradative processes are operative. For example, in pink grapefruit (*Citrus* × *paradisi,* Macf.) high temperatures (30–35°C) favor the accumulation of lycopene but not carotene; at low temperatures (5–15°C) the opposite is true.[63] However, in the tomato, temperatures above 30°C suppress the biosynthesis of lycopene but not β-carotene.[93]

Several plant growth regulators have also been shown to have a significant effect on the pigmentation in some harvested products. The use of ethylene for degreening citrus and banana fruits and stimulating the synthesis of carotenoids (tomato) has become a major commercial postharvest tool in many areas of the world. Ethylene is also known to stimulate the formation of anthocyanins in the grape when applied prior to harvest.[34] With floral crops, ethylene evolved from the harvested product or adjacent products dramatically accelerated pigment degradation, color fading, and senescence in many flowers (e.g., Vanda orchids). To prevent pigment degradation the use of chemicals that either inhibit the biosynthesis of ethylene (e.g., aminoethoxyvinylglycine) or impede its activity (silver thiosulfate) is increasing rapidly.

Other growth regulators, such as cytokinins, have a pronounced effect on the retention of chlorophyll.[47] The potential use of cytokinins to retard chlorophyll loss has, as a consequence, been tested in several green vegetables (broccoli, brussels sprouts (*Brassica oleracea,* L. Gemmifera group), celery (*Apium graveolens* var. *dulce,* Pers.), endive (*Cichorium endiva,* L.), and leaf lettuce (*Lactuca sativa,* L.)). In broccoli stored at 13°C, the storage life can be doubled with a single application of zeatin and dihydrozeatin.[25] While in many cases the results are positive, other postharvest techniques for color retention are commercially more acceptable.

VOLATILE COMPOUNDS

Classes of Volatile Compounds

Plants give off a wide range of volatile compounds, some of which are extremely important components of quality. Humans exhibit distinct patterns in the foods that they select to consume and flavor is known to be a primary criterion in this selection. Most flavors are comprised of a combination of both taste and odor. Since taste is generally thought to be limited to four basic sensations—sweet, sour, salty, and bitter—volatiles often play a very significant role in flavor. The four basic tastes can be contrasted with the potential perception of up to 10,000 distinct odors by the olfactory epithelium of humans. This difference in potential perception is also seen in the number of taste receptors for an individual, which are in the thousands, versus odor receptors, which are thought to be in the millions. Hence, aroma compounds are not only an extremely important part of taste but they also provide flavor with an almost unlimited potential diversity.

The importance of volatile compounds is not just limited to food products. The aroma of many ornamental and floral species compliments the visual perception of these and often is an important part of the plant's aesthetic quality. The contribution of plant volatiles may impart either positive characteristics as in the fragrance of gardenia flowers (*Gardenia jasminoides,* Ellis) or in the case of the carrion-flower (*Stapelia gigantea,* N.E. Br.) a distinctly undesirable sensation, at least from the standpoint of humans.

A primary requirement of plant volatiles is that they must be present in a gaseous or vapor state. Existence as a gas is essential for perception since the molecules must be able to reach the olfactory epithelium in the roof of the nasal passages. In addition, some degree of water solubility is essential.

The volatile components of postharvest products represent a diverse array of chemical compounds. These include esters, lactones, alcohols, acids, aldehydes, ketones, acetals, hydrocarbons and some phenols, ethers, and heterocyclic oxygen compounds. In the apple, 159 different volatiles have been isolated and identified (table 4-15). They include 20 acids, 28 alcohols, 71 esters, 26 carbonyls, 9 ethers and acetals, and 5 hydrocarbons.

Volatiles generally are present in very small amounts, often only a fraction of a part per million. In addition, of the large number of volatiles given off by a plant or plant part, typically only a very small number of compounds impart the characteristic aroma. In the apple, ethyl 2-methylbutyrate, although present in only very small amounts, is responsible for much of the characteristic aroma. These critical volatiles are called character impact compounds and plant products can be divided into the following four general groups based in part on the presence or absence of a character impact compound[67]:

1. Those whose aroma is composed primarily of one character impact compound.
2. Those whose aroma is due to a mixture of a small number of compounds, of which one may be a character impact compound.
3. Those whose aroma is due to a large number of compounds, none of which are character impact compounds and with careful combination of these components the odor can be reproduced.
4. Those whose aroma is made up of a complex mixture of compounds that cannot be reproduced.

Table 4-15. Volatile Compounds from Apple Fruit

Acids		
Formic	n-Pentenoic	n-Octanoic
Acetic	n-Hexanoic	Octenoic
n-Propionic	i-Hexanoic (from esters)	n-Nonanoic
n-Butyric	*trans*-2-Hexenoic	Nonenoic
i-Butyric	n-Heptanoic	n-Decanoic
n-Pentanoic	Heptenoic	Decenoic, etc.
i-Pentanoic	Benzoic	
Alcohols		
Methanol	2-Methylbutan-2-ol	n-Heptanol
Ethanol	2-Pentanol	2-Heptanol
n-Propanol	3-Pentanol	n-Octanol
i-Propanol	n-Hexanol	2-Octanol
n-Butanol	2-Hexanol	n-Nonanol

Table 4-15. (*Continued*)

i-Butanol	2-Methylpentan-2-ol	2-Nonanol
2-Butanol	*trans*-n-Hex-2-en-1-ol	n-Decanol, etc.
Pentanol	*cis*-n-Hex-3-en-1-ol	Geraniol
i-Pentanol	*trans*-n-Hex-3-en-1-ol	
2-Methylbutan-1-ol	n-Hex-1-en-3-ol	

Esters

Methyl formate	1-Decyl acetate	n-Pentyl 2-methylbutyrate
Ethyl formate	Methyl propionate	Methyl i-pentanoate
Propyl formate	Ethyl propionate	i-Pentyl i-pentanoate
i-Propyl formate	Propyl propionate	Methyl n-pentanoate
n-Butyl formate	n-Butyl propionate	Ethyl n-pentanoate
n-Pentyl formate	i-Butyl propionate	Propyl n-pentanoate
i-Pentyl formate	n-Hexyl propionate	n-Butyl n-pentanoate
Methyl acetate	Ethyl crotonate	Methyl n-hexanoate
Ethyl acetate	Methyl i-butyrate	Ethyl n-hexanoate
Propyl acetate	Ethyl i-butyrate	n-Butyl n-hexanoate
i-Propyl acetate	i-Butyl i-butyrate	n-Pentyl n-hexanoate
n-Butyl acetate	Pentyl i-butyrate	n-Hexyl n-hexanoate
i-Butyl acetate	Methyl n-butyrate	Ethyl phenacetate
t-Butyl acetate	Ethyl n-butyrate	Ethyl octanoate
n-Pentyl acetate	Propyl n-butyrate	n-Butyl octanoate
i-Pentyl acetate	i-Propyl n-butyrate	i-Butyl octanoate
2-Methyl-1-butyl acetate	n-Butyl n-butyrate	n-Pentyl octanoate
n-Hexyl acetate	i-Butyl n-butyrate	n-Hexyl octanoate
trans-2-Hexen-1-yl acetate	n-Pentyl n-butyrate	Ethyl nonanoate
cis-3-Hexen-1-yl acetate	i-Pentyl n-butyrate	Ethyl decanoate
Benzyl acetate	n-Hexyl n-butyrate	n-Butyl decanoate
2-Phenethyl acetate	Methyl 2-methylbutyrate	n-Pentyl decanoate
1-Octyl acetate	Ethyl 2-methylbutyrate	Ethyl dodecanoate
1-Nonyl acetate	Propyl 2-methylbutyrate	

Carbonyls

Formaldehyde	i-Pentanal	Furfural
Acetaldehyde	2-Methylbutanal	2-Hexanone
Propanal	2-Pentanone	2-Heptanone
Acetone	3-Pentanone	3-Heptanone
Butanal	Hexanal	4-Heptanone
i-Butanal	2-Hexenal	Acetophenone
2-Butanone	*cis*-3-Hexenal	Nonanal
2,3-Butanedione	Heptanal	7-Methyl-4-octanone
Pentanal	*trans*-2-Heptenal	

Ethers and acetals

1-Ethoxy-1-methoxyethane	1-Butoxy-1-ethoxyethane	1,1-Diethoxypentane
1,1-Diethoxyethane	1-Ethoxy-1-(2-methylbutoxy)ethane	2,4,5-Trimethyl-1,3-dioxolane
1-Ethoxy-1-propoxyethane	1-Ethoxy-1-hexoxyethane	
1,1-Diethoxypropane		

Hydrocarbons

Ethylene	Farnesene	2-Methylnaphthalene
Ethane	1-Methylnaphthalene	

Note: Data from Nursten.[67]

Table 4-16. Character Impact Compounds Responsible for the Characteristic Aroma of Selected Fruits and Vegetables

Fruits

Apple	2-Methylbutyrate, hexanal, 2-hexenal
Banana	Amyl esters, isoamyl acetate
Blueberries	*trans*-2-Hexenal, *trans*-2-hexenol, linalool
Grape	Methyl anthranilate, ethylacetate
Grapefruit	Limonene, terpenes, oxygenated terpenes
Lemon	Limonene, citral
Lime	Limonene, citral
Pear	Decadienoate esters
Orange	Sesquiterpene hydrocarbons, limonene
Raspberries	1-(p-Hydroxyphenyl)-3-butanone

Vegetables

Cabbage	Isothiocyanates
Celery	Dihydrophthalides, alkyliden phthalides, *cis*-3-hexen-1-yl pyruvate, diacetyl
Cucumber	2,6-Nonadienal
Mushrooms	1-Octen-3-ol, lenthionine
Onion	Disulfides, trisulfides, alkyl thiosulfonates
Parsley	Monoterpene hydrocarbons
Potato	2-Methoxy-3-ethyl pyrazine, 2,5-dimethyl pyrazine
Radish	4-Methylthio-*trans*-3-butenyl isothiocyanate
Red bean	Oct-1-ene-3-ol, hex-*cis*-3-enol
Soybean	Ethyl vinyl ketone

Note: Data from Salanke and Do.[80]

Examples of character impact compounds (CIC) are 2-methyl-3-ethylpyrazine in raw white potato tubers and ethylvinylketone in soybeans (*Glycine max,* (L.) Merrill) (table 4-16). Examples of foods whose aroma is made up of a small number of compounds are boiled white potatoes (2-methoxy-3-ethyl pyrazine (CIC) and methional), apple (2-methylbutyrate (CIC), hexanal and trans-2-hexenal), and boiled cabbage (*Brassica oleracea,* L. Capitata group) (dimethyl disulfide (CIC) and 2-propenylisothiocyanate).

Synthesis and Degradation

Due to the extremely wide range in types of compounds important in the aroma of harvested products, their biosynthesis is not discussed in detail. There are, however, several generalizations that can be made.

Volatile compounds that are important in the aroma of postharvest products are formed by one of three general means. Many are formed naturally by enzymes found within intact tissue. These would include nearly all of the odors from fresh fruits, vegetables, and flowers. While biosynthesis of many of these compounds has not been studied in detail, three major pathways are known to be important (fig. 4-35). These are the isoprenoid pathway, the shikimic acid pathway, and β-oxidation. The isoprenoid pathway contributes many of the terpenes (e.g., limonene), while the shikimic pathway provides

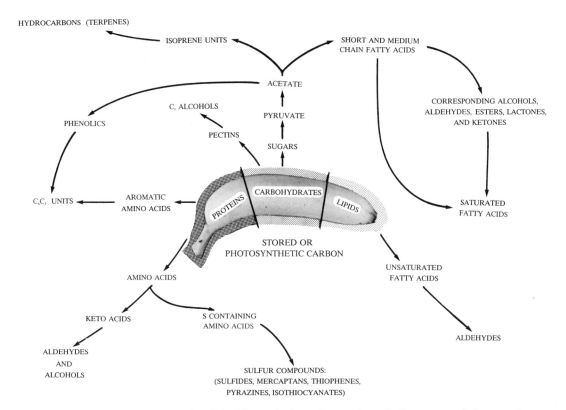

HYDROCARBONS (TERPENES)

ISOPRENE UNITS

SHORT AND MEDIUM
CHAIN FATTY ACIDS

ACETATE

C₁ ALCOHOLS

PYRUVATE

PHENOLICS

CORRESPONDING ALCOHOLS,
ALDEHYDES, ESTERS, LACTONES,
AND KETONES

PECTINS

SUGARS

C₆C₃ UNITS

AROMATIC
AMINO ACIDS

CARBOHYDRATES

PROTEINS   LIPIDS

SATURATED
FATTY ACIDS

STORED OR
PHOTOSYNTHETIC CARBON

UNSATURATED
FATTY ACIDS

AMINO ACIDS

KETO ACIDS

S CONTAINING
AMINO ACIDS

ALDEHYDES

ALDEHYDES
AND
ALCOHOLS

SULFUR COMPOUNDS:
(SULFIDES, MERCAPTANS, THIOPHENES,
PYRAZINES, ISOTHIOCYANATES)

Figure 4-35. A general schematic of the biosynthetic pathways for volatile aromas of plant products.

benzyl alcohol, benzaldehyde, and many of the volatile phenolic compounds. Beta-oxidation represents an important pathway for the production of volatiles throughout the oxidation of fatty acids. In table 4-17, a list of potential oxidation products for three fatty acids—oleic, linoleic, and linolenic—is given.

A second group of volatiles is produced enzymatically after damage to the tissue. Examples would be part of the aroma of cucumbers (*cis*-3-nonenal and hexanal) formed during disruption of the intact cells[27] or the formation of methyl and propyl disulfides in onions (*Allium cepa*, L.). Cellular disruption allows enzymes and substrates, previously sequestered separately within the cells, to interact. The production of the aroma of onion is perhaps the most thoroughly studied example of this type of volatile flavor production. Here the amino acids s-methyl-L-cysteine sulfoxide and s-n-propyl-L-cysteine sulfoxide are enzymatically degraded, forming the characteristic onion volatiles. Other amino acids may also represent precursors for volatiles, and as mentioned in the volatile production of undisturbed tissue, β-oxidation of fatty acids is also important in disrupted cells.

The third general means of flavor synthesis is through direct chemical reaction. This normally occurs with heating during processing or cooking. Since cooking alters the live product to a processed state, these volatiles are of less interest to postharvest physiologists. They can be important, however, when a particular postharvest handling practice alters the eventual flavor of a processed product.

Table 4-17. Oxidation Products of Three Unsaturated Fatty Acids

Oleic	Propanal
	Pentanal
	Hexanal
	Heptanal
	Nonanal
	2-Octenal
	2-Nonenal
	2-Decenal
Linoleic	Acetaldehyde
	Propanal
	Pentenal
	Hexanal
	2-Propenal
	2-Pentenal
	2-Hexenal
	2-Heptenal
	2-Octanal
	2-Nonenal
	2-Decenal
	Non-2,4-dienal
	Dec-2,4-dienal
	Undec-2,4-dienal
	Oct-1-en-3-ol
	2-Heptenal
Linolenic	Acetaldehyde
	Propanal
	Butanal
	2-Butenal
	2-Pentenal
	2-Hexenal
	2-Heptenal
	2-Nonenal
	Hex-1,6-dienal
	Hept-2,4-dienal
	Non-2,4-dienal
	Methyl ethyl ketone

Note: Data from Hoffman.[42]

In contrast to the biosynthesis of volatile compounds by plants, there has been much less interest in their degradation, due largely to the fate of the molecules once formed. Being volatile, most of these compounds simply dissipate into the atmosphere, eventually being degraded by biological, chemical, or photochemical reactions.

Postharvest Alterations

The volatiles produced by harvested products can be altered by a wide range of preharvest and postharvest factors. These include cultivar, maturity, season, production practices (e.g., nutrition), handling, storage, artificial ripening, and eventual method of preparation. Due to the importance of volatiles in the flavor quality of food crops, and aesthetic appeal of many ornamentals, care must be taken during the postharvest period to minimize undesirable changes.

Early harvest is known to have detrimental effects on the synthesis of the volatile constituents of many fruits. In the tomato, the production of volatiles increases with the development of the fruit and early harvest (breaker stage) with forced ripening (22–20°C) does not yield the same volatile profile as vine-ripened fruits (fig. 4-36). The concentrations of nonanal, decanal, dodecanal, neral, benzaldehyde, citronellyl propionate, citronellyl butyrate, geranyl acetate, and geranyl butyrate are higher in field-ripened than artificially ripened fruits. As tomatoes develop from a ripe to an overripe state, the concentrations of 2,3-butanedione, isopentyl butyrate, citronellyl butyrate, and geranyl butyrate increase while the concentrations of alcohols, aldehydes, acetates, and propionates generally tend to decrease. In nonclimacteric fruits (e.g., oranges) that do not ripen normally if picked at a preripe stage of development, the undesirable effect of early harvest on the flavor volatiles is even more pronounced.

Storage conditions and duration may also have a significant effect on the synthesis of volatiles after removal from storage. Apples stored under controlled atmosphere conditions (2% oxygen, 3.5°C) have been shown to have abnormal production of the critical flavor esters, butyl acetate and hexyl acetate, upon removal from storage. Similar findings have been reported for hypobaric storage of fruits and the exposure of some fruits to chilling injury.

Presently there is a very limited amount of information on the relationship between postharvest conditions and the beneficial or detrimental changes in the aroma of plant products. Much of the research to date has focused on determining the volatiles present and identifying specific character impact compounds. As more is learned about this important facet of postharvest biology, it will be possible to better control and perhaps even improve the aroma of many products.

PHENOLICS

Plant phenolics encompass a wide range of substances that have an aromatic ring with at least one hydroxyl group (table 4-18). Included are derivatives of these aromatic hydroxyl compounds due to substitutions, for example, the presence of O-methylation instead of hydroxyls on methylleugenol. In this group, common phenolics are the flavonoids, lignin, the hormone abscisic acid, the amino acids tyrosine and dihydroxyphenylalanine (DOPA), coenzyme Q, and numerous end products of metabolism. Phenolics represent one of the most abundant groups of compounds found in nature and are of particular interest in postharvest physiology because of their role in color and flavor. The concentration of phenolics varies widely in postharvest products. For example, in ripe fruits it ranges from very slight to up to 8.5% [persimmon (*Diospyros kaki,* L.f.)] of the dry weight (table 4-19).

Plant phenolics are generally reactive acidic substances that rapidly form hydrogen bonds with other molecules. Often they will interact with the peptide bonds of proteins and when the protein is an enzyme this interaction generally results in inactivation, a problem commonly encountered in the

Figure 4-36. Effect of stage of maturity and artificial ripening on tomato fruit volatiles (*after Shah et al.*[85]).

Peak[1]	Compound	Concentration (ppm in the fruit)		
		Artificially Ripened	Field Ripened	Overripe Red
2	2-Propanol	0.45	0.93	0.24
4	3-Methyl butanal	7.16	2.22	2.38
6	Propyl acetate	1.57	0.58	0.79
9	1-Hexanal	5.93	2.51	2.68
10	1-Butanol	1.12	0.35	0.76
11	2-Methyl-1-butanol	1.57	0.67	0.91
13	3-Pentanol	1.57	1.17	0.61
14	2-Methyl-3-hexol	7.94	2.51	2.68
17	3-Hexen-1-ol	0.78	1.28	0.79
18	Isopentyl tyrate	2.80	0.82	3.17
19	Isopentyl isovalerate	0.11	0.23	0.18
21	1-Nonanal	0.45	1.46	0.92
25	Benzaldehyde	0.22	1.17	1.40
26	1-Decanal	0.67	1.05	0.67
30	2-3-Butanedione	1.68	0.46	2.56
34	Citral b	1.56	5.95	3.66
35	1-Dodecanal	2.24	7.71	2.68
56	Linalyl acetate	2.80	2.28	1.22
57	Citronellyl propionate	5.82	17.87	7.14
58	Citronellyl butyrate	3.80	8.76	19.03
59	Geranyl acetate	2.78	2.92	2.38
60	Geranyl butyrate	1.68	2.80	4.02

[1]Peak numbers correspond to chromatogram.

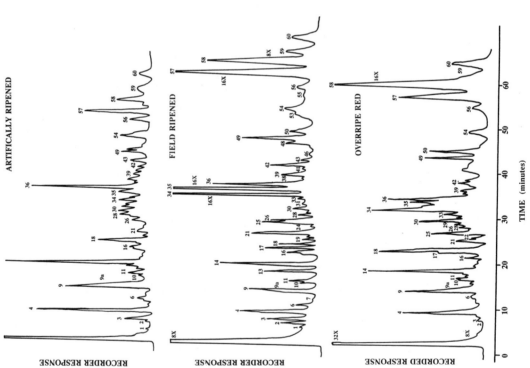

Table 4-18. Major Classes of Phenolics in Plants

Number of Carbon Atoms	Basic Skeleton	Class	Examples
6	C_6	Simple phenols	Catechol, hydroquinone
		Benzoquinones	2,6-Dimethoxybenzoquinone
7	$C_6 - C_1$	Phenolic acids	p-Hydroxybenzoic, salicylic
8	$C_6 - C_2$	Acetophenones	3-Acetyl-6-methoxybenzaldehyde
		Phenylacetic acids	p-Hydroxyphenylacetic
9	$C_6 - C_3$	Hydroxycinnamic acids	Caffeic, ferulic
		Phenylpropenes	Myristicin, eugenol
		Coumarins	Umbelliferone, aesculetin
		Isocoumarins	Bergenin
		Chromones	Eugenin
10	$C_6 - C_4$	Naphthoquinones	Juglone, plumbagin
13	$C_6 - C_1 - C_6$	Xanthones	Mangiferin
14	$C_6 - C_2 - C_6$	Stilbenes	Lunularic acid
		Anthraquinones	Emodin
15	$C_6 - C_3 - C_6$	Flavonoids	Quercetin, cyanidin
		Isoflavonoids	Genistein
18	$(C_6 - C_3)_2$	Lignans	Pinoresinol
		Neolignans	Eusiderin
30	$(C_6 - C_3 - C_6)_2$	Biflavonoids	Amentoflavone
n	$(C_6 - C_3)_n$	Lignins	
	$(C_6)_n$	Catechol melanins	
	$(C_6 - C_3 - C_6)_n$	Flavolans (condensed tannins)	

Source: After Harborne.[36]

study of plant enzymes. As a group, the phenols are susceptible to oxidation by the phenolases, which convert monophenols to diphenols and subsequently to quinones. In addition, some phenols are capable of chelating metals.

Phenolic compounds rarely occur in a free state within the cell; rather they are commonly conjugated with other molecules. Many exist as glycosides linked to monosaccharides or disaccharides. This situation is especially true of the flavonoids, which are normally glycosylated. In addition, phenols may be conjugated to a number of other types of compounds. For example, hydroxycinnamic acid may be found esterified to organic acids, amino groups, lipids, terpenoids, phenolics, and other groups, in addition to sugars. Within the cell, this state serves to render monophenols and diphenols less phytotoxic than when in the free state.

The general biological role of some phenolics in plants is readily apparent (e.g., pigments, abscisic acid, lignin, coenzyme Q) and some have been implicated as allelopathic agents, feeding deterrents, antifungal agents, and phytoalexins. However, for the majority of the phenolics in plants, their precise role remains in question.

Phenolics are commonly divided into three classes based on the number of phenol rings present. The simplest class includes the monocyclic phenols composed of a single phenolic ring. Common examples found in plants are

Table 4-19. Total Phenolic Content in Ripe Fruits

	Phenolic Content
Apple (*Malus sylvestris*, Mill.)	
Various cultivars	0.10–1.0 g $\cdot$ 100 g^{-1} fr. wt.
'Cox's Orange Pippin'	2.0–5.5 g $\cdot$ 100 g^{-1} dr. wt.
'Baldwin'	0.25 g $\cdot$ 100 g^{-1} fr. wt.
Cider apple 'Launette'	1.1 g $\cdot$ 100 g^{-1} fr. wt.
Cider apple 'Waldhofler'	0.45 g $\cdot$ 100 g^{-1} fr. wt.
Banana (*Musa* spp.)	0.53 g $\cdot$ 100 g^{-1} dr. wt.
Date (*Phoenix dactylifera*, L.)	0.5 g $\cdot$ 100 g^{-1} fr. wt.
Cherry (*Prunus cerasus*, L.)	
Montmorency	0.5 g $\cdot$ 100 g^{-1} fr. wt.
Grape (*Vitis* spp.)	
Riesling, cluster	0.95 g $\cdot$ 100 g^{-1} fr. wt.
'Tokay', cluster	0.48 g $\cdot$ 100 g^{-1} fr. wt.
'Muscat', skin	0.35 g $\cdot$ 100 g^{-1} fr. wt.
'Muscat', pulp	0.10 g $\cdot$ 100 g^{-1} fr. wt.
'Muscat', seed	4.5 g $\cdot$ 100 g^{-1} fr. wt.
Passion fruit (*Passiflora edulis*, Sims.)	1.4 mg $\cdot$ 100 g^{-1} fr. wt.
Peach (*Prunus persica*, (L.) Batsch.)	
Mixed cultivars	0.028–0.141 g $\cdot$ 100 g^{-1} fr. wt.
'Elberta'	0.069–0.180 g $\cdot$ 100 g^{-1} fr. wt.
'Elberta'	0.240 g $\cdot$ 100 g^{-1} fr. wt.
Pear (*Pyrus communis*, L.)	
'Muscachet'	0.4 g $\cdot$ 100 g^{-1} fr. wt.
Persimmon (*Diospyros kaki*, L.f.)	8.5 g $\cdot$ 100 g^{-1} dry wt.
Plum (*Prunus americana*, Marsh.)	
'Victoria', flesh	2.1 g $\cdot$ 100 g^{-1} dry wt.
'Victoria', skin	5.7 g $\cdot$ 100 g^{-1} dry wt.

Note: Data from van Buren.[95]

phenol, catechol, hydroquinone, and p-hydroxycinnamic acid. Dicyclic phenols such as the flavonoids have two phenol rings while the remainder tend to be lumped into the polycyclic or polyphenol class. The structures of several common phenols of each class are illustrated in figure 4-37. These general classes can be further divided into subclasses based on the number of carbon atoms and the pattern of the basic carbon skeleton of the molecule (table 4-18).

Biosynthesis of Phenols

Nearly all of the phenols are formed initially from phosphoenolpyruvate and erythrose 4-phosphate through shikimate in the shikimic acid pathway (fig. 4-38). The aromatic amino acid phenylalanine is a central intermediate that is deaminated and hydroxylated in the para position on the phenol ring, yielding p-hydroxycinnamic acid. As mentioned earlier in the section on pigments, malonate is essential in formation of the flavonoids. Three molecules of malonate in the form of malonyl-CoA combine with cinnamic acid (cinnamyl-

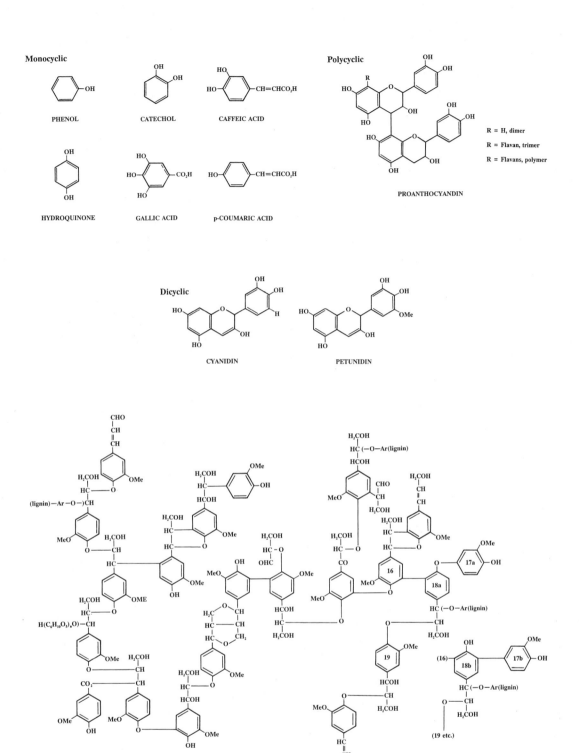

Figure 4-37. Structure of several common plant phenolics and the proposed structure of lignin *(the latter after Adler et al.[1]).*

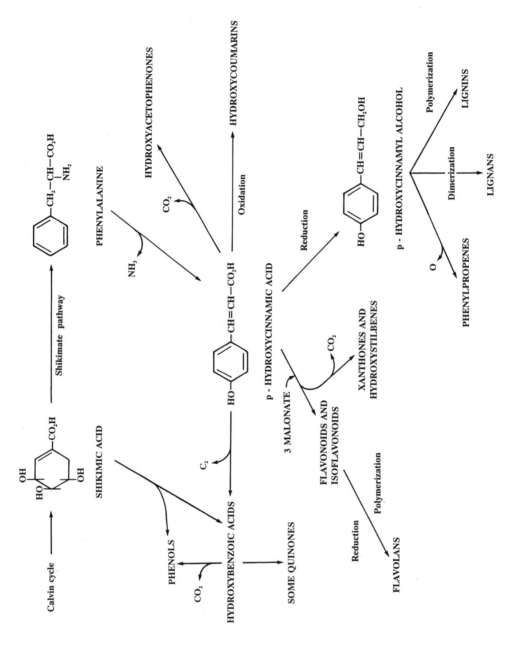

Figure 4-38. The biosynthesis of plant phenolics from shikimic acid and phenylalanine (*after Harborne*[36]).

CoA) to form a chalcone, which with ring closure gives the base structure for the flavonoids.

A number of physiological factors affect phenolic synthesis. These include light, temperature, and cellular carbohydrate, mineral, and water status. The response varies with the specific phenol and species in question. For example, the synthesis of anthocyanin is favored by low temperature while the synthesis of proanthocyanidin is repressed.

Decomposition of Phenols

For years many of the phenols were considered end products of metabolism, which was due in part to the inability to ascribe specific roles for these compounds within the cell. From isotope labeling studies, it is now known that phenols do not simply accumulate unchanged in the plant, but in fact are in a constant state of flux, undergoing synthesis, turnover, and degradation. The rate of turnover varies widely, from hours in certain floral pigments to, in some cases, weeks. The actual sequence of catabolism varies with the diverse number of types of phenolic compounds. One common feature, however, is eventual ring cleavage and opening. Higher plants have been shown to possess active aromatic ring-cleaving enzymes. With the simpler phenolics, β-oxidation appears to play a major role.

During the postharvest period of some plant products, changes in specific phenols are extremely important. They often entail the polymerization of the molecule into large, more complex substances.

Postharvest Alterations

Plant phenolics are of particular importance during the postharvest period due to their role in both flavor and color. While the majority of phenolics at the concentrations in which they are found in food products have no significant taste, several do substantially alter the flavor of certain products. The phenolic acids, when present in a sufficiently high concentration, are distinctly sour and some of the flavonoids in citrus (e.g., naringin) are bitter. In addition, phenolics in the immature fruits of several species are highly astringent. For example, immature fruits of banana (*Musa sapientum,* L.) contain 0.6% water-soluble tannins on a fresh-weight basis, the fruits of Chinese quince (*Cydonia sinensis,* Thouin)* contain 0.4%, carob beans (*Ceratonia siliqua,* L.) contain 1.7%, and persimmon fruits contain 2.0%.[60] These compounds are also found in other plant parts such as leaves, bark, and seed coats of some species. In the persimmon, this water-soluble tannin is composed of catechin, catechin-3-gallate, gallocatechin, gallocatechin-3-gallate, and an unknown terminal residue.[59] It is thought to be in the form of proanthocyanidin bound to similar neighboring subunits through carbon 4 of one unit to carbon 6 or 8 of another, forming a very large complex molecule. The protein-binding capacity of the tannin is so great that it is used to remove the protein from Japanese rice wine (sake), clarifying it.

---

* Syn. with *Chaenomeles sinensis,* Koehne.

During maturation and fruit ripening in the persimmon, the level of astringency and the concentration of water-soluble tannin diminishes, eventually giving fruit with an excellent flavor. This decrease in astringency is due to polymerization of the existing tannins, forming larger, water-insoluble molecules no longer capable of reacting with the taste receptors in the mouth. It has been known since the early part of this century that the astringency removal process could be substantially shortened by exposing the fruit to a high carbon dioxide atmosphere.[29] The rate of the induction response is temperature dependent; at 40°C the exposure may be as short as 6 hours. This initial induction process is followed by a period of astringency removal that occurs during a 3-day air storage period (30°C). It is thought that the anaerobic conditions, which block glycolysis, result in a small but significant buildup in acetaldehyde. Aldehydes react readily with phenols, resulting in cross-linking between neighboring molecules, hence producing the insoluble nonastringent form. This same basic aldehyde-phenol reaction is used in the production of Bakelite plastics, and when discovered in the early 1900s, it ushered in the plastics revolution.

Another important postharvest phenolic response is the discoloration (browning) of many products upon injury to the tissue. Injury may occur during harvest, handling, or storage, resulting in the breakdown of organizational resistance between substrates and enzymes within the cell. Common examples would be bruising of fruits or broken-end discoloration of snap beans. When browning occurs, constituent phenols are oxidized to produce a quinone or quinonelike compound that polymerizes, forming brown pigments. These unsaturated brown polymers are generally referred to as melanins or melaninodins. Among the compounds believed to be important as substrates are chlorogenic acid, neochlorogenic acid, catechol, tyrosine, caffeic acid, phenylalanine, protocatechin, and dopamine.

Research on phenolic discoloration has focused on the cellular constituents responsible for browning and handling and storage techniques that can be used to prevent browning. The success of attempts to correlate the concentration of total and specific phenols or the level of activity of the phenolase enzymes with discoloration varies widely depending on the tissue in question, and in some cases, depending on the cultivar studied. Several postharvest techniques, however, have been successful for inhibiting the browning of some products. For example, during the harvest of snap beans it is not uncommon for the ends of many of the bean pods to be broken. These brown fairly readily and upon processing yield a distinctly substandard product. The reactions involved in discoloration can be inhibited by exposing the beans to 7500–10,000 $\mu l \cdot l^{-1}$ sulfur dioxide for 30 seconds.[38] The use of controlled atmosphere storage and antioxidants has been successful with other products.

VITAMINS

Vitamins represent a group of organic compounds that are required in the diet in relatively small amounts for normal metabolism and growth. Plant products provide a major source for many of the vitamins required by humans. Exceptions would be vitamin B_{12}, which appears to be synthesized only by microorganisms, and vitamin D, obtained from the exposure of skin to ultra-violet irradiation. In plants, many of the vitamins perform the same biochemical function as they do in animal cells. As a consequence, most have a vital role in plant metabolism, in addition to being a source of vitamins for animals.

Although vitamins are required in only small amounts in the diet, deficiencies have been a serious problem throughout the history of man. Even today deficiencies of certain vitamins result in major nutritional diseases in many areas of the world. Insufficient vitamin A in the diet is perhaps the most common, especially in parts of Asia. Prolonged deficiency, particularly critical in young children, results in blindness.

Typically, vitamins are separated into two classes based on their solubility: the water-soluble vitamins (thiamine, riboflavin, nicotinic acid, pantothenic acid, pyridoxine, biotin, folic acid, and ascorbic acid), and those that are lipid-soluble (vitamins A, E, and K). Normally the lipid-soluble vitamins are stored in the body in moderate amounts; as a consequence, a consistent daily intake is not essential. The water-soluble vitamins, however, tend not to be stored and a fairly constant day-to-day supply is required.

Vitamins are needed in only small amounts in that most function in a catalytic capacity as coenzymes—organic compounds that participate in the function of an enzyme. The role of the water-soluble vitamins (principally coenzymes) is for the most part fairly well understood, whereas the precise metabolic function of the lipid-soluble vitamins is not yet clear.

Water-Soluble Vitamins

Thiamine

Structurally thiamine or vitamin B_1 is a substituted pyrimidine joined to a substituted thiazole group through a methylene bridge (fig. 4-39). It is water soluble and is relatively stable at pHs below 7.0. The vitamin is found widely in both plants and animals, in one of several forms. In plant tissues it is found most abundantly as free thiamine; however, it may also be found as mono-, di-, and triphosphoric esters and as mono- and disulfides in various biological materials. Thiamine functions in plants as the coenzyme thiamine pyrophosphate, which plays a major role in the glycolytic pathway (the decarboxylation of pyruvic acid), the tricarboxylic acid cycle (the decarboxylation of α-ketoglutaric acid), and the pentose phosphate pathway (as a carboxyl group transferring enzyme).

In contrast to some vitamins such as ascorbic acid, the concentration of thiamine in various plants and plant parts is fairly uniform, generally varying only within a 20-fold to 40-fold range. Typically, dried beans and peas contain approximately 700–600 $\mu g \cdot 100 \ g^{-1}$; nuts, 500–600 $\mu g \cdot 100 \ g^{-1}$; whole grain cereals, 300–400 $\mu g \cdot 100 \ g^{-1}$; and fruits and vegetables, 20–90 $\mu g \cdot 100 \ g^{-1}$. In addition to variation between species, the absolute concentration of thiamine in edible plant products varies somewhat with cultivar and growing conditions. After harvest, the vitamin is relatively stable during storage. Primary losses incurred are most commonly the result of cooking, due largely to the high water solubility of the molecules.

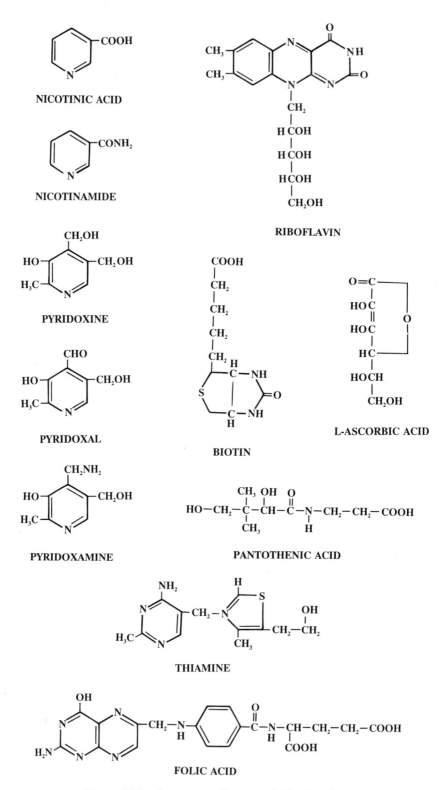

Figure 4-39. Structures of water-soluble vitamins.

Riboflavin

Riboflavin is a water-soluble derivative of D-ribose that contains an isoalloxazine ring (6,7-dimethyl-9-(D-1′-ribityl) isoalloxazine) (fig. 4-39). It is also known as vitamin B_2 and lactoflavin. Riboflavin or derivatives of riboflavin are found in all plants and many microorganisms; however, it is not synthesized by higher animals. Within plants it is found combined with other groups, largely as either flavin mononucleotide (FMN) or flavin adenine dinucleotide (FAD). This is also the case in most parts of the body of animals, although in the retina of the eye it is found in its free form. Flavin nucleotides act as prosthetic groups on oxidation-reduction enzymes. The isoalloxazine ring portion of the flavin nucleotide undergoes a reversible reduction to yield the reduced nucleotides $FMNH_2$ and $FADH_2$. Enzymes containing flavin nucleotides are essential for the oxidation of pyruvate and fatty acids and function in the electron transport system. Typically, the oxidized form of the molecule is colored yellow, red, or green, and the reduced form is colorless.

Leafy vegetables represent a relatively good source of riboflavin, although vegetables that are high in riboflavin (e.g., pimento pepper (*Capsicum annuum*, L. Grossum group), 0.46 mg·100 g^{-1} fresh weight; mushrooms, 0.30 mg·100 g^{-1}; lotus root (*Nelumbo nucifera*, Gaertn.), 0.22 mg·100 g^{-1}; salsify (*Tragopogon porrifolius*, L.), 0.22 mg·100 g^{-1}) are consumed in only small amounts in most diets.

Niacin

Niacin or nicotinic acid is found widely in both plants and animals either as the acid or amide (fig. 4-39). The name *nicotinic acid* comes from the role of the molecule as a component of the toxic alkaloid, nicotine, from tobacco. Nicotinic acid can be synthesized readily by animals if their diet contains sufficient protein, which is high in the amino acid tryptophan, the primary precursor of the vitamin. The coenzymes contain nicotinamide as an essential component. These are nicotinamide adenine dinucleotide (NAD) and nicotinamide adenine dinucleotide phosphate (NADP), known also as pyridine coenzymes. Both function as coenzymes in a large number of oxidation-reduction reactions catalyzed by what are known as pyridine-linked dehydrogenases. Most of the dehydrogenase enzymes are specific for either NAD or NADP, although several can utilize either form. In general, these reactions are reversible and are extremely important in many pathways within the cell.

Pyridoxine

Pyridoxine or vitamin B_6 is found in three forms—pyridoxine, pyridoxal, and pyridoxamine (fig. 4-39)—pyridoxine typically being converted to the latter two forms, which are more efficacious. The active coenzyme forms of the vitamin are the phosphate derivatives: pyridoxal phosphate and pyridoxamine phosphate. Pyridoxine coenzymes function in a wide range of important reactions in amino acid metabolism such as transamination, decarboxylation, and racemization reactions. Pyridoxal phosphate is also thought to be involved in the biosynthesis of ethylene, acting at the point of conversion of S-adenosyl methionine (SAM) to 1-aminocyclopropane-1-carboxylic acid (ACC). Ethylene synthesis is blocked at this step by 2-amino-4-aminoethoxy-*trans*-3-butenoic acid (AVG), which is a potent inhibitor of pyridoxal phosphate-mediated enzyme reactions.

The three forms of the vitamin are found widely distributed in both the plant and animal kingdom; the predominant form, however, varies between sources. In vegetables, pyridoxal is the predominant form. Cereals (0.2–0.4 mg $\cdot$ 100 g^{-1}) and vegetables (e.g., brussels sprouts, 0.28 mg $\cdot$ 100 g^{-1}; cauliflower (*Brassica oleracea*, L. Botrytis group), 0.2 mg $\cdot$ 100 g^{-1}; lima beans (*Phaseolus lunatus*, L.), 0.17 mg $\cdot$ 100 g^{-1}; spinach (*Spinacia oleracea*, L.), 0.22 mg $\cdot$ 100 g^{-1}) represent good sources of the vitamin whereas many of the fruits are quite low (e.g., apple, 0.045 mg $\cdot$ 100 g^{-1}; orange, 0.05 mg $\cdot$ 100 g^{-1}).

Pantothenic Acid

Pantothenic acid, formed from pantoic acid and the amino acid β-alanine, is found in limited quantities in most fruits and vegetables (fig. 4-39). The active form of the vitamin, coenzyme A (CoA), is synthesized from pantothenic acid in a series of steps. Coenzyme A functions as a carrier of acyl groups in enzymatic reactions during the synthesis and oxidation of fatty acids, pyruvate oxidation, and a number of other acetylation reactions within the cell.

Deficiencies of the vitamin in animals are rare; a limited amount of storage of the molecule does occur in the heart, liver, and kidneys. In diets with sufficient animal protein, most of the pantothenic acid is derived from this source. Dried peas and peanuts are considered good sources of the vitamin; walnuts, broccoli, peas, spinach, and rice are intermediate sources (0.5–2.0 mg $\cdot$ 100 g^{-1}); and onions, cabbage, lettuce, white potatoes, sweetpotatoes (*Ipomoea batatas*, (L.) Lam.), and most fruits are poor sources (0.1–0.5 mg $\cdot$ 100 g^{-1}).

Biotin

The vitamin biotin consists of fused imidazole and thiophene rings with an aliphatic side chain (fig. 4-39). Its structure, established in 1942, suggested the possible role of pimelic acid as the natural precursor of the molecule, which was later proven to be correct. Biotin is found widely in nature, usually in combined forms bound covalently to a protein through a peptide bond. When bound to a specific enzyme, it functions in carboxylation reactions. Here it acts as an intermediate in the transfer of a carboxyl group from either a donor molecule or carbon dioxide to an acceptor molecule. Examples of enzymes for which biotin acts as a carboxyl carrier are propionyl-CoA carboxylase and acetyl-CoA carboxylase.

Biotin is found widely distributed in foods and is also synthesized by bacteria in the intestine. As a consequence, deficiencies are extremely rare. When present they are normally associated with high intake of avidin, a protein found in raw egg whites that binds to the vitamin, making it unavailable. Legumes, especially soybeans (*Glycine max*, (L.) Merrill), represent an excellent plant source of biotin (61 μg·100 g^{-1} edible product). In addition, nuts such as peanuts (*Arachis hypogaea*, L.) (34 μg·100 g^{-1}), pecans (27 μg·100 g^{-1}), and walnuts (*Juglans regia*, L.) (37 μg·100 g^{-1}) and a number of vegetables are good sources (e.g., southern peas, 21 μg·100 g^{-1}; cauliflower, 17 μg·100 g^{-1}; mushrooms, 16 μg·100 g^{-1}), whereas most fruits and processed grains are consistently low in biotin.

Folic Acid

Folic acid is found widely distributed in plants, its name being derived from the Latin word *folium* for "leaf" from which it was first isolated. Structurally the molecule is composed of three basic subunits: (1) a substituted pteridine, (2) p-aminobenzoic acid, and (3) glutamic acid (fig. 4-39). The active coenzyme form of the vitamin is tetrahydrofolic acid, formed in a two-step reduction of the molecule. It functions as a carrier of one carbon units (e.g., hydroxymethyl-CH_2OH, methyl-CH_2, and formyl-CHO groups) when these groups are transferred from one molecule to another. These reactions are critical steps in the synthesis of purines, pyrimidines, and amino acids.

Folic acid is found widely in the plant kingdom and is also synthesized by microorganisms including intestinal bacteria. It is needed by humans and other animals in very small amounts (e.g., 0.4 mg · day^{-1} for humans) but is rapidly excreted from the body. Asparagus (*Asparagus officinalis*, L.), spinach, and dried beans are excellent sources of the vitamin; corn, snap beans, kale (*Brassica oleracea*, L. Acephala group), and many nuts are moderate sources (30–90 μg · 100 g^{-1}); and cabbage, carrots (*Daucus carota*, L.), rice (*Oryza sativa*, L.), cucumbers (*Cucumis sativus*, L.), white potatoes, sweetpotatoes, and most fruits are poor sources (0–30 μg · 100 g^{-1}).

Ascorbic Acid

Ascorbic acid is structurally one of the least complex vitamins found in plants. It is a lactone of a sugar acid (fig. 4-39) that is synthesized in plants from glucose or other simple carbohydrates. It was first isolated in crystalline form in 1923. In spite of years of research, the precise physiological function of ascorbic acid in plant and animal cells remains unclear. It is known to act as a cofactor in the hydroxylation of proline to hydroxyproline; however, other reducing agents can replace it.

Ascorbic acid is required in the human diet and only a small number of other vertebrates and is supplied primarily by fruits and vegetables, although a small amount is found in animal products such as milk, liver, and kidneys. In comparison with the other water-soluble vitamins in plants, ascorbic acid is found in relatively high concentrations. Guava (*Psidium guajava*, L.) (300 mg · 100 g^{-1} fresh weight), black currants (*Ribes nigrum*, L.) (210 mg · 100 g^{-1}), sweet peppers (125 mg · 100g^{-1}), and several greens (kale, collards (*Brassica oleracea*, L. Acephala group), turnips (*Brassica rapa*, L. Rapifera group), 120 mg · 100 g^{-1}) are excellent sources. The West Indian cherry is said to contain approximately 1300 mg · 100 g^{-1} fresh weight.[68]

Staples such as rice, wheat (*Triticum aestivum,* L.), corn, and many of the starchy tubers tend to be extremely low, however. Fruits have a distinct advantage in the diet in that they are often served raw. During cooking a significant portion of the ascorbic acid of many vegetables is lost. This loss is due primarily to leaching of the water-soluble vitamin out of the tissue and to oxidation of the molecule. Losses from leaching tend to be greater in leafy vegetables due to the surface area in contrast to bulkier products.

The concentration of ascorbic acid often varies with location within a specific plant part and between different parts on the same plant. For example, in many fruits the concentration in the skin is higher than in the pulp.

The concentration of ascorbic acid declines fairly rapidly in many of the more perishable fruits and vegetables after harvest. Losses are greater with increasing storage temperature and duration.

Lipid-Soluble Vitamins

Vitamin A

Vitamin A or retinol is an isoprenoid compound with a 6-carbon cyclic ring and an 11-carbon side chain (fig. 4-40). It is formed in the intestinal mucosa by cleavage of carotene. Of the numerous naturally occurring carotenoids, only ten have the potential to be converted into vitamin A and of these β-, α-, and γ-carotene are the most important. The presence of a β end group is essential for the formation of the molecule. Beta-carotene, which has two β end groups, has twice the potential vitamin A of α-carotene, which is composed of a β and ϵ end group. Cleavage appears to be due to the presence of the enzyme β-carotene-15,15′-dioxygenase, which oxidizes the central double bond; however, it is possible that other conversion mechanisms may be operative.

Vitamin A is extremely important in human nutrition in that its synthesis is dependent on carotene ultimately from plant sources. In contrast to ascorbic acid, only a small amount of vitamin A is needed in the diet. This amount ranges from 0.4–1.2 mg $\cdot$ day^{-1} depending on age and sex. Although it appears to be required in all of the tissues of the body, its general function in metabolism is not known, aside from its role in eyesight. A deficiency in young children results in permanent blindness, a common problem in many tropical areas of the world.

Since vitamin A per se is not present in plants, its potential concentration is measured in international units (IU), based on the concentration of α- and β-carotene in the tissue. One IU of vitamin A is equal to 0.6 μg of β-carotene or 1.2 μg of α-carotene. Leafy vegetables average approximately 5,000 IU $\cdot$ 100 g^{-1} fresh weight; and fruits are typically 100–500 IU $\cdot$ 100 g^{-1}, although the mango (*Mangifera indica,* L.) at 3,000 IU $\cdot$ 100 g^{-1} and papaya (*Carica papaya,* L.) at 2,500 IU $\cdot$ 100 g^{-1} are distinctly higher; while staple crops such as rice, peanuts, and cassava (*Manihot esculenta,* Crantz.) have virtually none. An exception would be the sweetpotato in which some of the high-carotene cultivars contain up to 14,000 IU of vitamin A per 100 g fresh weight.

The concentration of carotene is known to vary widely between species and cultivars, and less so due to production environmental conditions and

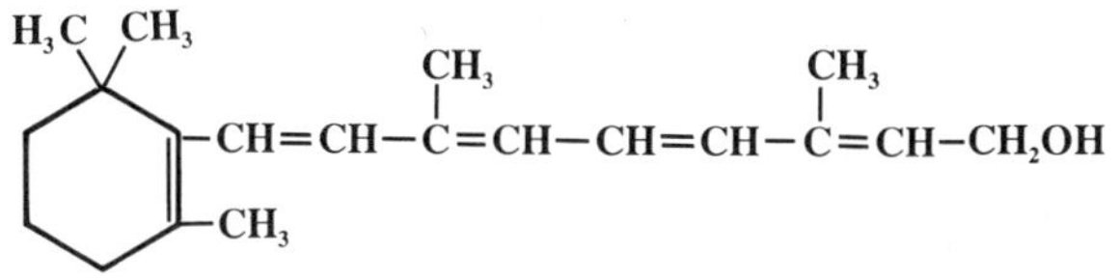

VITAMIN A

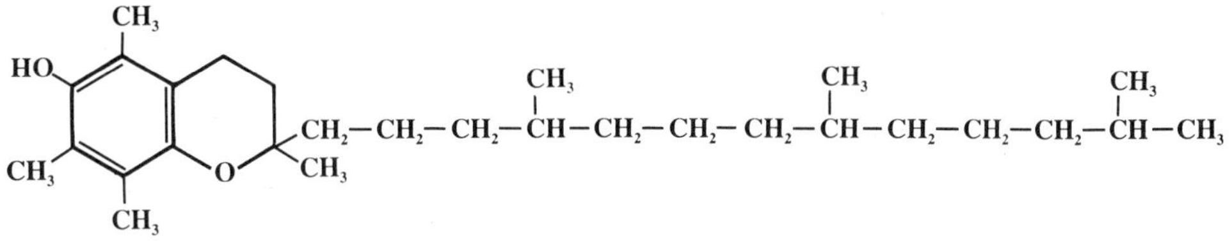

VITAMIN E

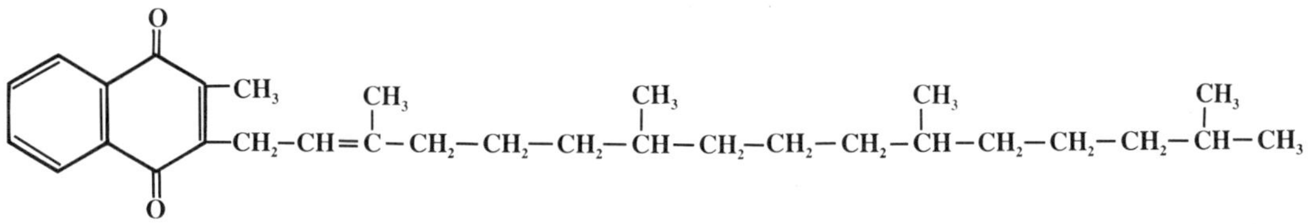

VITAMIN K

Figure 4-40. Structures of lipid-soluble vitamins.

cultural practices, although temperature and light are known to have significant effects.[19]

Vitamin E

Vitamin E or α-tocopherol is a molecule composed of a chromanol ring and a side chain formed from a phytol residue (fig. 4-40). In addition to α-tocopherol, β-, γ-, and δ-tocopherol are also found in photosynthetic plants, although α-tocopherol is the most active form as a vitamin.

The biological role of α-tocopherol in animals, as in plants, is unclear. Vitamin E deficiency results in a number of symptoms in test animals, one of which is infant mortality; hence the derivation of the name *tocopherol* from the Greek word *tokos* meaning "childbirth." Tocopherols are known to have antioxidant activity, which prevents the autooxidation of unsaturated lipids. As a consequence, one function may be the protection of membrane lipids.

In plants, α-tocopherol is found associated with the chloroplast membrane and is thought to also be present in mitochondria. It also appears to be located in various plastids (elaioplasts, aleuroplasts, amyloplasts, and chromoplasts).

Plant oils are excellent sources of tocopherols. Significant concentrations are found in wheat germ, corn, and pecan oils. Pecans contain up to 600 μg of tocopherol per gram of oil approximately 6 weeks prior to maturity. This amount declines to 100–200 μg $\cdot$ g^{-1} by maturation. In contrast to many other nuts (e.g., filberts (*Corylus avellana*, L.), walnuts, Brazil nuts (*Bertholletia excelsa*, Humb. & Bonpl.), almonds (*Prunus dulcis*, (Mill.) Webb.), chestnuts (*Castanea sativa*, Mill.), and peanuts), the γ-tocopherol isomer is found almost exclusively (>95%) in lieu of the α-tocopherol isomer.

The role of tocopherols in plants is thought to be related to their antioxidant properties. This is supported by the correlation between the concentration of tocopherols in pecans, which tend to deteriorate in storage due to the oxidation of their component lipids, and the length of time they can be successfully stored. In pecan oils with a constant linolate concentration, keeping time increased in a linear fashion up to 800 μg of tocopherol per gram of oil.[79] The germination of wheat seeds is also correlated with tocopherol content.[14] Tocopherols are also thought to function as a structural component of chloroplast membranes and may in some way function in the initiation of flowering of certain species.

Vitamin K

Vitamin K or phylloquinone is a lipid-soluble quinone that is in many ways structurally very similar to α-tocopherol. Both are cyclic compounds with a phytol residue side chain composed of isoprene units (fig. 4-40). Two forms of the vitamin, K$_1$ and K$_2$, are known. A deficiency of vitamin K impairs proper blood clotting in animals through the repressed formation of fibrin, the fibrous protein portion of blood clots. This in turn results in a tendency to hemorrhage. Aside from this specific function, its widespread occurrence in plants and microorganisms suggests a more general but presently undefined biological role. As a quinone, it could possibly function as an electron carrier.

In plants, phylloquinone is present in most photosynthetic cells; hence leafy green tissues represent an excellent dietary source of the vitamin. Plant parts that normally do not contain chlorophyll have little vitamin K. Likewise, mineral deficiencies that repress chlorophyll synthesis (e.g., Fe) also appear to decrease the concentration of vitamin K.

PHYTOHORMONES

Five groups of naturally occurring compounds, the phytohormones, are currently known to exist, each of which exhibits strong plant growth regulating properties. Included are ethylene, auxin, gibberellins, cytokinins, and abscisic acid (fig. 4-41); each is structurally distinct and active in very low concentrations within plants.

While each of the phytohormones has been implicated in a relatively diverse array of physiological roles in plants and detached plant parts, the precise mechanism in which they function is not yet known. During the postharvest period, ethylene is of major importance in that it is closely associated with the regulation of senescence in some products and the ripen-

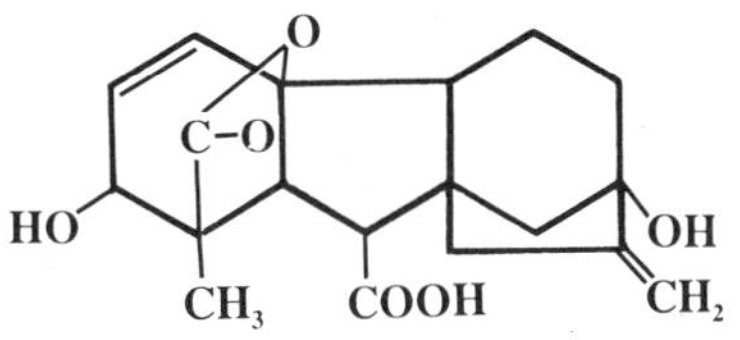

GIBBERELLIC ACID A₃

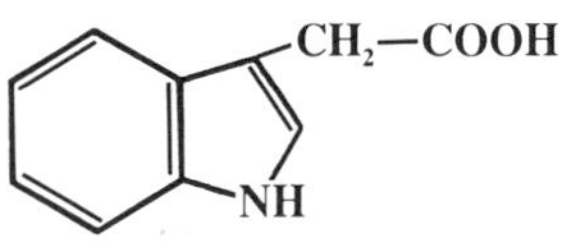

INDOLEACETIC ACID

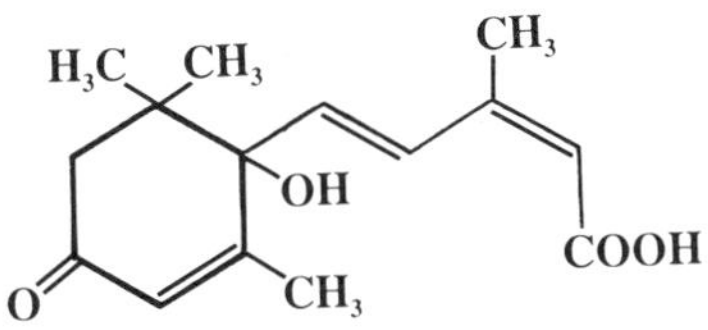

ABSCISIC ACID

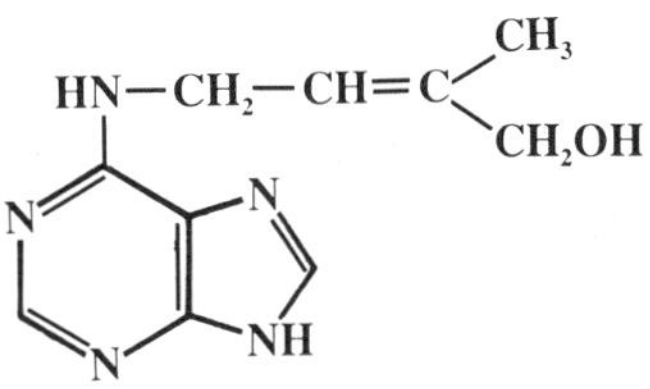

ZEATIN

$$CH_2{=}CH_2$$

ETHYLENE

Figure 4-41. Structures of plant hormones: ethylene, indoleacetic acid, abscisic acid, zeatin, and gibberellic acid A₃.

ing of many fruits. This section focuses primarily on the synthesis and deactivation of these molecules and changes in their concentration occurring during the postharvest period.

Classes, Synthesis, and Degradation of Phytohormones

Ethylene

Ethylene, being a gaseous hydrocarbon, is unlike the other naturally occurring plant hormones. Although ethylene was known to elicit such responses as geotropism and abscission early in this century, it was not until the 1960s that it began to be accepted as a plant hormone.

The effect of ethylene on plants and plant parts is known to vary widely. It has been implicated in ripening, abscission, senescence, dormancy, flowering, and other responses. Ethylene appears to be produced by essentially all living parts of higher plants, the rate of which varies with specific organ and tissue and their stage of growth and development. Rates of synthesis range from very low (0.04–0.05 $\mu l \cdot kg^{-1} \cdot hr^{-1}$) in blueberries (*Vaccinium* spp.) to extremely high (3,400 $\mu l \cdot kg^{-1} \cdot hr^{-1}$) in fading blossoms of *Vanda* orchids. Alterations in the rate of synthesis of ethylene have been found, in some cases, to be closely correlated with the development of certain physiological responses in plants and plant parts, for example, the ripening of climacteric fruits and the senescence of flowers.

Ethylene is synthesized from the sulfur-containing amino acid methionine that is first converted to s-adenosyl methionine (SAM) and then to the 4-carbon compound, 1-amino-cyclopropane-1-carboxylic acid (ACC) (fig. 4-42). During conversion to ACC, the sulfur-containing portion of the molecule, 5-methylthioadenosine, is cycled back to methionine via the formation of ribose and condensation with homoserine.

The final step in the synthesis pathway, the conversion of ACC to ethylene, is at present the least understood. The reaction requires the presence of oxygen and appears to represent the point where the ethylene synthesis pathway is inhibited by low oxygen conditions. The rate of synthesis of ethylene is known to be altered by a wide range of environmental factors. Oxygen concentration and temperature are two of the most important; when either are sufficiently low, synthesis is reduced. Stress (water, mechanical, and others) is known to stimulate ethylene synthesis, and under some conditions, markedly so.

Several potent inhibitors of ethylene synthesis have been found (rhizobitoxine and AVG, fig. 4-42) and were integral components in elucidating the pathway. Morris Lieberman first showed that fungal metabolites from *Rhizobium japonicum*, *Streptomyces* spp., and *Pseudomonas aeruginosa* inhibit the conversion of SAM to ACC. These unfortunately also inhibit other pyridoxal phosphate-requiring enzymes in plants and animals and as a consequence are of little commercial value for postharvest products that are to be consumed.

While ethylene appears to be synthesized in all cells, its precise site of synthesis within the cell is not yet known. Several lines of evidence point toward the enzyme being associated with the tonoplast. Vacuoles isolated from protoplasts were able to convert ACC to ethylene[32] and the enzyme exhibited stereospecificity[64] indicative of the ethylene-forming enzyme rather than a nonspecific conversion. Likewise, protoplasts that had their vacuoles removed (evacuolated) lost the capacity to produce ethylene from ACC; when the vacuoles were allowed to reform, synthesis was reinstated.[17]

Since ethylene is being continuously produced by plant cells, some mechanism is essential to prevent the buildup of the hormone within the tissue. Unlike other hormones, gaseous ethylene diffuses readily out of the plant. This passive emanation of ethylene from the plant appears to be the primary means of eliminating the hormone. During the postharvest period, techniques such as ventilation and hypobaric conditions help to facilitate this phenomenon by maintaining a high diffusion gradient between the interior of the product and the surrounding environment. A passive elimination system of this nature would imply that the internal concentration of ethylene is controlled largely by the rate of synthesis rather than the rate of removal of the hormone.

Ethylene may also be metabolized within the cell, decreasing the internal concentration. Products such as ethylene oxide and ethylene glycol have been found; however, their importance in regulating the internal concentration of ethylene in most species appears to be very minimal.

Auxin

The name *auxin,* from the Greek word "auxin" meaning to increase, is given to a group of compounds that stimulate elongation. Indoleacetic acid (IAA) (fig. 4-43*A*) is the prevalent form; however, recent evidence suggests that there are other indolic auxins naturally occurring in plants. Although auxin (indoleacetic acid) is found throughout the plant, the highest concentrations are localized in actively growing meristematic regions. It is found as both the free molecule and as inactive conjugated forms. When conjugated, auxin is metabolically bound to other low molecular weight compounds. This process appears to be reversible. The concentration of free auxin in plants ranges from 1 to 100 mg · kg^{-1} fresh weight. In contrast, the concentration of conjugated auxin has in some cases been shown to be substantially higher.

One striking characteristic of auxin is the strong polarity exhibited in its transport throughout the plant. Auxin is transported via an energy-dependent mechanism, basipetally away from the apical tip of the plant toward the base. This flow of auxin represses the development of axillary lateral buds along the stem, thus maintaining apical dominance. Movement of auxin out of the leaf blade toward the base of the petiole also appears to prevent leaf abscission.

Auxin has been implicated in the regulation of a number of physiological processes. For example, evidence for its role in cell growth and differentiation, fruit ripening, flowering, senescence, geotropism, abscission, apical dominance, and other responses has been given. The precise initial effect of the hormone that subsequently mediates this diverse array of physiological events is not yet known. During auxin-induced cell elongation, it is thought to act both through a rapid direct effect on an ATPase proton pump mechanism in the plasma membrane and a secondary effect mediated through enzyme synthesis.

The obvious similarity between the amino acid tryptophan and indoleacetic acid (fig. 4-43*A*) led to the initial proposal that tryptophan represented the precursor of the hormone. Subsequent tests with labeled tryptophan substantiated its role as precursor and helped to elucidate the specific steps involved in the degradation of the side chain of the amino acid. They include deamination, decarboxylation, and two oxidation steps, the precise sequence of which remains a subject of debate. Two general pathways are possible, one through indole-3-pyruvic acid and a second by way of tryptamine (fig. 4-43*B*).

In addition to these two primary means of auxin synthesis, the identification of several chlorinated auxins suggested the potential for alternate pathways. Here the chlorine atom is found on the benzene ring and appears to be added prior to alteration of the tryptophan side chain.

Auxin is active at very low levels in plant cells, and as a consequence, precise control over the internal concentration of the molecule is essential. As environmental conditions to which the plant is exposed change, relatively rapid and significant alterations in auxin concentration may be necessary. The concentration of auxin within a group of cells can be altered by: (1) the rate of synthesis of the molecule; (2) its rate of transport into or out of those cells; (3) the rate of breakdown of the molecule; and (4) the formation of conjugates or conversely auxin liberation from existing conjugates.

Interconversion and catabolism of indoleacetic acid can be mediated by both enzymatic (IAA oxidase, most probably a peroxidase) and nonenzymatic (e.g., H_2O_2 direct oxidation, light, ultraviolet radiation, and others)

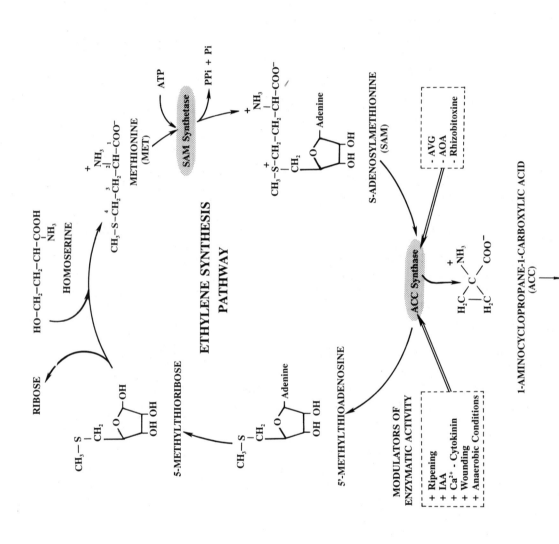

ETHYLENE SYNTHESIS
PATHWAY

1-AMINOCYCLOPROPANE-1-CARBOXYLIC ACID
(ACC)

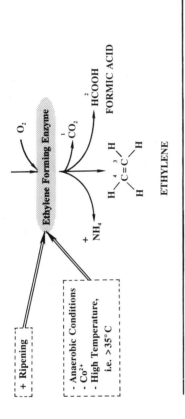

+ Ripening

- Anaerobic Conditions
- Co^{2+}
- High Temperature, i.e. >35°C

Ethylene Forming Enzyme

O_2

NH_4 +

CO_2 ¹

HCOOH ²

FORMIC ACID

$$H-C=C-H \quad (H,^4 C-^3 C, H)$$

ETHYLENE

INHIBITORS OF ETHYLENE SYNTHESIS

$$H-\overset{H}{\underset{NH_2}{C}}-\overset{H}{\underset{H}{C}}-O-C=C-\overset{H}{\underset{NH_2}{C}}-COOH$$

AMINOETHOXYVINYLGLYCINE (AVG)
[L-2-amino-4-(2-aminoethoxy)-<u>trans</u>-3-butenoic acid]

$H_2N-O-CH_2-COOH$

AMINOOXYACETIC ACID (AOA)

$$H-\overset{H}{\underset{HO}{C}}-\overset{H}{\underset{NH_2}{C}}-\overset{H}{\underset{H}{C}}-O-C=C-\overset{H}{\underset{NH_2}{C}}-COOH$$

RHIZOBITOXINE
[1-2-amino-4-(2-amino-3-hydroxypropoxy)-<u>trans</u>-3-butenoic acid]

Figure 4-42. Proposed pathway for the biosynthesis of ethylene from methionine (*after Yang*[102]). Also included are two potent inhibitors of ethylene synthesis (aminoethoxy-vinylglycine and rhizobitoxine) that act by inhibiting ACC synthase.

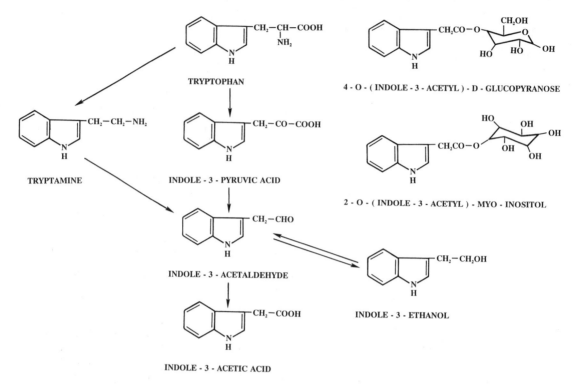

Figure 4-43A. Biosynthesis of indoleacetic acid from tryptophan via both the indole-3-pyruvic acid and tryptamine pathways.

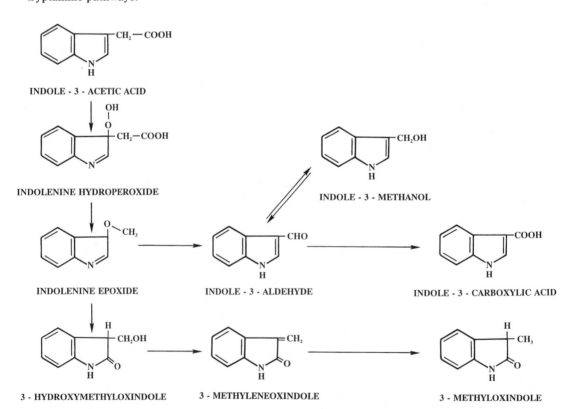

Figure 4-43B. Proposed pathway for oxidative degradation of indoleacetic acid.

means. The proposed pathway for indoleacetic acid oxidation is presented in figure 4-43. Peroxidases are found throughout the plant kingdom and some, in addition to exhibiting peroxidase activity, also appear to have the ability to oxidize auxin. Thus the endogenous concentration of auxin can be decreased by the action of these enzymes. Enzymatic control of the concentration of IAA, therefore, could represent a method of regulating certain physiological processes in which IAA is involved. In fact, the level of activity of IAA-degrading enzymes has been correlated with the development of specific responses (e.g., fruit ripening). This correlation on its own does not establish a cause and effect relationship, however.

Conjugation of auxin to other low molecular weight compounds represents a second means of modulating the concentration of the hormone within a cell. This process does not exclude the potential for reversibility; thus the reaction can be reversed to yield the free active forms. At present, there are three major groups to which auxin has been found to be bound, in each case through the carboxyl group of the hormone. These include a peptidyl IAA conjugate where auxin is linked to an amino acid through a peptide bond, glycosyl IAA conjugates where auxin is linked to a sugar through a glycosidic or an ester bond, and a myo-inositol conjugate where auxin is linked to myo-inositol through an ester bond (fig. 4-43).

Gibberellins

The gibberellins represent a group of acidic diterpenoids found in angiosperms, gymnosperms, ferns, algae, and fungi; they do not, however, appear to be present in bacteria. More than 68 different free and 16 conjugated gibberellins have been isolated, many of which represent intermediates in the synthesis pathway and lack hormonal activity. Typically the different gibberellins are designated with a number (e.g., GA_3, GA_4, GA_5, . . .) based on their chronological order of isolation and identification. While gibberellins have been shown to induce stem elongation and other responses [e.g., increase radial diameter in stems (conifers), induce flowering] their precise role in plants remains unknown. Often several gibberellins are found in the same plant.

The base molecule for the various forms is gibberellin, a 20-carbon diterpenoid (fig. 4-44). Some, however, are minus a methyl group and therefore have only 19 carbons (referred to as the $C_{19}-GA$'s). Individual gibberellins differ from each other in the oxidation state of the ring structure and the carbon and hydroxyl groups present. In plants and plant parts, gibberellins are also found as glycosides (typically of glucose) and other bound inactive forms. Several of these can be readily converted back to the free molecule upon hydrolysis.

Much of the data on the synthesis of gibberellins has come from studies on the fungus *Gibberella fujikuroi*. It appears that the same general pathway, at least until GA_{12}, is operative in higher plants. Since gibberellins are diterpenoids, metabolism of labeled precursors of the terpenoid biosynthesis pathway (e.g., acetyl CoA and mevalonate) was monitored to establish the proposed pathway (fig. 4-45). The first complete cyclic compound in the pathway is kaurene. Through a series of not clearly understood reactions,

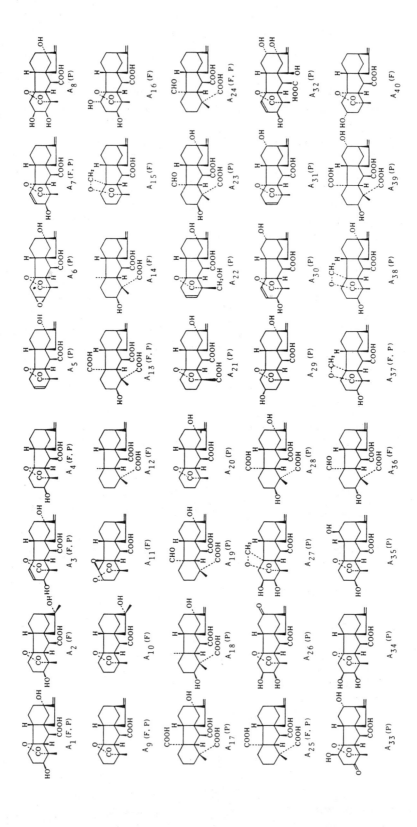

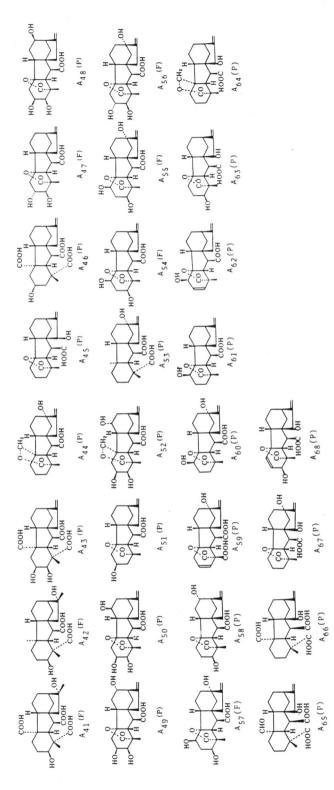

Figure 4-44. Structures of currently identified free gibberellins (*from Takahashi et al.*[90]). Letter adjacent to the number indicates whether of plant (P) or fungal (F) origin.

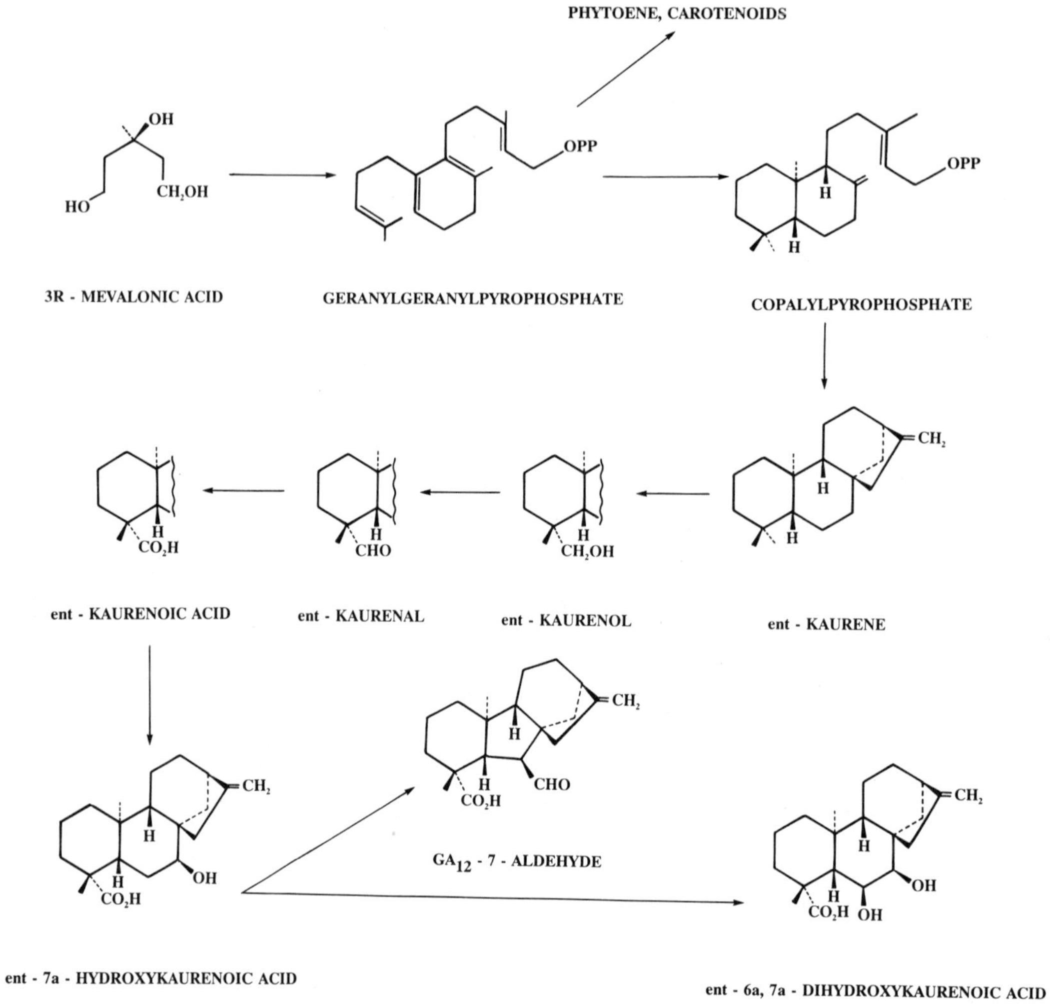

Figure 4-45. Pathway for 3R-mevalonic acid conversion to GA_{12}-7-aldehyde.

kaurene is then converted to various gibberellins. There appears to be some interconversion between specific gibberellins within plants. For example, GA^1 can be converted to GA_3 or GA_5, and subsequently to GA_8.

Conjugates may represent an important means of modulating the internal concentrations of gibberellins with the plant. When gibberellins are applied to plant tissue there is often a rapid conversion to the inactive glucoside form. In addition, although more stable than auxins, gibberellins can be degraded to inactive compounds and some evidence for compartmentalization has been found.

Gibberellins are synthesized in the apical leaf primordia, root tips, and developing seeds. The hormone does not exhibit the same strongly polar transport as seen with auxin, although in some species there is basipetal movement in the stem. In addition to being found in the phloem, gibberellins have also been isolated from xylem exudate, which suggests a more general, bidirectional movement of the molecule in the plant.

Several synthetic compounds have been shown to inhibit elongation of some plants, suggesting an antigibberellin activity. Commercial compounds such as AMO-1618 (2-isopropyl-4-dimethylamino-5-methylphenyl-1-piperidine-carboxylate methyl chloride), phosfon D (2,4-dichlorobenzyl-tributyl-phosphonium chloride), and cycocel or CCC ([2-chloroethyl]trimethyl ammonium chloride) have found widespread use in ornamental horticulture for producing more compact plants. Their use also appears to significantly modify the postharvest response of many of these plants.

AMO-1618 and phosfon D inhibit the normal synthesis of gibberellin within the plants through their inhibition of the enzyme kaurene synthase, which catalyzes the synthesis step between geranylgeranyl pyrophosphate to copalyl pyrophosphate. Cycocel also inhibits the synthesis of kaurene and therefore subsequently gibberellin, but its precise action is not yet known.

Cytokinins

Cytokinins are naturally occurring plant hormones that stimulate cell division. Initially they were called kinin; however, due to prior use of the name for a group of compounds in animal physiology, cytokinin (cyto kinesis or cell division) was adapted.

All of the naturally occurring cytokinins contain an N^6-substituted adenine moiety (fig. 4-46). Zeatin was the first naturally occurring cytokinin isolated and identified; however, since then a number of others have been identified. They are found as the base molecule: a riboside or a ribotide (presence of ribosyl group at the R_3 position).

The highest concentrations of cytokinins are found in embryos and young developing fruits, both of which are undergoing rapid cell division. The presence of high levels of cytokinins may facilitate their ability to act as a strong sink for nutrients. Cytokinins are also formed in the roots and are translocated via the xylem to the shoot. When in the leaf, however, the compounds are relatively immobile.

The precise mode of action of cytokinins is not known. While they do stimulate cell division, exogenous application is also known to cause several significant responses. When applied to detached leaves, cytokinins delay senescence; thus the rate at which degradative processes occur significantly decreases. This decrease is due in part to a facilitated movement of amino acids and other nutrients into the treated area. The site of response is localized to where the hormone is placed on the leaf, indicating little movement of cytokinin in the leaf. Considerable interest has been shown in this anti-senescence ability of cytokinins. Synthetic cytokinins such as N^6-benzyladenine have been applied to a number of postharvest products with varying degrees of success.

Other general effects of cytokinins on plants have been reported. These include (1) stimulating seed germination; (2) stimulating formation of seedless fruit; (3) breaking seed dormancy; (4) inducing bud formation; (5) enhancing flowering; (6) altering fruit growth; and (7) breaking apical dominance. These responses tend to only be found in certain species, and in some cases, cultivars, and are not widespread. It would appear, therefore, that they represent pharmacological responses rather than precise physiological roles

Figure 4-46. Naturally occurring cytokinins in plants (*after Sembdner et al.*[82]).

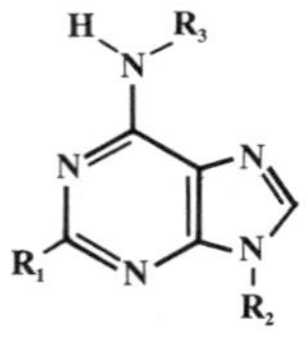

BASE STRUCTURE
OF CYTOKININ

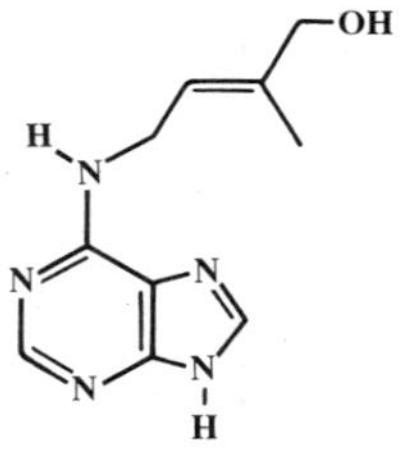

TRANS

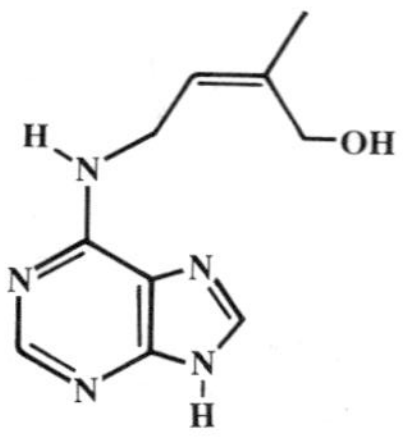

CIS

STERIO ISOMERS OF ZEATIN

SUBSTITUENTS R_1	R_2	R_3^*	TRIVIAL NAME	SYSTEMATIC NAME	ABBREVIATION
H	H	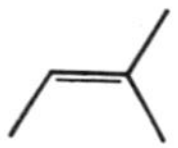	N^6 - (Δ^2 - ISOPENTENYL) ADENINE	6 - (3 - METHYLBUT - 2 - ENYLAMINO) - PURINE	i^6Ade
H	RIBOSYL	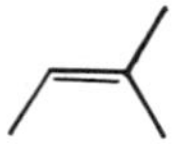	N^6 - (Δ^2 - ISOPENTENYL) ADENOSINE	6 - (3 - METHYLBUT - 2 - ENYLAMINO) - 9 - β - p - RIBOFURANOSYLPURINE	i^6A
H	H	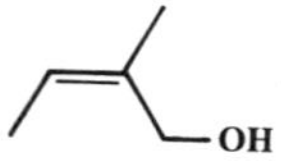	CIS - ZEATIN	6 - (4 - HYDROXY - 3 - METHYL - CIS - BUT - 2 - ENYLAMINO) - PURINE	c - io^6Ade
H	RIBOSYL	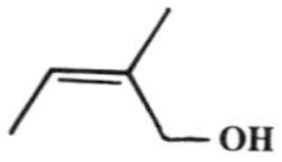	CIS - ZEATIN RIBOSIDE	6 - (4 - HYDROXY - 3 - METHYL - CIS - BUT - 2 - ENYLAMINO) - 9 - β - p - RIBOFURANOSYLPURINE	c - io^6A

		Structure	Trivial name	Chemical name	Abbreviation
H	H		TRANS - ZEATIN	6 - (4 - HYDROXY - 3 - METHYL - TRANS - BUT - 2 - ENYLAMINO) - PURINE	t - io⁶Ade
H	RIBOSYL		TRANS - ZEATIN RIBOSIDE	6 - (4 - HYDROXY - 3 - METHYL - TRANS - BUT - 2 - ENYLAMINO) - 9 - β - RIBOFURANOSYL - PURINE	t - io⁶A
H	H		DIHYDROZEATIN	6 - (4 - HYDROXY - 3 - METHYLBUTYLAMINO) - PURINE	H₂ - io⁶Ade
H	RIBOSYL		DIHYDROZEATIN RIBOSIDE	6 - (4 - HYDROXY - 3 - METHYLBUTYLAMINO) - 9 - β - p - RIBOFURANOSYLPURINE	H₂ - io⁶A
CH₃ - S	H			2 - METHYLTHIO - 6 - (3 - METHYLBUT - 2 - ENYLAMINO) - PURINE	ms² - i⁶Ade
CH₃ - S	RIBOSYL			2 - METHYLTHIO - 6 - (3 - METHYLBUT - 2 - ENYLAMINO) - 9 - β - p - RIBOFURANOSYLPURINE	ms² - i⁶A
CH₃ - S	H			2 - METHYLTHIO - 6 - (4 - HYDROXY - 3 - METHYL - CIS - BUT - 2 - ENYLAMINO) - PURINE	ms² - c - io⁶Ade
CH₃ - S	RIBOSYL			2 - METHYLTHIO - 6 - (4 - HYDROXY - 3 - METHYL - CIS - BUT - 2 - ENYLAMINO) - 9 - β - p - RIBOFURANOSYLPURINE	ms² - c - io⁶A
CH₃ - S	H			2 - METHYLTHIO - 6 - (4 - HYDROXY - 3 - METHYL - TRANS - BUT - 2 - ENYLAMINO) - PURINE	ms² - t - io⁶Ade
CH₃ - S	RIBOSYL			2 - METHYLTHIO - 6 - (4 - HYDROXY - 3 - METHYL - TRANS - BUT - 2 - ENYLAMINO) - 9 - β - p - RIBOFURANOSYLPURINE	ms² - t - io⁶A

*In all cases, N^6 is linked to the C - 1 of the isoprenoid side chain of the molecule in the plant. While these types of responses may not greatly expand the understanding of how cytokinins function in plants, some are of considerable interest in the commercial production and handling of agricultural plant products.

The biosynthesis of cytokinins is closely related to the metabolism of transfer RNA; both require the purine adenine. In addition, tRNA's have been shown upon hydrolysis to contain naturally occurring cytokinins and cytokininlike derivatives. Therefore, both tRNA bound and non-RNA cytokinins are found in plants.

Due to the presence of cytokinin in tRNA and the possible implication of the molecule at a very basic level of control in plants, considerably more interest has been displayed in the biosynthesis of the tRNA cytokinins than non-RNA cytokinins. Present evidence suggests that the cytokinin molecule is not incorporated into the tRNA molecule during its synthesis; rather the side chain (e.g., isopentenyl) is polymerized with adenine present in the chain. The isoprenoid group is formed through the mevalonate pathway.

The RNA-bound cytokinins, however, do not exhibit cytokinin-like activity while attached and do not appear to represent a source of free cytokinin, which is liberated upon the turnover (breakdown) of the tRNA. Its precise location of the tRNA molecule, adjacent to the 3' end of the anticodon that recognizes codons beginning with U (uridine), suggests that it may play a role in mRNA-tRNA ribosome recognition.

Non-RNA cytokinins appear to be synthesized in a pathway not involving tRNA directly. Little, however, of the specifics of this pathway is presently known past the formation of adenine. A general scheme for the interconversion of cytokinins forming the various derivatives has been proposed (fig. 4-47). The level of activity is affected by the structure of these derivatives. The length of the side chain, degree of side chain unsaturation, and stereochemistry of the double bond are important. Dihydroxyzeatin, without a double bond in the side chain, is only one tenth as active as zeatin. Both the *cis* and *trans* stereoisomers of the side chain's double bond are found (fig. 4-46); however, the *trans* forms appear to be much more active. *Cis* isomers are also found, primarily from RNA hydrolyzates.

Deactivation of cytokinins can occur through the conjugation of the molecule with a glycoside, giving an inactive compound. Cytokinins may also be degraded by the action of cytokinin oxidase, which cleaves the side chain. The adenine portion of the molecule is then metabolized as a substrate or oxidized.

Abscisic Acid

Abscisic acid, previously known as dormin and abscisin, is a naturally occurring growth inhibitor in plants. Chemically, it is a terpenoid that is structurally very similar to the terminal portion of many carotenoids (fig. 4-48). Both *cis* and *trans* isomers are possible; however, only the *cis* form, designated (+)-ABA, is active and is found almost exclusively in plants.

Abscisic acid is a potent growth inhibitor that has been proposed to play a regulatory role in such diverse physiological responses as dormancy, leaf and fruit abscission, and water stress. Typically the concentration within plants is between 0.01 and 1 ppm; however, in wilted plants the concentration may

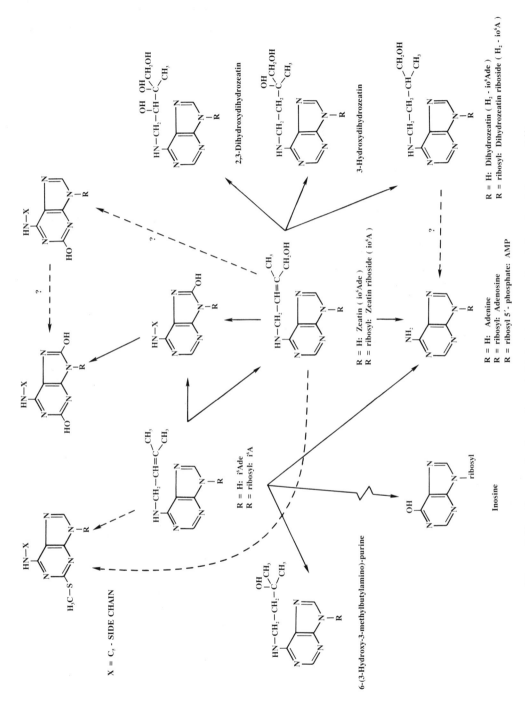

Figure 4-47. Proposed scheme for cytokinin interconversion (*after Sembdner et al.*[82]).

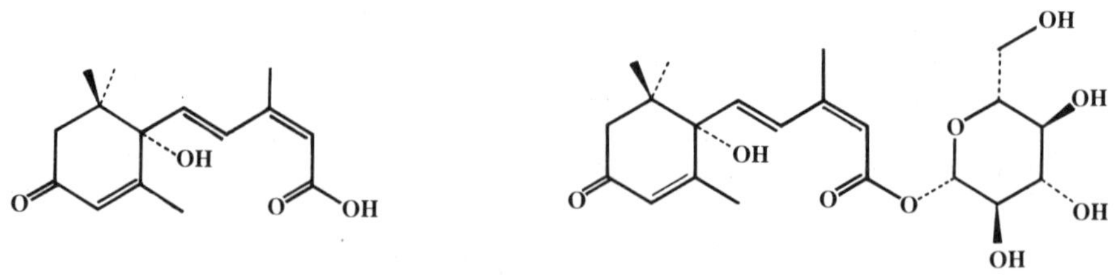

(+) - ABSCISIC ACID   (+) - ABSCISYL - *β* - D - GLUCOPYRANOSIDE

Figure 4-48.   Structures of abscisic acid and its inactive glucose conjugate.

increase as much as forty times. Abscisic acid is found in all parts of the plant; however, the highest concentrations appear to be localized in seeds and young fruits.

Two schemes for the synthesis of abscisic acid have been proposed. The first or direct method involves the formation of the C_{15} carbon skeleton of abscisic acid from three isoprene units derived from mevalonic acid (fig. 4-49). The precise series of steps have yet to be fully elucidated. A second or indirect method was initially suggested based on the close similarity between the terminal ends of certain carotenoids, for example, violaxanthin and abscisic acid. A lipogenase was subsequently isolated that would cleave these carotenoids, giving a range of compounds structurally similar to abscisic acid (e.g., xanthoxin). Exogenously applied xanthoxin was then shown to be converted to abscisic acid. In some cases, this indirect scheme of synthesis may occur; however, its importance appears to be minimal. The predominant means for abscisic acid synthesis is the direct scheme from mevalonic acid.

Degradation of abscisic acid or loss of activity occurs through two primary mechanisms: conjugation and metabolism (fig. 4-50). Abscisic acid rapidly forms an inactive conjugate with glucose. Glucosyl abscisate has been identified in a number of plants. Abscisic acid presumably may also form conjugates with other carbohydrates and types of compounds (i.e., proteins or lipids). It has been proposed that conjugation represents a means of interconversion between active and inactive forms of the molecule, thus controlling the internal concentration within the cell. Abscisic acid is also known to be rapidly metabolized by the plant, which results in a much less active derivative (e.g., phaseic acid) or inactive compounds.

Postharvest Alterations in the Concentration of Phytohormones

Due to the apparent effect of phytohormones on an array of physiological responses within the plant, the fate of these compounds after harvest is of considerable interest. Unfortunately, techniques for the measurement of phytohormones, especially auxin, gibberellins, cytokinins, and abscisic acid, are relatively complicated. In addition, few studies have monitored concurrently all five phytohormones, and as a consequence, a clear picture of gross postharvest alterations is not yet available. Typically, only one or occasion-

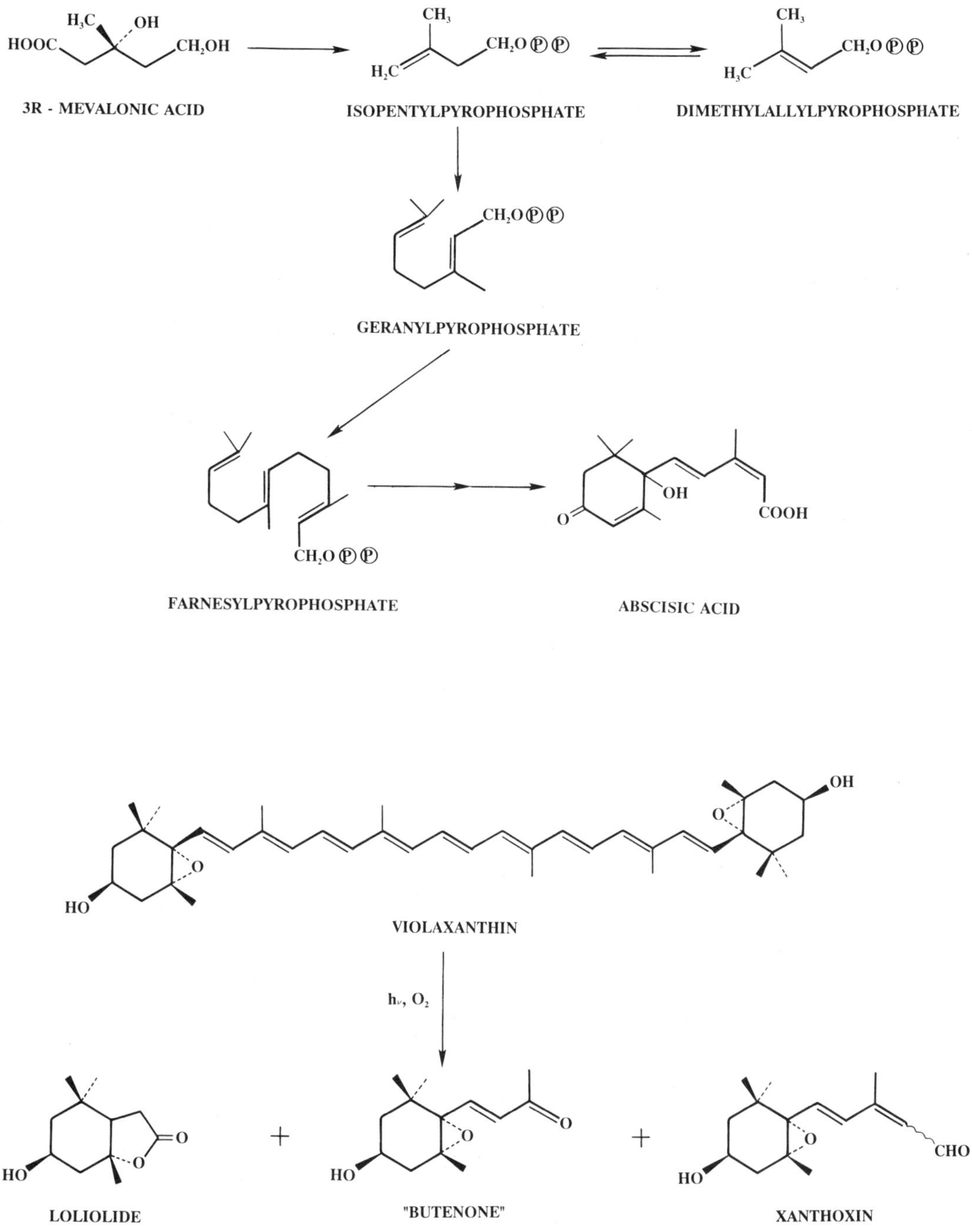

Figure 4-49. General pathway for the synthesis of abscisic acid from 3R-mevalonic acid (*top*) and a possible indirect scheme for the synthesis of abscisic acid from carotenoids (*bottom*). Violaxanthin is attacked by lipoxygenase, yielding xanthoxin and other products. Xanthoxin is then converted to abscisic acid. This does not appear to represent a significant pathway in vitro.

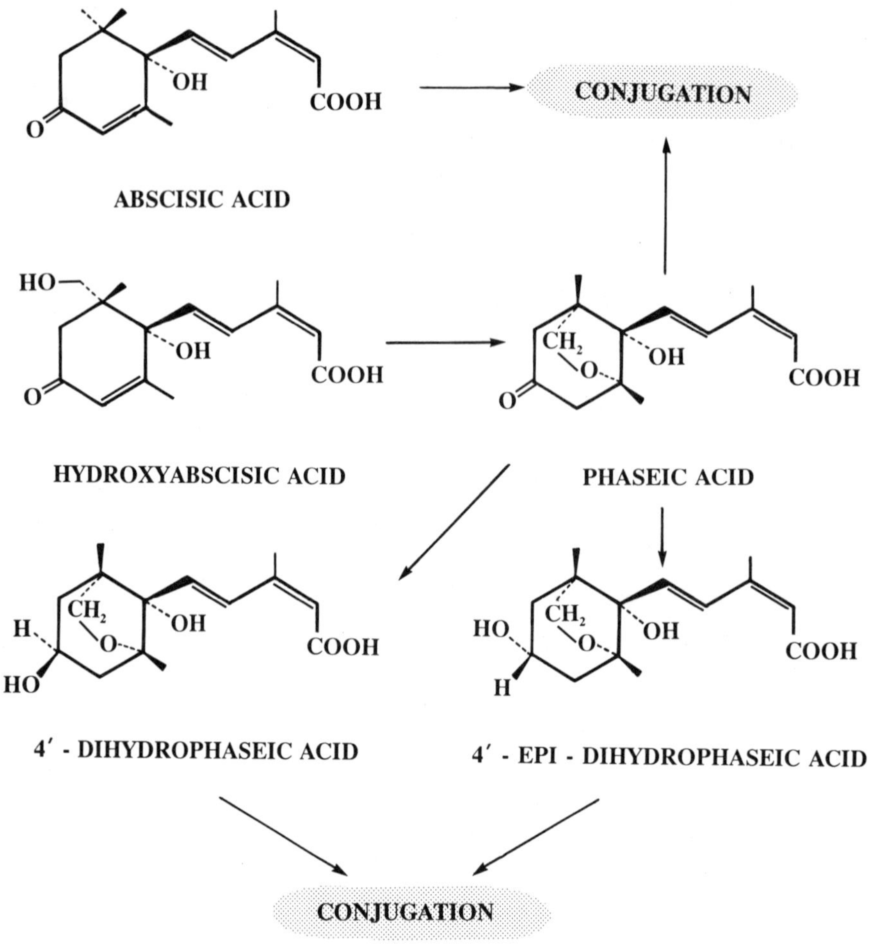

Figure 4-50. Conversion of abscisic acid to metabolites of low or no hormonal activity.

ally two phytohormones are assessed in a single study. More importantly, the interpretation of the results of most studies is complicated by isolation and/or quantification procedures used. Often internal standards are not utilized during isolation and bioassays are relied on to measure the relative activity of the impure isolates. Even when essential isolation and quantification prerequisites are met, it must be assumed that the hormone is not sequestered in a concentrated pool(s) somewhere within each cell (hence disruption during isolation would dilute the concentration) and that all the cell types making up an organ, such as a fruit, have equal amounts of the hormone. Both are rather dubious assumptions. Thus, even when a close correlation is found between the gross concentration and a specific physiological event, the precise meaning remains in question.

The phytohormone ethylene, being a gas that diffuses readily out of a tissue, has the distinct advantage that tissue disruption is not required for isolation and it can be analytically quantified relatively easily using gas chromatography at concentrations as low as several nl $\cdot$ l^{-1} of air. Because of this potential for accurate measurement and the apparent relative importance of the phytohormone in senescence and fruit ripening—two major postharvest phenomena of both physiological and commercial interest—the following discussion will focus to a large extent on ethylene.

Flower Senescence

Distinctive alterations in the concentration of phytohormones are found during the senescence of flowers whether attached or detached from the parent plant. The synthesis of ethylene rises sharply as the result of pollination and senescence in the carnation (*Dianthus caryophyllus,* L.) and other flowers. Exogenous application of ethylene accelerates senescence whereas factors that inhibit its synthesis or action delay senescence. The rise in ethylene production occurs first in the carnation stigma, followed by the ovary, receptacle, and lastly, the petal tissue. The chronological timing of the ethylene increase by the stigma would suggest that it is mediated by pollen tube penetration of the stigma rather than fertilization of the ovules. Thus, the initial response may be due partly to wounding.

With the increase in ethylene production upon pollination, there is a stimulation of ovary development with movement of carbohydrates out of the petals into the gynoecium. This transport effect can be stimulated without the occurrence of pollination by exposure of the flower to an appropriate concentration of ethylene. The close correlation between ethylene synthesis, ovary development, and flower senescence provides a strong argument for a natural role for ethylene in the flowering response.

Abscisic acid concentration also increases during carnation flower and rose (*Rosa* spp.) petal senescence; its exogenous application can accelerate the process. The relationship between increases in abscisic acid concentration and senescence are not clear. In rose petals, the increase in abscisic acid occurs several days after an increase in ethylene synthesis, while the opposite appears to be the case with the carnation flower. Thus, the nature of the interaction between these phytohormones is obscure.

Exogenous application of a synthetic cytokinin will extend the longevity of cut carnations.[16] In addition, the endogenous concentration of cytokinin increases as rose flowers begin to open but declines substantially thereafter.[61] Therefore, it has been suggested that cytokinins are important during the opening and maturation response of the flower.

Fruit Ripening

Climacteric fruits exhibit an exponential increase in the rate of ethylene emanation near the onset of ripening. During ripening of these fruits there is a large increase in the rate of respiration and carbon dioxide production. Since exogenous application of ethylene to preclimacteric fruits will induce ripening, the natural role of ethylene in inducing the ripening response has been a subject of considerable interest.

With many fruits exhibiting a respiratory climacteric, the increase in ethylene synthesis precedes increases in respiration; however, this sequence of events is not universal. With some species (e.g., mango and apple) the respiratory climacteric appears more or less simultaneously with, but not preceded by, an increase in ethylene synthesis. In feijoa (*Feijoa sellowiana,* O. Berg.), cherimoya (*Annona cherimola,* Mill.), and avocado fruits, the respiratory climacteric significantly precedes the increase in ethylene synthesis. In addition, the fruits' sensitivity to the internal concentration of ethylene may change as ripening approaches. Thus, while ethylene appears to be present in all ripening fruits, its concentration does not necessarily have to rise to initiate ripening.

Other phytohormones also appear to be important in the ripening response. Many avocado cultivars do not ripen while attached to the tree, suggesting that some inhibitor produced by the parent plant is operative. Likewise, when apple fruits are detached from the tree, the rate of ethylene synthesis increases.

Changes in auxin concentration have been proposed as part of a multihormonal scheme controlling ripening. This situation is supported by the retardation of ripening with the application of auxin and acceleration with an anti-auxin compound.[22,23] A decline in auxin concentration within some fruits as they approach the onset of ripening appears to be due to the action of IAA oxidase.[21] However, a decline in IAA does not appear to be a universal phenomenon (e.g., apple). Thus, there is insufficient evidence at this time to support IAA having a direct role in controlling the onset of ripening.

REFERENCES

1. Adler, E., S. Larsson, K. Lundquist, and G. E. Miksche. 1969. Acidolytic, alkaline, and oxidative degradation of lignin. *Int. Wood Chem. Symp. Abstr.,* 260p.
2. Argos, P., K. Pedersen, M. D. Marks, and B. A. Larkins. 1982. A structural model for maize zein proteins. *J. Biol. Chem.* **257:**9,984–9,990.
3. Badings, H. T. 1960. Principles of autooxidation processes in lipids with special regard to the development of off-flavors. *Neth. Milk Dairy J.* **14:**215–242.
4. Barrett, A., and D. Northcote. 1965. Apple fruit pectic substances. *Biochem. J.* **94:**617–627.
5. Bartley, I. M. 1976. Changes in the glucose of ripening apples. *Phytochemistry* **15:**625–626.
6. Biale, J. B. 1960. The postharvest biochemistry of tropical and subtropical fruits. *Adv. Food. Res.* **10:**293–354.
7. Biale, J. B., and R. E. Young. 1971. The avocado pear. In *The Biochemistry of Fruits and Their Products,* vol. 2, A. C. Hulme (ed.). Academic Press, New York, pp. 1–63.
8. Booth, V. H. 1963. Rapidity of carotene biosynthesis in *Narcissus. Biochem. J.* **87:**238–239.
9. Britton, G. 1976. Biosynthesis of carotenoids. In *Chemistry and Biochemistry of Plant Pigments,* vol. 1, T. W. Goodwin (ed.). Academic Press, London, pp. 262–327.
10. Carroll, D. E., M. W. Hoover, and W. B. Nesbitt. 1971. Sugar and organic acid concentrations in cultivars of Muscadine grapes. *J. Am. Soc. Hort. Sci.* **96:**737–740.
11. Cockshull, K. E., and A. P. Hughes. 1967. Distribution of dry matter to flowers in *Chrysanthemum morifolium. Nature* **215:**780–781.

12. Coggins, C. W., Jr., J. C. F. Knapp, and A. L. Richer. 1968. Postharvest softening studies of Deglet Noord dates: Physical, chemical, and histological changes. *Date Growers Rep.* **45**:3–6.
13. Colvin. J. R. 1980. *The Biochemistry of Plants,* vol. 3, J. Preiss (ed.). Academic Press, New York, 638p.
14. Dicks, M. 1965. Vitamin E content of foods and feeds for human consumption. *Wyo. Agric. Exp. Sta. Bull. 435,* 193p.
15. Edelman, J., and T. G. Jefford. 1968. The mechanism of fructosan metabolism in higher plants as exemplified in *Helianthus tuberosus. New Phytol.* **67**:517–531.
16. Eisinger, W. 1977. Role of cytokinins in carnation flower senescence. *Plant Physiol.* **59**:707–709.
17. Erdmann, H., R. J. Griesbach, R. H. Lawson, and A. K. Mattoo. 1989. 1-Amino-cyclopropane-1-carboxylic-acid-dependent ethylene production during re-formation of vacuoles in evacuolated protoplasts of *Petunia hybrida. Planta* **179**:196–202.
18. FAO. 1970. Amino-acid content of foods and biological data on proteins. *Nutritional Studies No. 24,* 275p.
19. Fernandez, M. C. C. 1954. The effect of environment on the carotene content of plants. In *Influence of Environment on the Chemical Composition of Plants.* South. Coop. Serv. Bull. 36, 198p.
20. Ferguson, C. H. R., and E. W. Simon. 1973. Membrane lipids in senescing green tissues. *J. Exp. Bot.* **24**:307–316.
21. Frenkel, C. 1972. Involvement of peroxidase and indole-3-acetic acid oxidase isozymes from pear, tomato, and blueberry fruit in ripening. *Plant Physiol.* **49**:757–763.
22. Frenkel, C., and R. Dyck. 1973. Auxin inhibition of ripening in Bartlett pears. *Plant Physiol.* **51**:6–9.
23. Frenkel, C., and N. F. Haard. 1973. Initiation of ripening in Bartlett pear with an antiauxin alpha (p-chlorophenoxy) isobutyric acid. *Plant Physiol.* **52**:380–384.
24. Frenkel, C., I. Klein, and D. R. Dilley. 1968. Protein synthesis in relation to ripening of pome fruits. *Plant Physiol.* **43**:1146–1153.
25. Fuller, G., J. A. Kuhnle, J. W. Corse, and B. E. Mackey. 1977. Use of natural cytokinins to extend the storage life of broccoli (*Brassica oleracea,* Italica group). *J. Am. Soc. Hort. Sci.* **102**:480–484.
26. Galliard, T. 1975. Degradation of plant lipids by hydrolytic and oxidative enzymes. In *Recent Advances in the Chemistry and Biochemistry of Plant Lipids,* T. Galliard and E. I. Mercer (eds.). Academic Press, London, pp. 319–357.
27. Galliard, T., D. R. Phillips, and J. Reynolds. 1976. The formation of *cis*-3-nonenal, *trans*-2-nonenal, and hexanal from linoleic acid hydroperoxide enzyme system in cucumber fruit. *Biochem. Biophys. Acta* **441**:181–192.
28. Garwood, D. L., F. J. McArdle, S. F. Vanderslice, and J. C. Shannon. 1976. Postharvest carbohydrate transformations and processed quality of high sugar maize genotypes. *J. Am. Soc. Hort. Sci.* **101**:400–404.
29. Gore, H. C., and D. Fairchild. 1911. Experiments on processing of persimmons to render them nonastringent. *U.S. Dept. Agric. Bur. Chem. Bull. 141,* 31p.
30. Gross, J., M. Gabai, and A. Lifshitz. 1971. Carotenoids in juice of Shamouti orange. *J. Food Sci.* **36**:466–473.
31. Guiraud, J. P., J. Daurelles, and P. Galzy. 1981. Alcohol production from Jerusalem artichokes using yeasts with inulinase activity. *Biotechnol. Bioeng.* **23**:1461–1466.
32. Guy, M., and H. Kende. 1984. Conversion of 1-aminocyclopropane-1-carboxylic acid to ethylene by isolated vacuoles of *Pisum sativum,* L. *Planta* **160**:281–287.
33. Habib, A. T., and H. D. Brown. 1956. The effect of oxygen and hydrogen ion concentration on color changes in processed beets, strawberries and raspberries. *Proc. Am. Hort. Sci.* **68**:482–490.

34. Hale, C. R., B. G. Coombe, and J. S. Hawker. 1970. Effect of ethylene and 2-chloroethylphosphoric acid on the ripening of grapes. *Plant Physiol.* **45**:620–623.

35. Hane, M. 1962. Studies on the acid metabolism of strawberry fruits. *Gartenbauwissenschaft* **27**:453–482.

36. Harborne, J. B. 1980. Plant phenolics. In *Secondary Plant Products,* E. A. Bell and B. V. Charlwood (eds.). Springer-Verlag, Berlin, pp. 329–402.

37. Harborne, J. B., and C. F. van Soumere (eds.). 1975. *The Chemistry and Biochemistry of Plant Proteins.* Academic Press, New York, 326p.

38. Henderson, J. R., and B. W. Buescher. 1977. Effects of SO_2 and controlled atmospheres on broken-end discoloration and processed quality attributes in snap beans. *J. Am. Soc. Hort. Sci.* **102**:768–770.

39. Hildebrand, D. F., and T. Hymowitz. 1982. Carotene and chlorophyll bleaching by soybeans with and without seed lipoxygenase-1. *J. Agric. Food Chem.* **30**:705–708.

40. Hobson, G. E. 1968. Cellulase activity during the maturation and ripening of tomato fruit. *J. Food Sci.* **33**:588–592.

41. Hobson, G. E. 1975. Protein redistribution and tomato fruit ripening. *Colloques Int. CNRS* **238**:265–269.

42. Hoffman, G. 1962. 1-Octen-3-ol and its relation to other oxidative cleavage products from esters of linoleic acid. *J. Am. Oil Chem. Soc.* **39**:439–444.

43. Hulme, A. C. 1972. The proteins of fruits: Their involvement as enzyme in ripening. A review. *J. Food Tech.* **7**:343–371.

44. Hulme, A. C., and L. S. C. Wooltorton. 1958. The acid content of cherries and strawberries. *Chem. Ind., London No. 22,* p. 659.

45. Impellizzeri, G., M. Piattelli, and S. Sciuto. 1973. Acylated betacyanins from *Drosanthemum foribundum. Phytochemistry* **12**:2,295–2,296.

46. Jacobi, G., B. Klemme, and C. Postius. 1975. Dark starvation and metabolism. IV. The alteration of enzyme activities. *Biochem. Physiol. Pflanz.* **168**:247–256.

47. Kefford, N. P., M. I. Burce, and J. A. Zwar. 1973. Retardation of leaf senescence by urea cytokinins in *Raphanus sativus. Phytochemistry.* **12**:995–1,003.

48. Kliewer, W. M. 1966. Sugars and organic acids of *Vitis vinifera. Plant Physiol.* (Lancaster) **41**:923–931.

49. Knee, M. 1988. Carotenol esters in developing apple fruit. *Phytochemistry* **27**:1,005–1,009.

50. Knee, M., and I. M. Bartley. 1981. Composition and metabolism of cell wall polysaccharides in ripening fruits. In *Recent Advances in the Biochemistry of Fruits and Vegetables,* J. Friend and M. J. C. Rhodes (eds.). Academic Press, New York, pp. 133–148.

51. Kolattukudy, P. E. 1980. Cutin, suberin and waxes. In *The Biochemistry of Plants,* vol. 4, P. K. Stumpf and E. E. Conn (eds.). Academic Press, New York, pp. 571–645.

52. Kolattukudy, P. E., and T. J. Walton. 1973. The biochemistry of plant cuticular lipids. *Prog. Chem. Fats Other Lipids* **13**:119–175.

53. Kollas, D. A. 1964. Preliminary investigation of the influence of controlled atmosphere storage on the organic acids of apples. *Nature* **204**:758–759.

54. Manners, D. J., and J. J. Marshall. 1971. Studies on carbohydrate-metabolizing enzymes. Part XXIV. The action of malted-rye alpha-amylase on amylopectin. *Carbohydr. Res.* **18**:203–209.

55. Marcus, A. (ed.). 1981. *The Biochemistry of Plants,* vol. 6, *Proteins and Nucleic Acids.* Academic Press, New York, 658p.

56. Marks, M. D., and B. A. Larkins. 1982. Analysis of sequence microheterogeneity among zein messenger RNAs. *J. Biol. Chem.* **257**:9,976–9,983.

57. Martin, C., and K. V. Thimann. 1972. The role of protein synthesis in the senescence of leaves. I. The formation of protease. *Plant Physiol.* **49**:64–71.

58. Martin, C., and K. V. Thimann. 1973. The role of protein synthesis in the senescence of leaves. II. The influence of amino acids on senescence. *Plant Physiol.* **50**:432–437.

59. Matsuo, T., and S. Itoo. 1978. The chemical structure of Kaki-tannin from immature fruit of the persimmon (*Diosysyros Kaki, L.*). *Agric. Biol. Chem.* **42**:1,637–1,643.

60. Matsuo, T., and S. Itoo. 1981. Comparative studies of condensed tannins from severed young fruits. *J. Jpn. Soc. Hort. Sci.* **50**:262–269.

61. Mayak, S., A. H. Halevy, and M. Katz. 1972. Correlative changes in phytohormones in relation to senescence processes in rose petals. *Physiol Plant.* **27**:1–4.

62. Mazliak, P. 1968. Chemistry of plant cuticles. In *Progress in Phytochemistry*, vol. 1, L. Reinhold and V. Liwschitz (eds.). J. Wiley and Sons, London, pp. 49–111.

63. Meredith, F. I., and R. H. Young. 1969. Effect of temperature on pigment development in Red Blush grapefruit and Ruby Blood oranges. In *Proceedings of the First International Citrus Symposium*, H. D. Chapman (ed.). University of California, Riverside, pp. 271–276.

64. Mitchell, T., A. J. R. Porter, and P. John. 1988. Authentic activity of the ethylene-forming enzyme observed in membranes obtained from kiwifruit (*Actinidia deliciosa*). *New Phytol.* **109**:313–319.

65. Money, R. W. 1958. Analytical data on some common fruits. *J. Sci. Food Agric.* **9**:19–20.

66. Money, R. W., and W. A. Christian. 1950. Analytical data on some common fruits. *J. Sci. Food Agric.* **1**:8–12.

67. Nursten, H. E. 1970. Volatile compounds: The aroma of fruits. In *The Biochemistry of Fruits and Their Products*, vol. 1, A. C. Hulme (ed.). Academic Press, New York, pp. 239–268.

68. Olliver, M. 1967. Ascorbic acid: Occurrence in foods. In *The Vitamins*, vol. 1, W. H. Sebrell, Jr. and R. S. Harris (eds.). Academic Press, New York, pp. 359–367.

69. Paillard, N. 1968. Analyse de l'arome de pommes de la variete 'Calville blanc' par chromatographie sur colonne capillaire. *Fruits d'Outre Mer.* **23**:393–387.

70. Piattelli, M., and L. Minale. 1964. Pigments of centrospermae. II. Distribution of betacyanins. *Phytochemistry* **3**:547–557.

71. Piattelli, M., L. Minale, and G. Prota. 1965. Pigments of centrospermae. III. Betaxanthins from *Beta vulgaris, L. Phytochemistry* **4**:121–125.

72. Pinsky, A., S. Grossman, and M. Trop. 1971. Lipoxygenase content and antioxidant activity of some fruits and vegetables. *J. Food Sci.* **36**:571–572.

73. Preiss, J., and C. Levi. 1980. Starch biosynthesis and degradation. In *The Biochemistry of Plants*, vol. 3, J. Preiss (ed.). Academic Press, New York, pp. 371–423.

74. Pressey, R. 1977. Enzymes involved in fruit softening. In *Enzymes in Food and Beverage Processing*, R. L. Ory and A. J. St. Angelo (eds.). ACS Symp. Ser. 47, American Chemical Society, Washington, D.C., pp. 172–191.

75. Proctor, M., and P. Yeo. 1973. *The Pollination of Flowers*. Collins, Glasgow, 418p.

76. Quackenbush, F. W. 1963. Corn carotenoids: Effects of temperature and moisture on losses during storage. *Cereal Chem.* **40**:266–269.

77. Richmond, A., and J. B. Biale. 1966. Protein synthesis in avocado fruit tissue. *Arch. Biochem. Biophys.* **115**:211–214.

78. Richmond, A., and J. B. Biale. 1967. Protein and nucleic acid metabolism in fruits. II. RNA synthesis during the respiratory rise of the avocado. *Biochem. Biophys. Acta* **138**:625–627.

79. Rudolph, C. J., G. V. Odell, and H. A. Hinrichs. 1971. Chemical studies on

pecan composition and factors responsible for the typical flavor of pecans. *41st Okla. Pecan Growers Assoc. Proc.*, pp. 63–70.

80. Salanke, D. K., and J. Y. Do. 1976. Biogenesis of aroma constituents of fruits and vegetables. *CRC Crit. Rev. Food Tech.* **8**:161–190.
81. Sasson, A., and S. P. Monselise. 1977. Organic acid composition of 'Shamouti' oranges at harvest and during prolonged postharvest storage. *J. Am. Soc. Hort. Sci.* **102**:331–336.
82. Sembdner, G., D. Gross, H.-W. Liehusch, and G. Schneider. 1980. Biosynthesis and metabolism of plant hormones. In *Hormonal Regulation of Development. I. Molecular Aspects of Plant Hormones*, J. MacMillan (ed.). Springer-Verlag, Berlin, pp. 281–444.
83. Senter, S. D., and R. J. Horvat. 1976. Lipids of pecan nutmeats. *J. Food Sci.* **41**:1,201–1,203.
84. Senter, S. D., and R. J. Horvat. 1978. Minor fatty acids from pecan kernel lipids. *J. Food Sci.* **43**:1,614–1,615.
85. Shah, B. M., D. K. Salunkhe, and L. E. Olsen. 1969. Effects of ripening processes on chemistry of tomato volatiles. *J. Am. Soc. Hort. Sci.* **94**:171–176.
86. Shine, W. E., and P. K. Stumpf. 1974. Fat metabolism in higher plants: Recent studies on plant oxidation systems. *Arch. Biochem. Biophys.* **162**:147–157.
87. Sterling, C. 1961. Physical state of cellulose during ripening of peach. *J. Food Sci.* **26**:95–98.
88. Steward, F. C., A. C. Hulme, S. R. Freiberg, M. P. Hegarty, J. K. Pollard, R. Rabson, and R. A. Barr. 1960. Physiological investigations on the banana plant. I. Biochemical constituents detected in the banana plant. *Ann. Bot.* **24**:83–116.
89. Swisher, H. E., and W. K. Higby. 1961. *Fruit and Vegetable Juice Processing Technology*, D. K. Tressler and M. A. Joslyn (eds.). AVI, Westport, Conn.
90. Takahashi, N., I. Yamaguchi, and H. Yamane. 1986. Gibberellins. In *Chemistry of Plant Hormones*, N. Takahashi (ed.). CRC Press, Boca Raton, Fla., pp. 57–151.
91. Tevini, M. 1977. Light, function, and lipids during plastid development. In *Lipids and Lipid Polymers in Higher Plants*, M. Tevini and H. K. Lichtenlhaler (eds.). Springer-Verlag, New York, pp. 121–145.
92. Tevini, M., and D. Steinmuller. 1985. Composition and function of plastoglobuli. II. Lipid composition of leaves and plastoglobuli during beech leaf senescence. *Planta* **163**:91–96.
93. Thomas, R. L., and J. J. Jen. 1975. Phytochrome-mediated carotenoids biosynthesis in ripening tomatoes. *Plant Physiol.* **56**:452–453.
94. Ulrich, R. 1970. Organic acids. In *The Biochemistry of Fruits and Their Products*, vol. 1, A. C. Hulme (ed). Academic Press, New York, pp. 452–453.
95. van Buren, J. 1970. Fruit phenolics. In *The Biochemistry of Fruit and Their Products*, vol. 1, A. C. Hulme (ed). Academic Press, New York, pp. 269–304.
96. Watt, B. K., and A. L. Merrill. 1963. Composition of foods—raw, processed, prepared. *USDA, ARS, Agric. Handb. 8*, 190p.
97. Weeks, W. W. 1986. Carotenoids. A source of flavor and aroma. In *Biogeneration of Aromas*, T. H. Parliment and R. Croteau (eds.). Am. Chem. Soc. Symp. Ser. 317, American Chemical Society, Washington, D.C., pp. 156–166.
98. Widdowson, E. M., and R. A. McCance. 1935. The available carbohydrate of fruits. Determination of glucose, fructose, sucrose and starch. *Biochem. J.* **29**:151–156.
99. Williams, M., G. Hrazdina, M. M. Wilkinson, J. G. Sweeny, and G. A. Iacobucci. 1978. High-pressure liquid chromatographic separation of 3-glucosides, 3,5-diglucosides, 3-(6-O-p-coumaryl)glucosides, and 3-(6-O-p-coumarylglucoside)-5-glucosides of anthocyanidins. *J. Chromatogr.* **155**:389–398.

100. Woodward, J. R. 1972. Physical and chemical changes in developing strawberry fruits. *J. Sci. Food Agric.* **23**:465–473.

101. Wyman, H., and J. K. Palmer. 1964. Organic acids in the ripening banana fruit. *Plant Physiol.* **39**:630–633.

102. Yang, S. F., and N. E. Hoffman. 1984. Ethylene biosynthesis and its regulation in higher plants. *Ann. Rev. Plant Physiol.* **35**:155–189.

ADDITIONAL READINGS

Abeles, F. B. 1973. *Ethylene in Plant Biology*. Academic Press, New York, 302p.

Addicot, F. T. (ed.). 1983. *Abscisic Acid*. Praeger, New York, 607p.

Akazawa, T., and I. Hara-Nishimura. 1985. Topographic aspects of biosynthesis, extracellular secretion, and intercellular storage of proteins in plant cells. *Ann. Rev. Plant Physiol.* **36**:441–472.

Bartholomew, E. T., and W. B. Sinclair. 1951. *The Lemon Fruit, Its Composition, Physiology and Products*. University of California Press, Berkeley, 163p.

Birch, G. G., and M. G. Lindley (eds.). 1987. *Developments in Food Flavors*. Elsevier, New York, 282p.

Bopp, M. (ed.). 1985. *Plant Growth Substances 1985*. Springer-Verlag, Berlin, 420p.

Brenner, M. L. 1981. Modern methods for plant growth substance analysis. *Ann. Rev. Plant Physiol.* **32**:511–538.

Brown, R. M., Jr. (ed.). 1982. *Cellulose and Other Natural Polymer Systems: Biogenesis, Structure, and Degradation*. Plenum, New York, 519p.

Castelfransco, P. A., and S. I. Beale. 1983. Chlorophyll biosynthesis: Recent advances and areas of current interest. *Ann. Rev. Plant Physiol.* **34**:241–278.

Chichester, C. O. (ed.). 1972. *The Chemistry of Plant Pigments*. Academic Press, New York, 218p.

Chinoy, N. J. (ed.). 1984. *The Role of Ascorbic Acid in Growth, Differentiation and Metabolism of Plants*. Kluwer Academic Pub., Hingham, Mass., 322p.

Chothia, C. 1984. Principles that determine the structure of proteins. *Ann. Rev. Biochem.* **53**:537–572.

Cohen, D. J., and R. S. Bandurski. 1982. Chemistry and psychology of the bound auxins. *Ann. Rev. Plant Physiol.* **33**:403–430.

Conn, E. E. 1980. Cyanogenic compounds. *Ann. Rev. Plant Physiol.* **31**:433–451.

Conn, E. E. (ed.). 1986. *The Shikimic Acid Pathway*. Plenum, New York, 347p.

Crozier, A. (ed.). 1983. *The Biochemistry and Physiology of Gibberellins,* 2 vols. Praeger, New York.

Crozier, A., and J. R. Hillman (eds.). 1984. *The Biosynthesis and Metabolism of Plant Hormones*. Cambridge University Press, Cambridge, 288p.

Dalling, M. J. (ed.). 1986. *Plant Proteolytic Enzymes,* 2 vols. CRC Press, Boca Raton, Fla.

Daussant, J., J. Mosse, and J. Vaughn (eds.). 1983. *Seed Proteins*. Academic Press, New York, 335p.

Davies, P. J. (ed.). 1987. *Plant Hormones and Their Role in Plant Growth and Development*. M. Nijhoff, Boston, Mass., 681p.

Dey, P. M., and R. A. Dixon (eds.). 1985. *Biochemistry of Storage Carbohydrates in Green Plants*. Academic Press, New York, 378p.

Duffus, C. M. 1984. *Carbohydrate Metabolism in Plants*. Longman, London, 183p.

Ellis, R. J. 1981. Chloroplast proteins: Synthesis, transport, and assembly. *Ann. Rev. Plant Physiol.* **32**:111–137.

Fuchs, Y., and E. Chalutz (eds.). 1984. *Ethylene: Biochemical, Physiological and Applied Aspects*. Kluwer Academic Pub., Hingham, Mass., 368p.

Fuller, G., and W. D. Nes (eds.). 1987. *Ecology and Metabolism of Plant Lipids*. American Chemical Society, Washington, D.C., 374p.

Galston, A. W., and T. A. Smith (eds.). 1985. *Polyamines in Plants*. Nijhoff/Junk Dordrecht, The Netherlands, 422p.

Goodwin, T. W. (ed.). 1988. *Plant Pigments*. Academic Press, New York, 362p.

Goodwin, T. W., and E. I. Mercer (eds.). 1983. *Introduction to Plant Biochemistry*. Pergamon, Oxford, England, 677p.

Gould, R. F. (ed.). 1977. Food proteins. *Adv. Chem. 160*, 312p.

Graebe, J. E. 1987. Gibberellin biosynthesis and control. *Ann. Rev. Plant Physiol.* **38**:419–465.

Green, P. J., O. Pines, and M. Inouye. 1986. The role of antisense RNA in gene regulation. *Ann. Rev. Biochem.* **55**:569–598.

Greppin, H., C. Penel, and T. Gaspar (eds.). 1986. *Molecular and Physiological Aspects of Plant Peroxidases*. University of Geneva, Switzerland, 468p.

Gross, J. 1987. *Pigments in Fruits*. Academic Press, London, 303p.

Hall, T. C., and J. W. Davies (eds.). 1979. *Nucleic Acids in Plants*, 2 vols. CRC Press, Boca Raton, Fla.

Harborne, J. B. (ed.). 1988. *The Flavonoids: Advances in Research*. Chapman and Hall, New York, 621p.

Harborne, J. B., and T. J. Mabry (eds.). 1982. *The Flavonoids: Advances in Research*. Chapman and Hall, London, 744p.

Harwood, J. L. 1988. Fatty acid metabolism. *Ann. Rev. Plant Physiol.* **39**:101–138.

Harwood, J. L. 1989. Lipid metabolism in plants. *Crit. Rev. in Plant Sciences* **8**:1–43.

Higgins, T. J. V. 1984. Synthesis and regulation of major proteins in seeds. *Ann. Rev. Plant Physiol.* **35**:191–221.

Jensen, U., and D. E. Fairbrothers (eds.). 1983. *Protein and Nucleic Acids in Plant Systematics*. Springer-Verlag, Berlin, 408p.

Kefford, J. F. 1970. *The Chemical Constituents of Citrus Fruits*. Academic Press, New York, 246p.

Kolattukudy, P. E. 1981. Structure, biosynthesis, and biodegradation of cutin and suberin. *Ann. Rev. Plant Physiol.* **32**:539–567.

Labavitch, J. M. 1981. Cell wall turnover in plant development. *Ann. Rev. Plant Physiol.* **32**:385–406.

Land, D. G., and H. E. Nursten (eds.). 1979. *Progress in Flavour Research*. Applied Science Pub., London, 371p.

Lawrence, B. M., B. D. Mookherjee, and B. J. Willis. 1986. *Flavors and Fragrances: A World Perspective*. Proc. of the 10th Intern. Cong. of Essential Oils, Fragrances and Flavors. Elsevier, New York, 1,104p.

Letham, D. S., and L. M. S. Palni. 1983. The biosynthesis and metabolism of cytokinins. *Ann. Rev. Plant Physiol.* **34**:163–197.

Lewis, D. H. (ed.). 1984. *Storage Carbohydrates in Vascular Plants: Distribution, Physiology and Metabolism*. Cambridge University Press, Cambridge, England, 284p.

Libert, B., and V. R. Franceschi. 1987. Oxalate in crop plants. *J. Agric. Food Chem.* **35**:926–938.

Maarse, H., and R. Belz. 1981. *Isolation, Separation and Identification of Volatile Compounds in Aroma Research*. Akademie-Verlag, Berlin, 290p.

Machlin, L. J. 1984. *Handbook of Vitamins: Nutritional, Biochemical, and Clinical Aspects*. Marcel Dekker, New York, 614p.

Mahato, S. B., S. K. Sarkar, and G. Poddar. 1988. Triterpenoid saponins. *Phytochemistry* **27**:3,037–3,067.

Margalith, P. Z. 1981. *Flavor Microbiology*. Charles C Thomas, Springfield, Ill., 309p.

Murphy, T. M., and W. F. Thompson. 1988. *Molecular Plant Development*. Prentice Hall, Englewood Cliffs, N.J., 222p.

Nes, W. D., G. Fuller, and L.-S. Tsai (eds). 1984. *Isopentenoids in Plants: Biochemistry and Function*. Marcell Dekker, New York, 596p.

Pharis, R. P., and R. W. King. 1985. Gibberellins and reproductive development in seed plants. *Ann. Rev. Plant Physiol.* **36**:517–568.

Platt, T. 1986. Transcription termination and the regulation of gene expression. *Ann. Rev. Biochem.* **55**:339–372.

Preiss, J. 1982. Regulation of the biosynthesis and degradation of starch. *Ann. Rev. Plant Physiol.* **33**:431–454.

Ribereau-Gayon, P. 1972. *Plant Phenolics*. Oliver and Boyd, Edinburgh, 254p.

Rivier, L., and A. Crozier. 1987. *Principles and Practice of Plant Hormone Analysis*, 2 vols. Academic Press, Orlando, Fla.

Robinson, T. 1980. *The Organic Constituents of Higher Plants: Their Chemistry and Interrelationships*. Cordus Press, North Amherst, Mass., 352p.

Rodricks, J. V. (ed.). 1976. Mycotoxins and other fungal related food problems. *Adv. Chem. 149,* 409p.

Roughan, P. G., and C. R. Slack. 1982. Cellular organization of glycerolipid metabolism. *Ann. Rev. Plant Physiol.* **33**:97–132.

Rubery, P. H. 1980. Auxin receptors. *Ann. Rev. Plant Physiol.* **32**:569–596.

Seib, P. A., and B. M. Tolbert (eds.). 1982. Ascorbic acid: Chemistry, metabolism, and uses. *Adv. Chem. 200,* 604p.

Sironval, C., and M. Browers (eds.). 1984. *Protochlophyllide Reduction and Greening*. Kluwer Academic Pub., Hingham, Mass., 408p.

Smith, T. A. 1985. Polyamines. *Ann. Rev. Plant Physiol.* **36**:117–143.

Spayd, S. E. (ed.). 1986. Naturally occurring toxins in horticultural food crops. *Acta Horticulturae 207,* 70p.

Stumpf, P. K., J. B. Mudd, and W. D. Nes (eds.). 1987. *The Metabolism, Structure, and Function of Plant Lipids*. Plenum Press, New York, 724p.

Swain, T., J. B. Harborne, and C. F. van Sumere (eds.). 1979. *Biochemistry of Plant Phenolics*. Plenum Press, New York, 651p.

Takahashi, N. (ed.). 1986. *Chemistry of Plant Hormones*. CRC Press. Boca Raton, Fla., 277p.

Thomson, W. W., and J. M. Whatley. 1980. Development of nongreen plastids. *Ann. Rev. Plant Physiol.* **31**:375–394.

Tucker, G. A., and J. A. Roberts (eds.). 1985. *Ethylene and Plant Development*. Butterworths, Boston, Mass., 416p.

van Sumere, C. F., and P. J. Lea (eds.). 1985. *The Biochemistry of Plant Phenolics*. Clarendon Press, Oxford, England, 483p.

Vikery, M. L. 1981. *Secondary Plant Metabolism*. University Park Press, Baltimore, 335p.

von Hippel, P. H., D. G. Bear, W. D. Morgan, and J. A. McSwiggen. 1984. Protein-nucleic acid interactions in transcription: A molecular analysis. *Ann. Rev. Biochem.* **53**:389–446.

Waller, G. R., and E. K. Nowacki. 1978. *Alkaloid Biology and Metabolism in Plants*. Plenum Press, New York, 281p.

Walton, D. C. 1980. Biochemistry and physiology of abscisic acid. *Ann. Rev. Plant Physiol.* **31**:453–489.

Yang, S. F., and N. E. Hoffman. 1984. Ethylene biosynthesis and its regulation in higher plants. *Ann. Rev. Plant Physiol.* **35**:155–189.

Zeevaart, J. A. D., and R. A. Creelman. 1988. Metabolism and physiology of abscisic acid. *Ann. Rev. Plant Physiol.* **39**:439–473.

5

DEVELOPMENT OF PLANTS AND PLANT PARTS

Plants and plant parts progress through a dynamic series of genetically controlled developmental processes terminating in their eventual senescence and death. Their development is the combination of both growth (an irreversible increase in size or volume accompanied by the biosynthesis of new protoplasmic constituents) and differentiation (qualitative changes in the cells) and can be viewed at either the whole plant or individual organ level.

Unlike animals, during the developmental period plants display a remarkable degree of variability in form that is strongly influenced by the environment in which they are grown. An animal, for example, a primate, will develop only four limbs regardless of variation in environment during the development period. Thus, in animals, morphological development is very tightly controlled. This situation is in sharp contrast with the tremendous diversity found in the plant kingdom. Responding to light, temperature, soil nutrient status, and other factors, two genetically identical plants may develop into structurally distinct mature plants. Environment, therefore, has a pronounced influence on the development of plants and plant parts, and this influence carries over into the postharvest period. Variations in composition and structure can significantly alter the way a product responds after harvest and as a consequence how it must be handled. If the physical and chemical changes occurring during the postharvest period are to be understood, it is essential that first there is an understanding of how the postharvest period fits into the entire developmental cycle of the plant.

The developmental period encompasses the entire time frame from the first initiation of growth to the eventual death of the plant or plant part. In the past, senescence was not generally accepted as part of development. It is now known that many of the initial processes of senescence are precisely regulated. Thus, the onset and initial developments in senescence do not represent a complete collapse of organization with ensuing chaos; rather this period represents a distinct portion of the overall developmental cycle of the plant or plant part.

In many crop plants this natural developmental cycle is interrupted prior to its completion by harvest. For example, lettuce plants are harvested very early in their developmental cycle; bean sprouts are harvested almost at the beginning. Grains and dried pulse crops, on the other hand, reach the end of their developmental cycle before being harvested. In its life cycle, a plant passes through a number of distinct developmental stages. These stages are closely synchronized with the development of the plant (internal control) and often with the environment in which the plant or plant part is held (external control). Many species have built-in control mechanisms that restrict a particular developmental step (e.g., germination, flowering). When the requirements for the control mechanism are satisfactorily met by an internal or external signal, development then proceeds.

SPECIFIC DEVELOPMENTAL STAGES

While there are numerous stages and substages in the developmental cycle of a plant, in this section the focus is on processes that can be and often are significantly modulated during the postharvest period. Developmental steps or stages such as dormancy, flowering, fruit ripening, abscission, and senescence are of critical importance for many postharvest products.

Dormancy

Dormancy represents a period of suspended growth and is common to many plants and plant parts (e.g., buds, tubers, seeds). In nature, dormancy serves to synchronize subsequent growth and development with desirable environmental conditions. As a consequence, many plants or plant parts are dormant during extended cold or dry periods, which are unfavorable for growth. In some seed-bearing species, the duration of dormancy within the seeds from a single parent plant may vary widely, which tends to enhance the long-term survival potential for the species. Other dormancy mechanisms appear to aid in the dispersal of the reproductive unit.

The postharvest period can dramatically affect dormancy. For many species of domesticated plants or plant parts, the storage environment is utilized to satisfy specific dormancy requirements. With the yam (e.g., *Dioscorea alata*, L. and others)[212] and the white potato (*Solanum tuberosum*, L.)[46] the dormancy period determines the length of time the tubers can be successfully stored. Thus, care must be taken to prolong the dormant period. For other species, it may be essential that the plant or plant propagule is not stored under conditions that will result in induced dormancy. Precise postharvest requirements, therefore, vary widely, depending on species and in some cases even cultivar in question.

Dormancy can be separated into three general classes based on the site or cause of the inhibition of growth.[119] Ecodormancy occurs when one or more factors in the basic growth environment are unsuitable for overall growth. These are external requisites such as moisture or temperature (more or less synonymous to the older dormancy term *quiescence*). Endodormancy occurs when an internal mechanism prevents growth even though the external conditions may be ideal. The initial reaction leading to growth control is a specific perception of an environmental or endogenous signal that occurs in the affected structure. Paradormancy is similar to endodormancy; however, the signal originates in or is initially perceived by a structure other than the one in which the dormancy is manifested.

The actual specific conditions for fulfillment of the requirements of the dormancy mechanism, thus allowing growth to proceed, vary widely. In the buds of many temperate perennials, sufficient exposure to cold temperatures above freezing and below a specific maximum is required. Many seeds also respond to low temperatures. Holding seeds in cold, moist conditions (stratification) has been a widely used agricultural practice for centuries. Other species fulfill their dormancy requirements with environmental signals such as temperature fluctuations (*Lycopus europaeus,* L.), light quality (certain lettuce cultivars), photoperiod (*Vertonica persica,* Poir.), or moisture (many desert species). The dormancy mechanism may respond to the summation of an external signal (e.g., hours of chilling) or it may require several types of signals simultaneously (e.g., photoperiod and temperature) to ensure against aberrant environmental conditions inadvertently triggering the continuation of growth.

Flowering

The ability to reproduce is a unifying and essential characteristic of all organisms. Although there are a number of reproductive strategies found in the plant kingdom, sexual reproduction by way of flower and seed production is one of the most common. Many flowering species have evolved beautiful and elaborate floral appendages that appear to facilitate successful reproduction. Because of their beauty, many of these, in turn, have been domesticated by man, adding immeasurably to the aesthetic quality of life.

Flowering represents a distinct stage in the overall developmental cycle of most plants. Therefore, the factors controlling flowering, from initiation to anthesis and eventually senescence, are of considerable interest to plant scientists. For postharvest physiologists, species in which the flowering process is in some way modulated during storage are of particular interest. In this section, the relationship between flowering and storage is examined.

Cut flowers held in refrigerated storage are often the first products associated with the postharvest physiology of floral crops. Considerable research has been devoted to expanding the longevity of cut flowers both during and after storage. Although cut flowers represent a substantial part of the total volume of flowers and flowering products handled, the range of products is actually much broader. For example, many flowers are sold attached to the parent plant [e.g., potted chrysanthemums (*Chrysanthemum morifolium,* Ramat.)]. The handling techniques utilized for intact plants and plant propagules may vary distinctly from those used for detached flowers of the same species. In some cases, flowering may be induced during storage with subsequent floral development occurring after storage (e.g., flowering bulbs such as tulips (*Tulipa* spp.) or Easter lilies (*Lilium longiflorum,* Thumb.)). For some biennial species, it may be essential to store the plants or plant propagules under conditions that will not induce flowering.

Flower Induction

Flowering can be separated into two distinct physiological processes: induction and development. Flower induction is known to be controlled or modulated by four primary mechanisms. These include vernalization, photoperiodism, thermoperiodism, and the stage of morphological development of the plant. Some species are influenced by only one of these mechanisms whereas the flowering of others is controlled by two or more. Vernalization is the promotion of flowering by low-temperature preconditioning of plants.[163] Individual species may be receptive to vernalization at a very specific growth stage or throughout development. For example, seeds of many biennials or winter annuals can be induced to flower by increasing their moisture content to approximately 40% and maintaining a low temperature. The actual flowering process does not proceed, however, until germination and subsequent development of the plant. Intact plants, such as *Matthiola*, are induced to flower by a 2- to 3-week cold period. Bulbs of Dutch iris (*Iris* spp.),[100] tulip,[164] hyacinths (*Hyacinthus* spp.), muscari (*Muscari botryoides*, Mill.),[174] and Easter lilies require a cold period for flower induction.[53]

The minimum length of the cold period and the effective temperature range vary with species. With many bulbs, part of the postharvest storage period may be utilized for flower induction. Proper timing of vernalization allows synchronizing flowering with specific sales periods. For many biennial species, it may be essential to store the plants or plant propagules under conditions that will not result in flower induction. Exposure to low temperatures at the transplant stage can be sufficient to cause premature flowering after field planting and loss of a significant portion of the crop.[66,141]

Photoperiodism, the length of the daily light and dark periods, provides a means by which many species synchronize the timing of their reproductive phase during the growing season. Flowering plants can be divided into three general groups based on photoperiodic response: short-day plants, long-day plants, and day-neutral plants. For a few very sensitive species, extremely short exposures to the proper day–night time sequence are sufficient to induce flowering. Storage conditions of some species, then, may be modified to enhance or conversely prevent flower induction depending on needs.

The influence of daily temperature fluctuations on plant growth and development is known as thermoperiodism. The effects of thermoperiodism on flowering tend to be more quantitative than inductive. Optimum temperature fluctuations may result in more flowers per plant and other growth effects rather than an all or nothing response such as found with photoperiod or vernalization. Thermoperiodism may be a diurnal phenomenon or a seasonal phenomenon. In the tulip, the optimum temperature varies during the growth of the plant. Therefore, thermoperiodism is much more important during periods of plant growth rather than during periods of suspended growth in storage.

The stage of development of the plant also affects the flowering response; this situation, however, varies widely in the plant kingdom. Some species (e.g., *Pharbitis nil*, Choisy) will flower under optimum inductive conditions at the cotyledonary stage of development. Other species require longer periods, extending in length from days to years. Several species of bamboo require approximately 50 years before flowering and then do so only once.

Flower induction then can be modulated or controlled by several environmental parameters and the stage of development of the plant. In some instances, the storage period for a specific crop can be utilized to induce flowering. This is the case with many bulbs. It may also be essential to store plants that have already been induced to flower under conditions where the flowering stimulus is not diminished or eliminated. When poststorage flowering is not desirable, storage conditions must be selected that will prevent induction.

Flower Development

After receiving the floral induction stimulus that results in the conversion of the vegetative meristem into a floral one, the meristem undergoes a series of developmental changes. These changes can be grouped into the following general stages: cell division, cell elongation and flower maturation, anthesis, and senescence. The initial stage involves intense cell division with the sequential development of the individual floral parts. The floral parts are initiated by periclinal division of cells, usually found deeply beneath the protoderm of the apical meristem. Generally they develop in a distinct sequence, arising from the outside inward, that is, sepals → petals → stamen → pistil (fig. 5-1). The duration and timing of the cell division phase vary widely

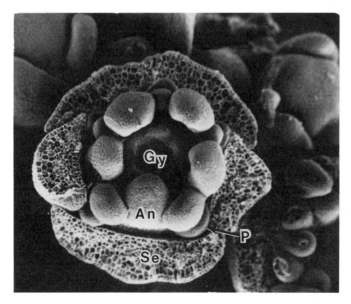

Figure 5-1. Most perfect flowers develop their parts in a distinct sequence, forming whorls from the outside to the inside, that is, sepals (Se) → petals (P) → androecium (An) → gynoecium (Gy), as illustrated by this geranium (*Pelargonium × hortorum*, Bailey) flower bud. The androecium is composed of the anthers and filaments and the gynoecium is made up of the stigma, style, and ovary. (*Photograph courtesy of H. Y. Wetzstein.*)

between individual floral parts. For example, the petals and stamen often cease cell division much earlier than the carpels, which may continue to divide in some species even after fruit set. The initial organogenesis of the floral parts has been widely studied by plant anatomists.[181] These studies tend to focus on the initial developmental stage of each floral part through completion of cell division.

The cell division phase of flower development is followed by cell expansion. The timing and duration of the cell expansion phase vary widely between the individual floral parts as well as due to species and environmental conditions. Each flower part appears to go through the same general sequence of developmental events as the complete flower, that is, cell division, cell expansion, maturation, and senescence. As with flower induction and cell division, the cell expansion phase is also strongly modulated by environmental factors. Unlike induction and division, however, cell expansion and subsequent flower opening stages have been little studied. These stages are of critical importance during postharvest handling and storage.

During the cell expansion phase, the individual floral parts develop, often approaching their final size. The rate at which development occurs varies widely with species. In many flowers, petal growth proceeds very rapidly just prior to opening (fig. 5-2) with a differential elongation rate between the interior and exterior portion of the flower (petal or perianth leaf) mediating the actual opening response.[151] Elongation induced flower opening differs from leaf movements that are generally turgor driven.

Anthesis, especially in species with large brightly colored flowers, represents one of the most spectacular phenomena in the plant kingdom. The actual flower opening response, however, varies widely. Many flowers open and remain open until death, the duration of which is dependent on species

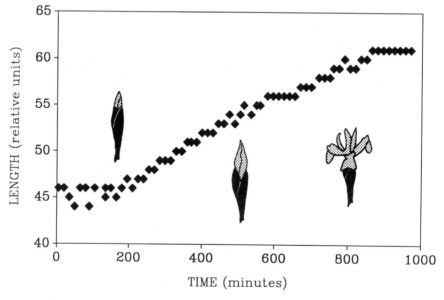

Figure 5-2. The rapid elongation of the petals is the driving force in flower opening in many species. This elongation phenomenon is illustrated by the increase in bud length of iris flowers during anthesis *(after Reid and Evans[168])*.

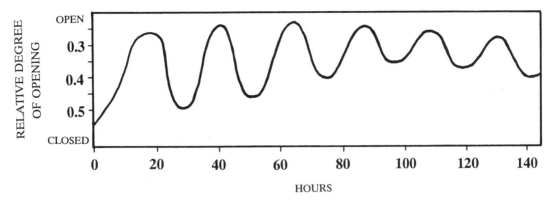

Figure 5-3. Although virtually all of the flower species that are used as cut flowers in the floral trade remain open after anthesis, a few exhibit distinct opening and closing cycles. This cycling is illustrated by the movement of *Kalanchoë blossfeldiana*, v. Poellnitz. flowers *(after Engelmann et al.[69])*.

and environmental conditions. Others open and close at certain times of the day (fig. 5-3), the number of cycles depending on the longevity of the flower. The former tend to be the most commonly used species in the floral trade. The actual opening response may be modulated by light or dark rhythm (fig. 5-4), temperature, and moist conditions.

Petal and, in some cases, sepal pigmentation are critical factors in flower quality, and like other events in the development of flowers, pigmentation can be affected by environmental conditions. The timing of pigment synthesis relative to other developmental events in the flower varies with species. Most flowers develop their full complement of pigments prior to opening. In some cases, however, major changes in pigments occur after anthesis. For example, the flowers of *Victoria amazonica*, Sowerby* are pure white the night of anthesis. Pigments are synthesized while the flowers are closed on the subsequent day, and upon reopening, the flowers are red their second night of bloom.[72]

In nature, pollination follows anthesis in the normal sequence of developmental events. Pollination sets off a dramatic series of chemical and morphological changes within the flower. In some species, pollination greatly accelerates the senescence of certain flower parts (e.g., petals and stamen) and as a consequence can be undesirable. For example, pollination results in rapid flower fading in orchids[12] and petunia (*Petunia × hybrida*, Hort.)[80] and abscission in the snapdragon (*Antirrhinum majus*, L.), sweet peas (*Lathyrus odoratus*, L.), and other flowers (fig. 5-5). (Flower senescence is discussed later in this chapter.)

Fruit Ripening

As fruits near the end of their growth phase, a series of qualitative and in some cases quantitative transformations occur with ripening. The fruit of agricultural products varies widely in both chemical composition and physi-

---

* Synonymous to *V. regia*, Lindl.

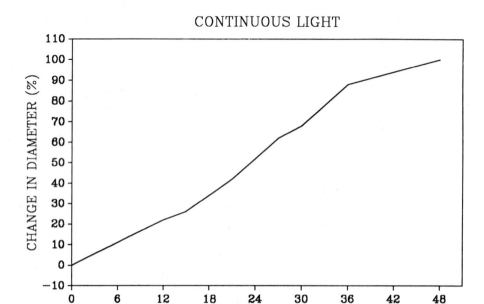

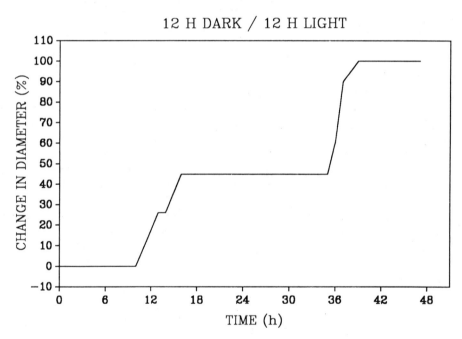

Figure 5-4. The environment used for holding flowers after harvest can have a pronounced effect on bud opening. This figure illustrates the effect of continuous light (*top*) versus a 12-hour dark/light cycle (*bottom*) on the opening of cut rose flowers (*after Evans and Reid*[70]).

Figure 5-5. Pollination by insects (right) can trigger flower senescence and abscission in some species (e.g., snapdragon, *Antirrhinum majus,* L.).

cal structure. These fruits, however, may be divided into two very general classes based on the structure of their fruit wall: parenchymatic or fleshy fruits (e.g., banana (*Musa* spp.), citrus, apple (*Malus sylvestris,* Mill.), tomato (*Lycopersicon esculentum,* Mill.)) and sclerenchymatic or dry fruits (e.g., rice (*Oryza sativa,* L.), wheat (*Triticum aestivum,* L.), peas (*Pisum sativum,* L.), corn (*Zea mays,* L.), nuts). With fleshy fruits, the fruit tissue rather than the actual reproductive organ, the embryo, represents the object of commercial importance.* In nature, the fruit tissue in fact has only an indirect role in the perpetuation of the species. With most dry fruits, the

---

* There are a few exceptions, for example, almond and coffee.

individual seeds (fertilized mature ovule) rather than the fruit is the article of postharvest interest. Many one-seeded dry indehiscent fruits, however, are regarded as seeds in a functional sense, a unit of dissemination.

In fleshy fruits, ripening refers to the transition period between maturation of the fruit and senescence. During ripening, fleshy fruits undergo a series of distinct, and in some cases, very dramatic changes in their physical and chemical condition. With dried fruits, however, ripening is less well defined and may refer, depending on species, to all the final stages of development and maturation. For example, the ripening period in wheat is divided into six stages: watery ripe, milky ripe, mealy ripe, waxy ripe, fully ripe and dead ripe.[150] During the watery ripe stage cell division in both the endosperm and embryo is still occurring. Similar, but not necessarily identical, stages have been ascribed to many other dried fruits.

Ripening, therefore, describes distinctly different events in fleshy and dry fruits. Thus, the concept of ripening and the changes occurring during this time period will be treated separately.

Fleshy Fruits

Botanically a fruit is a mature ovary that contains one or more seeds and may include accessory floral parts. In fleshy fruits, the fruit wall, which may consist of the ovary wall or the ovary wall fused with noncarpellary tissue(s), is much developed. Depending on the type of fruit, the entire ovary wall or various parts differentiate into the fleshy parenchymatic tissue. The individual "fruit" of commerce may be formed from a single, enlarged ovary (simple fruit, e.g., avocado (*Persea americana,* Mill.)), a number of ovaries belonging to a single flower (aggregate fruit, e.g., strawberry (*Fragaria* × *Ananassa,* Duchesne)) or enlarged ovaries of several flowers including accessory parts fused to form the fruit (multiple fruit, e.g., pineapple (*Ananas comosus,* (L.) Merrill)) (table 5-1).

Table 5-1. Classification of Fruits

Dry Fruits		*Fleshy Fruits*		
Indehiscent	*Dehiscent*	*Simple*	*Aggregate*	*Multiple*

Grains
  wheat, rice
	Legume
	  pea, bean
		Berry
		  tomato
			Berry
			  strawberry,
			  raspberry
				Berry
				  mulberry,
				  pineapple

Nuts
  chestnut, pecan
	Silique
	  mustard
		Drupe
		  peach		

Caryopsis
  corn
	Capsule
	  snapdragon
		Pome
		  apple		

Achene
  sunflower
	Follicle
	  columbine
		Hesperidium
		  orange		

Samara maple, elm				
Schizocarp celery				

While immature the parenchymatic tissues that comprise the main body of the fruit have a firm texture and the cells retain their protoplasts. At the transition phase following maturation, the fruit undergoes a series of generally nonreversible qualitative changes that render it attractive for consumption. These diverse physical and chemical alterations are termed *ripening* in fleshy fruits, and lead to the senescence phase of development, culminating in the ultimate death of the fruit.[29] Changes in texture, storage materials, pigments, and flavor components may occur. Due to the great diversity in the origin and composition of the parenchymatic tissue and the number of ovaries that comprise the individual fruit, considerable diversity is found in the ripening behavior between species.

Fruits vary widely in the length of time required to ripen once the ripening process begins and in their longevity after ripening before the tissue deteriorates to an unacceptable level. One major responsibility of postharvest physiologists and technologists is to develop the information and methods needed to maximize the duration of the period between ripening and deterioration.

Fruits of different species vary in their ability to ripen once detached from the parent plant. Many fruits must be harvested only when fully ripe (e.g., grape (*Vitis vinifera*, L.), cherry (*Prunus avium*, L.), bramble fruits) in that they are not capable of ripening after detachment (nonclimacteric fruits, table 5-2). These fruits do undergo many physical and chemical changes after harvest; however, these alterations are largely degradative and generally do not enhance the products' aesthetic qualities. Other fruits can be harvested unripe but if properly handled will undergo normal ripening even though detached from the plant (climacteric fruit). Examples of these include apples, bananas, and tomatoes. These fruits soften, undergo hydrolytic conversions in storage materials, and synthesize the pigments and flavors associated with a ripe fruit. With the exception of certain cultivars of several species (e.g., the avocado), detachment is not a prerequisite for ripening; each will ripen while attached to the parent plant.

This latter class of fruits, which can ripen normally after harvest, has been widely studied in that many offer a much greater degree of storage and marketing flexibility. If harvested unripe and held under conditions that prevent ripening, the ripening process can then be induced, allowing synchronization of ripening and marketing. The quality attributes that make up the aesthetic appeal of the ripe fruit are not present during the storage period and as a consequence are not subject to loss. Thus, consistently high-quality fruit can be marketed over extended periods. Many of the fruits that are available in retail stores during periods well beyond their normal field production time are of this latter class.

A relatively large number of fruits that fall into both of the preceding classes are marketed and utilized in an unripe state. Ripening results in undesirable changes that in fact decrease the aesthetic quality of the fruit. Many of the fruits that are eaten as vegetables are examples (e.g., bitter melon (*Momordica charantia*, L.), cucumbers (*Cucumis sativus*, L.), squash (*Cucurbita pepo*, L.), okra (*Abelmoschus esculentus*, (L.) Moench.)). Normally these fruits are harvested in an immature or mature unripe condition. Inhibiting the ripening process, therefore, may be essential during the postharvest period.

Table 5-2. Classification of Fleshy Fruits According to Their Respiratory Pattern

Common Name	Scientific Name	Reference
Climacteric Fruits		
Apple	*Malus sylvestris*, Mill.	28
Apricot	*Prunus armeniaca*, L.	28
Avocado	*Persea americana*, Mill.	28
Banana	*Musa* spp.	28
Biriba	*Rollinia deliciosa*, Safford	31
Bitter melon	*Momordica charantia*, L.	102
Blueberry, highbush	*Vaccinium corymbosum*, L.	98
Blueberry, lowbush	*Vaccinium angustifolium*, Ait.	98
Blueberry, rabbiteye	*Vaccinium ashei*, Reade	124
Breadfruit	*Artocarpus altilis*, (Parkins.) Fosb.	31
Cantaloupe	*Cucumis melo*, L. Cantalupensis group	126
Cherimoya	*Annona cherimola*, Mill.	28
Chinese gooseberry	*Actinidia chinensis*, Planch.	158
Corossol sauvage	*Rollinia orthopetala*, A. DC.	30
Feijoa	*Feijoa sellowiana*, O. Berg.	28
Fig, common	*Ficus carica*, L.	129
Guava, 'Purple Strawberry'	*Psidium littorale*, var. *longipes*, (O. Berg.) Fosb.	9
Guava, 'Strawberry'	*Psidium littorale*, Raddi.	9
Guava, 'Yellow Strawberry'	*Psidium littorale*, var. *littorale*, Fosb.	9
Guava	*Psidium guajava*, L.	9
Honeydew melon	*Cucumis melo*, L. Inodorus group	156
Kiwi (see Chinese gooseberry)		
Mammee-apple	*Mammea americana*, L.	8
Mango	*Mangifera indica*, L.	28
Papaw	*Asimina triloba*, (L.) Dunal.	28
Papaya	*Carica papaya*, L.	28
Passion fruit	*Passiflora edulis*, Sims.	28
Peach	*Prunus persica*, (L.) Batsch.	28
Pear	*Pyrus communis*, L.	28
Persimmon	*Diospyros kaki*, L.f.	166
Plum	*Prunus americana*, Marsh.	28
Sapote	*Casimiroa edulis*, Llave.	28
Soursop	*Annona muricata*, L.	31
Tomato	*Lycopersicum esculentum*, Mill.	28
Watermelon	*Citrullus lunatus*, (Thunb.) Mansf.	142
Nonclimacteric Fruits		
Blackberry	*Rubus* spp.	124
Cacao	*Theobroma cacao*, L.	31
Cashew	*Anacardium occidentale*, L.	31
Cherry, sour	*Prunus cerasus*, L.	39
Cherry, sweet	*Prunus avium*, L.	28
Cucumber	*Cucumis sativus*, L.	28
Grape	*Vitis vinifera*, L.	28
Grapefruit	*Citrus* × *paradisi*, Macfady	28
Java plum	*Syzygium cumini*, (L.) Skeels	9
Lemon	*Citrus limon*, (L.) Burm. f.	28
Litchi	*Litchi chinensis*, Sonn.	8
Mountain apple	†*Syzygium malaccense*, (L.) Merrill & Perry	9

Table 5-2. (*continued*)

Common Name	Scientific Name	Reference
Olive	*Oiea europaea*, L.	134
Orange	*Citrus sinensis*, (L.) Osbeck.	28
Pepper	*Capsicum annuum*, L.	177
Pineapple	*Ananas comosus*, (L.) Merrill	28
Rose apple	‡*Syzygium jambos*, (L.) Alston	9
Satsuma mandarin	*Citrus reticulata*, Blanco	166
Star apple	*Chrysophyllum cainito*, L.	157
Strawberry	*Fragaria* × *Ananassa*, Duchesne	28
Surinam cherry	*Eugenia uniflora*, L.	9
Tree tomato	*Cyphomandra betacea*, (Cav.) Sendtu.	159

* Originally cited as *Eugenia cumini*, (L.) Druce.
† Originally cited as *Eugenia malaccensis*, L.
‡ Originally cited as *Eugenia jambos*, L.

CHANGES THAT OCCUR WITH RIPENING

During ripening, fleshy fruits undergo major changes in their chemical and physical state. These changes represent a wide spectrum of synthetic and degradative biochemical processes, many that occur concurrently or sequentially within the fruit, although not all. Table 5-3 lists a cross-section of these alterations. Those that represent changes in quality attributes of the fruit can be grouped into three general categories: (1) textural changes, (2) changes in pigmentation, and (3) changes in flavor.

The induction and development of ripening are tightly controlled steps in the overall developmental cycle of a fruit. During this period, specific enzymes are synthesized or activated, triggering or accelerating specific metabolic events. For example, in some fruit (e.g., apples), protein synthesis increases dramatically with the beginning of the ripening response.[97] These proteins are thought to represent enzymes required for ripening.

Softening

Softening is one of the most significant quality alterations consistently associated with the ripening of fleshy fruits. Alterations in texture affect both the edibility of the fruit and the length of time the fruit may be held. In many of the fruits that are consumed in an unripe state (e.g., cucumbers, squash), softening may be detrimental. In others, it is an essential component in the development of optimum quality. As a consequence, when feasible during the postharvest handling and storage period, an attempt is made to maximize control over textural changes, whether it is to prevent, synchronize, or accelerate the process.

Once the softening process is initiated, the rate of textural change is a function of the type of fruit and the conditions under which the product is held. Often acceptable flesh texture represents a very narrow range that can be rapidly exceeded, diminishing the quality of the product. With the exception of some turgor-mediated textural alterations, softening in most fruits represents an irreversible process once it is initiated. This situation does not

Table 5-3. Physical and Chemical Alterations That Occur During the Ripening of Fleshy Fruits

	Important Quality Attributes
1. Seed maturation	

2. Changes in pigmentation
  a. degradation of chlorophyll
  b. unmasking of existing pigments
  c. synthesis of carotenoids
  d. synthesis of anthocyanins
	Color

3. Softening
  a. changes in pectin composition
  b. possible alterations in other cell wall components
  c. hydrolysis of storage materials
	Texture

4. Changes in carbohydrate composition
  a. starch conversion to sugar
  b. sugar interconversions
	Flavor

5. Production of aromatic volatiles	
6. Changes in organic acids	
7. Fruit abscission	
8. Changes in respiration rate	
9. Changes in the rate of ethylene synthesis	
10. Changes in tissue permeability	

11. Changes in proteins
  a. quantative
  b. qualitative
    1. enzyme synthesis	

12. Development of surface waxes	

Note: The order does not represent the sequence of occurrences during ripening.

necessarily mean that all synthetic processes enhancing wall rigidity cease; rather the degradative reactions proceed at a more rapid rate.

The texture of fleshy fruits is affected by the composition of their cell walls and cellular constituents and their degree of hydration. During ripening, enzymatically mediated degradative changes in the cell walls have been found in most fruits. The enzymes may be either synthesized, activated, or a combination of both, at or near the onset of the ripening process. In some fruits, such as the banana, hydrolysis of stored carbohydrates within the cells also results in a significant effect on texture. Changes in cellular osmotic properties are generally not associated with normal textural changes during ripening; however, in many fruits desiccation-induced osmotic alterations represent a potentially critical path for undesirable textural changes that may occur throughout the postharvest period.

The parenchymatic tissue of fleshy fruits is composed of polyhedral or elongated cells with considerable intercellular space between their cell walls. The primary cell wall consists of cellulose microfibrils embedded in a matrix of other polysaccharides and protein. Between adjacent cells, the walls are separated by the middle lamella, which is rich in pectic compounds.[154] Collectively the cell wall and middle lamella are composed of cellulose, hemicellulose, pectin, and protein. Most primary cell walls of the edible portion of fleshy fruits are unlignified and seldom have secondary cell walls present. The cell wall and middle lamella of the tomato fruit consists of cellulose (17%), pectin (22%), protein (17%), araban-galactan (21%), and xylose plus glucose (13–23%).[214] Hydrolysis of these cell wall components into their component monomers yields a cross-section of compounds. Amino acids present are characteristic of wall proteins; xylose and mannose of hemicellulose polymers; galactose, galacturonic acid, and arabinose of pectic compounds; and glucose of cellulose.

Enzymatic-mediated degradative changes in the composition of the cell walls, resulting in textural alterations, could be through attack of any of the primary polymers. Research by several laboratories has been directed toward establishing which polymers are altered significantly as softening develops and what types of structural changes are occurring.

Although cellulose makes up a major portion of the composition of cell walls, the role of cellulase* in textural changes during ripening remains in question. Substantial increases in cellulase activity have been reported during ripening[18,171,191] and increases in avocado cellulase activity precede increases in polygalacturonase activity by about 3 days.[15] Thus, there is a positive correlation between cellulase activity and softening. Further experiments on the avocado, however, have not substantiated a role for cellulase in cell wall modification leading to softening.[86] Cellulase isolated from avocado did not solubilize the cellulosic components from avocado cell walls. Likewise, alterations in the hemicellulosic component of the cell wall during ripening tend to be relatively insignificant in the softening of most fruits. While a decrease in the size of hemicellulose molecules has been demonstrated with ripening (tomato), there is no evidence for a significant role in textural changes.[94]

In contrast to cellulose and hemicellulose, very marked changes in the pectin fraction[95] found predominantly in the middle lamella occur with ripening and these changes parallel softening (fig. 5-6). Soluble pectins increase during this period while insoluble pectin declines. In unripe peaches (*Prunus persica,* (L.) Batsch.), approximately 25% of the pectin fraction is water soluble; upon ripening, this amount increases to approximately 70% of the total.[155]

Enzymatic-mediated changes in the insoluble pectin fraction, resulting in solubilization, may occur through the cleavage of linkages between the pectin molecule and other cell wall components or through the direct hydrolysis of the pectin molecule. The former appears to occur in the apple, whereas hydrolysis is the predominant mechanism in a relatively wide cross-section of other fruits.

The hydrolysis of pectin molecules involves the action of two types of enzymes: pectinesterases and polygalacturonases. Pectinesterase (also referred to as pectin methylesterase) catalyzes the hydrolysis (removal) of methyl esters, proceeding in a linear manner down the pectin molecule,

---

* Cellulase is often a complex of enzymes, for example, β-glucosidase, exo-$\beta(1\rightarrow4)$-glucanase, and two endocellulases were identified in tomato.[191]

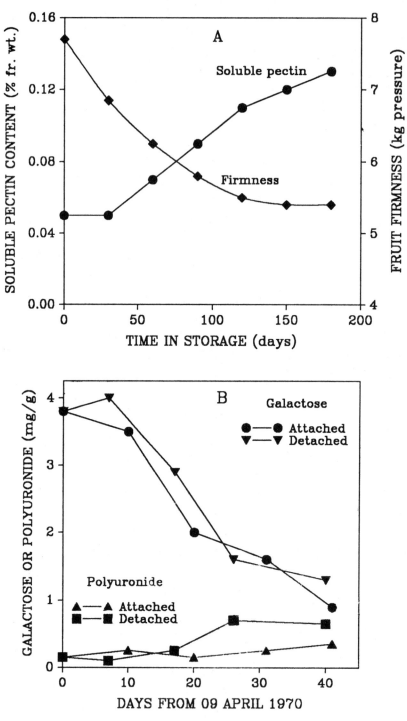

Figure 5-6. During fruit ripening there are pronounced changes in texture. In many fruits, softening involves degrading the large pectin molecules found in the middle lamella that bind together the walls of neighboring cells. *A:* As the size of the pectin molecules decreases, their water solubility increases as illustrated here in 'Delicious' apples (*Malus sylvestris,* Mill.) *(after Gerhardt and Smith[79])*. *B:* A decline in cell wall galactose content and an increase in polyuronides are also indicative of pectin hydrolysis *(after Knee[111])*.

leaving free carboxyl groups. While this process does not result directly in softening, it must precede further degradation of the molecule by polygalacturonase.

The polygalacturonases are the only known pectolytic enzymes found in fruits. Two polygalacturonases are generally present: an endopolygalacturonase that attacks the pectin molecule at various sites within the chain and an exopolygalacturonase that sequentially removes galacturonic acid residues from the end of the molecule. Of the two, endopolygalacturonase is much more important with regard to physical changes in texture. Cleavage of a pectin molecule in the center yields two molecules, one half the size of the original, greatly increasing their solubility. However, removal of a terminal subunit by exopolygalacturonase results in only a very slight alteration in the properties of the pectin molecule. Both mediate changes in solubility; however, the two enzymes differ greatly in the rate in which they may effect change. The levels of the two polygalacturonases vary with species and within species. For example, apples appear to lack endopolygalacturonase but it is present in pears (*Pyrus communis,* L.), peaches, strawberries, tomatoes, and many other fruit.[160] Freestone peaches, which soften extensively during ripening, have high activities of both endo- and exopolygalacturonase. Clingstone cultivars, however, have low activities of endopolygalacturonase and display much less softening and pectin solubilization with ripening.[161]

Even with the very marked correlations that have been reported between polygalacturonase activity and fruit softening, the precise role of polygalacturonase in softening is far from clear. Recent findings with transgenic plants have renewed the debate on the relative importance of polygalacturonase. When transgenic tomato plants were produced in which the level of polygalacturonase synthesis in the fruit was substantially reduced (approximately 90%), fruit softening was not significantly different from normal nontransformed fruit.[188] Likewise, when polygalacturonase gene transcription was added to a mutant line that lacks the enzyme, the subsequent cell wall degradation was not sufficient for softening.[81] These results have pointed toward several possible conclusions: (1) The importance of polygalacturonase in cell wall degradation is currently overrated; (2) relatively small amounts of polygalacturonase may be sufficient to instigate softening; and (3) polygalacturonase gene expression and subsequent transcription is sufficiently complex (as indicated by data of Sheehy et al.[184]) and poorly understood to warrant definite conclusions at this time.

Hydrolytic changes in the cellular constituents during ripening may also affect fruit texture. These changes, however, tend to be much less significant than concurrent alterations in cell wall structure and are often difficult to separate from the latter.

Alteration in cellular hydration is known to result in marked changes in the texture of plant products. Often, the percent water loss needed before realizing undesirable textural changes is quite small. Although changes in the rate of transpiration have been found during the ripening of fruits such as the banana,[187] changes in osmotic conditions do not appear to be a significant part of the normal softening process during ripening. Only small changes in hydration normally occur during ripening when fruits are held under proper relative humidity conditions. With improper handling, however, water loss may yield significant losses in textural quality during the postharvest period.

Hydrolytic Conversion of Storage Materials

Many fruits that have the potential to ripen after being detached from the plant undergo significant changes in their storage form of carbon during this process. Starch, in many fruits, is the prevalent carbon storage compound and it undergoes hydrolytic conversion during ripening, yielding free sugar. In other fruits, carbon may be stored as lipids or organic acids that may or may not be altered during ripening.

The banana fruit represents an excellent example of hydrolytic alterations during ripening (fig. 5-7). The edible mesocarp tissue contains 20% to 30% starch when the fruit is at a mature green stage. During ripening, the starch concentration decreases to only 1% to 2% while the concentration of sugar increases from 1% to 14–15%.[220] Along with changes in the textural properties of the tissue, this conversion of starch to sugar enhances the palatability of the fruit.

The rate and extent of hydrolysis of carbon storage compounds during ripening vary widely between fruit of different species. With the banana, there is a dramatic conversion of starch to sugar and this conversion generally occurs over a relatively short period of time (e.g., 8–10 days).[204] In the avocado, however, lipids represent the predominant storage form of carbon. These compounds do not appear to represent a respiratory source of carbon during the ripening process.

In fruits that ripen normally only when attached to the parent plant there is often a significant increase in sugars during this period. In this case, the sugars are not derived via the hydrolysis of carbon compounds stored within the fruit. Rather, these sugars are composed of carbon transported from the

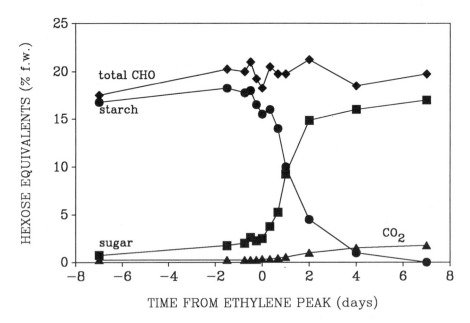

Figure 5-7. In banana (*Musa* spp.) fruit there is a very rapid hydrolysis of starch and a corresponding increase in sugar concentration during ripening (*after Beaudry et al.*[22]). In other fruits (e.g., oranges, *Citrus sinensis,* (L.) Osbeck.) changes in composition may occur at a much slower rate.

parent plant into the fruit during the ripening process. As a consequence, the accumulation of sugar is dependent on attachment; fruit removal prior to ripening prevents further accumulation. Although postharvest softening will generally occur in these fruits if prematurely harvested, they do not develop the full complement of other aesthetic and edible quality characteristics associated with a ripe fruit.

Changes in Pigmentation

Changes in the coloration of fruits during ripening are often spectacular. In nature, these changes facilitate the natural dispersal of the seeds of many species by attracting animals.

With fruits consumed by humans, color changes that occur during the ripening period are often used as an index to the degree of ripeness. Thus, the timing of harvest for some fruits (e.g., tomato) may be determined using fruit color. Color is also the primary criterion used by consumers in determining ripeness of many fruits. Thus, color changes during ripening and storage are of primary importance. It is necessary to be cognizant, therefore, of the conditions that mediate both desirable and undesirable changes in the color after harvest.

Alteration in the coloration of fruits normally involves the loss of chlorophyll and either the synthesis of other pigments such as carotenoids and anthocyanins[217] (fig. 4-34) and/or the unmasking of these pigments formed earlier in the development of the fruit. Some fruits do, however, retain their green coloration throughout the ripening period, for example, avocado, kiwi (*Actinidia chinensis*, Planch.), honeydew melons (*Cucumus melo*, L.), some apple cultivars.

The timing, rate, and extent of change in fruit color vary widely between different species and cultivars of the same species. The timing of color change can be assessed relative to the time of harvest or to the actual ripening of the fruit. Many red apple cultivars develop much of their coloration during the final stages of development prior to ripening. Synthesis of pigments may continue, however, throughout the ripening period. A number of climacteric fruits have the potential to develop their normal coloration after removal from the parent plant (e.g., banana, tomato). Many fruits, however, only develop their normal coloration while attached to the parent plant. If picked prior to ripening, these fruits may lose much of their chlorophyll but will seldom develop more than a small portion of their normal complement of pigments.

Changes in fruit color may or may not coincide with the development of the other quality criteria associated with ripening. With apples, color development does not closely parallel the respiratory climacteric. Color, therefore, is not generally an acceptable means of assessing ripeness of this fruit. There is, however, a relatively close association between color changes and ripening in climacteric fruits such as the banana and bitter melon and nonclimacteric fruits such as the cherry, blueberry (*Vaccinium* spp.), and strawberry.

The rate of color change also varies widely. In many fruits there is a relatively slow but steady synthesis of the pigments making up the final coloration of the product. In others this change is compressed into a relatively short time period, giving a spectacular effect. For example, the fruit of the bitter melon when exposed to ethylene can proceed from green to bright orange in a 24-hour period.[102] Likewise, the color change from green to yellow in the banana can be compressed into several days with appropriate treatment. Many of these very rapid changes in coloration are mediated to a large extent by the rapid degradation of chlorophyll and the exposure of existing pigments that were previously masked rather than de novo pigment synthesis alone (fig. 5-8).

Most color changes in fruit are associated with a decrease in the concentration of chlorophyll molecules in the chloroplasts. The chloroplasts are transformed during ripening through extensive changes in their internal membranes and the synthesis of carotenoids (yellow to red in color) into chromoplasts. The loss in chlorophyll is mediated through an increase in the activity of the enzyme chlorophyllase, which degrades the molecule.*[125] Concurrent synthesis of carotenoids does not occur in all fruit, however. Significant color changes may be mediated in some fruits through the degradation of chlorophyll and the exposure of preexisting carotenoids. For example, during the transformation of three cultivars of banana from green to yellow during ripening, the mean chlorophyll content of the peel decreased from 77 $\mu g \cdot g^{-1}$ to 0 $\mu g \cdot g^{-1}$ while the carotenoid concentration increased from 10.5 $\mu g \cdot g^{-1}$ to only 11.3 $\mu g \cdot g^{-1}$.[56]

Unlike chlorophyll and carotenoids, which are sequestered in chloroplasts or chromoplasts, anthocyanins accumulate in the vacuoles and are responsible for the pink, red, purple, and blue colors of fruits. Cells with high concentrations of anthocyanins may be found distributed throughout the entire fleshy portion of the fruit (e.g., sweet cherry cultivars) or found in abundance only in the epidermal or subepidermal tissues of the fruit (e.g., cranberry, *Vaccinium macrocarpon*, Ait.[178]; apples; plums, *Prunus domestica*, L.; and pears[52]) (fig. 5-9). Additional details on the synthesis and degradation of pigments are covered in chapter 4.

Color alterations within a fruit during ripening are affected by a number of factors. As mentioned previously, the time of harvest may significantly alter the development of the normal complement of pigments in many fruits. Light, temperature, and oxygen concentration may also have a pronounced effect on color development. Light is not essential for the synthesis of carotenoids and in some cases [peach and apricot (*Prunus armeniaca*, L.)] color development has been shown to be greater in the absence of light. Although not an obligatory requirement for synthesis, light has also been shown to enhance the synthesis of carotenoids in tomatoes harvested prior to ripening. Light in the blue portion of the spectrum stimulated carotenoid synthesis while red light accelerated chlorophyll degradation.[99]

Anthocyanin synthesis is markedly stimulated by light in many fruits (e.g., apple). Red apple cultivars allowed to mature in total darkness remain green.[186] The light-mediated synthesis appears to involve two photoreactions: a low-energy reversible phytochrome controlled reaction and a high-energy reaction. Thus, both light quantity and quality are important in anthocyanin development.[118]

---

* Two other enzymes (chlorophyll oxidase and peroxidase) have been shown to degrade chlorophyll.[3,132]

Figure 5-8. In some cases (e.g., banana, *Musa* spp.), the degradation of chlorophyll unmasks largely preexisting pigments, accounting for the rapid change in fruit color (ripening stages: I = green; II = green-yellow; III = yellow-green; IV = yellow) *(after Gross and Flugel[83]).*

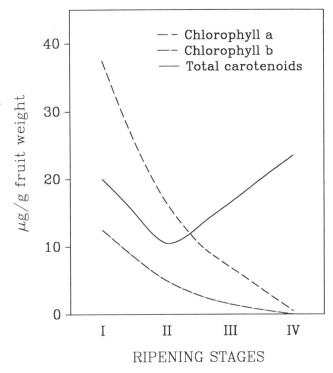

Temperature can affect both the rate of synthesis of specific pigments and their final concentration within the fruit. The optimum and maximum temperature for synthesis of a specific pigment varies between species. For example, lycopene synthesis in the tomato is inhibited above 30°C,[82] whereas in the watermelon (*Citrullus lanatus,* (Thunb.) Mansf.) synthesis is not prevented until the fruit temperature rises above 37°C.[203] Oxygen is essential for carote-

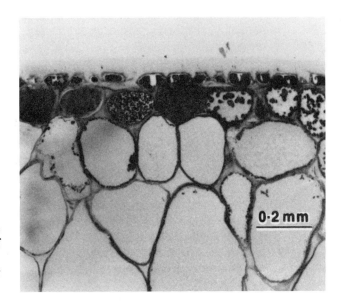

Figure 5-9. Anthocyanin pigments are found in the outer two layers of cells of cranberry (*Vaccinium macrocarpon,* Ait.) fruit (cv. 'Pilgrim') *(after Sapers et al.[178]).*

noid synthesis[203] and increasing the oxygen concentration enhances the synthesis of these pigments.[55]

Changes in Flavor

Flavor is generally considered to be comprised of two major attributes: taste and odor. Taste is perceived by taste buds in the mouth, while odor is detected by olfactory receptors in the nose. Fruit flavor is primarily a composite of sugars, acids, and volatile compounds and with ripening there are very dramatic changes in the flavor of most fruits.

Predominant alterations in taste during ripening center largely around changes in sugars and organic acids. With fruits that must ripen while attached to the parent plant, sugars increase via translocation of sucrose from the leaves. Upon arrival, sucrose in grapes is hydrolyzed by invertase, forming glucose and fructose. The total sugar concentration (as measured by soluble solids) in white Riesling grapes begins to accumulate rapidly with the onset of ripening (fig. 5-10). In some climacteric fruits, although not all, changes in internal sugars represent products derived from the hydrolysis of carbohydrate reserves within the tissue. For example, the sugar concentration in the banana increases 12-fold to 15-fold during ripening (fig. 5-8). Increases in free sugars are due to hydrolysis of starch reserves by α-amylase and β-amylase and/or starch phosphorylase. The activity of these enzymes increases markedly during the ripening of many fruits. Lipids may also be hydrolyzed and converted to sugars.

Some fruits that do not accumulate significant reserve carbohydrates (e.g., honeydew melons) will also ripen after harvest. Their internal sugar concentration is dependent on accumulation of sugars prior to harvest; thus, early harvest can greatly compromise final quality.

Changes in acidity are also important in the development of the characteristic taste in many fruits (fig. 5-11). Although there are a number of organic acids found in plants, generally only one or two accumulate in the fruit of a species. For example, malic acid predominates in apple, banana, and cherry; citric acid in oranges (*Citrus sinensis*, (L.) Osbeck.), lemons (*Citrus limon*, (L.) Burm. f.), and currants (*Ribes nigrum*, L.); citric and malic acids in tomato and gooseberry (*Ribes grossularia*, L.); and malic and tartaric acids in the grape.

During ripening, there is a decrease in organic acids in most fruits. This loss is due largely to the utilization of these compounds as respiratory substrates and as carbon skeletons for the synthesis of new compounds during ripening. The decrease in total acidity in the grape tends to coincide with the onset of ripening and the accumulation of sugars (fig. 5-11). The concentration of organic acids does not, however, decline in all fruits during ripening. In the banana, there is a significant increase in the concentration of malic acid and a decrease in pH (e.g., 5.4 to 4.5).[148]

The aroma of a fruit is an extremely important quality criterion and as fruits ripen there is an increase in the rate of synthesis of these volatile compounds (fig. 5-12). Over 200 different compounds have been identified emanating from ripe banana fruits. Similar numbers of volatile compounds have been identified in various other fruits[147] (for additional details see chapter 4). Only a relatively small number of the total complement of volatile

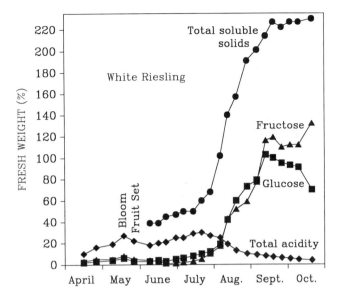

Figure 5-10. Sugars begin to accumulate very rapidly with the onset of ripening in grapes, *Vitis vinifera*, L. *(after Kliewer[110]).*

compounds, however, tends to make up the characteristic aroma perceived for a specific fruit.

Fruits vary considerably in the concentration and type of volatile compounds they produce. In the banana, odor concentrates between 65 ppm and 338 ppm were collected whereas the strawberry yielded only 5–10 ppm. In addition, ability to perceive the odor of specific compounds varies widely. For example, vanillin can be sensed at a concentration of 0.000,001 mg $\cdot$ m^{-3} of air while diethyl ether can be detected by humans only when the concentration has reached 1 mg $\cdot$ m^{-3} or greater.

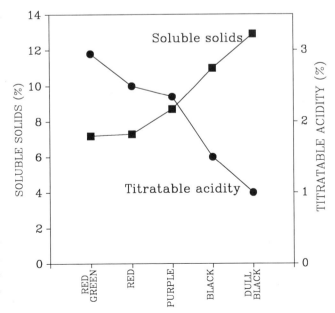

Figure 5-11. Sweetness (as indicated by soluble solids) and acidity are important components in the taste of many fruits. Both commonly change with the onset of ripening, illustrated by blackberry (*Rubus macropetalus,* Dougl. ex Hook.) fruit *(after Walsh et al.[207]).*

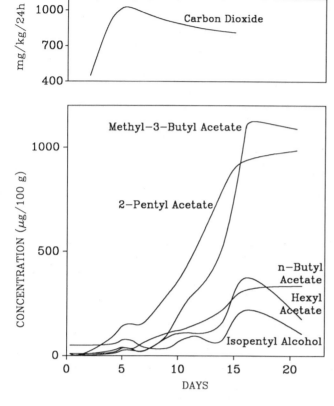

Figure 5-12. In addition to taste, perceived by the taste buds in the mouth, aromatic compounds are extremely important components of the flavor of edible products. The synthesis of critical volatiles typically increases markedly with ripening, illustrated by changes in volatile compounds evolving from banana fruit *(after Drawert[59])*.

There are distinct quantitative and qualitative differences in the composition of the volatiles produced by individual fruits as they ripen. More than 85 compounds are known to emanate from ripe peaches, none of which exhibit a characteristic peachlike aroma by themselves. Rather, the aroma of a ripe peach appears to be comprised of gamma- and delta-lactones, esters, aldehydes, benzyl alcohol, and d-limonene.[58] The concentration of these compounds increases as the fruit ripens. If picked at a preripe stage of development (hard-mature) and artificially ripened at room temperature, the production of these volatiles is markedly depressed. This loss in volatiles appears to be due to the fact that many of the aromatic compounds are apparently synthesized in the leaves and translocated into the fruit.[92] Even with fruits (e.g., tomato) that ripen "normally" after detachment, the characteristic odor may be significantly diminished, although still within an acceptable range.

In addition to stage of maturity at harvest, handling and storage conditions may also significantly alter the synthesis of volatiles. Fruits held in controlled atmosphere and hypobaric storage are often incapable of synthesizing normal quantities of these compounds upon removal. This inability does not appear to be due to an inhibition of the overall ripening process since with apples chlorophyll degradation and ethylene synthesis proceed normally.[87]

CHANGES IN THE RESPIRATORY RATE OF
CERTAIN FRUITS

A number of fruits exhibit a spectacular increase in the rate of respiration, called the respiratory climacteric, during ripening. During the final stages of maturation, fruit respiration declines to a very low level (the preclimacteric minimum, see fig. 3-25 in chapter 3). This decline is followed by a tremendous increase in the rate of respiration (as much as four to five times) with the onset of ripening. The rate of respiration eventually peaks (climacteric peak) and then declines during the postclimacteric period. Climacteric fruits are also distinguished from fruits not exhibiting this burst in respiration (nonclimacteric) in that they increase their rate of synthesis of the hormone ethylene in response to exposure to low levels of the hormone (termed autocatalytic production of ethylene).

The respiratory climacteric was first studied in detail by Kidd and West.[107] Since this early work, postharvest physiologists have had a fascination with the respiratory climacteric that has translated into an impressive accumulation of research data.

The respiratory climacteric does not occur in all fruits, however. In fact, the fruits of most plants do not appear to display a climacteric. Rather, a relatively low, consistent rate of respiration is maintained in these fruits during ripening (fig. 3-25). Many nonclimacteric fruits do not have large respiratory energy reserves that can be hydrolyzed as respiratory substrates during ripening. Respiratory substrates in these fruits are supplied by the parent plant throughout the ripening process. While lacking the respiratory climacteric and autocatalytic production of ethylene, nonclimacteric fruits obviously do ripen. What fundamental differences then exist in the ripening mechanism between climacteric and nonclimacteric fruits? It is currently believed that these two classes of fruits differ primarily in the rate in which the events of ripening proceed. At present, however, the climacteric/nonclimacteric classification remains useful.

The list of climacteric fruits that have been identified is generally longer than that of nonclimacteric fruits. However, it is somewhat misleading in that it is much easier to establish that the respiratory climacteric exists in a species than to prove its nonexistence. Thus, there are probably substantially more nonclimacteric fruits than climacteric ones.

Why has there been such an interest in the respiratory climacteric among postharvest physiologists? From a scientific basis, the respiratory climacteric represents a convenient map or timetable on which other changes occurring during ripening can be placed and related to each other. Early work also indicated that environmental factors that inhibited the respiratory climacteric likewise inhibited ripening. Thus, fruit could be held in storage and induced to ripen as needed for individual studies. This situation allowed scientists to greatly extend the time period in which the ripening process could be studied. In addition, many of the important fruit crops (e.g., apple, pear, banana, tomato) are also climacteric. Both the level of grower interest and support for research on these crops enhanced their attractiveness as research projects.

Commercially, our ability to harvest many climacteric fruits when unripe, store them for extended periods under conditions that prevent ripening, and subsequently ripen them greatly expanded the length of time these fruits were available fresh for consumption. Sales were no longer limited to just the harvest season and yearly consumption was similarly expanded.

The timing of the respiratory climacteric relative to optimum eating quality of the fruit varies with species. In pear fruits, the climacteric peak more or less coincides with this optimum. In the apple and banana it occurs slightly before and in the tomato it occurs well before optimum ripeness.

Individual species also vary in the speed in which the climacteric proceeds and the maximum rate of respiration at the climacteric peak[27] (fig. 3-26). The climacteric occurs quite rapidly in the banana and breadfruit, at intermediate rates in apple and mango (*Mangifera indica*, L.), slowly in the fig (*Ficus carica*, L.), and not at all in nonclimacteric fruits, for example, lemon, grape, and strawberry. At the climacteric peak of the breadfruit (*Artocarpus altilis*, (Parkins.) Fosb.) and cherimoya (*Annona cherimola*, Mill.),[31,116] respiration is proceeding at 170–180 mg $CO_2 \cdot kg^{-1} \cdot hr^{-1}$; in the mango[43] the rate is approximately 80 mg $CO_2 \cdot kg^{-1} \cdot hr^{-1}$; and in the apple and tomato[127,190] the rate is 10–20 mg $CO_2 \cdot kg^{-1} \cdot hr^{-1}$.

CONTROL OF RIPENING

Ripening represents a complex of controlled synthetic and degradative changes, many of which appear to be biochemically independent of one another (table 5-3). A number of the general changes associated with ripening have been described previously. What are the essential elements controlling the ripening process once triggered?

A source of respiratory substrate is required to provide the energy to drive the reactions occurring during ripening. This substrate also provides carbon skeletons for many of the new compounds that are formed. In many fruits, the respiratory substrate comes from the parent plant as the fruit ripens while still attached. In others, it is recycled from stored carbon within the fruit.

The importance of respiration to ripening was shown by using inhibitors of the metabolic production of energy (e.g., dinitrophenol, arsenite).[25,130] When energy production is inhibited, ripening is normally also inhibited.

Also required for ripening are quantitative and qualitative changes in the normal complement of enzymes. These enzymes bring about the characteristic synthetic and degradative changes (softening, flavor, pigmentation) that occur during ripening. The need for de novo synthesis of enzymes during ripening was illustrated with the incorporation of radioactive amino acids into fruit proteins. Incorporation increased substantially from the preclimacteric period to the early climacteric rise but dropped drastically at the climacteric peak.[169] If an inhibitor of protein synthesis was added, softening and pigmentation changes were terminated without affecting the climacteric rise in respiration.[76] This situation suggested that the respiratory rise was not linked to the synthesis of new enzymes, whereas many of the ripening changes apparently are.

INITIATION OF RIPENING

Plant hormones appear to play an extremely important, yet unexplained, role in the developmental processes of plants. Intuitively, then, it would be expected that plant hormones also play a role in one of the final developmental stages, the ripening of fruits. Early work, long before ethylene was known to be a plant hormone, indicated that exogenous ethylene could dramatically induce the ripening of climacteric fruits. Ethylene was in fact first referred to as the "ripening hormone."

As many fruits approach physiological maturity, there is a distinct increase in their susceptibility to ethylene-induced ripening.[113,135] Climacteric fruits removed from the parent plant generally ripened more readily than similar fruit that remained attached. In some cultivars of the avocado, ripening is completely blocked while the fruit remains on the tree. Substances transported from the parent plant into the fruit apparently modulate its susceptibility to ripening. Therefore, factors (possibly hormones) other than ethylene appear to be operative in the ripening process.

The hormone ethylene is synthesized and evolves from the cells of all fruits throughout their growth and development. During the ripening phase of climacteric fruit, ethylene appears to assume a much more dominant regulatory role. This ethylene associated with ripening appears to represent a separate system from the normal background levels of ethylene synthesized by the plant. The two sources of ethylene have been designated system 1 and system 2.[137] System 1 ethylene is found throughout development in both climacteric and nonclimacteric fruits, and system 2 ethylene is activated in climacteric fruits* during ripening.

It has been known for a number of years that exogenous application of ethylene to climacteric fruit would induce ripening.[112] In fact, in the mid-1940s, a book describing the use of ethylene to ripen tomatoes, a climacteric fruit, was published.[90] The discovery of ethylene as the causal agent led from the observation that fruits held in rooms warmed by oil heaters ripened much earlier than those warmed by other means.† Sufficient ethylene was given off in the exhaust fumes to initiate the ripening process. Later research established that ethylene also evolved from the fruits themselves, the rate of production being much greater in ripe than in nonripe fruits.[33,114] Hence, ethylene emanating from ripe fruits would initiate ripening of unripe fruits if they were stored together. Nonclimacteric fruits also produce ethylene; however, only system 1 ethylene and the fruits' response to the hormone differs from climacteric fruits (fig. 5-13). While exposure of the fruit to ethylene does increase respiration it does not initiate the ripening process in nonclimacteric fruits. Other differences between the two fruit types center around their respiratory response to applied ethylene and their synthesis of the hormone (table 5-4).

When climacteric fruits are exposed to ethylene at a sufficient concentration and length of time, the respiratory climacteric of the fruit is initiated and respiration increases, often dramatically. The fruits' sensitivity to ethylene is greatest just prior to the respiratory rise. After respiration has been stimulated, additional ethylene has no stimulatory effect. Above a threshold level of ethylene, the respiratory increase is independent of the concentration of ethylene to which the fruit is exposed. If the ambient ethylene is removed after the respiratory rise has begun, the response is irreversible and the respiratory rate continues to increase.

---

* And certain other tissues (e.g., climacteric flowers).

† For a review of the chronological sequence of ethylene use in agriculture, see Kays and Beaudry.[101]

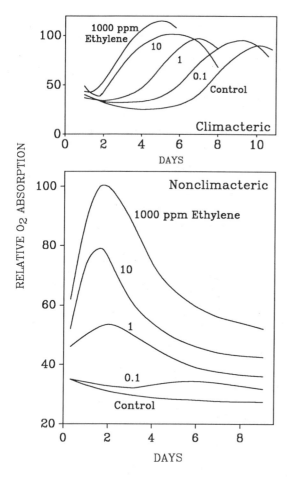

Figure 5-13. The respiratory rate of both climacteric *(top)* and nonclimacteric *(bottom)* fruit increases when exposed to an external source of ethylene; however, the response differs in several ways. For example, the increase in respiration (as indicated by oxygen absorption) of nonclimacteric fruit is concentration dependent, whereas increases in ethylene concentration only shift the time of the respiratory increase in climacteric fruits *(after Biale[29])*. See table 5-4 for other differences in response of climacteric and nonclimacteric fruit to exogenous ethylene.

Table 5-4. Differences Between Climacteric and Nonclimacteric Fruits in the Synthesis of Ethylene and Their Response to Exogenous Ethylene

	Climacteric	*Nonclimacteric*
Response to exogenous application of ethylene	Stimulates respiration only prior to respiratory rise	Stimulates respiration throughout postharvest life
Magnitude of respiratory response	Independent of ethylene concentration	Dependent on ethylene concentration
Reversibility of ethylene-mediated respiratory rise	Irreversible	Reversible, dependent on continued exposure
Autocatalytic production of ethylene	Present	Absent
Endogenous concentration of ethylene	Highly variable, ranging from low to very high	Low

In nonclimacteric fruits, ethylene exogenously applied can stimulate the rate of respiration throughout the postharvest life of the tissue. The magnitude of the response is dependent on the concentration of ethylene to which the fruit is exposed (peak respiratory rates are generally proportional to the logarithm of the ethylene concentration applied). Continued high respiration is dependent on the continuous exposure to ethylene. Removal of the exogenous ethylene results in a resumption of the respiratory rate found prior to exposure.

Climacteric and nonclimacteric fruits also differ in two additional ways (table 5-4). Climacteric fruits exhibit autocatalytic production of ethylene (ethylene-induced ethylene synthesis) whereas in nonclimacteric fruits it is absent. Furthermore, in nonclimacteric fruits the endogenous concentration of ethylene is generally quite low, and remains so throughout the ripening period. In climacteric fruits, the internal concentration is highly variable, being quite low prior to ripening and very high during the respiratory climacteric.

The application of ethylene to a number of climacteric fruits is known to initiate the early onset of ripening. The minimum concentration of ethylene and the length of exposure required to accelerate ripening vary with different fruits. For example, 10 μl · l^{-1} of ethylene is required for the induction of ripening in the avocado whereas only 1 μl · l^{-1} is sufficient for the banana.[29] These differences may in part reflect differences in the resistance of the tissue to the diffusion of ethylene. Therefore, determination of the internal concentration of ethylene is essential since it represents the physiologically active concentration. The internal concentration can be calculated using Fick's law of diffusion[44]:

$$\frac{ds}{dt} = -x\,\frac{A'D\,(C_{in} - C_{out})}{T}$$

where ds/dt expresses the rate of transport, D is the diffusion coefficient, C_{in} and C_{out} are the concentrations of the gas within and outside the fruit, T is the thickness of the barrier to diffusion, A' is the surface area of the fruit, and x is the fraction of the surface area through which gaseous exchange occurs. The internal concentration of ethylene at the onset of normal ripening in the persimmon (*Diospyros kaki,* L. f.) is 0.44 μl · l^{-1},[166] 0.1–1.0 μl · l^{-1} in the banana,[28] and 3.0 μl · l^{-1} in honeydew melons.[156]

In addition to variation between types of fruits to the minimum concentration of ethylene required to initiate ripening, fruits of the same cultivar change in their sensitivity to ethylene as they approach the time of natural ripening. In 'Anjou' pears, the minimum concentration and time required to initiate ripening decreased from a 14-day exposure of 1 μl of ethylene · l^{-1} of air in fruits that were 57% mature, to 0.5 μl · l^{-1} for 15 days when 71% mature, to 0.2 μl · l^{-1} for 17 days when 86% mature.[209] The resistance of young fruit to ripening by exogenous ethylene has been documented in a number of other species.

If ethylene does in fact act as the signal turning on ripening in climacteric fruits, it was initially felt that the internal concentration of ethylene should increase prior to the first signs of ripening within the fruit (generally the onset of the respiratory climacteric). In some species this circumstance is in fact the case. The synthesis of ethylene by banana (fig. 5-14),[43] and cantaloupe (*Cu-*

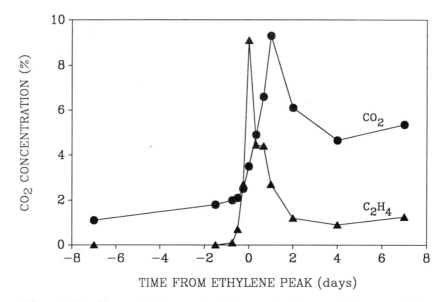

Figure 5-14. Changes in the rate of ethylene synthesis precede the onset of the respiratory climacteric in some fruits, as illustrated in banana (*Musa* spp.) *(after Beaudry et al.[22]).* In other climacteric fruit (e.g., mango (*Manifera indica*, L.) and feijoa (*Feijoa sellowiana*, O. Berg.)), however, the increase in ethylene synthesis may occur after the increase in respiration.

cumis melo, L.),[126] avocado,[43] and honeydew melon[156] precedes the onset of the respiratory climacteric. In the mango[43] and the feijoa (*Feijoa sellowiana*, O. Berg.),[166] however, the increase in ethylene synthesis occurs after the initial onset of the respiratory climacteric. In these fruits the concentration of ethylene already present appears to be sufficient to stimulate ripening; therefore, an increase in internal ethylene need not necessarily precede the climacteric. The timing of the onset of ripening in these fruits must then be controlled by factors other than ethylene (e.g., changes in the responsiveness of the tissue to the hormone).

It is now known that a number of biochemical events at the genomic level precede this increase in ethylene synthesis in climacteric fruit (e.g., the synthesis of new mRNA's that appear to code for enzymes required in ripening). What then is the role of ethylene if the events leading to the initiation of ripening are set into motion prior to an increase in ethylene synthesis? Since fruits are comprised of a diversity of cell types, and these cells are in varying stages of maturity, it seems unlikely that the decision to initiate ripening would be made in all of the cells simultaneously. Rather, the onset of ripening in climacteric fruit probably occurs in one area of the fruit and this decision spreads, through the diffusion of ethylene from these cells, triggering the ripening of the entire organ. Thus, the role of ethylene in climacteric fruit may be more accurately described as one of synchronizing ripening. A synchronizing role for ethylene is supported by the fact that when preclimacteric banana are wounded in one relatively small area of the fruit, ripening is often initiated.

Auxins, principally indole compounds, have been isolated and identified in various fruits. The principal form is IAA, with other compounds probably representing intermediates, breakdown products, storage forms, or detoxification products. Obtaining a clear picture of the role of IAA in ripening has been hampered by the inability to accurately assess the concentration of the hormone within the plant. Unlike ethylene, where changes in the internal concentration and rate of synthesis can be monitored externally in undisturbed cells, the assessment of auxin, cytokinin, abscisic acid, and gibberellic acid has in the past required cellular disruption, which greatly complicates matters. A number of assumptions must now be made, for example, there is no loss, or conversely, no gain in the concentration of the hormone during extraction; there are not active versus latent pools of hormone within a single cell that are combined when the tissue is homogenized. New immunological, nuclear magnetic resonance, x-ray, and nucleotide probe techniques are beginning to make progress in solving some of these problems, although a clear picture has yet to emerge.

Generally, auxin activity has been found to be low in fruits during the latter stages of development[50,193,202](see review by McGlasson et al.[135]). Exogenous application of synthetic auxins and auxin breakdown products has been used to artificially alter the internal concentration of auxin and then assess changes in the ripening response of fruit. Like extraction, exogenous treatment is also open to the production of artifacts that often produce contradictory responses and confuse interpretation.

Synthetic auxins (e.g., NAA, 2,4-D) when applied in appropriate concentrations generally have been found to delay the onset of ripening and the development of senescence.[16] In some climacteric fruits, however, auxin application stimulates the synthesis of ethylene, advancing the onset of ripening. Based on studies with auxins, antiauxins, and oxidative degradation products of IAA, Frenkel[74,75,77] has proposed that ripening may be in part modulated by a deficiency in auxin with ethylene evolution being partially dependent on the oxidative degradation of IAA by IAA oxidase (i.e., peroxidases). However, in apple fruit there is a threefold to fourfold increase in IAA concentration prior to the rapid rise in ethylene synthesis, thus not supporting the hypothesis that IAA is an inhibitor to ripening.[144]

The relationship between gibberellic acid, cytokinin, and abscisic acid and fruit ripening is even more obscure. The concentration of gibberellins is generally high in developing fruits but decreases with maturation. Exogenous application of GA_3 to fruits is known to delay several of the characteristic changes occurring with ripening. Both the loss of chlorophyll and the increase in carotenoids in oranges are delayed.[49] Regreening (chlorophyll synthesis) is actually stimulated. Application of GA_3 has also been shown to delay softening in several fruits (e.g., oranges,[122] prunes (*Prunus domestica,* L.),[162] and apricots,[1] apparently in part through a suppression of polygalacturonase activity).[16]

Like gibberellic acid, the concentration of cytokinin is high in many fruits during initial development, declining markedly with maturation. However, it is not always the case. In olive fruits, cytokinin activity is at its highest during ripening,[185] suggesting that cytokinin does not in fact regulate the onset of ripening. Although exogenous applications of cytokinins have been shown to cause changes in both plastid and nonplastid pigments, these responses may simply be pharmacological effects.

Dried Fruits*

Like their fleshy counterparts, dried fruits such as grains, legumes, and nuts also have a segment of their development that is commonly, although not universally, referred to as the ripening phase. The precise time period and the actual physical and chemical changes that are occurring are, however, very poorly defined. For example, ripening is seen as the final developmental stage in wheat plants, beginning after the seed has developed significantly. The ripening period is separated into six distinct stages (watery ripe, milky ripe, mealy ripe, waxy ripe, fully ripe, and dead ripe for harvest) beginning well after the logarithmic period of cell division. In rice, however, the ripening period has been cited as encompassing the entire developmental period between flowering and harvest, including both cell division and enlargement. This situation is in sharp contrast to soybeans (*Glycine max,* (L.) Merrill) where the term *ripening* is very seldom used.† Thus, while both fleshy and dried fruits are said to ripen, and there are distinct parallels in the ripening phase between the two fruit types, the two differ in several very important ways. For example, during the ripening of dried fruits, physical and chemical changes occurring within the seed rather than in fleshy accessory tissue surrounding the seed are of primary interest. Likewise, alterations in dried fruits during ripening tend to be those that complete the maturation process of the reproductive propagule, enhancing its reproductive potential. In fleshy fruits, ripening changes center primarily on the edible accessory tissues rather than on the seed or seeds per se. These changes, in general, only indirectly promote the fruit or seed reproduction potential, through enhancing dispersal rather than increasing germination potential or seedling vigor.

What then is the ripening period of dried fruits? Ripening is seen as those changes that enhance the reproductive potential of the seeds that generally begin toward the end of the logarithmic period of dry weight accumulation. In most dry fruits, these changes are not dependent on a continued input of carbon from the parent plant for their occurrence. For example, removal of seeds that have not completed their normal dry weight accumulation and maturation phases, if handled properly (controlled water loss and temperature), will undergo ripening.[6] Thus, once attaining a certain developmental stage, ripening can occur independent of carbon input by the parent plant, although the continued flow of carbon from the parent plant may be beneficial to overall developmental processes.

SEED DEVELOPMENT

Seed development can be separated into four distinct stages during which specific developmental events predominate. The stages are (1) fertilization,

---

* The term *dried fruit* is used here to denote grains, legumes, and nuts rather than dried fleshy fruits such as raisins or prunes.

† In soybean the term *maturation,* when used, encompasses those changes occurring during the final stages of seed development.

(2) cell division, (3) seed filling, and (4) ripening (fig. 5-15). In the second stage, once fertilization has taken place, there is a tremendous increase in synthetic activity as many of the structural components of the seed are formed. It is a period of rapid cell division and organelle development. Stage 2 is followed by a period in which photosynthates are translocated into the young cells and reserve forms of carbon and nitrogen (e.g., starch, lipids, storage proteins) are synthesized. The seed-filling period is characterized by a rapid increase in dry weight and a gradual decline in seed moisture content (fig. 5-15). Seed filling is followed by ripening where the seed progresses through a series of physical and chemical alterations essential for the production of a propagule capable of germination and growth. Ripening is characterized by changes that are independent of carbon loading into the seed. Under normal conditions it generally begins at or near the end of the logarithmic period of dry weight accumulations, though the ripening process can be initiated earlier with certain treatments. During ripening, seed fresh weight is at or near its peak and subsequently declines.

With each developmental stage, there are no distinct points at which one stops and the next begins. Rather there is often considerable overlap as one developmental stage declines in activity and the subsequent one proceeds. It is in part due to a gradual transition in developmental processes within a cell but also this overlap is due to the fact that not all of the cells within a seed are identical; they may vary in type, age, and degree of development. Likewise, within a fruit with multiple seeds, the individual seeds may vary in age.[65]

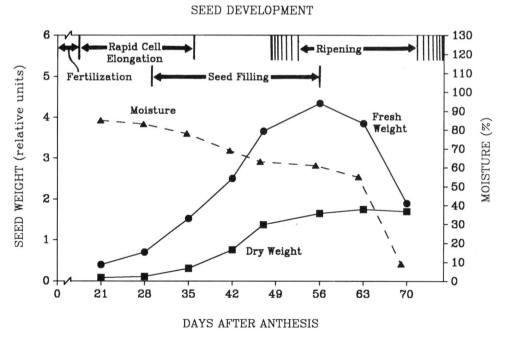

SEED DEVELOPMENT

Figure 5-15. Seed developmental stages in relation to changes in moisture content and fresh and dry weight after anthesis (data from soybean, *Glycine max*, (L.) Merrill) (*after Rosenburg and Rinne*[175]).

Typically, seeds at the base of the ear or pod are the least mature, and their ripening also lags behind.*

While it has been common to delineate the end of the ripening phase as occurring at harvest, this distinction is usually not warranted.† Often the biochemical and physical changes occurring that are required to produce a seed capable of germination are not complete when the grower decides it is time for harvest.[64] The term *after-ripening* is sometimes used to encompass these changes. After-ripening often may represent the final stages of the ripening process, which for a number of species do not terminate with harvest. Barton[19] lists periods of dry storage during which germination potential was completed, ranging from two weeks (*Lepidium virginicum,* L.) to several years (*Sporobolus cryptandrus,* Gray).

RIPENING

The focus in this section is on the final stage in the development of the seed: ripening. The length of the ripening period varies considerably between individual species and in some cases may differ significantly within a species. For example, germination potential, used as a measure of the completion of the developmental process, differs considerably in the seed of wheat cultivars 'Holdfast' and 'Atle' of the same age and weight.[211] Seed of 'Holdfast' harvested 7 weeks after anthesis had 80–90% germination and similar seed of 'Atle' had only 7–10% germination.

MORPHOLOGICAL ALTERATIONS DURING SEED RIPENING

As *Phaseolus lunatus,* L. seeds approach the ripening stage, the cotyledons comprised largely of parenchyma cells are well vacuolated, contain chloroplasts with well-developed grana, have starch grains, and have their polysomes associated with the endoplasmic reticulum (fig. 5-16A).‡[109] During ripening the cotyledons undergo significant changes in chloroplast structure with the dismantling of the internal membrane and the disappearance of the grana. The loss of polysomes (fig. 5-16*B*) during this period appears to coincide with a sharp decline in protein synthesis (largely storage proteins) during this final stage of development. Similarly there is a disappearance of golgi bodies. For many seeds, chlorophyll degradation occurs during the ripening period.

CHEMICAL CHANGES DURING RIPENING

Seeds, especially those in the legume family, contain large quantities of proteins that accumulate largely during the seed-filling period.[23] Typically the bulk of the protein component is made up of a relatively small number of

---

* The opposite is true in maize.

† A similar situation can be seen in fleshy fruits, especially for climacteric species that almost exclusively instigate or continue their ripening process after harvest depending on stage of maturity at harvest.

‡ Klein and Pollock[109] term what is herein referred to as the seed-filling stage as *ripening* and ripening as the *maturation stage*.

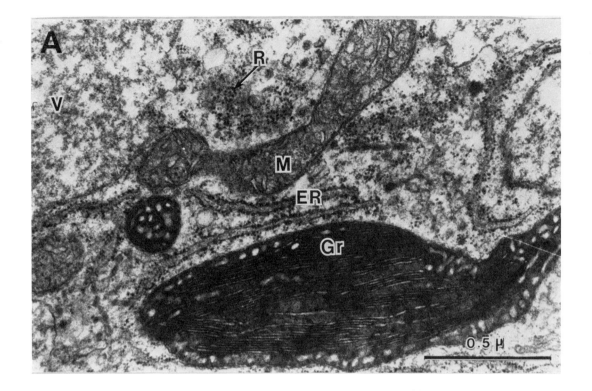

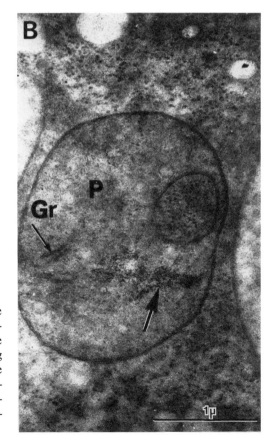

Figure 5-16. *A:* Electron micrograph of immature lima bean cotyledon cells (75% moisture). The ribosomes occur in clusters (polysomes) around the rough endoplasmic reticulum. *B:* During ripening (35% moisture) they appear singly and there are significant changes in the density of the mitochondria matrix and plastids *(from Klein and Pollock[109])*. (M, mitochondria; ER, endoplasmic reticulum; Gr, grana; R, ribosomes; P, protein body.)

individual polypeptides, collectively referred to as storage proteins.*[10,88,89] In soybean, two storage proteins (conglycinin and glycinin) constitute approximately 70% of the seed protein at maturity and 30% to 40% of the total seed weight. There are temporal differences in the accumulation of the protein fractions during development.[51] Messenger RNAs for glycinin and conglycinin can be first detected 14–18 days after pollination.[139] However, during the final developmental period these mRNAs were no longer detected, indicating a cessation of storage protein synthesis. A similar decline in storage protein synthesis is seen in *Ricinus communis,* L.[106]

While storage protein synthesis draws to a close during the ripening period, other proteins, presumably those required for ripening and/or subsequent germination, continue to be synthesized.[105] This situation is illustrated by the change in number of soluble protein bands (fig. 5-17) and changes in abundance of RNAs occurring during development (fig. 5-18).[63]

The decline in moisture content during the ripening period plays an extremely important role in redirecting the metabolism of the seed. Metabolism shifts from a developmental mode to one dedicated to preparing the propagule for eventual germination. The importance of dehydration in this shift has been demonstrated using immature seeds excised from the parent plant. If the seeds are kept hydrated after excision, germination is inhibited.[5] However, if a controlled moisture-loss treatment is imposed, the seeds undergo the necessary biochemical alterations required for subsequent germination upon rehydration. Moisture loss, whether occurring naturally in the field or simulated artificially, has been linked to the inhibition of synthesis of certain proteins and the induction of new proteins.[175]

A number of other biochemical alterations occur during ripening, although most have not been adequately characterized. In soybean seed, there are significant changes in the activity of specific enzymes controlling carbohydrate interconversion during the latter stages of development. For example, galactinol synthase activity increases, often peaking around the time the seed reaches its maximum dry weight,[179] and appears to be important in the conversion of sucrose to raffinose and stachyose.

Seed chlorophyll content also declines. The concentration of total chlorophyll recedes rapidly as soybean seeds approach 60–70% moisture[194] and this decline tends to coincide with a rapid decline in seed respiratory rate (fig. 5-19). In pecan kernels, there is a linear decline in respiratory rate on a logarithmic scale with seed moisture content (from 13% to 3%).[21] For many seeds there is a distinct transition to a low level of respiration around the seed moisture content that is typically considered safe for harvest. Collectively, ripening changes enhance the seeds' potential for subsequent growth.

Abscission

The shedding of plant parts is a natural process, affording plants a number of survival and evolutionary advantages. Shedding allows plants to remove

---

* The amino acid sequences for 13 globulin storage proteins have been delineated. Existing evidence indicates that all of the globulin storage proteins in flowering plants descended from two genes that existed at the beginning of angiosperm evolution.[40]

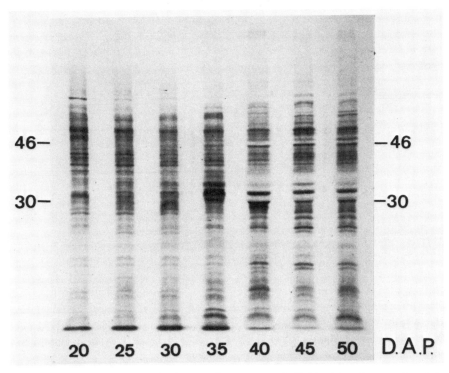

Figure 5-17. Changes in soluble proteins, seen as individual bands on polyacrylamide fluorographs, with the development of *Ricinus communis,* L. endosperms after pollination (*from Kermode and Bewley*[105]). (DAP, days after pollination.)

injured, infected, or senescent organs, recycle nutrients, adjust leaf/flower/fruit numbers when under stress conditions, disperse seeds, and other benefits. Most shedding responses are precipitated by physical and chemical changes occurring in a specialized abscission zone. Abscission per se represents the separation of cells, tissue or organs from the remainder of the plant at one of these zones. During the developmental phase, naturally occurring abscission responses are part of normal growth under non-stress conditions. After harvest, however, abscission responses generally are detrimental to product quality. Examples of this would be the abscission of florets in cut and potted flowers,[47] leaves of foliage plants,[131] and food crops such as cabbage (*Brassica oleracea,* L. Capitata group),[136] celery (*Apium graveolens* var. *dulce,* Pers.), and similar vegetables. As a consequence, considerable research effort has been directed toward the inhibition of abscission.

Abscission zones are comprised of specialized cells that are only one to three cells wide but they transverse a major portion of the cross-sectional area of the organ (fig. 5-20). Prior to the onset of abscission, the cells in the abscission zone are relatively undistinguishable from neighboring cells with the exception of being somewhat smaller. The presence and location of the abscission zone on a plant is genetically controlled.

A wide range of plant parts is abscised by various species. The most commonly shed parts are leaves, flowers, fruits, and seeds; however, branches, bark, roots, spines, and other parts may be shed, depending on the

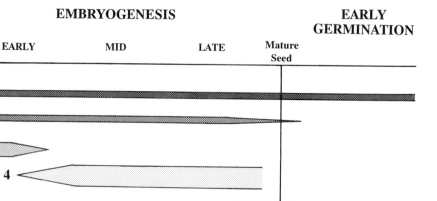

Figure 5-18. Changes in the timing of synthesis of specific groups of mRNAs and their relative abundance (vertical width of band) during seed development and the early stages of germination in *Gossypium hirsutum*, L. *(after Dure[63])*.

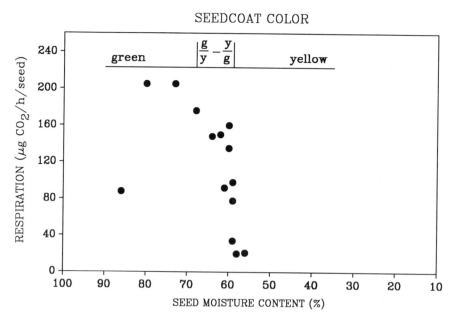

Figure 5-19. Changes in soybean (*Glycine max*, (L.) Merrill) seed respiration (cv. 'Fiskeky V') in relation to changes in moisture content and seedcoat color *(after TeKrony et al.[194])*.

species. Leaf abscission can involve individual leaves, individual leaflets on compound leaves, and in a limited number of species, entire leaf branchlets (e.g., *Larix*). Entire flowers abscise in response to stress conditions and individual floral parts (e.g., calyx, petals, stamen) abscise after anthesis. In some species, petal abscission occurs very quickly after pollination (e.g., *Digitalis, Clarkia,* and *Antirrhinum*). Others shed their petals at varying intervals after anthesis, for example, in some species of *Linum* and *Geranium* petal abscission is during the afternoon of the day of anthesis; for *Gossypium hirsutum,* L. the day after anthesis; and for *Eschscholzia californica,* Cham., the fifth day after anthesis.[7]

Individual fruits may be shed due to the lack of pollination, due to adjustment of the fruit load on the plant, or due to physical injury to the organ or subtending branch. Seeds abscise from the placenta of most fruits. In dry, dehiscent fruits, abscission allows the propagule to be dispersed from the

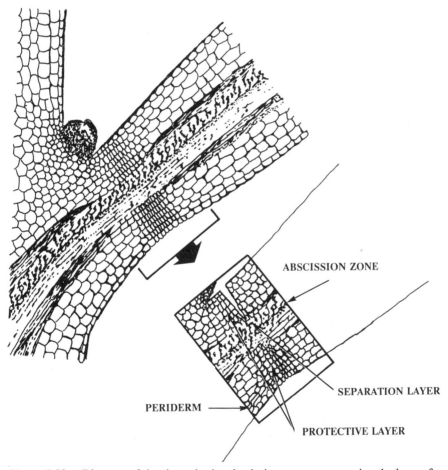

Figure 5-20. Diagram of the tissue in the abscission zone transversing the base of a leaf petiole. Abscission zone cells are typically smaller and less differentiated than neighboring cells. As the chemical and physical processes leading to abscission develop (box), the juncture between neighboring cells within the zone weakens and subsequently separates. A layer of cells then seals the leaf scar, protecting the plant from pathogen invasion and water loss *(after Addicott[7])*.

parent plant via a diverse array of species-dependent mechanisms. The actual dehiscence process represents a modification of abscission. Rather than detachment, dehiscence involves the opening of a structure, with the separation occurring along a defined zone of cells.

The location of the abscission zone is adjacent to the site of detachment. For simple leaves, it is normally at the base of the petiole. Fruit abscission zones may be between the ovary and receptacle, the ovary and pedicel, the nut and receptacle, or other locations depending on the species.

Shedding involves two processes: (1) changes in the physical and chemical condition of the cells within the abscission zone, which results in a progressive weakening of the juncture; and (2) mechanical separation of the plant part. After the induction of the abscission process there is a lag phase during which there is no change in the physical strength of the abscission zone. Following the lag phase, the abscission zone undergoes a progressive weakening, referred to as a decline in break strength. Physical forces are important in the eventual mechanical separation of the part. In addition to gravitational forces, wind, rain, ice accumulation, and other factors often facilitate final separation.

Physical and chemical changes in the abscission zone that precipitate the loss in break strength are energy dependent and mediated through the synthesis of enzymes required for abscission. During the lag phase, RNA and protein synthesis occur,[2] and if blocked, there is a marked inhibition in the subsequent weakening process. Enzymatic hydrolysis of the middle lamella and the primary cell wall reduce the strength of the molecules holding the cells together along the separation zone. Fracture typically follows a line through the middle lamella between neighboring cells.

During the lag phase, the enzyme cellulase (β-1, 4-glucan-4-glucan hydrolase) begins to be synthesized. It exists in two forms (isozymes), only one of which increases during abscission.[121] Cellulase is secreted from the cells bordering the fracture plane into the middle lamella region where wall breakdown is most pronounced. Ethylene is thought to be involved in the secretion process, since its removal from the atmosphere slows the rate of weakening of the abscission zone.[54] The activity of other enzymes, for example, pectinase and hemicellulase, may also be important, although at present, data supporting a significant role for each are very limited.

Environmental and other factors that accelerate abscission (e.g., mineral deficiency, drought, low light, pollination, lack of pollination, pollution) have been well documented. However, how these signals are translated into the induction of the abscission response is not well understood. The importance of plant hormones in this process has long been a popular supposition supported by a series of research reports since the 1930s. Both IAA and ethylene appear to be intimately involved in the abscission process, although the role of IAA was the center of considerable debate for a number of years, since it was found that application of IAA could both retard and accelerate abscission. This apparent paradox was explained by the ability of IAA to both induce ethylene synthesis and cause changes in the sensitivity of cells within the abscission zone to ethylene.[4] IAA application early in the abscission process (stage 1) resulted in the abscission zone being maintained in an ethylene-insensitive condition; however, if application was delayed until stage 2, application triggered an increase in ethylene synthesis that accelerated abscission. In nature, the cells in the abscission zone are thought to require sensitizing (stage 1), which involves a reduction in juvenility factors such as IAA.[2] Once accomplished, ethylene then (stage 2) induces the synthesis of mRNAs necessary for the formation of cell wall-degrading enzymes.

During the postharvest period, ethylene, low-light intensity, and water stress are the most common causes of abscission in ornamental plants. Physiologically active levels of ethylene may be produced by the plants themselves[103] or may come from an external source (e.g., internal combustion engines, improperly adjusted gas heaters). Preventing losses caused by ethylene-mediated abscission after harvest has been approached three ways: (1) avoidance of exposure to exogenous sources of ethylene, (2) treatments to inhibit the synthesis of ethylene by the plant, and (3) treatments to reduce the sensitivity of the plant to ethylene.

Proper ventilation and other management practices (e.g., sanitation, proper product selection for mixed storage, use of ethylene scrubbers) should be routinely used to prevent the concentration of ethylene from reaching a detrimental level. Although the inhibition of ethylene synthesis with various compounds [e.g., aminoethoxyvinylglycine (AVG)] may be beneficial, it does not protect the plant material from other sources of ethylene. The use of silver ions, a potent inhibitor of ethylene binding,[26] has proven to be an extremely effective way of preventing abscission in many ornamentals.*[167] When formulated as silver thiosulfate, it is relatively nonphytotoxic and stable while readily translocated within the plant. An earlier technique was the application of synthetic auxins, which was found to be effective for the inhibition of leaf abscission of harvested English holly (*Ilex aquifolium*, L.)[170] and several other ornamental foliages. For some food crops, high levels of carbon dioxide, a competitive inhibitor of ethylene binding,[45] can reduce ethylene-mediated abscission.

Senescence

Living organisms share several universal traits, one being a finite existence. All organisms eventually die. The question thus becomes not will they die but when will they die? In the plant kingdom the life expectancy of plants and plant parts varies widely, ranging from only minutes to hundreds of years.† When moving live, harvested plant material from the producer to the eventual consumer, there is innate concern not just with the impending death of the product but often even more so with the general degradative changes in the product that lead to eventual death. Interest in these changes arises from the fact that many of these alterations diminish product quality. The objective during the postharvest period is not to prevent the eventual death of the product, or for that matter to extend the storage life to its theoretical maxi-

---

* It should not be used for food crops.

† One of the classic experiments on the survival of buried seeds was started by W. J. Beal in 1879. At 5- and later 10-year intervals, samples were unearthed and germinated,[108] demonstrating longevities of up to 100 years in some species.

mum duration; economic constraints make the latter an unrealistic commercial option. Thus, if an adequate supply of lettuce is being harvested for the market every week of the year, there is little incentive to maintain lettuce in storage for 6 months even though it may be theoretically possible.

Many of the changes that occur after harvest, especially those in highly perishable products, are part of the process of senescence. Senescence can be defined as a series of endogenously controlled deteriorative changes that result in the natural death of cells, tissues, organs, or organisms. It differs from aging, which entails changes that accumulate over time without reference to death as an eventual consequence.[138] Nonliving objects, therefore, can age. In living organisms, aging is not a direct cause of death but may increase the probability of death by decreasing product resistance to stress.

Senescence is an integral part of the normal developmental cycle of plants and can be viewed on a cellular, tissue, organ, or organism level.[120,146] The death of certain cells represents part of the normal development of plants. An example can be seen in single-celled root hairs formed just behind the growing tip of the root that have a relatively short life expectancy.[143] One of their primary roles is anchorage of the root, allowing the root tip to overcome the physical resistance encountered as it penetrates further into the soil. Physical resistance of sufficient magnitude has been shown to result in an increase in ethylene synthesis by the root,[104] which in turn appears to enhance the development of root hairs. As the root elongates, the existing root hairs are no longer in a strategic position for anchoring the tip; thus, new root hairs are formed and the older ones senesce. Similar patterns of death have been documented for various other individual cells (e.g., fiber and sclerid cells).

Groups of cells may also senesce while the cells around them remain alive. An example would be the thin layer of cells making up abscission and dehiscence zones. These cells undergo a form of programmed senescence. Although the cells are not dead at the time of separation, many of the physical and chemical changes are similar to those occurring in cells during senescence.[145]

Organ senescence is most often encountered during the postharvest period. Many handling and storage techniques have been designed to retard the rate of development of senescence in organs such as fruits, flowers, leaves, roots, and tubers. In nature, organ senescence allows the removal of plant parts that have already fulfilled their biological function. It eliminates the maintenance costs (energy) to the parent plant for that organ, plus may allow for recycling part of the nutrients out of the organ, increasing the plant's overall efficiency. During development of an agricultural product such as a fruit, the plant part goes through a series of developmental stages, several of which involve the death of certain organs or tissues. For example, at anthesis a simple fruit is comprised of the immature ovary and the various flower parts. Within a week or so after flower opening and pollination, most of the floral parts (petals, pistil, stigma, style), having fulfilled their biological function, senesce and drop to the ground (fig. 5-21). With the exception of the seeds, the tissues making up the fruit in turn senesce. Thus, of the thousands of cells that made up the fruit, nearly all have by now died. Those remaining are largely concentrated in the cotyledons, which with germination will also senesce. Thus, organ senescence and death are integral parts of the normal cycle from seed to seed.

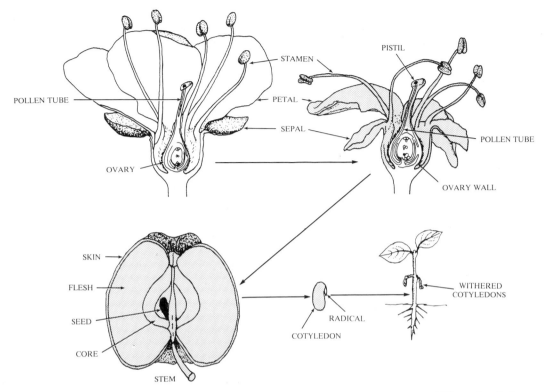

Figure 5-21. Organ senescence and death represent an integral part of the normal life cycle of plants. Flower pollination, for example, sets off a dramatic series of developmental events within the ovary. Coupled to pollination is the onset of senescence in the floral parts that are no longer essential (e.g., petals, stamen, stigma, and style). Carbon and nitrogen compounds may be recycled out of these organs into the developing ovaries. The fruit tissue subsequently decomposes, leaving the seeds, which in turn undergo senescence and death of a major portion of their cells.

If senescence represents degradative changes that lead to death of cells, is there a genetically controlled programmed sequence that once triggered instigates the senescence syndrome, or does senescence simply reflect the inability of the tissues to maintain homeostasis?*[173] Cells are dynamic homeostatic systems that have the ability to undergo repair. For example, harvested fruit sensitive to chilling injury can counteract low-temperature injury during storage if exposed to periods of intermittent warming.[24] Likewise, DNA repair is a routine phenomenon.[206] Is senescence in harvested products simply a situation where the cells loose their ability to maintain themselves or at least repair themselves at a rate equal to that at which degradative changes are occurring? Both situations, a programmed induction of senescence and a loss of homeostasis, appear to occur. A programmed onset of senescence has evolved in some species (e.g., monocarpic plants) and organs (fruit ripening, flower petal senescence, leaf abscission), which confers a competitive advan-

---

* Homeostasis is the tendency to maintain internal stability via a coordinated response to the disruption of normal condition or function.

tage to the plant (e.g., nutrient conservation, enhanced propagule dispersal). Petal senescence in flowers is a striking example of programmed senescence. Once activated, many of the constituents within the petals are recycled into nonsenescing neighboring tissue. The signal to activate the senescence program may reside within the tissue (e.g., fruit ripening) or come from an external stimuli (e.g., leaf abscission due to insufficient light).

In the alternative situation, the tissue undergoes a series of senescence changes that are designed to prolong the viability and integrity of the cells. Repair reactions essential for the maintenance of homeostasis require energy and carbon skeletons; however, for many products, harvest terminates the external acquisition of these requisites. Thus, postharvest requirements must be met by recycling components from within the product. At the cellular level the progressive dismantling of cellular components is involved. As might be anticipated, there is a fairly distinct sequence of priority in cellular components for dismantling. The mitochondria, needed for continued energy production, the nucleus, and the plasma membrane are maintained until the cell nears a final organizational collapse and death.

A programmed onset to recycling of nutrients out of cells and the intracellular recycling of constituents to maintain homeostasis both exhibit many similarities (e.g., sequential dismantling of cellular components) (fig. 5-22). In the former, however, the object is to maximize the recycling of constituents out of the cell whereas in the latter recycling is largely to maintain the condition of the cell. Thus, differences would be anticipated in rate between both scenarios and the level of synchronization between neighboring cells.

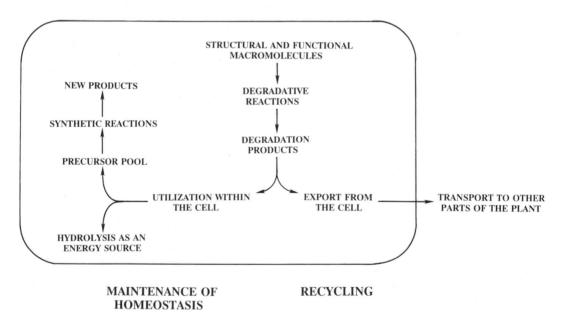

MAINTENANCE OF   RECYCLING
HOMEOSTASIS

Figure 5-22. Scheme for recycling cellular macromolecules in harvested products. The resulting degradation products can be exported from the cell when the senescence syndrome is programmed for recycling or utilized within the cell for energy and carbon skeletons needed to maintain homeostasis. Some utilization within the cell is also required for recycling. The degradative reactions are enhanced by (1) induction and/or activation of hydrolytic enzymes, and (2) decompartmentation of macromolecules within the cell (*after Sexton and Woolhouse[183]*).

Chemical and Ultrastructural Changes During Senescence

In contrast to the once-held view that senescence was simply an organizational collapse of the cell, it is now known that senescence is initially a tightly controlled developmental step in which there is normally a highly ordered sequence of events. The idea of a controlled series of events is underscored by the fact that senescence is an active process, requiring energy. The disruption of energy availability, such as through depressed respiration via low oxygen,[216] retards the rate of development of senescence. Likewise, senescence is dependent on gene activation and synthesis of new mRNAs (fig. 5-23) and probably the inactivation of others.[133,149,218] Certain proteins are required since the inhibition of protein synthesis by cycloheximide also inhibits senescence (fig. 5-24).[115] Cycloheximide inhibits protein synthesis on 80s ribosomes found largely in the cytoplasm. In contrast, the inhibition of protein synthesis with chloramphenicol, which acts on 70s ribosomes found in the plastids and mitochondria, does not inhibit senescence.[215] Thus, the control of senescence appears to reside in the nuclear genome rather than in DNA found in individual organelles.

Total protein content, while characteristically declining in senescing leaves, does not necessarily decline in all senescing organs. During fruit ripening total protein concentration has been shown to remain essentially unchanged (e.g., avocado,[32] banana,[205] citrus[122]), or in some cases, it increases (apple,[97] pear,[85] cantaloupe[176]). In each case, the concentration of certain enzymes increases. These characteristically are hydrolytic in nature and are involved in dismantling large molecules. In fruit, increases in ribonuclease, invertase, acid phosphatase, α-1,3-glucanase, α-1,4-glucanase, cellulase, and polygalacturonase have been reported; in leaves, protease, ribonuclease, invertase, acid phosphatase, and α-1,3-glucanase increase. The two organs differ in the absence of increases in proteases in fruit and cell wall-softening enzymes in senescing leaves.[42]

CHANGES WITHIN THE CELLS DURING THE EARLY STAGES OF SENESCENCE

The cellular membrane system (plasma membrane, endoplasmic reticulum, vacuolar membrane, the highly specialized thylakoid membranes of the chloroplasts, etc.) represent selective barriers to the movement of compounds within and between cells. Control via access is one important means of regulation for a number of biochemical and physiological events within the cell. During senescence, however, there is a progressive loss of membrane integrity, and as a consequence, regulatory control.[196] As might be anticipated, the chemistry and structure of the various membranes within the cell differ, which in turn affects the operative senescence processes and their rate.

One common but highly significant alteration in the membranes occurring with senescence is a change in fluidity. With the development of senescence, fluidity decreases and the membranes become more rigid.[197] This situation can alter the activity of enzymes that are associated with the membrane and receptors on the membrane. Likewise, a change in membrane structure from

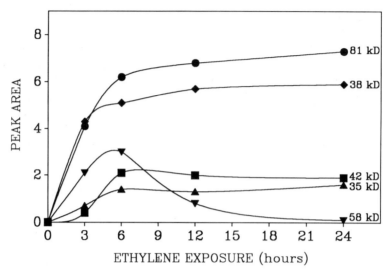

Figure 5-23. Changes in the relative abundance of selected mRNAs in carnation (*Dianthus caryophyllus,* L.) petals with time after the onset of senescence mediated via exposure to ethylene (7.5 $\mu l \cdot l^{-1}$) (*after Woodson and Lawton*[218]). Thus, genes for specific proteins appear to be turned on during the early stages of senescence and their duration of transcription varies.

a liquid crystalline to a membrane with the presence of gel phase domains results in leakiness and a loss of compartmentalization.[17]

A relatively consistent event in the senescence of chlorophyll-containing harvested plant products is the degradation of chlorophyll and the corresponding transition in color of the product.* While chlorophyll degradation is

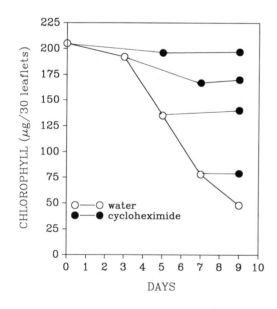

Figure 5-24. Inhibition of chlorophyll degradation in *Anacharis* leaflets using cycloheximide, an inhibitor of protein synthesis (*after Makovetski and Goldschmidt*[128]).

* Depending on the product, the loss of chlorophyll can unmask a range of pigments and therefore colors.

in most products closely tied with senescence, its degradation does not appear to be an essential requisite for senescence to proceed. The nonobligatory nature of this relationship is quite dramatically illustrated in the leaves of a mutant genotype of fescue (*Festuca pratensis,* Huds.) in which death occurs without chlorophyll degradation.[195]

Chlorophyll loss occurs in tandem with significant changes in the chloroplast per se that undergo a sequential series of alterations in their ultrastructure.[78] Ultrastructural changes center on the thylakoids, the internal membrane system within the chloroplast and the site of the light reactions in photosynthesis (photosystem I and II). Initially the stroma thylakoids lose their integrity, which is followed by swelling and disintegration of the grana thylakoids.[96] Interestingly, the double membrane envelope that encloses the chloroplast does not begin to lose its integrity until very late in the senescence of the chloroplast, which allows the dismantling process to be reversed until quite late in the degradative sequence. Regreening is a common occurrence in leaves and some fruits exposed to inductive conditions. Likewise, retention of the outer membrane system would account for the ability of the chloroplasts in some products (e.g., tomato fruit) to be transformed into other types of plastids.

CHANGES WITHIN THE CELLS AS THEY APPROACH DEATH

In keeping with the idea of an orderly dismantling of the cell, it would be anticipated that cellular structures and organelles that are absolutely essential would be retained until the cell enters the very final stages of senescence. The mitochondrial,[67,172] plasmalemma, nuclear, and vacuolar membranes persist to the very late stages of senescence. Regardless of whether the cells are programmed to recycle their components outward to other parts of the plant, or to utilize the hydrolyzed components within the cell, these structures are essential.

Environmental Factors Modulating the Rate of Senescence

Stress can significantly modulate the rate of senescence of harvested plants and plant parts. In some cases, stress can induce the onset of senescence. Environmental stresses such as temperature (high and low), composition of the gas atmosphere surrounding the product, water deficit or excess, pathogens, herbivores, irradiation, mechanical damage, mineral imbalances, salinity, and air pollutants have been shown to accelerate senescence in various organs. Some environmental stresses (e.g., light deprivation) can trigger the onset of senescence.[38] During the postharvest period, mild thermal and gas stresses (refrigerated and controlled atmosphere storage) are routinely used to decrease the rate of development of senescence.

Endogenous Regulators of Senescence

Both ethylene and abscisic acid have been shown to stimulate sen senescence-associated processes in a variety of organs. Exogenou tion of ethylene has long been known to stimulate the ripening an cence of climacteric fruit, senescence in flowers, leaf abscission, and other senescence processes. For example, the leaves on harvested branches of holly can be stored in the dark for extended periods of time without significant losses; however, exposure to ethylene induces rapid senescence.

Treatments that inhibit the synthesis (e.g., AVG) or action (e.g., silver ions) of ethylene tend to delay but not prevent senescence (e.g., fruit ripening[208]). Likewise, lowering the internal concentration of ethylene using hypobaric conditions has also been shown to delay senescence.[57] As with ethylene, application of abscisic acid promotes senescence-related processes such as chlorophyll loss in leaves,[68] increased protein degradation and decreased synthesis,[84] and alterations in membrane structure.[91] The precise endogenous role of ethylene and abscisic acid as promoters of senescence is not yet clear.

In contrast to ethylene and abscisic acid, cytokinin and calcium ions have been shown to delay senescence. In general, exogenous application of cytokinin or synthetic forms of cytokinin at the appropriate concentration and timing has been shown to delay senescence in many tissues.[200] In addition, a decline in internal concentration of cytokinins coincides with the onset and development of senescence in some tissues (e.g., petals[201] and leaves[199]). In fruits, the transition of chloroplasts to non-chlorophyll-containing chromoplasts is accompanied by a decrease in endogenous cytokinin concentration.[135]

Exogenous Ca^{2+} application delays senescence in a number of organs (e.g., leaves, fruits).[71,123,153] This effect is thought to be related to the intercellular role of Ca^{2+} in stabilizing the cell wall and external surface of the plasmalemma.[71] Within the cell the concentration of Ca^{2+} is substantially lower than that found in the extracellular region and its role appears to be much more complex. Within the cell, Ca^{2+} is thought to act as part of an information transduction system where extracellular signals are translated into changes in metabolism.[152]

The precise role of ethylene, abscisic acid, cytokinin, and Ca^{2+} in the promotion or inhibition of senescence and their interaction among themselves and with other endogenous growth regulators is not presently known. As the understanding of this complex relationship improves, so should the ability to manipulate these factors to retard senescence.

MATURATION

During development, the plant passes through a series of distinct but often overlapping stages. With monocarpic plants, development begins with germination, passing through the juvenile stage and progressing to maturity and finally senescence. Maturity, as viewed from the natural reproductive biology of the plant, is generally considered to be the stage of development where the plant is capable of shifting from vegetative to reproductive growth. In agriculture, however, maturity is much more arbitrary. Generally, maturity is seen as a stage of development superimposed on the plant or plant part relative to human needs. The object in question is considered mature when it meets the requirements for harvest (i.e., harvestable maturity). This stage does not necessarily imply that the product meets the maturity requirements for immediate utilization. Many products are sufficiently mature for harvest but not for utilization. For example, apples or bananas that are to be held in storage for considerable time are harvested prior to having developed sufficiently for immediate consumption. With proper handling, they will continue to develop after harvest, reaching an acceptable level of culinary perfection.

Harvest maturity (fig. 5-25) varies widely with the plant product involved. Plants that are sold or consumed at the seedling stage reach a harvestable maturity very early in the natural development cycle of the plant. Plant parts such as inflorescence (artichokes, cauliflower) or partially developed fruits (cucumbers, bitter melon, sweet corn) progress through a significant portion of their developmental cycle. Fruits such as apples, bananas, and citrus are nearly fully developed, and most nut and seed crops are fully developed at harvest. Harvestable maturity, therefore, can occur throughout the developmental cycle, with the precise time varying with the product in question.

Harvestable maturity for a number of crops occurs over a relatively wide time frame. For example, with cassava (*Manihot esculenta,* Crantz) or taro (*Colocasia esculenta,* (L.) Schott) this time period may be several months in duration. With *Alocasia macrorrhiza,* (L.) G. Don., the giant taro, the period of harvestable maturity extends for several years without an appreciable loss of quality. Many crops, however, have relatively short, precise time periods for harvest and exceeding this period results in the impairment of quality or product loss. The length of this period ranges from weeks (e.g., oranges) to days (e.g., apples) to hours within a single day (e.g., pollen, gherkins (*Cucumis sativus,* L.)).

The timing and duration of this period of harvestable maturity can be modulated by a number of factors. For example, cultivar can have a significant influence. Environmental conditions during development can also have a pronounced effect on the timing and length of the period of harvestable maturity. Substantial losses are encountered each year due to environmentally induced alterations in the maturation time period.

Harvesting at the proper stage of maturity is essential for optimum quality and often for the maintenance of this quality after harvest. When then is the optimum point in the period of harvestable maturity for harvest? Within a given crop, optimum maturity is a highly subjective determination. One critical variable is who in the production–harvest–storage–marketing–utilization chain determines the criteria. Since the needs at each step in the chain may vary, the criteria utilized will often also vary. Optimum harvest maturity for the grower is a function of both product and marketing conditions. When can the crop be harvested to maximize profits? When supply is low and price is high, lettuce is often harvested very early in the normal harvestable maturity time period. Thus, for an individual grower, optimum maturity may vary with each successive crop. Growers must also consider how the crop is to be harvested. Optimum harvest maturity for a crop of tomatoes to be harvested using a destructive once-over mechanical harvest is going to be substantially later than when multiple hand harvests are used.

The method in which the product is to be handled after harvest may also influence what is considered the optimum degree of maturity for harvest. Apples that are to be held in cold storage for extended periods are generally harvested at a less mature stage than those destined for the immediate fresh market. The optimum maturity for spinach that is to be hand harvested for

Developmental Stages

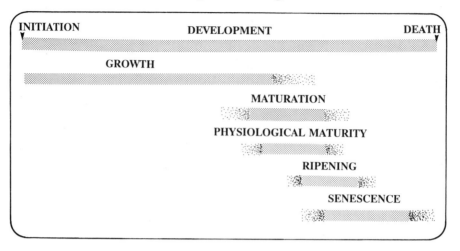

Harvestable Maturity

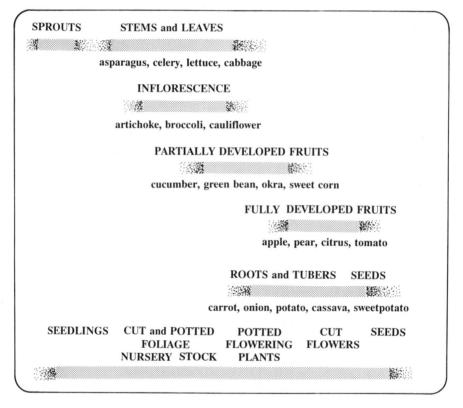

Figure 5-25. The timing of harvestable maturity within the normal developmental cycle of a plant varies widely between individual products. Some (e.g., bean sprouts) have just begun development when they reach a harvestable maturity whereas others (e.g., grains, seeds) complete the cycle from seed to seed *(after Watada et al.[210])*.

Adam

Figure 5-26. Optimum maturity is often a highly subjective determination, varying with who in the production–harvest–storage–marketing–utilization chain makes the decision and how it is made *(from Basset[20]).*

fresh market often does not coincide with spinach (*Spinacia oleracea,* L.) that is to be processed. Likewise, spinach that is to be processed and packed as a chopped product can be harvested later than spinach packed with the leaves intact.

Optimum harvest maturity is not a fixed point in the developmental cycle of a plant or plant part. Optimum harvest maturity varies depending on the criteria utilized to determine it. These criteria may vary substantially between the various individuals in the production–harvest–storage–marketing–utilization chain and the constraints under which they are operating at that particular time (fig. 5-26).

How then do we determine when optimum harvestable maturity has been reached? With most crops, optimum maturity is determined by specific physical and/or chemical characteristics of the plants or plant part to be harvested. Implicit in the use of single or multiple physical or chemical characteristics to determine optimum maturity is that changes in the selected parameter correlate with the attainment of the general composite of quality characteristic of the product.

Physical Measurements of Maturity

Physical attributes such as size, color, or texture are commonly used to determine crop maturity. With individual crops, these physical characteristics may be very specific. For example, muskmelon (*Cucumis melo,* L. Reticulatus group) maturity is determined by the ease at which the stem separates from the fruit. Although very closely related, totally different criteria are used for honeydew melons (*Cucumis melo,* L. Inodorus group). Processing spinach is harvested shortly after a hollow area develops within the interior of the base of the plant where the stem and taproot meet. With cut roses, optimum maturity for harvest varies with individual cultivar; yellow cultivars generally can be cut at a slightly tighter bud stage than pink or red cultivars. The latter are allowed to develop until one or two of the outer petals begin to unfurl and the sepals are either at a right angle to the stem axis or pointing downward.

Maturity criteria based on physical characteristics can be made either objectively or subjectively. Experienced growers commonly determine harvest date based on a subjective assessment of the crop's physical characteristics. Although lacking precise analytical measurements, many have developed the ability to determine with a relatively high degree of precision the optimum maturity for a specific crop. Not all growers, however, are equally proficient at determining maturity. In addition, with some crops it is not possible to accurately determine optimum maturity with a subjective evaluation. Thus, objective or analytical measurements of maturity have progressively become more widely used, and if properly carried out, they tend to be highly consistent.

Analytical assessment of maturity requires using a standard instrument to measure some physical characteristic that is known to change in relation to the maturation of the product in question. For example, the maturation of peas is determined by the force required to shear the seed. Several instruments have been developed for this purpose (e.g., Tenderometer, Maturometer). Flesh firmness is commonly used to determine the maturity of apples. As the fruit matures and ripens, dissolution of the middle lamella of the cell wall results in softening of the edible accessory tissue. Firmness is measured using the resistance to penetration of the tissue by a plunger of a standard shape and size. Several relatively inexpensive pressure testers are commercially available (e.g., Magness-Taylor, University of California, and Effegi pressure testers).[41] Recent work has been directed toward developing nondestructive techniques for measuring firmness.[48,140]

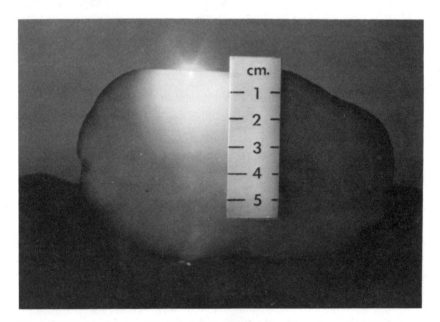

Figure 5-27. As illustrated with this longitudinal section of a potato (*Solanum tuberosum*, L.) tuber, part of the light striking a plant product moves into the tissue. The light can be reflected from the surface (surface reflectance), move into the tissue a short distance, and then exit (body reflectance), pass entirely through the sample (transmitted), or be absorbed. (*Photograph courtesy of G. G. Dull.*)

Optical measurements may also be used to assess the maturity of many crops. Normally removing a representative sample of fruit and analyzing it under laboratory conditions is required. Color changes for many fruits are associated with ripening and may be used to determine the optimum maturity for harvest. Thus, the simplest types of equipment measure the surface color of the sample.[73]

An individual ray of light striking the product can undergo one of four possible fates (fig. 5-27). It may be reflected from the surface of the product without entering the surface cells such as light striking a polished metal surface. Generally only a very small amount of light is reflected from the surface of the product (surface reflectance); most of the light enters the product. The direction of much of the light entering most products is altered by interactions with internal constituents. Some of the light entering moves back out of the sample in the general direction of entry or along the sides (body reflectance). Some of the light passes entirely through the sample (transmitted light) while a major portion is generally absorbed (absorbed light). When light, composed of a cross-section of wavelengths in the visible spectrum, enters a plant product, pigments absorb specific wavelengths. For example, chlorophyll absorbs red light (660 nm) and blue light (400 nm) but very little in the green region of the spectrum. The green color perceived for chlorophyll is due to this nonabsorbed light (body reflectance) moving back out of the tissue. Wavelengths of the body reflectance that were not absorbed can then be readily measured.[61] Likewise absorbed wavelengths can be determined by the difference in the spectral composition of the light entering versus leaving the sample. The quantity of light absorbed of a specific wavelength is correlated with the quantity of a specific absorbing pigment, which is in turn used as an index of the degree of maturation or ripening.

Analysis of transmitted light (light that passes through the sample) allows the measurement of internal color characteristics and/or the presence of certain internal disorders. For example, the presence of hollow heart in potato tubers can be nondestructively detected with transmitted light.[34] Since the disorder is found in the interior of the tuber, and the exterior is devoid of any physical manifestations, dissection was the only means of detection prior to optical analysis. Likewise, near-infrared light can be used to nondestructively determine the soluble solids in cantaloupes.[60] The use of light, therefore, represents a very powerful tool for determining not only stage of maturity but also quality of harvested products. The equipment required to determine maturity using light, however, is generally considerably more expensive than that required for pressure or shear measurements.

Chemical Measurements of Maturity

During maturation many plant products undergo distinct chemical alterations that are correlated with maturity. Thus, it is possible to use either a subjective or objective measure of these changes to determine maturity. The most common chemical parameters measured are water, soluble solids, specific sugars, starch, and acidity. Numerous other compounds or groups of compounds, however, can be or are used for specific crops.

Subjective evaluations of the basic chemistry of the product are made by feel, taste, and aroma. The concentration of water, the predominant chemical in many fleshy plant products, is often judged subjectively by the feel of the product. The moisture content of many dried seeds and nuts is commonly estimated in this manner and often with reasonable accuracy.

Taste and odor are also utilized as subjective measures of chemical composition. We can perceive four basic taste sensations (sweet, sour, salty, and bitter) and up to an estimated 10,000 distinct odors. Thus, sweetness (sugars), sourness (acidity), and aroma are sometimes estimated in this manner. Numerous factors can, however, significantly alter our estimation of maturity using these parameters, and as a consequence, they are less commonly used than subjective physical measurements.

Objective measurements of maturity are widely used for many crops. Water content can be quickly measured by weighing the sample before and after drying. Instruments such as the Ohaus Moisture Determination Balance simplify this process; however, only one sample can be dried at a time. Water content is extremely important in the postharvest handling and storage of seed and nut crops that must be relatively dry at harvest. Soluble solids or specific sugars are also used extensively (e.g., processing grapes). Soluble solids can be determined on samples of juice squeezed from the tissue using a refractometer or hydrometer. Acidity may also be readily measured by titration. In some crops (e.g., *Citrus*) the ratio of soluble solids to titratable acidity more accurately reflects the proper maturity for harvest.

The conversion of starch to sugar is also used as an index of maturity in some products. The reaction of starch with iodine produces a bluish to purple color and the pattern of staining and intensity can be used as a measure of the amount of starch remaining in the tissue.

QUALITY

In harvested plant products, quality is the composite of those characteristics that differentiate individual units of the product and have significance in determining the units' degree of acceptability to the user.[13,117] These characteristics or components of quality differ between types of products, where within the production–storage–marketing–utilization chain they are assessed, and between the individuals assessing the quality.

Components of Quality

The ultimate consumers attempt to assess the general quality of a product at the time of purchase. Generally the assessment represents a composite of several characteristics of the product that they have found or that they believe are good indices of overall quality. These components of quality can be separated into two major classifications: nutritional and sensory. Nutritional components are primarily applicable to plant products that are to be consumed; however, nutrition may also strongly modulate the quality of other products, for example, those used as reproductive propagules (seeds, small plants) or for their ornamental value (foliage plants). Sensory criteria include appearance, texture, aroma, and taste. Each of these criteria is not necessarily applicable to all products. Taste is not a quality consideration for ornamental flowers while odor may or may not be, depending on the species and/or cultivar.

Quality, therefore, is assessed from the relative values of several characteristics considered together (seldom is only a single parameter used). While it is possible to develop a precise list of what factors make up the quality of a particular product, ascertaining their relative importance is difficult. Not all characteristics contribute equally to the overall quality of the product. Likewise, the value of one parameter may change relative to other parameters. For some products, numerical scores have been assigned for individual characteristics where each represents part of a total numeric value. Schemes have been developed where the composite score of all components represents the grade of the product. Unfortunately, even when the maximum possible value for each individual component of quality is weighed relative to its estimated contribution, the scores are rarely additive; the composite decreases the amount of information available. For example, a very low score for one quality component (e.g., number of flowers on a potted chrysanthemum plant) may render the product unacceptable to the consumer while lowering the overall grade only slightly. Likewise, one quality characteristic may or may not be closely tied to another. Apple taste and aroma are important quality components but they only significantly improve the product's overall quality when the fruit's textural properties are at or near optimum.

Size

The size of individual units of a product can significantly affect consumer appeal, handling practices, storage potential, market selection, and final use. With many agricultural products, consumers discriminate based on size. Often exceedingly small and/or large individual units are considered undesirable. Small water chestnuts are less desirable due to the greater losses incurred during peeling and the additional labor required. In contrast, large peas are generally considered lower in quality than small or *petit pois*. Seldom are large apples selected for long-term storage in that they are much more susceptible to postharvest physiological disorders such as internal breakdown. Size may also determine what markets are available to the producer for the sale of the product. Early-season peaches grown in Georgia must be at least 4.1 cm in diameter to be shipped out of the state. Likewise, large sweetpotatoes (*Ipomoea batatas,* (L.) Lam.) are used for processing or animal feed rather than for fresh market sales.

Size may be determined by one of three general means: (1) dimension (length, width, diameter, or circumference), (2) weight, or (3) volume. In some instances, multiple measurements of size on a single product are utilized. For example, the Economic Commission for Europe uses both stem diameter and length for grading asparagus (*Asparagus officinalis,* L.). Standards for trees and shrubs in the United States (American Association of Nurserymen) include such measurements as overall height, trunk diameter, height of branching, number of branches, and either diameter and depth of ball or size of the container.

Shape and Form

Shape (the general outline of the product) and form (the arrangement of individual parts) are also important components of the overall quality of many postharvest products. Shape can be determined precisely using specific measurements and/or their mathematical relationships; however, more often than not it is ascertained subjectively. Shape is an important factor in distinguishing between individual cultivars (e.g., carrots (*Daucus carota*, L.), pears, apples, pecans (*Carya illinoinensis*, (Wang.) C. Koch.)). Likewise, unacceptable products are often eliminated based on their lack of conformity to a predetermined shape. Thus, like size, shape and/or form can alter acceptability, potential markets, and final use. With ornamental products, form in addition to shape is an important quality criterion. The number, placement, and orientation of individual branches, flowers, and leaves may be important in ascertaining quality. For example, the British Standard Institution recognizes 11 different forms of fruit trees.

Both shape and form can be altered by a number of factors, of which species, cultivar, and production conditions are particularly important. Improper postharvest handling and storage practices may also significantly affect shape and form. Improper sleeving of poinsettias (*Euphorbia pulcherrima*, Willd ex Klotz.) and leaf shedding (e.g., *Ficus benjamina*, L.) due to low light are examples of situations where shape and form can be seriously compromised.

Color

The color of an agricultural product probably contributes more to the assessment of quality than any other single factor. Consumers have developed distinct correlations between color and the overall quality of specific products. Tomatoes should be red; bananas should be yellow. Reverse this (i.e., yellow tomatoes and red bananas) and it would be difficult to give the products away, even if their quality were superior. Hence, on the first visual assessment of the quality of a product, color is critical.

Color is a function of the light striking the product, the differential reflection of certain wavelengths, and visual perception of those wavelengths. This light can

- Be reflected from the surface (surface reflectance)
- Penetrate the surface and then be reflected from the surface (body reflectance)
- Be absorbed (absorbed light)
- Move completely through the product (transmitted)

The perceived color is due to the absorption of some wavelengths and the reflection of others in the visible portion of the electromagnetic spectrum

(380–760 nm). Wavelength ranges for reflected light are

Blue: 400–500 nm
Yellow: 550–600 nm
Red: 600–700 nm

If there is equal reflection of all wavelengths, the sensation is white, and equal absorption of all wavelengths gives black. Which wavelengths are absorbed is determined primarily by the pigmentation of the product (see chapter 4).

Color can be described by three basic properties: (1) hue, the actual color that is a function of the dominant wavelength reflected; (2) lightness, the amount of light reflected (it depends not only on the product but also on the intensity of light from the source); and (3) saturation, the portion of the total light having a given wavelength. Perception of the product's color can be altered by changing any of these three properties.

Products with smooth, polished surfaces tend to be shiny whereas those with irregular surfaces are flat in color. The absence of sheen is due to the irregular surface reflecting the light at different angles. A bright luster is added to some products (e.g., apples, oranges) by waxing and polishing the surface. Perception of color can also be altered by the quality of the light striking the product. Artificial lights that do not display a spectrum that closely coincides with that of sunlight distort the perception of product color.

While color is used as a primary criterion to assess the general quality of many products, quality and color do not necessarily correlate closely with each other. In some cases, the association between what is perceived as optimum color and optimum quality is not at all valid. For example, a number of orange cultivars have fruit that are quite green when at their peak of quality. Since most consumers believe that oranges should be orange, the marketability of green fruit is much diminished. As a consequence, when destined for the fresh market these fruits are either dyed or gassed with ethylene to remove the chlorophyll pigments imparting the green color. In pecans, a dark brown kernel coloration is considered undesirable. This supposition is based on the gradual increase in darkness and rancidity that occurs with age. However, some cultivars of excellent quality are quite dark at optimum harvest. In both cases, color does not accurately reflect the true quality of the product.

A number of preharvest and postharvest factors can affect the color of harvested products. For example, cultivar may have a tremendous effect on color and/or color stability after harvest. Apple cultivars have been selected with red, yellow, and green fruit. Likewise, a wide range of carnation (*Dianthus caryophyllus,* L.) cultivars have been selected because of their flower color. Color diversity is also extensive in flowering ornamentals where flower color is an extremely important component of overall quality.

The color of many products changes during development. Banana fruit make the transition from green to yellow upon ripening, indicating an increase in acceptability for consumption. Conversely, the outer leaves of brussels sprouts change from green to yellow as they begin to deteriorate. Maturity, therefore, may have a significant effect on product color. Additional preharvest and postharvest factors that can alter the color of agricultural products include nutrition, moisture content, season, weather, improper handling, chilling injury, and physical damage.

Condition and Absence of Defects

Condition is a somewhat nebulous quality consideration that appears to encompass a wide range of the properties of the product in question. Our assessment of condition may include general visible quality parameters such as color, shape, and freedom from defects. It may also include considerations less easy to define such as freshness, cleanliness, and maturity. Freshness often includes the general physical condition of the product, for example, wilting of lettuce or shriveling of fruits. Other characteristics of freshness may be more elusive. For example, aroma may be lost or an undesirable aroma may be gained as product freshness deteriorates. Oil seed crops and nuts develop a rancid odor with the loss of freshness. Freshness may also carry a time connotation, that is, being recently harvested. Considerable emphasis is now placed on the purchase of fresh flowers, fruits, and vegetables.

Variation is an inherent factor in the production of agricultural products. Because of variation, some portion of the total of each harvested commodity will deviate in what is considered optimum for one or more quality components (shape, size, color, etc.). These products display quality defects, the presence of a fault that is undesirable, which prevents them from being optimal in quality. Defects can be classified under six general headings:

1. Biological factors: pathological, entomological, animal
2. Physiological factors: physiological disorders, nutritional imbalance, maturity
3. Environmental factors: climate and weather, soil, water supply
4. Mechanical damage
5. Extraneous matter: growing medium, vegetable matter, chemical residue
6. Genetic aberrations

These include both production and postproduction causes of defects. Of primary interest from the context of postharvest physiology are postproduction causes of quality loss. The most important of these from the standpoint of losses incurred are harvest and postharvest mechanical damage and the occurrence of insect and pathogen problems. In addition, it is possible to exert a much greater control over some causes of defects than others. For example, the weather during production can be little influenced. After harvest, however, it is possible to precisely control the temperature and humidity to which many products are exposed. Understanding the causes for defects, especially those that can be influenced, often allows taking the necessary steps to prevent their occurrence.

Taste

The taste of an edible product is perceived by specialized taste buds on the tongue. Although there are a great many tastes, most appear to primarily represent combinations of four dominant chemical sensations: sweet, sour, bitter, and salty. Of these, sweet and sour predominate, with bitterness being important in some products. Saltiness, on the other hand, is seldom a factor in fresh products. Aroma or odor, however, is not limited to edible plant products as a quality component and is therefore treated separately. Taste represents one of the quality attributes consumers try to correlate with visual parameters of the product (e.g., maturity, color, cultivar). The need to correlate results from taste being seldom a direct quality consideration in deciding whether or not to purchase a product. Rather, the assessment of taste is made after purchase.

Sweetness due to sugars and sourness from organic acids are dominant components in the taste of many fruits. In some cases, wholesale purchases are made based on sugar concentration (as indicated by soluble solids) or sugar/acid ratio. Likewise, bitterness may be an important component of the taste of some products (e.g., grapefruit, bitter melon). Many phenolic compounds, in varying levels of polymerization, are major contributors to bitterness. In some products, bitterness is considered a desirable attribute whereas in others it is undesirable and is avoided with cultivar selection (e.g., cucumber) and/or maturity. Of the many factors that can affect the taste quality of a product, ripeness, maturity, cultivar, irrigation, and fertilization are especially important.

Odor

Odor is perceived through the chemical stimulation of sites on the olfactory epithelium. In contrast to the four primary taste sensations, a trained person can distinguish more than 10,000 distinct odors, some in very minute quantities (e.g., 10^{-9} mg $\cdot$ m^{-1}). Unlike taste, odor may be a primary quality criterion assessed in deciding whether or not to purchase a product. The aroma of *Gardenia jasminoides,* Ellis flowers or other aromatic ornamentals often contributes significantly to the decision to purchase the product.

The chemical compounds that make up the aromatic properties of plant products need to be volatile, generally at the temperature at which the product is utilized, and exhibit at least some degree of water solubility. In recent years, considerable research has been directed toward isolating and identifying the volatile compounds produced by plants (for additional details, see chapter 4). Most volatile compounds do not contribute to the characteristic odor of a product due to their lack of a distinguishable odor and/or being in such minute quantities as to be insignificant. Generally only a small number of volatile compounds are important in the aroma. Likewise, volatile compounds may also make up distinctly undesirable odors. These are especially evident when some products are improperly handled after harvest (e.g., the exposure of brussels sprouts (*Brassica oleracea,* L. Gemmifera group) to anaerobic conditions) or held for too long. Other factors that can influence both desirable and undesirable odors include species, cultivar, maturity, and ripening.

Although it is possible to precisely measure the concentration of many of the desirable and undesirable compounds that impart the characteristic odor of a specific product, it is in fact rarely done in assessing quality. Generally, odor is determined subjectively by simply smelling the product.

Texture

The myriad of agricultural plant products display a seemingly unlimited array of textural characteristics. These may be external textural properties, for example, the surface geometry of an ice plant (*Mesembryanthemum crystallinum,* L.) leaf, or those internal properties that are of critical importance in edible products. Texture is comprised of those properties of a product that can be appraised visually or by touch. With edible products, textural properties may also be assessed by skin and muscle senses in the mouth. Textural properties can be divided into three major classifications with subclasses within each[192]:

1. Mechanical characteristics
   a. Hardness: the force necessary to attain a given deformation
   b. Cohesiveness: the strength of internal bonds of the product
   c. Viscosity: the rate of flow per unit force
   d. Elasticity: the speed at which the material returns to its original shape after deformation
   e. Adhesiveness: the work required to overcome the attraction between food and mouth
   f. Brittleness: the force required to fracture a product
   g. Chewiness: the energy required to masticate a solid food until it can be swallowed
   h. Gumminess: the energy required to masticate a semisolid food until it can be swallowed
2. Geometrical characteristics
   a. Particle size and shape
   b. Particle shape and orientation
3. Other characteristics
   a. Moisture content
   b. Lipid content

The internal textural properties of plant products are due to the composition of the cells and the structure of these cells and their supporting tissues. The structural properties may be due primarily to the cell walls per se or also to non-cell-wall materials such as storage carbohydrates. Turgor pressure is an extremely important parameter affecting texture in many fleshy products. Physical and chemical differences in structure result in the many dissimilarities found between species and products, and within individual products.

Texture may be measured both subjectively and objectively. We subjectively scrutinize the texture of an apple when we bite into it. The textural property of apples represents one of the single most important components of overall quality. Their texture could also be determined using a more objective method. In recent years, a number of instruments have been developed to objectively measure discrete differences in textural properties between and within individual products. These include shear and compression instruments and instruments that assess texture indirectly by determining the moisture, fiber, or lipid content of the product.

As with other quality attributes, a number of factors can affect the texture of plant products. Among these, maturity is one of the most important for a significant cross-section of products. Products such as asparagus, beans, and peas become fibrous or harden with advancing maturity and many of the fleshy fruits (e.g., peach, apple, pear) soften. The handling and storage conditions to which many products are exposed after harvest may also significantly alter their textural properties. The loss of water due to improper control of the relative humidity to which the product is exposed can result in serious textural quality losses. These losses may be reversible as in leaf lettuce, or largely irreversible (e.g., apples). Exposure of some products to chilling temperatures, even for short periods, may result in textural alterations. For example, holding sweetpotatoes at temperatures below 10°C can result in a condition called hardcore where the center of the root becomes woody and inedible. Several other factors that can significantly affect the textural properties of harvested products are cultivar and agronomic practices such as nutrition and irrigation.

Other Components of Quality

Not all of the critical components of quality for all products are described by the before-mentioned characteristics. Some plant products, due to their end use or unique nature, require specialized quality criteria. For example, with products that are to be used as propagules (e.g., seeds, seedlings, transplants), viability, vigor, and purity are of critical importance. With seeds, the freedom from seed-borne diseases is also a consideration.

Nutritional quality may be important for some food crops. For example, losses in vitamins during handling and storage are known to occur and may be significant. Although there is a growing interest in nutrition, it is at present rarely utilized by consumers as a quality criterion in discriminating between possible choices within a given commodity.

Measurement of Quality

Most commercially grown crops are graded for quality at one or more points between harvest and final retail sales. For many products, precisely defined quality standards (grades) have been established and are enforced by various government agencies or grower organizations. This assessment of quality may be subjective or objective in nature, depending on the technique utilized. For most crops more than one criterion is utilized when assessing quality. For example, a subjective visual evaluation may be made by trained personnel for surface characteristics such as shape, disease, insect damage, discoloration, and wilting. This type of evaluation is limited to characteristics that can be assessed visually. Subsequently, the product may then be screened using one of many possible automated objective measurements. These include characteristics such as size, weight, specific gravity, mechanical properties, spectrophotometric properties, and chemical analyses. Since the methods for subjective quality evaluation are highly varied, we shall focus on the more objective measures of quality.

Size

Size separation of individual product units may be accomplished using a number of techniques depending on the product in question. Screens of perforated plates or wire cloth of various hole sizes and shapes are used for separation of small products such as seeds. Size separation may also be made using belts with holes, cups, rollers, or diverging rollers. For example, Chinese water chestnuts (*Eleocharis dulcis,* (Burm. f.) Trin. ex Henschel) are graded for size using two diverging rollers placed in a V configuration. The corms are fed onto the narrow end of the rollers; as they proceed toward the base, the distance between the rollers becomes greater. Eventually the chestnut drops onto one of several conveyers beneath, each collecting a representative size classification.

Weight

As with size, a variety of techniques have been developed for the separation of products based on their weight. Often weight is a more precise measure for separation than size (diameter) in that weight generally varies with the cube of the diameter of regularly shaped products. Typically a cup or pocket holding a single unit moves along a path where the weight required to drop the product into its appropriate class changes with distance. The actual weight required to activate the tripping device is generally adjustable so that the machine may be used for more than one product.

Shape

Instrumentation for shape separation has been developed for some plant products. Mechanical means of this type of quality assessment is especially prevalent in the grading of seeds. For example, disc, cylinder, incline belt, and spiral separators may be used to select for or against length or roundness. Separators may also be used for cleaning seed lots. More recently, the widespread availability of computers and the development of computer imaging technology have stimulated the development of automated shape analysis.[182] It is currently possible to compute the (1) area, (2) perimeter, (3) maximum length, (4) width, (5) centroid, (6) horizontal and vertical fret, (7) curvature, (8) circularity, and (9) form factor of harvested products.[93] The techniques for sorting products such as tomatoes,[180] apples,[165] sweetpotatoes,[219] and other agricultural products have been developed. Interfacing computer vision with robotics will have a pronounced effect on grading of harvested products in the future.

Specific Gravity

Separation by specific gravity generally entails the use of a solution with a controlled specific gravity. A specific gravity is selected that allows the heavier products to sink or the lighter products to float to the surface. Specific gravity has been shown to correlate with the quality of processed toma-toes.[213] When it is not desirable to wet the product, a continuous stream of air may be used for separation. However, in addition to the specific gravity, this medium also separates based on the aerodynamic characteristics of the product, which may significantly influence the actual separation. Airstream separation is especially useful for seeds where for some species it may be correlated with vigor.[189] Air is also extensively used for the removal of foreign material from the product.

Specific gravity can also be measured using spectrophotometry. Near infrared light is used to separate intact potatoes into relatively precise dry matter categories.[64]

Mechanical Properties

A number of quality tests have been developed based on the mechanical properties of the product. Typically these instruments measure either the resistance to shear or penetration.* Several shear instruments have been developed to determine the quality of peas[11] and broadbeans[14] harvested fresh for processing (e.g., the Texturetester and Tenderometer). The Texturetester (Food Technology Corporation) allows the measurement of shear values for a large cross-section of fleshy products. A later, more sophisticated instrument, the Instron, greatly expanded the characterization of the mechanical properties of harvested products. This instrument is used almost exclusively for research purposes, however, rather than routine quality assessment.

Pressure testers or penetrometers that measure the resistance of the product to penetration by a standard plunger are used for some fruit crops (e.g., apples, peaches, pears). The most common of these are the Magness-Taylor, Effegi, and U.C. Pressure testers. Mechanical properties may also be assessed using the ability of the product to conduct sonic or ultrasonic waves. Although generally nondestructive, these techniques have proven to be less precise and are generally used infrequently.

Spectrophotometric Properties

The use of light in quality evaluation has increased substantially in recent years due to the amount of information that can be derived, the speed at which it is accomplished, and its nondestructive nature. Initial electronic sorters selected or rejected individual product units based on surface color, relative brightness, or the intensity of specific wavelengths reflected. Two early instruments were the Gardner and the Hunter color difference meters. Adaptations of this principle were used for automated separation of various fruits based on external color. Subsequently, the development of the spectro-computer (Biospect) has allowed for correlation of changes in a monochromatic light source striking the product with internal quality attributes. Instru-

---

* See Bourne[41] for a list of commercially available instruments.

ments have now been developed that will identify potato tubers with hollow heart (an internal disorder that exhibits no external symptoms),[34] screen onion bulbs (*Allium cepa,* L.)[36] and potato tubers[60] for dry matter percent, screen papayas (*Carica papaya,* L.) for internal color,[35] screen honeydew and cantaloupes for soluble solids,[62] and screen apple fruit for water core.[37] It is probable that scientists will soon be able to nondestructively measure the internal concentration of specific sugars within harvested products.

Surface Characteristics

Air trapped on the surface of the product alters the rate at which the product settles in a solution. Chemicals are added to the solution that enhance the trapping. Generally this technique is used for separating damaged fruits, which tend to have more air adhere to their surface and thus float.

Chemical Analyses

Individual products can be measured for a cross-section of chemical characteristics, for example, moisture, sugar, soluble solids, acidity, pH, impurities, rancidity, fiber. With the exception of spectrophotometric measurements, chemical analyses tend to be destructive. As a consequence, normally only a small representative sample is measured. This information is then used to infer the condition of the large quantity from which the sample was drawn. Therefore, chemical analyses are not used to determine the quality of each individual unit of the product.

Other Techniques

Various other techniques for assessing quality utilizing thermal, electrical, or radiant energy have been tested. As nondestructive techniques, these typically require more energy, and as a consequence, they typically cause greater tissue damage than spectrophotometric techniques.

Quality Standards

While most individuals have a general subjective set of quality standards for the products with which they are familiar, these tend to be highly arbitrary. What is acceptable quality to one consumer may be totally unacceptable to another. Likewise, what is considered to be of acceptable quality to the seller may not coincide with that of the buyer. Individual quality criteria are also subject to variation due to changes in supply and demand. When a particular product is scarce, a certain level of quality may be acceptable whereas the same product would be very much unacceptable if the supply were abundant. The disadvantages of not having a well-defined set of standards to characterize quality were evident even in the Roman era. The Emperor Diocletian established standards for many food products.[198]

The advantages of standards are twofold. First, standards protect the purchaser by providing a uniform quality product. It is especially important when purchases are made without having seen the product. The buyer knows from past experience what he or she will receive. Second, standards typically bring about a general upgrading of the overall quality of a particular commodity. Producers are aware of the economic advantages of providing a top-quality product and thus have the incentive to alter their production practices to achieve as high a standard as possible.

Standards are established by individual countries or groups of countries (e.g., the European Economic Community); by provinces, states, counties, or municipal authorities within individual countries; and by grower organizations. There is an increasing impetus for worldwide standards for many agricultural products. These standards are being established by the Codex Alimentarius Commission in France.

Compliance with standards may be voluntary or mandatory, depending on the organization establishing them and the authorization under which the organization operates. Most standards established or operated by government organizations are generally mandatory.

An important requisite for quality standards is a clear, concise, and accurate description of each criterion used to monitor quality and the acceptable range for each criterion within a specific grade. For example, the upper and lower diameter limits for each grade of lemon and orange fruits are defined as well as the ranges for other quality components. Rigorous characterization of grades leads to uniformity in grading between individuals, locations, and time during the season.

REFERENCES

1. Abdel-Gawad, H., and R. J. Romani. 1974. Hormone-induced reversal of color change and related respiratory effects in ripening apricot fruit. *Plant Physiol.* **32**:161–165.
2. Abeles, F. B. 1968. Role of RNA and protein synthesis in abscission. *Plant Physiol.* **43**:1,577–1,586.
3. Abeles, F. B., L. J. Dunn, P. Morgans, A. Callahan, R. E. Dinterman, and J. Schmidt. 1988. Induction of 33-kD and 60-kD peroxidases during ethylene induced senescence of cucumber cotyledons. *Plant Physiol.* **87**:609–615.
4. Abeles, F. B., and B. Rubinstein. 1964. Regulation of ethylene evolution and leaf abscission by auxin. *Plant Physiol.* **39**:963–969.
5. Adams, C. A., M. C. Fjerstad, and R. W. Rinne. 1983. Characteristics of soybean seed maturation: Necessity for slow dehydration. *Crop Sci.* **23**:265–267.
6. Adams, C. A., and R. W. Rinne. 1981. Seed maturation in soybeans (*Glycine max*, L. Merr.) is independent of seed mass and of the parent plant, yet is necessary for production of viable seeds. *J. Exp. Bot.* **32**:615–620.
7. Addicott, F. T. 1982. *Abscission.* University of California Press, Berkeley, 369p.
8. Akamine, E. T., and T. Goo. 1978. Respiration and ethylene production in mammee apple (*Mammea americana*, L.). *J. Am. Soc. Hort. Sci.* **103**:308–310.
9. Akamine, E. T., and T. Goo. 1979. Respiration and ethylene production in fruits of species and cultivars of *Psidium* and species of *Eugenia. J. Am. Soc. Hort. Sci.* **104**:632–635.
10. Altschul, A. M., L. Y. Yatsu, R. L. Ory, and E. M. Engleman. 1966. Seed proteins. *Ann. Rev. Plant Physiol.* **17**:113–136.

11. Anthistle, M. J. 1961. The composition of peas in relation to texture. *Fruit Veg. Cann. Quick Freeze Res. Assoc. Sci. Bull. No. 4.*

12. Arditti, J., and H. Flick. 1976. Post-pollination phenomena in orchid flowers. VI. Excised floral segments of Cymbidium. *Am. J. Bot.* **63:**201–211.

13. Arthey, V. D. 1975. *Quality of Horticultural Products.* Butterworths, London, 228p.

14. Arthey, V. D., and C. Webb. 1969. The relationship between maturity and quality of canned broadbeans (*Vicia faba,* L.). *J. Food Technol.* **4:**61–74.

15. Awad, M., and R. E. Young. 1979. Postharvest variation in cellulase, polygalacturonase, and pectin methylesterase in avocado (*Persea americana* Mill. cv. Fuerte) fruits in relation to respiration and ethylene production. *Plant Physiol.* **64:**306–309.

16. Babbitt, J. K., M. J. Powers, and M. E. Patterson. 1973. Effects of growth-regulators on cellulase, polygalacturonase, respiration, color, and texture of ripening tomatoes. *J. Am. Soc. Hort. Sci.* **98:**77–81.

17. Barber, R. F., and J. E. Thompson. 1980. Senescence-dependent increase in permeability of liposomes prepared from cotyledon membranes. *J. Exp. Bot.* **31:**1,305–1,313.

18. Barnes, M. F., and B. J. Patchett. 1976. Cell wall degrading enzymes and the softening of senescent strawberry fruit. *J. Food Sci.* **41:**1,392–1,395.

19. Barton, L. V. 1965. Seed dormancy: General survey of dormancy types in seeds. In *Handbook of Plant Physiology,* vol. 15a(2), W. Rukland (ed.). Springer-Verlag, Berlin, pp. 699–720.

20. Basset, B. 1984. Adam, 7-21 (cartoon). Universal Press Syndicate.

21. Beaudry, R. M., J. A. Payne, and S. J. Kays. 1985. Variation in the respiration of harvested pecans due to genotype and kernel moisture level. *HortScience* **20:**752–754.

22. Beaudry, R. M., R. F. Severson, C. C. Black, and S. J. Kays. 1989. Banana ripening: Implications of changes in glycolytic intermediate concentrations, glycolytic and gluconeogenic carbon flux, and fructose 2,6-bisphosphate concentration. *Plant Physiol.* **91:**1,436–1,444.

23. Beevers, L., and R. Poulson. 1972. Protein synthesis in cotyledons of *Pisum sativum,* L. I. Changes in cell-free amino acid incorporation capacity during seed development and maturation. *Plant Physiol.* **49:**476–481.

24. Ben-Arie, R., S. Lavee, and S. Guelfat-Reich. 1970. Control of woolly breakdown of 'Elberta' peaches in cold storage by intermittent exposure to room temperature. *Proc. Am. Soc. Hort. Sci.* **95:**801–802.

25. Ben-Yehoshua, S. 1964. Respiration and ripening of discs of the avocado fruit. *Plant Physiol.* **17:**71–80.

26. Beyer, E. M. 1976. A potent inhibitor of ethylene action in plants. *Plant Physiol.* **58:**268–271.

27. Biale, J. B. 1950. Postharvest physiology and biochemistry of fruits. *Ann. Rev. Plant Physiol.* **1:**183–206.

28. Biale, J. B. 1960. Respiration of fruits. In *Encyclopedia of Plant Physiology,* vol. 12(2), W. Ruhland (ed.). Springer, Berlin, pp. 536–592.

29. Biale, J. B. 1964. Growth, maturation, and senescence in fruits. *Science* **146:**880–888.

30. Biale, J. B. 1976. Recent advances in postharvest physiology of tropical and subtropical fruits. *Acta Hort.* **57:**179–187.

31. Biale, J. B., and D. E. Barcus. 1970. Respiratory patterns in tropical fruits of the Amazon Basin. *Trop. Sci.* **12:**93–104.

32. Biale, J. B., and R. E. Young. 1971. The avocado pear. In *The Biochemistry of Fruits and Their Products,* 2 vols., A. C. Hulme (ed). Academic Press, London, pp. 65–105.

33. Biale, J. B., and R. E. Young. 1981. Respiration and ripening in fruits—retrospect and prospect. In *Recent Advances in the Biochemistry of Fruits and Vegetables,* J. Friend and M. J. C. Rhodes (eds.). Academic Press, New York, pp. 1–39.

34. Birth, G. S. 1960. A nondestructive technique for detecting internal discoloration in potatoes. *Am. Potato J.* **37:**53–60.

35. Birth, G. S., G. G. Dull, J. B. Magee, H. T. Chan, and C. G. Cavaletto. 1984. An optical method for estimating papaya maturity. *J. Am. Soc. Hort. Sci.* **109:**62–66.

36. Birth, G. S., G. G. Dull, W. T. Renfroe, and S. J. Kays. 1985. Nondestructive spectrophotometric determination of dry matter in onion. *J. Am. Soc. Hort. Sci.* **110:**297–303.

37. Birth, G. S., and K. L. Olsen. 1964. Nondestructive detection of watercore in delicious apples. *Proc. Am. Soc. Hort. Sci.* **85:**74–84.

38. Biswal, U. C., and B. Biswal. 1984. Photocontrol of leaf senescence. *Photochem. Photobiol.* **39:**875–879.

39. Blanpied, G. D. 1972. A study of ethylene in apple, red raspberry and cherry. *Plant Physiol.* **48:**627–630.

40. Borroto, K., and L. Dure III. 1987. The globulin seed storage proteins of flowering plants are derived from two ancestral genes. *Plant Mol. Biol.* **8:**113–131.

41. Bourne, M. C. 1980. Texture evaluation of horticultural crops. *HortScience* **15:**51–57.

42. Brady, C. 1988. Nucleic acid and protein synthesis. In *Senescence and Aging in Plants,* L. D. Noodén and A. C. Leopold (eds.). Academic Press, New York, pp. 147–179.

43. Burg, S. P., and E. A. Burg. 1962. Role of ethylene in fruit ripening. *Plant Physiol.* **37:**179–189.

44. Burg, S. P., and E. A. Burg. 1965. Gas exchange in fruits. *Physiol. Plant.* **18:**870–884.

45. Burg, S. P., and E. A. Burg. 1967. Molecular requirements for the biological activity of ethylene. *Plant Physiol.* **42:**144–152.

46. Burton, W. G. 1957. The dormancy and sprouting of potatoes. *Food Sci. Abstr.* **29:**1–12.

47. Cameron, A. C., and M. S. Reid. 1983. Use of silver thiosulfate to prevent flower abscission from potted plants. *Sci. Hort.* **19:**373–378.

48. Chen, P., and Z. Sun. 1989. Nondestructive methods for quality evaluation and sorting of agricultural products. 4th *Int. Conf. Properties of Agric. Materials Proc.* Rostock, GDR.

49. Coggins, C. W., Jr., and H. Z. Hield. 1958. Gibberellin on orange fruit. *Calif. Agric.* **12**(9):11.

50. Coombe, B. G. 1960. Relationship of growth and development to changes in sugars, auxin, and gibberellins in fruit of seeded and seedless varieties of *Vitis vinifera. Plant Physiol.* **35:**241–250.

51. Danielson, C. E. 1952. A contribution to the study of the synthesis of reserve proteins of ripening pea seeds. *Acta Chem. Scand.* **6:**149–159.

52. Dayton, D. F. 1966. The pattern and inheritance of anthocyanin distribution in red pears. *Proc. Am. Soc. Hort. Sci.* **89:**110–116.

53. DeHertogh, A. A. 1988. *Holland Bulb Forcer's Guide,* 4th ed. Netherlands Flower-Bulb Institute, New York, 369 p.

54. De la Fuente, R. K., and A. C. Leopold. 1969. Kinetics of abscission in the bean leaf petiole explant. *Plant Physiol.* **44:**251–254.

55. Denisen, E. L. 1951. Carotenoid content of tomato fruits. I. Effect of temperature and light. II. Effects of nutrients, storage and variety. *Iowa State Coll. J. Sci.* **25:**549–574.

56. Desai, B. B., and P. B. Deshpande. 1975. Chemical transformations in three cultivars of banana (*Musa paradisica* Linn.) fruit stored at 20°C. *Mysore J. Agric. Sci.* **9**:634–643.

57. Dilley, D. R. 1977. Hypobaric storage of perishable commodities—fruits, vegetables, flowers, and seedlings. *Acta Hort.* **62**:61–70.

58. Do, J. Y., D. K. Salunkhe, and L. E. Olson. 1979. Isolation, identification and comparison of the volatiles of peach fruit as related to harvest maturity and artificial ripening. *J. Food Sci.* **34**:618–621.

59. Drawert, F. 1975. Formation des aromes a différents stades de l'évolution du fruit; enzyme intervenant dans cette formation. In *Facteurs et Régulation de la Maturation des Fruits,* No. 238. CNRS, Paris, pp. 309–319.

60. Dull, G. G., G. S. Birth, and R. G. Leffler. 1989. Use of near infrared analysis for the nondestructive measurement of dry matter in potatoes. *Am. Pot. J.* **66**:215–225.

61. Dull, G. G., G. S. Birth, and J. B. Magee. 1980. Nondestructive evaluation of internal quality. *HortScience* **15**:60–63.

62. Dull, G. G., G. S. Birth, D. A. Smittle, and R. G. Leffler. 1989. Near infrared analysis of soluble solids in intact cantaloupe. *J. Food Sci.* **54**:393–395.

63. Dure, L. 1985. Embryogenesis and gene expression during seed formation. *Oxford Surv. Plant Mol. Cell Biol.* **2**:179–197.

64. Eckerson, S. 1913. A physiological and chemical study of after-ripening. *Bot. Gaz.* **55**:286–299.

65. Egli, D. B., J. E. Leggett, and J. M. Wood. 1978. Influences of soybean seed size and position on the rate and duration of filling. *Agron. J.* **70**:127–130.

66. Eguchi, T., T. Matsumura, and T. Koyama. 1963. The effect of low temperatures on flower and seed formation in Japanese radish and Chinese cabbage. *Proc. Am. Soc. Hort. Sci.* **82**:322–331.

67. Eisenburg, B. A., and G. L. Staby. 1985. Mitochondrial changes in harvested carnation flowers (*Dianthus caryophyllus,* L.) during senescence. *Plant and Cell Physiol.* **26**:829–837.

68. El-Antably, H. M. M., P. F. Wareing, and J. Hillman. 1967. Some physiological responses to D. Abscisin (Dormin). *Planta* **73**:74–90.

69. Engelmann, W., I. Eger, A. Johnsson, and H. G. Karlsson. 1974. Effect of temperature pulses on the petal rhythm of *Kalanchoe:* An experimental and theoretical study. *Int. J. Chronobiol.* **2**:347–358.

70. Evans, R. Y., and M. S. Reid. 1985. Control of petal expansion during diurnal opening of roses. *Acta Hort.* **181**:55–63.

71. Ferguson, I. B. 1984. Calcium in plant senescence and fruit ripening. *Plant Cell Environ.* **7**:477–489.

72. Floren, G. 1941. Untersuchungen über Blütenfärbmuster und Blütenfärbunger. *Flora* **135**:65–100.

73. Francis, F. J. 1980. Color quality evaluation of horticultural crops. *HortScience* **15**:58–59.

74. Frenkel, C. 1972. Involvement of peroxidase and indole-3-acetic acid oxidase isozymes from pear, tomato, and blueberry fruit in ripening. *Plant Physiol.* **49**:757–763.

75. Frenkel, C. 1975. Role of oxidative metabolism in the regulation of fruit ripening. In *Facteurs et Régulation de la Maturation des Fruits,* No 238. CNRS, Paris, pp. 201–209.

76. Frenkel, C., I. Klein, and D. R. Dilley. 1968. Protein synthesis in relation to ripening of pome fruits. *Plant Physiol.* **43**:1,146–1,153.

77. Frenkel, C., and M. K. Mukai. 1984. Possible role of fruit cell wall oxidative activity on ethylene evolution. In *Ethylene. Biochemical, Physiological and Applied Aspects,* Y. Fuchs and F. Chalutz (eds.). Martinus Nijhoff/W. Junk, The Hague, pp. 303–316.

78. Gepstein, S. 1988. Photosynthesis. In *Senescence and Aging in Plants*, L. D. Noodén and A. C. Leopold (eds.). Academic Press, New York, pp. 85–109.

79. Gerhardt, F., and E. Smith. 1946. Physiology and dessert quality of Delicious apples as influenced by handling, storage and simulated marketing practice. *Proc. Wash. State Hort. Assoc.* **1,945**:151–172.

80. Gilissen, L. J. W. 1977. Style controlled wilting of the flower. *Planta* **133**:275–280.

81. Giovannoni, J. J., D. Della Penna, A. B. Bennett, and R. L. Fischer. 1989. Expression of a chimeric polygalacturonase gene in transgenic *rin* (ripening inhibitor) tomato fruit results in polyuronide degradation but not fruit softening. *Plant Cell* **1**:53–63.

82. Goodwin, T. W., and M. Jamikorn. 1952. Biosynthesis of carotenes in ripening tomatoes. *Nature* **170**:104–105.

83. Gross, J., and M. Flugel. 1982. Pigment changes in peel of the ripening banana *Musa cavendish. Gartenbauwis* **47**:62–64.

84. Grossmann, K., and J. Jung. 1982. Pflanzlicher Seneszenzvorgange. *Z. Acker-Pflanzenbau* **151**:149–165.

85. Hansen, E. 1967. Ethylene-stimulated metabolism of immature 'Bartlett' pears. *Proc. Am. Soc. Hort. Sci.* **91**:863–867.

86. Hatfield, R., and D. J. Nevins. 1986. Characterization of the hydrolytic activity of avocado cellulase. *Plant Cell Physiol.* **27**:541–552.

87. Hatfield, S. G. S., and B. D. Patterson. 1975. Abnormal volatile production by apples during ripening after controlled atmosphere storage. In *Facteurs et Régulation de la Maturation des Fruits*, No. 238. CNRS, Paris, pp. 57–62.

88. Hill, J. E., and R. W. Breidenbach. 1974. Proteins of soybean seeds. I. Isolation and characterization of the major components. *Plant Physiol.* **53**:742–746.

89. Hill, J. E., and R. W. Breidenbach. 1974. Proteins of soybean seeds. II. Accumulation of the major components during seed development and maturation. *Plant Physiol.* **53**:747–751.

90. Hills, L. D., and E. H. Haywood. 1946. *Rapid Tomato Ripening*. Faber and Faber, London, 143p.

91. Ho, T. D. 1983. Biochemical mode of action of abscisic acid. In *Abscisic Acid*, F. T. Addicott (ed). Praeger, New York, pp. 147–169.

92. Horvat, R. J., and G. W. Chapman. 1990. Comparison of volatile compounds from peach fruit and leaves (cv. Monroe) during maturation. *J. Agric. Food Chem.* **38**:1442–1444.

93. Howarth, M. S., and W. F. McClure. 1987. Agricultural product analysis by computer vision. *Am. Soc. Agric. Eng. Paper No. 87-3043*.

94. Huber, D. J. 1983. Polyuronide degradation and hemicellulose modifications in ripening tomato fruit. *J. Am. Soc. Hort. Sci.* **108**:405–409.

95. Huber, D. J. 1983. The role of cell wall hydrolases in fruit softening. *Hort. Rev.* **5**:169–219.

96. Huber, D. J., and D. W. Newman. 1976. Relation between lipid changes and plastid ultra-structural changes in senescing and regreening soybean cotyledons. *J. Exp. Bot.* **27**:490–511.

97. Hulme, A. C. 1954. Studies in the nitrogen metabolism of apple fruit. *J. Exp. Bot.* **5**:159–172.

98. Ismael, A. A., and W. T. Kender. 1969. Evidence of a respiratory climacteric in highbush and lowbush blueberry fruit. *HortScience* **4**:342–344.

99. Jen, J. J. 1974. Influence of spectral quality of light on pigment systems of ripening tomatoes. *J. Food Sci.* **39**:907–910.

100. Kamerbeek, G. A., and J. J. Beijer. 1964. Vroege bloei van Iris 'Wedgwood'. *Meded. Dir. Turinb.* **27**:598–604.

101. Kays, S. J., and R. M. Beaudry. 1987. Techniques for inducing ethylene effects. *Acta Hort.* **210**:77–116.

102. Kays, S. J., and M. J. Hayes. 1978. Induction of ripening in the fruits of *Momordica charactia*, L. by ethylene. *Trop. Agric.* **55**:167–172.

103. Kays, S. J., C. A. Jaworski, and H. C. Price. 1976. Defoliation of pepper transplants in transit by endogenously evolved ethylene. *J. Am. Soc. Hort. Sci.* **101**:449–451.

104. Kays, S. J., C. W. Nicklow, and D. H. Simons. 1974. Ethylene in relation to the response of roots to physical impedance. *Plant and Soil* **40**:565–571.

105. Kermode, A. R., and J. D. Bewley. 1985. The role of maturation drying in the transition from seed development to germination. II. Postgerminative enzyme production and soluble protein synthetic pattern changes within the endosperm of *Ricinus communis*, L. seeds. *J. Exp. Bot.* **36**:1,916–1,927.

106. Kermode, A. R., D. J. Gifford, and J. D. Bewley. 1985. The role of maturation drying in the transition from seed development to germination. III. Insoluble protein synthetic pattern changes within the endosperm of *Ricinus commurus*, L. seeds. *J. Exp. Bot.* **36**:1,928–1,936.

107. Kidd, F., and C. West. 1930. Physiology of fruit. I. Changes in the respiratory activity of apples during their senescence at different temperatures. *Proc. Roy. Soc.* (*London*) **106B**:93–109.

108. Kivilaan, A., and R. S. Bandurski. 1981. The one-hundred-year period for Dr. Beal's seed viability experiment. *Am. J. Bot.* **68**:1,290–1,292.

109. Klein, S., and B. M. Pollock. 1968. Cell fine structure of developing lima bean seeds related to seed desiccation. *Am. J. Bot.* **55**:658–672.

110. Kliewer, W. M. 1967. The glucose-fructose ratio of *Vitis vinifera* grapes. *Am. J. Enol. Vitic.* **18**:33–41.

111. Knee, M. 1973. Polysaccharide changes in cell walls of ripening apples. *Phytochemistry* **12**:1,543–1,549.

112. Knee, M. 1985. Evaluating the practical significance of ethylene in fruit storage. In *Ethylene and Plant Development*, J. A. Roberts and G. A. Tucker (eds.). Butterworths, London, pp. 297–315.

113. Knee, M., S. G. S. Hatfield, and W. J. Bramlege. 1987. Response of developing apple fruits to ethylene treatment. *J. Exp. Bot.* **38**:972–979.

114. Knee, M., F. J. Proctor, and C. J. Dover. 1985. The technology of ethylene control: use and removal in post-harvest handling of horticultural commodities. *Ann. Appl. Biol.* **107**:581–595.

115. Knypl, J. S., and W. Mazurczyk. 1971. Arrest of chlorophyll and protein breakdown in senescing leaf discs of kale by cycloheximide and vanillin. *Curr. Sci.* **40**:294–295.

116. Kosiyachinda, S., and R. E. Young. 1975. Ethylene production in relation to the initiation of respiratory climacteric in fruit. *Plant and Cell Physiol.* **16**:595–602.

117. Kramer, A., and B. A. Twigg. 1970. *Quality Control for the Food Industry*, vol. 1, AVI, Westport, Conn., 556p.

118. Ku, P. K., and A. L. Mancinelli. 1972. Photocontrol of anthocyanin synthesis. *Plant Physiol.* **49**:212–217.

119. Lang, G. A. 1987. Dormancy: A new universal terminology. *HortScience* **22**:817–820.

120. Leopold, A. C. 1961. Senescence in plant development. *Science* **134**:1,727–1,732.

121. Lewis, L. N., and J. E. Varner. 1970. Synthesis of cellulase during abscission of *Phaseolus vulgaris* leaf explants. *Plant Physiol.* **46**:194–199.

122. Lewis, L. N., C. W. Coggins, Jr., C. K. Labanauskas, and W. M. Dugger, Jr. 1967. Biochemical changes associated with natural and gibberellin A_3 delayed senescence in the navel orange rind. *Plant Cell Physiol.* **8**:151–160.

123. Lieberman, M., and S. Y. Wang. 1982. Influence of calcium and mangesium on ethylene production by apple tissue slices. *Plant Physiol.* **69**:1,150–1,155.

124. Lipe, J. A. 1978. Ethylene in fruits of blackberry and rabbiteye blueberry. *J. Am. Soc. Hort. Sci.* **103**:76–77.

125. Looney, N. E., and M. E. Patterson. 1967. Chlorophyllase activity in apples and bananas during the climacteric phase. *Nature* **214**:1,245–1,246.

126. Lyons, J. M., W. B. McGlasson, and H. K. Pratt. 1962. Ethylene production, respiration and internal gas concentrations in cantaloupe fruits at various stages of maturity. *Plant Physiol.* **37**:31–36.

127. Lyons, J. M., and H. K. Pratt. 1964. Effect of stage of maturity and ethylene treatments on respiration and ripening of tomato fruits. *Proc. Am. Soc. Hort. Sci.* **84**:491–500.

128. Makovetski, S., and E. E. Goldschmidt. 1976. A requirement for cytoplasmic protein synthesis during chloroplast senescence in the aquatic plant *Anacharis canadensis. Plant Cell Physiol.* **17**:859–862.

129. Marei, N., and J. C. Crane. 1971. Growth and respiratory response of fig (*Ficus carica*, L. cv. Mission) fruits to ethylene. *Plant Physiol.* **48**:249–254.

130. Marks, J. P., R. Bernlohr, and J. P. Varner. 1957. Esterification of phosphate in ripening fruit. *Plant Physiol.* **32**:259–262.

131. Marousky, F. J., and B. K. Harbough. 1979. Interactions of ethylene, temperature, light and carbon dioxide on leaf and stipule abscission and chlorosis in *Philodendron scandens* spp. *oxycardium. J. Am. Soc. Hort. Sci.* **104**:876–880.

132. Martinoia, E., M. J. Dalling, and P. Matile. 1982. Catabolism of chlorophyll: Demonstration of chloroplast-localized peroxidative and oxidative activities. *Z. Pflanzenyphysiol.* **107**:269–279.

133. Matile, P., and F. Winkenbach. 1971. Function of lysosomes and lysosomal enzymes in senescing corolla of the morning glory (*Ipomoea purpurea*). *J. Exp. Bot.* **122**:759–771.

134. Maxie, E. C., P. B. Catlin, and H. T. Hartman. 1960. Respiration and ripening of olive fruits. *Proc. Am. Soc. Hort. Sci.* **75**:275–291.

135. McGlasson, W. B., N. L. Wade, and I. Adato. 1978. Phytohormones and fruit ripening. In *Phytohormones and Related Compounds: A Comprehensive Treatise*, vol. 2, D. S. Letham, P. B. Goodwin, and T. J. V. Higgins (eds.). Elsevier/North-Holland, Amsterdam, pp. 447–493.

136. McKeown, A. W., E. C. Lougheed, and D. P. Murr. 1978. Compatability of cabbage, carrots, and apples in low pressure storage. *J. Am. Soc. Hort. Sci.* **103**:749–752.

137. McMurchie, E. J., W. B. McGlasson, and I. L. Eaks. 1972. Treatment of fruit with propylene gives information about the biogenesis of ethylene. *Nature* **237**:235–236.

138. Medawar, P. B. 1957. *The Uniqueness of the Individual.* Methuen, London, 191p.

139. Meinke, D. W., J. Chen, and R. W. Beachy. 1981. Expression of storage-protein genes during soybean seed development. *Planta* **153**:130–139.

140. Merideth, F. I., R. G. Leffler, and C. E. Lyons. 1988. A firmness detector for peaches using impact forces analysis. *Am. Soc. Agric. Eng. Paper No. 88-6570.*

141. Miller, J. C. 1929. A study of some factors affecting seed-stalk development in cabbage. *Cornell Univ. Agric. Exp. Sta. Bull. 488,* 46p.

142. Mizano, S., and H. K. Pratt. 1973. Relations of respiration and ethylene production to maturity in the watermelon. *J. Am. Soc. Hort. Sci.* **98**:614–617.

143. Molisch, H. 1938. *The Longevity of Plants*, H. Fullington (trans.). Science Press, Lancaster, Penn., 226p.

144. Mousdale, D. M. A., and M. Knee. 1981. Indolyl-3-acetic acid and ethylene levels in ripening apple fruits. *J. Exp. Bot.* **32**:753–758.

145. Noodén, L. D. 1988. The phenomena of senescence and aging. In *Senescence and Aging in Plants,* L. D. Noodén and A. C. Leopold (eds.). Academic Press, New York, pp. 1–50.

146. Noodén, L. D., and J. W. Thompson. 1985. Aging and senescence in plants. In *Handbook of the Biology of Aging*, C. E. Finch and E. L. Schneider (eds.). Van Nostrand Reinhold, New York, pp. 105–127.

147. Nursten, H. E. 1970. Volatile compounds: The aroma of fruits. In *The Biochemistry of Fruits and Their Products*, A. C. Hulme (ed.). Academic Press, New York, pp. 239–268.

148. Palmer, J. K. 1971. The banana. In *The Biochemistry of Fruits and Their Products*, vol. 2, A. C. Hulme (ed). Academic Press, London, pp. 65–105.

149. Pech, J. C., and R. J. Romani. 1979. Senescence in pear *Pyrus communis* fruit cells cultured in a continuously renewed auxin-deprived medium. *Plant Physiol.* **63**:814–817.

150. Peterson, R. F. 1965. *Wheat. Botany, Cultivation and Utilization*. Leonard Hill, London, 442p.

151. Pfeffer, W. 1873. *Physiologische Untersuchungen*. W. Engelmann, Leipzig, Germany, 216p.

152. Poovaiah, B. W. 1988. Calcium and senescence. In *Senescence and Aging in Plants*, L. D. Noodén and A. C. Leopold (eds.). Academic Press, New York, pp. 369–389.

153. Poovaiah, B. W., and A. C. Leopold. 1973. Deferral of leaf senescence with calcium. *Plant Physiol.* **52**:236–239.

154. Porter, K. R., and R. D. Machado. 1959. Studies on the endoplasmic reticulum. VI. Its form and distribution during mitosis in cells of onion root tip. *J. Biophys. Biochem. Cytol.* **7**:167–180.

155. Postlmayer, H. L., B. S. Luk, and S. J. Leonard. 1956. Characterization of pectin changes in freestone and clingstone peaches during ripening and processing. *Food Technol.* **10**:618–625.

156. Pratt, H. K., and J. D. Goeschl. 1968. The role of ethylene in fruit ripening. In *Biochemistry and Physiology of Plant Growth Substances*, F. Wightman and G. Setterfield (eds.). Runge, Ottawa, pp. 1,295–1,302.

157. Pratt, H. K., and D. B. Mendoza. 1980. Fruit development and ripening of the star apple (*Chrysophyllum cainito* L.). *HortScience* **15**:721–722.

158. Pratt, H. K., and M. S. Reid. 1974. Chinese gooseberry: Seasonal patterns in fruit growth and maturation, ripening, respiration and the role of ethylene. *J. Sci. Food. Agric.* **25**:747–753.

159. Pratt, H. K., and M. S. Reid. 1976. The tamarillo: Fruit growth and maturation, ripening, respiration, and the role of ethylene. *J. Sci. Food. Agric.* **27**:399–404.

160. Pressey, R. 1977. Enzymes involved in fruit softening. *Am. Soc. Chem. Symp. Ser.* **47**:172–191.

161. Pressey, R., and J. K. Avants. 1978. Difference in polygalacturonase composition of clingstone and freestone peaches. *J. Food Sci.* **43**:1,415–1,423.

162. Proebsting, E. L. 1968. Early Italian prune quality can be improved. *Proc. Wash. State Hort. Assoc.*, pp. 188–191.

163. Purvis, O. N. 1961. The physiological analysis of vernalisation. In *Encylopedia of Plant Physiology*, vol. 16, W. Rukland (ed.). Springer-Verlag, Berlin, pp. 76–122.

164. Rees, A. R. 1977. The cold requirements of tulip cultivars. *Scientia Hort.* **7**:383–389.

165. Rehkugler, G. E., and J. A. Throop. 1986. Apple sorting with machine vision. *Am. Soc. Agric. Eng. Trans.* **29**:1,388–1,397.

166. Reid, M. S. 1975. The role of ethylene in the ripening of some unusual fruits. In *Facteurs et Régulation de la Maturation des Fruits*, No. 238. CNRS, Paris, pp. 177–182.

167. Reid, M. S. 1985. Ethylene and abscission. *HortScience* **20**:45–50.

168. Reid, M. S., and R. Y. Evans. 1985. Control of cut flower opening. *Acta Hort.* **181**:45–54.

169. Richmond, A., and J. B. Biale. 1966. Protein and nucleic acid metabolism in fruits: I. Studies of amino acid incorporation during the climacteric rise in respiration of the avocado. *Plant Physiol.* **41**:1,247–1,253.

170. Roberts, A. N., and R. L. Ticknor. 1970. Commercial production of English holly in the Pacific Northwest. *Am. Hort.* **49**:301–314.

171. Roe, B., and J. H. Bruemmer. 1981. Changes in pectic substances and enzymes during ripening and storage of 'Keitt' mangos. *J. Food Sci.* **46**:186–189.

172. Romani, R. 1978. Long term maintenance of mitochondria function in vitro and the course of cyanide-insensitive respiration. In *Plant Mitochondria*, G. Ducet and C. Lance (eds.). Elsevier/North Holland, Amsterdam, pp. 3–10.

173. Romani, R. J. 1987. Senescence and homeostasis in postharvest research. *HortScience* **22**:865–868.

174. Roozen, F. M. (ed.). 1980. *Forcing Flowerbulbs.* International Flower-Bulb Centre, Hillegom, The Netherlands.

175. Rosenburg, L. A., and R. W. Rinne. 1986. Water loss as a prerequisite for seedling growth in soybean seeds (*Glycine max,* L. Merr.). *J. Exp. Bot.* **37**: 1,663–1,674.

176. Rowan, K. S., W. B. McGlasson, and H. K. Pratt. 1969. Changes in adenosine pyrophosphates in cantaloupe fruit ripening normally and after treatment with ethylene. *J. Exp. Bot.* **20**:145–155.

177. Salveit, M. E. 1977. Carbon dioxide, ethylene, and color development in ripening mature green bell peppers. *J. Am. Soc. Hort. Sci.* **102**:523–525.

178. Sapers, G. M., S. B. Jones, and G. T. Maher. 1983. Factors affecting the recovery of juice and anthocyanin from cranberries. *J. Am. Soc. Hort. Sci.* **108**:246–249.

179. Saravitz, D. M., D. M. Pharr, and T. E. Carter, Jr. 1987. Galactinol synthesis activity and soluble sugars in developing seeds of four soybean genotypes. *Plant Physiol.* **83**:185–189.

180. Sarkar, N., and R. R. Wolfe. 1985. Future extraction techniques for sorting tomatoes by computer vision. *Am. Soc. Agric. Eng. Trans.* **28**:970–974, 979.

181. Sattler, R. 1973. *Organogenesis of Flowers: A Photographic Text-Atlas.* University of Toronto Press, Toronto, 207p.

182. Schatzki, T. F., A. Grossman, and R. Young. 1983. Recognition of agricultural objects by shape. *IEEE Trans. on Pattern Anal. Artif. Intell.* **5**:645–653.

183. Sexton, R., and H. W. Woolhouse. 1984. Senescence and abscission. In *Advanced Plant Physiology*, M. B. Wilkins (ed.). Pitman, London, pp. 469–497.

184. Sheehy, R. E., M. Kramer, and W. R. Hiatt. 1988. Reduction of polygalacturonase activity in tomato fruit by antisense RNA. *Nat. Acad. Sci. (USA) Proc.* **85**:8,805–8,809.

185. Shulman, Y., and S. Lavee. 1976. Endogenous cytokinins in maturing Manzanillo olive fruits. *Plant Physiol.* **57**:490–492.

186. Siegelman, H. W. 1964. Physiological studies on phenolic biosynthesis. In *Biochemistry of Phenolic Compounds*, J. B. Harborne (ed.). Academic Press, New York, pp. 437–456.

187. Simmonds, N. W. 1966. Fruit biochemistry. In *Bananas*. Longman, London, pp. 223–225.

188. Smith, C. J. S., C. F. Watson, J. Ray, C. R. Bird, P. C. Morris, W. Schuch, and D. Grierson. 1988. Antisense RNA inhibition of polygalacturonase gene expression in transgenic tomatoes. *Nature* **334**:724.

189. Smittle, D. A., R. E. Williamson, and J. R. Stansell. 1976. Response of snap beans to seed separation by aerodynamic properties. *HortScience* **11**:469–471.

190. Smock, R. M. 1972. Influence of detachment from the tree on the respiration of apples. *J. Am. Soc. Hort. Sci.* **97:**509–511.

191. Sobotka, F. E., and D. A. Stelzig. 1974. An apparent cellulase complex in tomato (*Lycopersicon esculentum* L.) fruit. *Plant Physiol.* **53:**759–763.

192. Szczesnick, A. S. 1963. Classification of external characteristics. *J. Food Sci.* **28:**385–389.

193. Takahashi, N., I. Yamaquchi, T. Kono, M. Igoshi, K. Hirose, and K. Suzuki. 1975. Characterization of plant growth substances in *Citrus unshiu* and their change in fruit development. *Plant Cell Physiol.* **16:**1,101–1,111.

194. TeKrony, D. M., D. B. Egli, J. Balles, T. Pfeiffer, and R. J. Fellows. 1979. Physiological maturity in soybean. *Agron. J.* **71:**771–775.

195. Thomas, H., and J. L. Stoddart. 1975. Separation of chlorophyll degradation from other senescence processes in leaves of a mutant genotype of meadow fescue (*Festuca pratensis*). *Plant Physiol.* **56:**438–441.

196. Thompson, J. E. 1988. The molecular basis for membrane deterioration during senescence. In *Senescence and Aging in Plants,* L. D. Noodén and A. C. Leopold (eds.). Academic Press, New York, pp. 51–83.

197. Thompson, J. E., S. Mayak, M. Shinitzky, and A. H. Halevy. 1982. Acceleration of membrane senescence in cut carnation flowers by treatment with ethylene. *Plant Physiol.* **69:**859–863.

198. Townshend, F. 1967. Food standards. In *Quality Control in the Food Industry,* vol. 1, S. M. Herschdoerfer (ed.). Academic Press, London, pp. 185–365.

199. Van Staden, J. 1976. Season changes in the cytokinin context of *Ginkgo biloba* leaves. *Plant Physiol.* **38:**1–5.

200. Van Staden, J., E. L. Cook, and L. D. Noodén. 1988. Cytokinins and senescence. In *Senescence and Aging in Plants,* L. D. Noodén and A. C. Leopold (eds.). Academic Press, New York, pp. 281–328.

201. Van Staden, J., and G. G. Dimalla. 1980. The effect of silver thiosulphate preservative on the physiology of cut carnations. II. Influence on endogenous cytokinin. *Z. Pflangenphysiol.* **99:**19–26.

202. Vendrell, M. 1969. Reversion of senescence: Effects of 2,4-dichloro-phenoxyacetic acid and indoleacetic acid on respiration, ethylene production, and ripening of banana fruit slices. *Aust. J. Biol. Sci.* **22:**601–610.

203. Vogele, A. C. 1937. Effect of environmental factors upon the color of the tomato and the watermelon. *Plant Physiol.* **12:**929–955.

204. von Loesecke, H. W. 1950. *Bananas.* Interscience, New York, 189p.

205. Wade, N. L., P. B. H. O'Connell, and C. J. Brady. 1972. Content of RNA and protein of the ripening of banana. *Phytochemistry* **11:**975–979.

206. Walker, G. C. 1985. Inducible DNA repair systems. *Ann. Rev. Biochem.* **54:**425–475.

207. Walsh, C. S., J. Popenoe, and T. Solomos. 1983. Thornless blackberry is a climacteric fruit. *HortScience* **18:**482–483.

208. Wang, C. Y., and W. M. Mellenthin. 1977. Effect of aminothoxy analog of 2-rhizobitoxine on ripening of pears. *Plant Physiol.* **59:**546–549.

209. Wang, C. Y., W. M. Mellenthin, and E. Hausen. 1972. Maturation of 'Anjou' pears in relation to chemical composition and reaction to ethylene. *Am. Soc. Hort. Sci. J.* **97:**9–12.

210. Watada, A. E., R. C. Herner, A. A. Kader, R. J. Romani, and G. L. Staby. 1984. Terminology for the description of developmental stages of horticultural crops. *HortScience* **19:**20–21.

211. Wellington, P. S. 1956. Studies on the germination of cereals. I. The germination of wheat grains in the ear during development, ripening, and after-ripening. *Ann. Bot.* **20:**105–120.

212. Wickham, L. D., H. C. Passam, and L. A. Wilson. 1984. Dormancy responses to postharvest application of growth regulators in *Dioscorea* species. 2. Dormancy

responses in ware tubers of *D. alata* and *D. esculeuta*. *J. Agric. Sci.* **102**:433–436.

213. Williams, J. W., and W. A. Sistrunk. 1979. Effects of cultivar, irrigation, ethephon, and harvest date on the yield and quality of processing tomatoes. *J. Am. Soc. Hort. Sci.* **104**:435–439.

214. Williams, K. T., and A. Bevenue. 1954. Some carbohydrate components of tomato. *J. Agric. Food. Chem.* **2**:472–474.

215. Wollgiehn, R., and B. Parthier. 1964. Der Einfluss des Kinetins auf den NS-und Protein-Stoffwechsel in abgeschnittenen mit Hemmstoffen behandelten Tabakblättern. *Phytochemistry* **3**:241–248.

216. Wood, J. G., and D. H. Cruicleshawk. 1944. The metabolism of starving leaves. 5. Changes in amounts of some amino acids during starvation of grass leaves and their bearing on the nature of the relationship between proteins and amino acids. *Aust. J. Exp. Biol. Med. Sci.* **22**:111–123.

217. Woodard, J. R. 1972. Physical and chemical changes in developing strawberry fruits. *J. Sci. Food Agric.* **23**:465–473.

218. Woodson, W. R., and K. A. Lawton. 1988. Ethylene-induced gene expression in carnation petals. Relationship to autocatalytic ethylene production and senescence. *Plant Physiol.* **87**:498–503.

219. Wright, M. E., J. H. Tappen, and F. E. Sistler. 1985. The size and shape of typical sweet potatoes. *Am. Soc. Agric. Eng. Paper No. 85-6017*, pp. 207–225.

220. Young, R. E., S. Salminen, and P. Sornsrivichai. 1975. Enzyme regulation associated with ripening in banana fruit. In *Facteurs et Régulations de la Maturation des Fruits*, No. 238. CNRS, Paris, pp. 271–280.

ADDITIONAL READINGS

Abeles, F. B. 1973. *Ethylene in Plant Biology.* Academic Press, New York, 302p.

Addicott, F. T. 1981. *Abscission.* University of California Press, Berkeley, 369p.

Alavoine, F., M. Crochon, J. Fallot, P. Moras, and J. C. Pech. 1988. *La qualité gustative des fruits. Méthodes pratiques d'analyse.* Antony, France, Centre National du Machinisme Agricole, du Ginie Rural, des Eaux et des Forets, 70p.

Arthey, V. D. 1975. *Quality of Horticultural Products.* Butterworth, London, 228p.

Bradbeer, J. W. 1988. *Seed Dormancy and Germination.* Chapman and Hall, New York, 146p.

Brady, C. J. 1987. Fruit ripening. *Ann. Rev. Plant Physiol.* **38**:155–178.

Brenner, M. L. 1981. Modern methods for plant growth substance analysis. *Ann. Rev. Plant Physiol.* **32**:511–538.

Chadwick, C. M., and D. R. Garrod. 1986. *Hormones, Receptors, and Cellular Interactions in Plants.* Cambridge University Press, Cambridge, England, 375p.

Conn, E. E. 1980. Cyanogenic compounds. *Ann. Rev. Plant Physiol.* **31**:433–451.

Dalling, M. J. (ed.). 1986. *Plant Proteolytic Enzymes,* 2 vols. CRC Press, Boca Raton, Fla.

Davies, P. J. 1987. *Plant Hormones and Their Role in Plant Growth and Development.* M. Nijhoff, Boston, 681p.

Dennis, F. G., Jr. 1987. Two methods of studying rest: Temperature alteration and genetic analysis. *HortScience* **22**:820–824.

Goodenough, P. W., and R. K. Atkin. 1981. *Quality in Stored and Processed Vegetables and Fruits.* Academic Press, New York, 398p.

Grace, J. (ed.). 1977. *Plant Responses to Wind.* Academic Press, New York, 389p.

Grierson, D. 1987. Senescence in fruits. *HortScience* **22:**859–862.

Gross, J. 1987. *Pigments in Fruits*. Academic Press, New York, 303p.

Halevy, A. H. 1987. Recent advances in postharvest physiology of carnations. *Acta Hortic.* **216:**243–254.

Huber, S. C. 1986. Fructose 2,6-bisphosphate as a regulatory metabolite in plants. *Ann. Rev. Plant Physiol.* **37:**439–466.

Hultin, H. O., and M. Milner (eds.). 1978. *Postharvest Biology and Biotechnology*. Food and Nutrition Press, Westport, Conn. 462p.

Joas, J. 1987. Quelques observation a propos du circuit de distribution de la banane Antillaise (cv. Cavendish) et des principaux facteurs définissant la qualité du fruit. *Fruits* **42**(9):493–504.

Kelly, M. O., and P. J. Davies. 1988. The control of whole plant senescence. *Crit. Rev. Plant Sci.* **7:**139–173.

Klambt, D. (ed.). 1986. *NATO Advanced Research Workshop on Plant Hormone Receptors*. Springer-Verlag, Berlin, 319p.

Kuhlemier, C., P. J. Green, and N. Chua. 1987. Regulation of gene expression in higher plants. *Ann. Rev. Plant Physiol.* **38:**221–257.

Labavitch, J. M. 1981. Cell wall turnover in plant development. *Ann. Rev. Plant Physiol.* **32:**385–406.

Lang, G. A. 1987. Dormancy: A new universal terminology. *HortScience* **22:**817–820.

Leshem, Y. Y., A. H. Halevy, and C. Frenkel. 1986. *Processes and Control of Plant Senescence*. Elsevier, New York, 216p.

Lipton, W. J. 1987. Senescence in leafy vegetables. *HortScience* **22:**854–859.

Mayak, S. 1987. Senescence of cut flowers. *HortScience* **22:**863–865.

Mayer, A. M., and A. Poljakoff-Mayber. 1982. *The Germination of Seeds*. Pergamon, Oxford, 211p.

Monselise, S. P., and R. Goren. 1987. Preharvest growing conditions and postharvest behavior of subtropical and temperate-zone fruits. *HortScience* **22:**1,185–1,189.

Noodén, L. D., and A. C. Leopold (eds.). 1988. *Senescence and Aging in Plants*. Academic Press, New York, 526p.

Osborne, D. J. 1989. Abscission. *Crit. Rev. Plant Sci.* **8:**103–129.

Purohit, S. S. (ed.). 1985. *Hormonal Regulation of Plant Growth and Development*. Kluwer Academic, Dordrecht, The Netherlands, 412p.

Rivier, L., and A. Crozier. 1987. *Principles and Practice of Plant Hormone Analysis*, 2 vols. Academic Press, Orlando, Fla.

Roberts, J. A., and R. Hooley. 1988. *Plant Growth Regulators*. Chapman and Hall, New York, 190p.

Romani, R. J. 1987. Senescence and homeostasis in postharvest research. *HortScience* **22:**865–868.

Salunkhe, D. K. 1974. *Storage, Processing, and Nutritional Quality of Fruits and Vegetables*. CRC Press, Cleveland, 166p.

Satter, R. L., and A. W. Galston. 1980. Mechanisms of control of leaf movements. *Ann. Rev. Plant Physiol.* **32:**83–110.

Sharples, R. O., and D. S. Johnson. 1987. Influence of agronomic and climactic factors on the response of apple fruit to controlled atmosphere storage. *HortScience* **22:**763–766.

Shibaoka, A. (ed.). 1978. *Controlling Factors in Plant Development*. Botanical Society of Japan, Tokyo, 277p.

Smith, H. 1982. Light quality, photoperception, and plant strategy. *Ann. Rev. Plant Physiol.* **33:**481–518.

Sweeney, B. M. 1987. *Rhythmic Phenomena in Plants*. Academic Press, New York, 172p.

Thimann, T. V. (ed.). 1980. *Senescence in Plants*. CRC Press, Boca Raton, Fla., 276p.

Thomas, H., and J. L. Stoddart. 1980. Leaf senescence. *Ann. Rev. Plant Physiol.* **31**:83–111.

Tobin, E. M., and J. Silverborne. 1985. Light regulation of gene expression in higher plants. *Ann. Rev. Plant Physiol.* **36**:569–593.

Tucker, G. A., and J. A. Roberts (eds.). 1985. *Ethylene and Plant Development.* Butterworths, Boston, 416p.

White, P. L., and N. Selvey (eds.). 1974. *Nutritional Qualities of Fresh Fruits and Vegetables.* Futura, New York, 186p.

STRESS IN
HARVESTED PRODUCTS

NATURE OF STRESS IN RELATION TO
HARVESTED PRODUCTS

Plants have evolved over many millennia, adapting to a wide range of environmental conditions. Relative to the tremendous environmental diversity encountered over the Earth's surface, the range of conditions in which plants can survive is remarkably wide. The precise conditions and potential duration of exposure that a plant can withstand before the onset of serious injury and death, however, vary substantially between species. For example, many species of tropical origin [e.g., citrus, banana (*Musa* spp.)] sustain significant damage from low temperatures that are considered mild for innumerable temperate species. Likewise, different parts of the same plant may display a significant disparity in the precise conditions under which injury occurs.

Within this broad range of survival conditions, there is a relatively narrow span of conditions under which the plant or plant part functions at its optimum. When the plant is exposed to conditions outside the boundaries of this narrow optimum span, it becomes subject to an element of stress. Although now the subject of several books, the precise definition of stress relative to harvested biological material remains somewhat nebulous. Generally, stress is seen as any environmental factor that is capable of inducing a potentially injurious strain in a living system. More specifically, a stress is an external factor (or succession of factors) of such magnitude that it tends to disrupt the normal physiological processes of the organism.

A stress, therefore, interrupts, restricts, or accelerates normal metabolic processes and does so in an adverse or negative manner. The extent of injury sustained is determined by the severity of the stress, the length of time the plant is exposed, and the plant's general constitutive resistance to the stress. Based on this definition, however, it is evident that most recommended postharvest environmental conditions represent stresses, that is, conditions that interrupt, restrict, or accelerate normal metabolic processes in an adverse or negative manner. While the storage of an apple (*Malus sylvestris,* Mill.) fruit at 0°C represents by this definition a significant stress, it also represents the optimum storage temperature for preservation of the fruit. Therefore, implicit within the definition of stress is the question of whether stress is defined relative to the plant per se or relative to the eventual use for the plant or plant part.

From the postharvest physiologist's position, stress is an external factor that will result in undesirable changes in quality if the plant or plant part is exposed to it for sufficient duration or at a sufficient intensity. Therefore, while recommended apple storage conditions represent a stress to the apple fruit, they represent to the postharvest physiologist the optimum conditions for the maintenance of the product's quality.

While attached to the parent plant, an individual plant part is continually replenished with energy, nutrients, hormones, water, and other requisites. This availability of essential substances enhances the plant parts' ability to withstand environmental stresses and recover from the strains produced. Plant products that are severed from the parent plant at harvest, however, are deprived of this continual replenishing process, which in turn can significantly alter the plant product's ability to withstand certain stresses and recover from the strain incurred.

Harvested plant products, then, are not only subjected to a significant mechanical stress when severed from the parent plant but they are subjected to a progressive series of stresses during the postharvest period. As a consequence, the postharvest period can be seen as a time of stress management. In this context, stress is defined relative to the end use of the product.

The importance of stress management during the postharvest period necessitates understanding how plants or plant parts respond to stress. Stress results in a strain, any physical or chemical change produced by the stress, of which there are two types.[146] An elastic strain is reversible upon the removal of the stress whereas a plastic strain is irreversible and produces a permanent injury to the tissue. Physical manifestations of a strain may be seen, for example, as changes in the physical shape of the cell or the cessation of protoplasmic streaming. A shift in metabolism would be an example of a chemical change precipitated by a stress.

Stress can produce an injurious effect on a tissue in one of three principal ways. When a single stress is imposed, a direct, indirect, or secondary stress injury may occur. A direct stress injury is the result of direct plastic strain and is often characterized by a relatively rapid manifestation of the injury (e.g., freeze damage). An indirect stress injury is due to an elastic strain that if imposed for a sufficient duration produces an indirect plastic strain. A secondary stress injury is due to the induction of a secondary stress, which in turn causes the injurious strain. High-temperature-induced desiccation is an example of a secondary stress injury (i.e., high temperature increases evaporation, which results in a water stress from which the injury is derived).

Plants have an inherent adaptive potential that they use to resist stress. This resistance to stress can occur via two possible means. The plant can avoid the stress by preventing or decreasing the penetration of the stress into its tissues (stress avoidance). This avoidance results in an increase in the amount of stress required to produce a given strain. The plant may also respond by decreasing or eliminating the strain caused by the stress upon entering the tissue (stress tolerance). This tolerance results in a decrease in the strain produced by a specific stress.

During the postharvest period, there are a number of potential stresses to which the plant material may be exposed. These include: temperature, water, gas, radiation, chemical, mechanical, gravitational, herbivory, and pathological stresses. It is desirable, therefore, to understand not only the effect of these stresses on the plant and its components of quality but also the response of the plant to stress.

TEMPERATURE STRESS

High-Temperature Stress

High temperature represents perhaps the most critical single factor in the maintenance of quality of harvested plant products. With intact plants, the conversion of water from a liquid state to a gas and its loss from the plant provides the primary means for maintaining the plant's temperature within a biologically safe range. Because of evaporation (transpiration) sweetpotato plants (*Ipomoea batatas,* (L.) Lam.) growing under conditions where the surface temperature of the soil exceeded 50°C during the day were able to maintain their maximum leaf temperature under 33°C.[144]

Control of temperature by transpiration, however, necessitates a continual replenishment of the water lost. When plant parts are severed from the root system at harvest, or prevented in some other way from replenishing this water, the plant loses its potential to repress a buildup in temperature. As a consequence, the surrounding atmospheric temperature, thermal energy from the sun striking the product, and metabolic heat produced by the product itself often result in a progressive increase in product temperature. If the temperature is not reduced, significant losses in quality occur. Consequently, one of the initial primary concerns for many plant products after harvest is the removal of heat. Postharvest physiologists and agricultural engineers have established for many products the effect on quality of prolonged exposure to high temperature, the most desirable method for removal of heat, and the optimum temperature for subsequent storage.

In addition to a decrease or inhibition of transpiration after harvest, plant parts severed from the parent plant have the availability of minerals, carbohydrates, hormones, and other vital substances radically altered. As a consequence, the response potential of the product to high temperature may be likewise altered. In many cases, detached postharvest products appear to be more vulnerable to high-temperature stress than intact plants.

The upper limit of plants to high temperatures varies with species, cultivar, stage of development, condition, and plant part in question. Some thermophilic bacteria appear to be able to grow and develop at temperatures up to the boiling point of water.[21] In contrast, apple and tomato (*Lycopersicon esculentum,* Mill.) fruits have temperature maxima of 49–52°C and 45°C, respectively.[107] Generally tissues that are in a state of rest are more stable than those that are actively growing. Likewise, low moisture content is advantageous (fig. 6-1). Dry wheat grains (approximately 9% moisture) can withstand temperatures of approximately 90°C whereas moist grains succumb at much lower temperatures.[91]

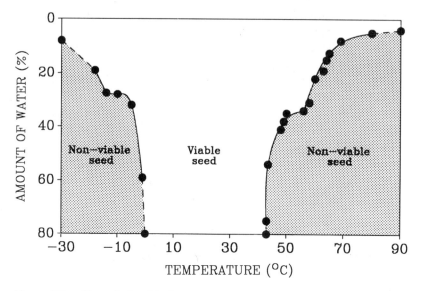

Figure 6-1. The relationship between corn (*Zea mays*, L.) grain moisture content and death caused by temperature extremes. High moisture increases the susceptibility of grain to both low- and high-temperature stress (*after Robbins and Petsch[203]; low-temperature data from Kisselbach and Ratcliff[136]*).

The injury sustained by the plant or plant part due to high-temperature stress is time dependent (fig. 6-2). Generally plant products withstand periods of very high temperature for a relatively short duration much better than they withstand exposure to somewhat lower temperatures for longer periods of time. The temperature at which a product will die varies inversely with exposure time, with time being an exponential function.

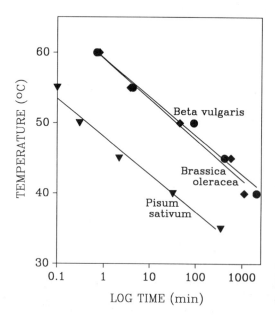

Figure 6-2. The relationship between temperature and the log of the duration of exposure required to cause death (*data from Collander[46]*).

The actual injury incurred by the plant due to high-temperature stress can be seen at several levels. For example, primary direct injury such as membrane damage, lipid liquidification, and protein and nucleic acid denaturation may be seen. The integrity of membrane structure and function is essential for normal cellular function[15]; thus, its disruption can mediate widespread secondary responses. Both membrane disruption and protein denaturation are injuries caused by acute exposures to high temperatures. The effects of chronic exposures, however, are not understood as well. Injury may be due to changes in the rate of breakdown of metabolites, metabolic imbalances, or other factors.

Primary indirect effects of high-temperature stress can be seen as the inhibition of pigment synthesis, the formation of surface burns (e.g., sun scald of grape (*Vitis vinifera*, L.), cherry (*Prunus avium*, L.), and tomato fruits) and lesions and protein breakdown. In tomato fruits, high temperatures can significantly alter the synthesis of pigments during ripening. At 32–38°C, only yellow pigmentation develops with the synthesis of lycopene, the principal red pigment being inhibited.[251] The softening of tomato fruits during ripening may also be inhibited by high temperatures,[95] possibly due to decreased activity of pectolytic enzymes.[186] Bananas stored at 40°C failed to ripen[270] while at 35°C the production of volatiles was greatly reduced.[271] The composite effect of high temperatures during the postharvest period is a reduction in longevity of the product after harvest and/or reduced quality.

Secondary injuries such as desiccation are also common. High temperatures can greatly increase the rate of transpiration. This increase is due to both a direct effect on the diffusion constant of water and an increase in the vapor pressure gradient between the product and its surrounding environment. A leaf temperature of 5°C above the atmospheric temperature is equivalent to an increase in the vapor pressure gradient caused by a 30% reduction in the surrounding relative humidity.[53] Thus, when *Ficus benjamina*, L. plants were held at 39°C for 4 to 12 days, leaf desiccation occurred, greatly decreasing the quality of the plants.[47]

Plants respond to heat stress through heat avoidance and/or changes in heat tolerance. Avoidance may occur through a decrease in respiratory rate, the synthesis of compounds that increase the reflectance of radiant energy, an increase in transpiration, or other means. The plant may effect changes in heat tolerance through changes in its chemical composition or moisture content. For example, many species are known to respond to heat stress by synthesizing a new set of proteins, called heat shock proteins, while repressing the synthesis of most normal proteins.[131] These proteins, formed during exposure to a 39–41°C treatment, provided thermal protection to a subsequent heat shock treatment of 45–48°C, which would have otherwise been lethal.[130] There appears to be a rapid synthesis of mRNAs for these proteins that can be detected 3 to 10 minutes after exposure to the high temperature.

The single-celled nonphotosynthesizing eukaryote *Tetrabymena pyriformis* can alter the fatty acid composition of the lipid portion of its membranes in response to high- or low-temperature stress,[84] allowing the organism to adapt rapidly to changes in environment. Since the fluidity of the lipid component of the membrane increases with increasing temperature, it can significantly alter membrane function and integrity. Specific enzymes, synthesized in response to thermal stress, replace certain fatty acids in the lipid portion of the membrane. This process decreases the unsaturated to saturated fatty acid ratio, which in turn decreases the fluidity of the membrane at high temperatures.[60] How widely this heat stress survival mechanism is found in the plant kingdom and its importance in postharvest plant products has yet to be established.

Low-Temperature Stress: Chilling Injury

Many species of tropical and subtropical origin and some of temperate origin are injured by temperatures in the 0–20°C range. Chilling stress occurs in these species at nonfreezing temperatures and in some cases temperatures well above freezing. This phenomenon is especially important in the postharvest handling and storage of plant products in that the use of low temperature is the most effective means of extending the storage life of many products.

The degree of chilling injury incurred by a plant or plant part depends on the temperature to which it is exposed, the duration of exposure, and the species sensitivity to chilling temperatures (fig. 6-3). The lower the temperature to which a product is exposed below its threshold chilling temperature, the greater the severity of the eventual injury.[248] The rate of development of injury symptoms in storage is also generally decreased with temperature; however, upon removal to nonchilling conditions the full manifestation of the stress becomes apparent (fig. 6-4). Likewise, the longer the duration of exposure to temperatures below the threshold, the greater the injury. Only a few hours of exposure of banana fruit to chilling temperatures is sufficient to cause injury, whereas weeks or months may be required for some cultivars of apple and grapefruit (*Citrus × paradisi,* Macfady.).

The sensitivity of a plant or plant part to chilling stress varies due to a number of factors, of which species, cultivar, plant part, and morphological and physiological condition at time of exposure are of critical importance.[77] Some of the more common chill-sensitive postharvest products are apples, asparagus (*Asparagus officinalis,* L.), avocado (*Persea americana,* Mill.), beans (*Phaseolus* spp.), banana, cranberry (*Vaccinium macrocarpon,* Ait.), cassava (*Manihot esculenta,* Crantz.), various species of citrus, corn (*Zea mays* var. *rugosa,* Bonaf.), cucumber (*Cucumis sativus,* L.), eggplant (*Solanum melongena,* L.), okra (*Abelmoschus esculentus,* (L.) Moench.), mango (*Mangifera indica,* L.), melons, nectarines (*Prunus persica* var. *nucipersica,* (Suckow) Schneid.), papaya (*Carica papaya,* L.), peach (*Prunus persica,* (L.) Batsch.), pepper (*Capsicum annuum,* L.), pineapple (*Ananas comosus,* (L.) Merrill), plum (*Prunus domestica,* L.), some cultivars of potato (*Solanum tuberosum,* L.), pumpkin (*Cucurbita* spp.), sweetpotato, yam (*Dioscorea* spp.), and numerous ornamentals. The critical threshold temperature below which chilling stress occurs varies widely between these species. For example, banana fruit are injured by temperatures in the 12–13°C range; mango, 10–13°C; lime (*Citrus aurantiifolia,* (Christm.) Swingle.), muskmelon (*Cucumis melo,* L. Reticulatus group), and pineapple, 7–10°C; cucumber (*Cucumis sativus,* L.), eggplant, and papaya, 7°C; and apples* and oranges

---

* Sensitivity varies with cultivar.

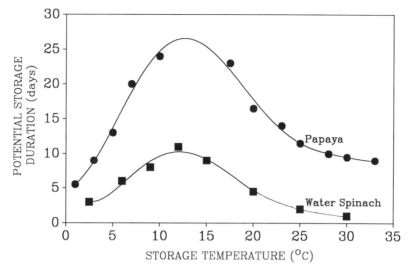

Figure 6-3. The relationship between storage temperature and the maximum potential storage duration for papaya fruit (*Carica papaya*, L.) and water spinach (*Ipomoea aquatica*, Forsk.) *(after Paull[190])*.

(*Citrus sinensis*, (L.) Osbeck.), 5°C. Fruit ripening of some species can significantly alter chilling sensitivity. Avocado and tomato fruit tend to be very sensitive when at the preclimacteric stage of development but lose sensitivity as they ripen. Chilling sensitivity increases to a maximum that coincides with the climacteric peak, declining rapidly thereafter as the fruit ripens further.[141] Likewise, susceptibility to chilling injury can also be modulated by production conditions[80] and mineral nutrition.[219] Evidence also suggests that organelles within the cell vary in their susceptibility.[40]

Chilling stress and injury does not just occur during storage. Chilling temperatures may be encountered in the field, during handling or transit,

Figure 6-4. Effect of duration of exposure to chilling temperatures on the respiratory rate of okra fruit *(after Ilker[114])*. The longer the exposure to chilling (●), the greater the respiration rate of the product once transferred to room temperature (■). Elevated respiration is indicative of damage to the tissue.

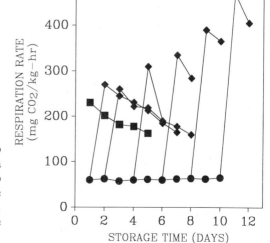

during wholesale distribution, in the retail store, and in the home. Inadvertently placing bananas in the home refrigerator is a common cause of injury. For most chill-sensitive products, quality losses are generally slower in development than with the banana but no less dramatic in the end.

Direct stress injury, the primary response to a chilling temperature, is generally considered to be physical in nature. Chilling temperatures are thought to result in changes in the physical properties of the cellular membranes that result in a series of possible indirect injuries or dysfunctions.[235] A transition occurs in the fluidity of the membranes that is thought to coincide with the threshold chilling temperature in at least some chill-sensitive species.[199] Rather than a uniform alteration across the membrane, fluidity changes probably occur in microdomains within the membrane.[193] Alternatively, low temperatures may have a direct effect on specific critical proteins, resulting in all or part of the damage.

After sufficient exposure of species sensitive to chilling, changes in the membranes result in a number of possible secondary responses, for example, loss of membrane integrity, leakage of solutes (fig. 6-5), loss of compartmentation, and changes in enzyme activity (fig. 6-6).[257] These secondary changes lead to the eventual manifestation of chilling injury symptoms. Specific physical and chemical symptoms vary widely between the various chill-sensitive plants. Injury may be in the form of surface lesions (e.g., cucumber fruits[69]) (fig. 6-7), inhibition of ripening (e.g., tomatoes[208]), discoloration (e.g., banana fruit[119]) (fig. 6-7), apple scald,[265] low-temperature breakdown,[266] inhibited growth, decay, and wilting.

Dysfunctions resulting from primary molecular changes induced by chilling temperatures can be repaired and/or reversed in some species if the tissue is returned to nonchilling conditions before permanent injury occurs. By using temperature cycling (also known as intermittent warming), 'Bramley's Seedling' apples were stored at 0°C using 5-day exposures to 15°C at intervals during the storage period, without the development of the character-

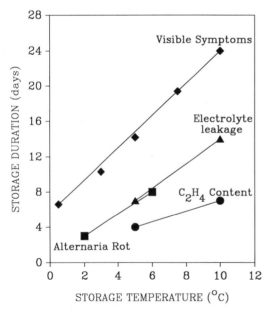

Figure 6-5. The relationship between storage duration and temperature on incipient chilling injury symptoms in papaya fruit (*Carica papaya*, L.) (*after Paull[190]*). Symptoms include electrolyte leakage, fruit cavity ethylene accumulation, Alternaria rot, and visible signs of injury.

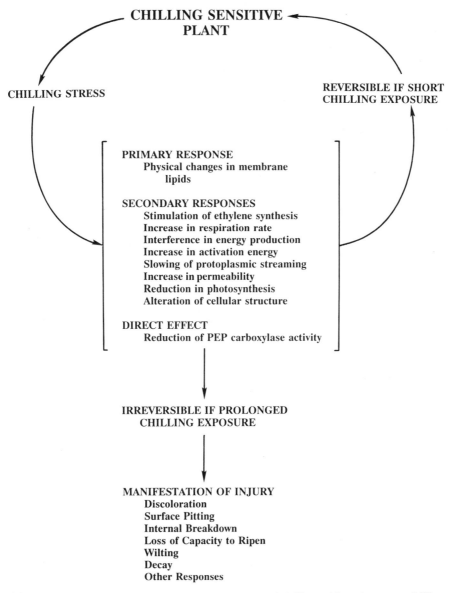

Figure 6-6. A simplified scheme of the responses of chill-sensitive plants to a chilling stress *(after Wang[257]).*

istic low-temperature breakdown at 0°C.[225] Similar improvements with return to nonchilling temperatures have been shown for citrus, cucumbers, peppers, and several other crops. The potential for reversal varies with species, condition of the product, length of time under chilling conditions, and temperature to which the product is exposed.

In addition to direct and indirect stress, chilling temperatures may also lead to secondary stress injury that can result in quality loss, increased susceptibility to decay and death. When the root systems of sensitive plants (e.g., tobacco (*Nicotiana tabacum,* L.), corn (*Zea mays,* L.), sugarcane (*Saccharum officinarum,* L.)) were exposed to temperatures just above

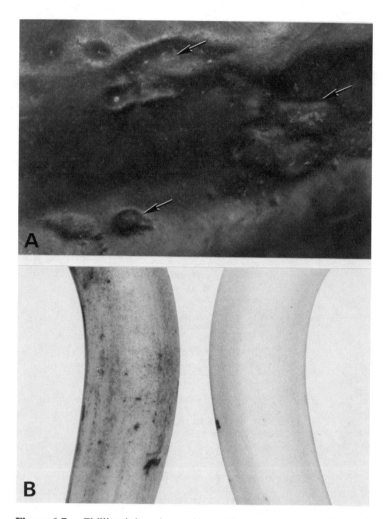

Figure 6-7. Chilling injury is manifested in an array of physical and chemical symptoms. For example, in cucumber (*Cucumis sativus,* L.) *(A)* injury is seen as localized surface pitting, while bananas (*Musa* spp.) *(B)* display a generalized surface discoloration (*left*).

freezing, the plants subsequently wilted.[180] If exposed for sufficient time, desiccation-induced death occurs, apparently due to the inability of the plant to absorb water.[127] Water loss and wilting may also occur with exposure of the aerial portion of the plant to chilling temperatures. In some species, the stomata are unable to close and the plant's ability to transport water is impaired.

As with high-temperature stress, *Tetrahymena pyriformis* displays a remarkable adaptive mechanism, responding to low temperatures via rapid alterations in membrane physical and chemical characteristics.[60] Acyl groups within the membrane lipids undergo selective replacement, increasing the level of unsaturation with a commensurate increase in fluidity. Chill-sensitive products may lack this adaptive mechanism or the rate of membrane alteration may be too slow to circumvent injury. Alternatively, injury may be through a nonlipid mechanism (e.g., specific proteins). Regardless of the mechanism resulting in injury, chill-sensitive tissues display undesirable alterations in quality when exposed to low-temperature stress. Aside from avoiding chilling temperatures, few techniques have been successful for circumventing chilling injury. Perhaps the most successful method has been intermittent warming of peaches and nectarines, especially when coupled with controlled atmosphere storage.[3] There has been some indication that the development of low-temperature injury in apples can be reduced using lower storage relative humidities.[216] Likewise, postharvest dips of apples in $CaCl_2$ solutions have been shown to reduce low-temperature breakdown of 'Jonathan' apples.[217]

Low-Temperature Stress: Freezing Injury

Postharvest plant products vary in the temperature at which they freeze and the damage incurred due to freezing.[172] For most fleshy flowers, fruits, and vegetables, freezing occurs one to several degrees below the freezing point of water (table 6-1). This depression in freezing point is due to the presence of solutes within the aqueous medium of the cell. With most of these products, freezing has disastrous consequences on quality. Generally, even brief exposure to freezing temperatures renders the product useless, especially for fresh market sales. Many nonfleshy products (e.g., seeds and nuts) and some acclimatized ornamentals can withstand temperatures well below freezing with little or no damage. For example, wheat seeds (10.6% moisture[156]) and *Ranunculus* tubers (9% moisture[13]) have been shown to withstand temperatures of $-196°C$ with no loss in survival. However, the same tissues similarly exposed at an elevated moisture content (25% and 30–50%, respectively) were killed. It is evident, therefore, that the ability of a tissue to withstand freezing temperatures is strongly dependent on its moisture level. Tissues that are very low in moisture tend to have the small amount of water remaining present in a ''bound'' state. This ''bound'' water is not converted to ice crystals even at very low temperatures.

Some hydrated cells—cells that would normally be killed by freezing temperatures—may be frozen without loss of viability if the freezing process occurs under the appropriate conditions.[159] Ice crystals may be formed extracellularly or intracellularly depending on the rate of cooling. Normally extracellular ice forms initially, occurring in the intercellular spaces outside of the cells. As the crystals grow in size, the cells contract and may eventually collapse. Contraction is caused by a redistribution of water within the tissue. When ice crystals begin to form in the extracellular medium, the solutes within the water are excluded, increasing the chemical potential in the unfrozen solution. This situation results in an osmotic imbalance between the intracellular and extracellular water, which is resolved by the movement of intracellular water outward through the plasma membrane. Freezing of extracellular water, therefore, results in an increase in the solute concentration of both the extracellular and intracellular solutions. Water movement continues until the chemical potential of the unfrozen water is in equilibrium with that of the ice.

Table 6-1. Freezing Points of Selected Fruits, Vegetables, Cut Flowers, Florist Greens and Bulbs, Corms, Rhizomes, Tubers, and Roots of Ornamentals and the Killing Point of Young and Mature Roots of Selected Woody Ornamentals

	Highest Freezing Point*	
	°C	°F
Fruits		
Apple	− 1.1	30.0
Apricot	− 1.1	30.1
Avocado	− 0.3	31.5
Banana	− 0.6	31.0
Berries		
Blackberry	− 0.8	30.5
Blueberry	− 1.3	29.7
Cranberry	− 0.9	30.4
Currant	− 1.0	30.2
Dewberry	− 1.3	29.7
Gooseberry	− 1.1	30.0
Loganberry	− 1.3	29.7
Raspberry	− 1.1	30.0
Strawberry	− 0.8	30.6
Cherimoya	− 2.2	28.0
Cherry, sour	− 1.7	29.0
Cherry, sweet	− 1.8	28.8
Date	− 15.7	3.7
Fig, fresh	− 2.4	27.6
Grapefruit	− 1.1	30.0
Grape, Vinifera	− 2.2	28.1
Grape, American	− 1.3	29.7
Lemon	− 1.1	30.0
Lime	− 1.6	29.1
Loquat	− 1.9	28.6
Mango	− 0.9	30.3
Nectarine	− 0.9	30.4
Olive, fresh	− 1.4	29.4
Orange, California and Arizona	− 1.3	29.7
Orange, Florida and Texas	− 0.8	30.6
Papaya	− 0.9	30.4
Peach	− 0.9	30.3
Pear	− 1.6	29.2
Persimmon, Japanese	− 2.2	28.1
Pineapple	− 1.0	30.2
Plum and prune	− 0.8	30.5
Pomegranate	− 3.0	26.6
Quince	− 2.0	28.4
Sapote	− 1.9	28.6
Starapple	− 1.2	29.9
Tamarind	− 3.7	25.4
Tangerine, Temple orange, and related citrus fruits	− 1.1	30.1
Vegetables		
Artichoke, globe	− 1.2	29.9
Asparagus	− 0.6	30.9

Table 6-1. *(continued)*

	Highest Freezing Point*	
	°C	°F
Bean, green, or snap	− 0.7	30.7
Bean, lima	− 0.6	31.0
Beet, bunched	− 0.4	31.3
Beet, topped	− 0.9	30.3
Broccoli, sprouting	− 0.6	30.9
Brussels sprouts	− 0.8	30.5
Cabbage, early	− 0.9	30.4
Cabbage, late	− 0.9	30.4
Carrot, mature (topped)	− 1.4	29.5
Carrot, immature (topped)	− 1.4	29.5
Cauliflower	− 0.8	30.6
Celeriac	− 0.9	30.3
Celery	− 0.5	31.1
Collard	− 0.8	30.6
Corn, sweet	− 0.6	30.9
Cucumber	− 0.5	31.1
Eggplant	− 0.8	30.6
Endive and escarole	− 0.1	31.9
Garlic, dry	− 0.8	30.5
Horseradish	− 1.8	28.7
Kale	− 0.5	31.1
Kohlrabi	− 1.0	30.2
Leeks, green	− 0.7	30.7
Lettuce	− 0.2	31.7
Melons		
Cantaloupe (3/4-slip)	− 1.2	29.9
Cantaloupe (full-slip)	− 1.2	29.9
Casaba	− 1.1	30.1
Crenshaw	− 1.1	30.1
Honeydew	− 0.9	30.3
Persian	− 0.8	30.5
Watermelon	− 0.4	31.3
Mushroom	− 0.9	30.4
Okra	− 1.8	28.7
Onion (dry) and onion salts	− 0.8	30.6
Onion, green	− 0.9	30.4
Parsley	− 1.1	30.0
Parsnip	− 0.9	30.4
Pea, green	− 0.6	30.9
Pepper, sweet	− 0.7	30.7
Potato, early-crop	− 0.6	30.9
Potato, late-crop	− 0.6	30.9
Pumpkin	− 0.8	30.5
Radish, spring	− 0.7	30.7
Rhubarb	− 0.9	30.3
Rutabaga	− 1.1	30.1
Salsify	− 1.1	30.0
Spinach	− 0.3	31.5
Squash, winter	− 0.8	30.5
Squash, summer	− 0.5	31.1

See page 351 for footnotes.

Table 6-1. *(continued)*

	Highest Freezing Point*	
	°C	°F
Sweetpotato	− 1.3	29.7
Tomato, mature-green	− 0.6	31.0
Tomato, firm-ripe	− 0.5	31.1
Turnip	− 1.1	30.1
Turnip greens	− 0.2	31.7
Water chestnuts	− 2.8	27.0
Watercress	− 0.3	31.4
Nuts		
Chestnuts, Chinese	− 5.3	22.5
Chestnuts, European	− 1.7	28.9
Coconut	− 0.8	30.6
Peanut	− 2.1	28.3
Pecan	− 6.7	19.9
Walnut, Persian	− 5.5	22.1
Ornamentals		
Cut flowers		
Acacia	− 3.6	25.6
Anemone	− 2.1	28.2
Aster, China	− 0.9	30.3
Camellia	− 0.7	30.7
Carnation	− 0.7	30.8
Chrysanthemum	− 0.8	30.5
Columbine	− 0.5	31.1
Cornflower	− 0.6	31.0
Daisy, Shasta	− 1.1	30.0
Delphinium	− 1.6	29.2
Feverfew	− 0.6	30.9
Gardenia	− 0.3	31.4
Heath	− 1.8	28.7
Hyacinth	− 0.3	31.4
Iris, bulbous	− 0.8	30.6
Lily	− 0.5	31.1
Narcissus (daffodils)	− 0.1	31.8
Orchid	− 0.3	31.4
Peony, tight buds	− 1.1	30.1
Poinsettia	− 1.1	30.1
Ranunculus	− 1.7	28.9
Rose (in preservative)	− 0.4	31.2
Rose (in dry packed)	− 0.4	31.2
Snapdragon	− 0.9	30.4
Stock	− 0.4	31.3
Sweet pea	− 0.9	30.4
Violet, sweet	− 1.8	28.8
Florist greens		
Asparagus (Plumosus)	− 3.3	26.0
Dracaena	− 1.6	29.1
Eucalyptus	− 1.8	38.8

Table 6-1. *(continued)*

	Highest Freezing Point*	
	°C	°F
Ferns		
Dagger and Wood ferns	− 1.7	28.9
Holly	− 2.8	27.0
Huckleberry	− 2.9	26.7
Ivy, English	− 1.2	29.9
Magnolia	− 2.8	27.0
Mistletoe	− 3.9	25.0
Mountain-laurel	− 2.4	27.6
Podocarpus	− 2.3	27.9
Rhododendron	− 2.4	27.6
Salal, lemon leaf	− 2.9	26.8
Bulbs, corms, rhizomes, tubers, and roots		
Amaryllis	− 0.7	30.8
Begonia	− 0.5	31.1
Caladium fancy-leaved	− 1.3	29.7
Calla	− 2.5	27.5
Dahlia	− 1.8	28.7
Gladiolus	− 2.1	28.2
Gloxinia	− 0.8	30.5
Hyacinth	− 1.5	29.3
Lily, Longiflorium (Easter)	− 1.7	28.9
Narcissus	− 1.3	29.6
Tulip (for forcing)	− 2.4	27.6
Tulip (for outdoors)	− 2.4	27.6

	Killing Temperature			
	Young Roots†		50% of Root System‡	
	°C	°F	°C	°F
Woody Ornamentals in Containers				
Acer palmatum Thunb. cv. Atropurpureum			− 10.0	14.0
Buxus sempervirens, L.	− 3	26.6	− 9.4	15.0
Cornus florida, L.	− 6	21.2	− 6.7	20.0
Cotoneaster adpressa praecox, Bois & Berth			− 12.2	10.0
Cotoneaster dammeri, Schneid	− 5	23.0		
Cotoneaster dammeri cv. Skogsholmen	− 7	19.4		
Cotoneaster horizontalis, Decne.			− 9.4	15.0
Cotoneaster microphyllus, Lindl.	− 4	24.8		
Cryptomeria japonica D. Don.			− 8.9	16.0
Cytisus × *praecox* Bean.			− 9.4	15.0
Daphne cneorum, L.			− 6.7	20.0
Euonymus alata, Sieb. cv. Compacta	− 7	19.4		
Euonymus fortunei cv. Argenteo-marginatus			− 9.4	15.0
Euonymus fortunei cv. Colorata			− 15.0	5.0
Euonymus fortunei, Hand.-Mazz. cv. Carrieri			− 9.4	15.0

Table 6-1. *(continued)*

	Young Roots†		50% of Root System‡	
	°C	°F	°C	°F
Euonymus kiautschovica, Loes (*E. patens*)	− 6	21.2		
Hedera helix, L. cv. Baltica			− 9.4	15.0
Hypericum spp., L.	− 5	23.0		
Ilex cornuta, Lindl. & Paxt. cv. Dazzler	− 4	24.8		
Ilex crenata Thunb. cv. Helleri.	− 5	23.0		
Ilex crenata Thunb. cv. Hetzi.			− 6.7	20.0
Ilex crenata Thunb. cv. Stokes.			− 6.7	20.0
Ilex crenata, Thunb. cv. Convexa			− 6.7	20.0
Ilex cv. Nellie Stevens	− 5	23.0		
Ilex cv. San Jose	− 6	21.2		
Ilex glabra, Gray			− 9.4	15.0
Ilex opaca, Ait.	− 5	23.0	− 6.7	20.0
Ilex × meserveae Hu. cv. Blue Boy	− 5	23.0		
Juniperus conferta, Parl.	− 11	12.2		
Juniperus horizontalis cv. Douglasii			− 17.8	0
Juniperus horizontalis Moench. cv. Plumosa	− 11	12.2	− 17.8	0
Juniperus squamata, D. Don. cv. Meyeri	− 11	12.2		
Kalmia latifolia, L.	− 9	15.8		
Koelreuteria paniculata, Laxm.	− 9	15.8		
Leucothoe fontanesiana, Sleum.	− 7	19.4	− 15.0	5.0
Magnolia stellata, Maxim.	− 6	21.2	− 5.0	23.0
Magnolia × soulangeana, Soul.			− 5.0	23.0
Mahonia bealei, Carr.	− 4	24.8		
Pachysandra terminalis, Sieb. & Zucc.			− 9.4	15.0
Picea glauca, Voss.			− 23.3	− 10.0
Picea omorika, Purkyne.			− 23.3	− 10.0
Pieris floribunda, Benth. & Hook.			− 15.0	5.0
Pieris japonica, D. Don.	− 9	15.8	− 12.2	10.0
Pieris japonica, D. Don. cv. Compacta			− 9.4	15.0
Potentilla fruticosa, L.			− 23.3	− 10.0
Pyracantha coccinea, Roem.			− 7.8	18.0
Pyracantha coccinea, Roem. cv. Lalandei	− 4	24.8		
Rhododendron carolinianum, Rehd.			− 17.8	0
Rhododendron catawbiense, Michx. cv. Roseum Elegans	− 11	12.2		
Rhododendron catawbiense, Pursh.			− 17.8	0
Rhododendron Exbury, Hybrid	− 8	17.6		
Rhododendron cv. Gibraltar (an Exbury Hyb. azalea)			− 12.2	10.0
Rhododendron cv. Hino Crimson	− 7	19.4		
Rhododendron cv. Hinodegiri (azalea)			− 12.2	10.0
Rhododendron cv. Purple Gem	− 9	15.8		
Rhododendron P.J.M. Hybrids			− 23.3	− 10.0
Rhododendron prunifolium	− 7	19.4		
Rhododendron schlippenbachii, Maxim.	− 9	15.8		
Stephanandra incisa, Zabel. cv. Crispa	− 8	17.6		
Taxus × media, Rehd. cv. Hicksii.	− 8	17.6		
Taxus × media, Rehd. cv. Nigra.			− 12.2	10.0

Table 6-1. *(continued)*

	Killing Temperature			
	Young Roots†		50% of Root System‡	
	°C	°F	°C	°F
Viburnum carlesii, Hemsl.			− 9.4	15.0
Viburnum plicatum, Thunb. v. tomentosum	− 7	19.4		
Vinca minor, L.			− 9.4	15.0

Note: Data from Havis,[99] Lutz and Hardenburg,[158] Studer et al.,[240] Whiteman,[262] and Wright.[268]
* Highest reading of all cultivars tested.
† Measurements made on artificially acclimatized plants.[240]
‡ Havis.[99]

When the rate of cooling is relatively slow, the efflux of water through the plasma membrane is rapid enough that excessive supercooling of the intracellular medium does not occur and intracellular ice crystals do not form. If, however, cooling is quite rapid, the rate of transfer of water may not be sufficient. The results are excessive supercooling of the intracellular solution and generally ice formation within the cell (figure 6-8). This type of ice formation causes foliage sunscalding of evergreen ornamentals during the winter.[261] The plants are not impaired by the normal slow-freezing process during the winter; for example, some species may withstand temperatures of −87°C. However, when in very bright, direct sunlight, the leaf tissue may thaw even though the surrounding air temperature is well below freezing. Intermittent blockage of the sunlight by clouds can result in very rapid cooling (8–10°C · min^{-1}) and refreezing of the leaf, causing intercellular ice formation. Thus, the same plant that can withstand very low winter temperature extremes may be killed by temperatures of only a few degrees below freezing (e.g., −10°C).

After thawing, freeze-injured tissue exhibits a flaccid, water-soaked appearance (fig. 6-9) with damage to the plasma membrane as a central feature. The rapid occurrence of the response suggests that injury is not due to a metabolic dysfunction such as that occurring with chill injury. Injury is thought to occur when the cooling rate is relatively rapid and intercellular ice forms. Injury can also occur when the cooling rate is relatively slow due to concentrating or dehydrating of the intercellular solution.[166,167] Manifestation of freezing injury may be due to (1) expansion-induced lysis or splitting of the plasma membrane during warming; (2) loss of osmotic responsiveness during cooling; (3) altered osmotic behavior during warming; and (4) intracellular ice formation.[235]

Many plants, and in some cases plant parts, may be acclimatized under appropriate conditions, greatly decreasing their susceptibility to freeze injury. Based on the central role of the plasma membrane in damage symptomology, it is inferred that acclimatization must therefore involve cellular alterations in the membrane that enhance its ability to withstand freezing temperatures. This ability may be due to changes that confer an altered permeability of the membrane to water and/or an increased stability of the membrane.

Figure 6-8. Freeze damage in the woody ornamental *Magnolia grandiflora,* L. In the intact plant *(A),* injury (right) is seen as discoloration and altered leaf and shoot orientation after thawing. Leaves *(B)* display darkening (right) and, in many species, a water-soaked appearance.

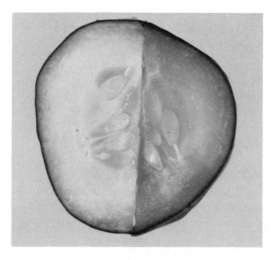

Figure 6-9. Freeze damage in cucumber (*Cucumis sativus,* L.) fruit (right half). Upon thawing, frozen tissue of succulent species typically becomes translucent, placid, and darker in color.

WATER STRESS

Water stress is a universal problem in the production of agricultural plant products. Even under optimum soil moisture conditions, plants may undergo a cyclic daily occurrence of mild but significant water stress. Likewise, water stress also occurs from excessive moisture, a common occurrence in high-rainfall areas.

Water stress begins when the tissue's moisture content, intracellular or extracellular, deviates from the optimum. When the strain is due to a loss of water, the cell's turgor pressure drops below its maximum value. This drop has a pronounced effect on growth rate in that the expansion of new cells is largely controlled by turgor pressure. The economic consequences of this loss of growth (yield) each year due to water stress are astronomical.

The potential for water stress does not stop at harvest; rather in most postharvest products the stress potential increases sharply. When plants are actively growing, they are able to maintain a dynamic equilibrium with the environment between water uptake by the root system and loss due to transpiration. When individual parts are severed from the parent plant at harvest, however, their ability to replace water lost through transpiration is eliminated, making them much more susceptible to water stress. The absence of this water replacement potential is obvious in products such as fleshy fruits and vegetables severed from the parent plant and its root system at harvest. In some cases (e.g., carrot (*Daucus carota,* L.), parsnip (*Pastinaca sativa,* L.)), a portion of the root system per se is the plant part harvested; however, it is removed from its source of moisture. With the exception of decapitated plant parts, in which transpirational losses can be readily replaced without the presence of the root system, harvested fleshy products are dependent on their existing internal moisture supply after harvest. The conservation of this internal water, therefore, is a primary concern during the postharvest period.

Container-grown ornamentals represent a somewhat similar situation. Here the container's growing medium also contains moisture that comprises part of the plant's limited resources. The potential contribution of container water during the postharvest period is in part a function of the medium's percent moisture at the time of harvest and its volume. Containerized ornamentals have the distinct advantage over many harvested products, however, in that moisture losses can be readily replenished if applied in time.

Water deficit stress can be described using cell water potential as an index (for a more detailed discussion of cell water relations, see chapter 7). Hsiao[106] has separated water stress into three somewhat arbitrary classes based on cell water potential. Mild stress occurs when the cell water potential is lowered by only a few bars.* It represents a small loss of turgor pressure. Moderate stress occurs between a few bars and -12 to -15 bars, and mediates a sufficient decline in turgor pressure to result in the wilting of leaves of most species. Severe stress is found below -15 bars when the cells are subjected to serious dehydration and mechanical stress.

---

* Bars are a unit of pressure (e.g., 1 bar = 10^5 pascals or 0.987 atmospheres) used to measure water potential. Water moves from areas of high potential to areas of low potential.

Water deficit stress leads to a direct dehydration strain within the cells that in turn mediates a number of indirect strains (e.g., metabolic alterations, changes in enzyme activation, altered ion flux). Using leaves as a model, the effect of water stress in the functioning of the tissue is illustrated in table 6-2. Cell growth appears to be the most sensitive plant process to water stress, with drops in cell water potential of less than one bar being sufficient to decrease the rate of cell elongation. A decline of this magnitude does not, however, stop growth. In fact, some detached products (e.g., cabbage (*Brassica oleracea,* L. Capitata group) heads) will continue to grow in storage, producing stems of considerable length. In this case, there is not a net increase in weight but in fact a decline, with requisite water and nutrients being recycled from existing sources within the head.

Closely following growth in the degree of sensitivity to water deficit stress are cell wall and protein synthesis, protochlorophyll formation, and nitrate reductase levels, followed by changes in the concentration of abscisic acid and cytokinin, stomatal closure, and a sequential drop in the rate of photosynthesis. Changes in each of these parameters occur at relatively mild levels of stress (i.e., <6 bars). At higher levels of stress (cell water potentials of -10 to -20 bars), changes in respiration and proline and sugar accumulation occur.

Harvested plant products vary widely in their ability to withstand water deficiency stress. Primary factors include species, surface-to-volume ratio, surface characteristics, and stage of development. Seed, spores, bulbils, and other reproductive bodies of many species can be air-dried without loss of viability. For example, Kentucky bluegrass (*Poa pratensis,* L.) seed is capa-

Table 6-2. Generalized Sensitivity to Water Stress of Plant Processes or Parameters

Process or Parameter Affected	*Sensitivity to Stress** Very Sensitive → Relatively Insensitive; Reduction in Tissue Water Potential to Affect Process			*Remarks*
	0 Bar	*10 Bar*	*20 Bar*	
Cell growth				
Wall synthesis				Fast-growing tissue
Protein synthesis				Fast-growing tissue
Protochlorophyll formation				Etiolated leaves
Nitrate reductase level				
ABA accumulation				
Cytokinin level				
Stomatal opening				Depends on species
CO$_2$ assimilation				Depends on species
Respiration				
Proline accumulation				
Sugar accumulation				

Source: After Hsiao.[105]

* Length of the horizontal lines represents the range of stress levels within which a process becomes first affected. Dot-shaded lines signify deductions based on more tenuous data.

ble of withstanding a moisture content of approximately 0.1%[98] whereas other seeds die at much greater moisture levels. Soybean (*Glycine max*, (L.) Merrill) seed can be readily air-dried when mature with no loss in viability. When harvested at a slightly less than mature stage, rapid drying is lethal. If, however, the seeds are held under the appropriate temperature and relative humidity conditions for a sufficient period, and are then allowed to slowly decrease in moisture content, viability is maintained. Apparently the seeds are capable of undergoing the essential physical and chemical changes needed to maintain viability even though prematurely severed from an external source of moisture and nutrients.

In contrast to mature reproductive structures, most fruits, vegetables, and flowers decline rapidly in quality with only minor losses in moisture (fig. 6-10). Generally the loss of only 5–10% moisture renders a wide range of products unsalable (table 6-3).[126,191,205] There is generally a loss of crispness and turgidity, and for green vegetables, green coloration.

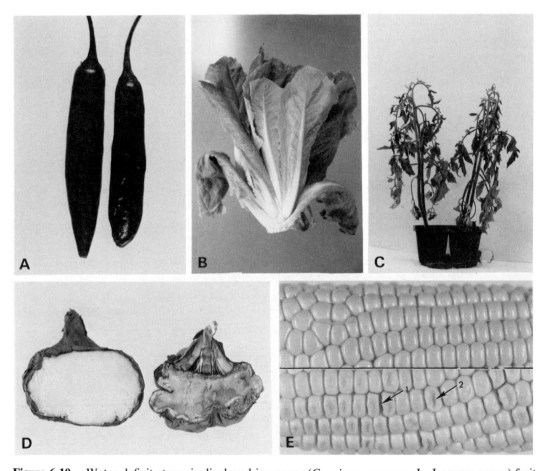

Figure 6-10. Water deficit stress is displayed in pepper (*Capsicum annuum*, L. Longum group) fruit *(A, right);* leaf lettuce (*Lactuca sativa*, L.) *(B);* tomato (*Lycopersicon esculentum*, Mill.) transplants *(C);* Chinese water chestnut (*Eleocharis dulcis*, (Burm. f.) Trin. ex Henschel) corm *(D, right);* and sweet corn (*Zea mays* var. *rugosa*, Bonaf.) *(E, bottom).* A shriveled or wilted appearance is common in most succulent products undergoing a water deficit stress. In sweet corn *(E, bottom),* dehydration is first seen as separation between individual kernels (1) and indentation on the top of the individual kernels (2).

Table 6-3. Maximum Acceptable Loss of Water from Selected Harvested Products Before the Commodity Is Considered Unsalable

	Maximum Acceptable Loss (%) *
Apple	5
Asparagus	8
Bean, broad	6
Bean, runner	5
Beetroot, storing	7
Beetroot, bunching, with leaves	5
Blackberries, Bedford Giant	6
Brussels sprouts	8
Cabbage, Primo	7
Cabbage, January King	7
Cabbage, Decema	10
Carrot, storing	8
Carrot, bunching, with leaves	4
Cauliflower, April Glory	7
Celery, white	10
Cucumber, Femdam	7
Leeks, Musselburgh	7
Lettuce, Unrivalled	5
Lettuce, Kordaat	3
Lettuce, Kloek	3
Onion, Bedfordshire Champion	10
Orange	5
Parsnip, Hollow Crown	7
Potato, maincrop	7
Potato, new	7
Peas in pod, early	5
Peas in pod, maincrop	5
Pepper, green	7
Raspberries, Malling Jewel	6
Rhubarb, forced	5
Spinach, Prickly True	3
Sprouting broccoli	4
Strawberries, Cambridge Favourite	6
Sweet corn	7
Tomato, Eurocross BB	7
Turnip, bunching with leaves	5
Watercress	7

Note: Data from Kaufman et al.[126] Pieniazeh,[191] and Robinson et al.[205]
* Percent of original fresh weight.

The rate of development of water deficit stress also varies widely with different products. Organs such as potatoes and onions that are capable of long-term storage lose water less readily than peppers and avocados and much less readily than immature okra and leaf lettuce (*Lactuca sativa*, L.). Leaf vegetables with large surface-to-volume ratios are particularly vulnerable to the rapid loss of water.

The loss of moisture results in a reduction in the fresh weight of the harvested product, which when sold on a weight basis is translated into a loss in value. Much more significant losses in value, however, are incurred due to undesirable quality changes. In addition to losses in turgidity and firmness, there are other symptoms and secondary stresses depending on the product in question. Water deficit stress can result in the discoloration of flowers, leafy vegetables, and some fruits (e.g., oranges); loss in flavor and aroma (citrus fruits); decline in nutritional value due to losses in vitamins; increased susceptibility to chilling injury and other physiological disorders; increased incidence of pathogen invasion; and accelerated rate of senescence.

Plants and plant parts vary widely in their ability to withstand water deficit stress. Many intact plants exposed to periods of low water availability become more resistant to subsequent stress, a process called acclimitization or drought hardening. Some species have also evolved adaptive mechanisms such as increased cuticle formation, shedding of leaves, rapid stomatal closure, and shape and size alterations to combat drought situations.

Excess moisture can also result in water stress conditions in harvested products. With intact plants, excessive moisture replaces the gas phase of the soil or root media with water, which leads to several possible secondary stresses. Excess water may cause leaching of nutrients from the plant, or more importantly, it may impose a gas stress on the plant (see later in this chapter for additional details). The latter can be due to insufficient molecular oxygen, excessive carbon dioxide, or excessive production of the hormone ethylene. These changes may result in increased invasion by pathogens. One common flooding symptom in many plants is the wilting of the aboveground portion of the plant. Wilting is caused by a water deficit stress due to impaired uptake of water by the root system.

Seeds and other postharvest plant parts may also sustain stress due to excessive moisture. With seeds, the optimal moisture level for storage is much lower than that for the normal germination, growth, development, and reproduction cycle. As a consequence, low moisture is a desirable condition for long-term quality maintenance in most seeds. (There are, of course, notable exceptions, e.g., the seed of *Acer saccharinum*, L. will not withstand even air drying.) In most seeds, however, increases in moisture above the optimum for storage result in undesirable changes in quality, which from a postharvest standpoint represent a stress (fig. 6-11).

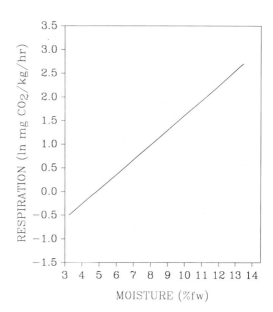

Figure 6-11. The relationship between water content in pecan (*Carya illinoensis,* (Wang.) C. Koch) seeds and respiratory activity *(after Beaudry et al.[12]).*

Different seeds have different optimum moisture levels. Generally the moisture content that leads to a significant increase in respiration coincides with the level at which spoilage begins to occur in storage. The most common cause for increases in seed moisture after drying is storage at an excessively high relative humidity. Under these conditions, the moisture content of the grain increases until an equilibrium is reestablished between the seed and its storage environment. Seed moisture level also has a pronounced effect on the activity of storage pathogens. For example, fungal attack of corn stored at 18% moisture was significantly greater after 7.5 months of storage than that stored at 14–15.5% moisture.[41] Likewise, the maximum acceptable storage moisture content varies with temperature. Higher moisture content seeds must be stored at lower temperatures to prevent deterioration.

Water-excess stress is also a common occurrence in other postharvest products when moisture is allowed to remain or collect on the surface of the product. Perhaps the most common situation occurs when the temperature of the product is below the dew point of the water vapor in the surrounding air. Condensation forms on the surface of the product, which impairs gas diffusion and greatly enhances the potential for pathogen invasion. In some products, the damage incurred is relatively minor whereas in others (e.g., Chinese chestnuts (*Castenea mollissima,* Blume.)) it is devastating. Even in Roman times great care was taken to prevent this occurrence.

GAS COMPOSITION

Stress during the postharvest period imposed by the gas atmosphere within the product can have a pronounced effect on quality. A gas stress that slows down the normal maturation and senescence phases of the life cycle of the plant or plant part can be a positive attribute and is a basic component in the handling and storage of certain crops. Of primary concern during storage is the concentration of oxygen, carbon dioxide, and ethylene within the tissue. Postharvest biologists have found that by imposing an elastic strain, one that while altering the normal rate of metabolic events does not produce a permanent injury, they can greatly extend the time interval in which quality can be maintained in certain products. In some cases, the altered gas environment is beneficial not so much because of direct changes in the metabolism of the produce per se than because of the imposition of stress injury or death to other organisms (e.g., insects, rodents) that may otherwise compromise the quality of the stored product. A consistent risk of imposing an elastic strain on stored plant products is the occurrence of a plastic strain that results in an irreversible injury. Exceedingly low oxygen or high carbon dioxide or ethylene may under the appropriate conditions have disastrous consequences on quality.

Oxygen Stress

With decreasing availability of molecular oxygen to low oxygen affinity enzymes, there is a decline in the general metabolism of the cells. Large changes in metabolic rate do not occur, however, before reaching relatively low environmental concentrations of oxygen (i.e., <5%). The critical factor is the actual concentration of oxygen within the cells. Internal oxygen concentration is controlled by product resistance to oxygen diffusion, rate of utilization of the molecule, and the differential in oxygen partial pressure between the product exterior and interior. Due to differences in the rate of use and diffusion, there is considerable variation in the optimum external oxygen concentration for various products. Large dense products (e.g., sweetpotato storage roots) generally have a higher external oxygen requirement than smaller and/or less dense products. Likewise, some products must be stored at higher temperatures than others to prevent chilling injury, thus increasing their oxygen utilization rate. Examples of recommended oxygen concentrations for several postharvest products under their optimum storage conditions are 'Bramley's Seedling' apples, 11–13% O_2 (3–4°C, 8–10% CO_2); 'Golden Delicious' apples, 3% O_2 (2.5°C, 5% CO_2); 'William's Bon Chrétien' pear (*Pyrus communis,* L.), 2% O_2 (0°C, 2% CO_2); blackcurrent (*Ribes nigrum,* L.), 5–6% O_2 (2–4°C, 40–50% CO_2); and red cabbage, 3% O_2 (0°C, 3% CO_2).[78,238]

Low oxygen stress can be separated into three general classes based on the strain produced. Severe stress occurs when anaerobic conditions are reached within the product. This stress produces a plastic strain that generally results in a rapid and irreversible decline in product quality if continued for a sufficient duration (fig. 6-12, *top*). Moderate stress occurs at oxygen concentrations above those required for anaerobiosis; however, it produces a plastic strain that can considerably impair the quality of the product. Particularly notable are undesirable changes in flavor or aroma. Mild stress produces an elastic strain that does not result in injury and in some products it can greatly increase longevity and maintain quality (fig. 6-12, *bottom*). Not all components of quality are similarly affected. However, since we are concerned with the composite of all quality parameters, the optimum oxygen concentration is above the level that results in a decline in any critical quality component. It should be noted that optimum oxygen conditions are not necessarily economically viable. Although small improvements in longevity can be achieved for a number of postharvest products with mild oxygen stress, the additional expense seldom warrants its use.

Under very low internal oxygen conditions, the terminal electron acceptor in the electron transport system, cytochrome oxidase, ceases to function. This enzyme has a high affinity for oxygen and as a consequence inhibition does not occur until the internal oxygen concentration is less than 0.2%. When inhibition does occur, $NADH_2$ can no longer be oxidized to NAD and cycled back to the tricarboxylic acid cycle. As a consequence, the tricarboxylic acid cycle is also inhibited. The glycolytic pathway is not blocked, however, leading to a buildup of acetaldehyde and ethanol that are phytotoxic to the cell. As these metabolites accumulate, they disrupt cellular organization (e.g., ethanol may act upon cellular membranes),[137] which eventually leads to death.

Anaerobic conditions in most products result in an increase in respiratory rate. Increased respiration is caused by the low energy yield: 2 ATPs per molecule of glucose versus 36 when under aerobic conditions. Since the cell still requires energy when anaerobic, and the efficiency of production is much lower, respiration must be significantly accelerated to meet even minimal requirements.

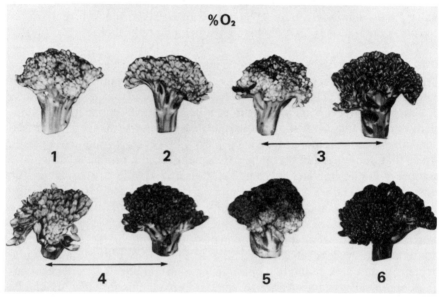

Figure 6-12. Low oxygen conditions can be either beneficial or detrimental, depending on the product, oxygen concentration, temperature, and other factors. The detrimental effect of too little oxygen on stored brussels sprouts (*Brassica oleracea*, L. Gemmifera group) is seen as internal discoloration *(top)*. Both a positive and negative effect of low oxygen on stored broccoli (*Brassica oleracea*, L. Italica group) florets is displayed in the bottom photograph. Inadequate oxygen results in loss of chlorophyll and subsequent lightening of the florets while low but adequate oxygen minimizes chlorophyll degradation. (Arrows indicate the range in symptoms.) *(Photographs courtesy of W. J. Lipton[150].)*

Susceptibility to detrimental low oxygen stress is related to the nature of the product: its size, stage of development, anatomy, morphology, and general condition. Considerable variation in the susceptibility to low oxygen stress can be seen between cultivars. For example, the injury sustained by the storage roots of sweetpotato cultivars exposed to flood-induced low oxygen conditions varied widely but did not correlate closely with the absolute alcohol concentration attained or the activity of alcohol dehydrogenase, the enzyme controlling its synthesis.[39] Other factors, such as differences in ability to metabolize accumulated ethanol upon return to aerobic conditions, or the extent of damage during the lag period before the onset of catabolism, may be more important in determining the extent of injury incurred.

Low oxygen conditions after harvest often develop when air exchange is inadequate. Poor air exchange may be due to excessively tight packaging (especially when product temperature is high), harvesting and bulk storage of grains when their moisture level has not been sufficiently reduced, over-watering of containerized plants, prolonged submergence in or coverage with water, as well as improperly controlled gas atmosphere storage conditions. Low oxygen stress may also occur prior to harvest, with the majority of the injury symptoms developing during the postharvest period. Sweetpotato storage roots exposed to flooding prior to harvest are predisposed to spoilage in storage.

The extent of the development of low oxygen injury is highly dependent on the duration of the exposure to anaerobic conditions. Generally, the longer the exposure, the greater the accumulation of stress metabolites and subsequent injury. However, since the reactions for the conversion of pyruvate to ethanol are reversible upon return to aerobic conditions, most plants, if not extensively injured, are capable of recycling the metabolites formed, thus preventing further injury.[55]

Plant products exposed to anaerobic conditions sustain significant losses in quality, and if exposure is sufficiently long, death will occur. Injury symptomology and subsequent quality losses vary considerably between various products. Anaerobic conditions can result in the formation of undesirable flavors and odors, altered texture, discoloration, increased incidence of physiological disorders, changes in composition, as well as other undesirable changes.[79,135,151,152,227]

Mild oxygen stress is utilized to decrease the metabolic rate of a number of harvested plant products, decreasing the rate at which quality is lost and increasing the product's functional life expectancy. Unlike severe and moderate low-oxygen stresses, the effects of mild stress are beneficial. Examples of several beneficial effects of mild oxygen stress after harvest are decreases in the rate of softening, weight loss, compositional changes, and pigmentation loss. With climacteric fruits, low oxygen storage conditions are used to retard the onset of ripening.[134]

Harvested plant products can also be injured by excess oxygen in the postharvest environment. Several studies have monitored the effects of extremely high oxygen concentrations (i.e., up to 100%); however, the absence of their occurrence outside of the laboratory makes them of only theoretical value.[9,132] High oxygen stress may possibly occur under ambient storage conditions for some products that are produced in low oxygen environments, for example, aquatic crops such as lotus root (*Nelumbo nucifera,* Gaertn.) or Chinese water chestnut (*Eleocharis dulcis,* (Burm. f.) Trin. ex Henschel). However, this area of research has been little explored.

Carbon Dioxide Stress

Carbon dioxide may also impose a significant stress on harvested products. The response produced may be beneficial or detrimental to product quality depending on the nature of the product in question, the concentration of carbon dioxide within the tissue, the duration of exposure, the internal oxygen concentration, and other factors. Increased levels of carbon dioxide have been shown to significantly decrease the respiratory rate of a number of products. The molecules' primary action appears to be on the kinetics of a reversible reaction within the tricarboxylic acid cycle catalyzed by succinate dehydrogenase. As the rate of the enzyme is inhibited, the concentration of succinate increases, often to toxic levels.[109,139] Other enzymes in the cycle may also be affected but to a lesser extent.[181] Carbon dioxide inhibition may also lead to the accumulation of toxic quantities of acetaldehyde and ethanol,[221] a disorder referred to as *zymasis* in early research on carbon dioxide toxicity. At carbon dioxide concentrations above 20%, zymasis leads to acetaldehyde poisoning due to the inhibition of alcohol dehydrogenase.[243]

Plant products vary widely in their susceptibility to carbon dioxide injury. Considerable variation can be found between individual cultivars,[108,133] with fruit maturity,[51,58] and with location within the product. For example, the nonchlorophyll-containing rib portion of lettuce (cv. 'Crisp Head') is much more susceptible to carbon dioxide injury than sections of the leaf containing chlorophyll.[150]

Injury from excessively high carbon dioxide levels is seen in many products as an increased incidence of various internal and external physiological disorders. These include brown heart[133,227] and core flush of apples,[145] internal browning of cabbage,[115] low-temperature breakdown of apples,[77] brown stain of lettuce,[237] discoloration of mushrooms,[226] calyx discoloration and internal browning of bell peppers,[182] surface blemishes on tomatoes, and pitting of broad bean (*Vicia faba*, L.) pods.[245] Quality losses may also be due to accelerated softening,[151] the formation of off-flavors [e.g., strawberries (*Fragaria* × *Ananassa*, Duchesne),[50] broccoli (*Brassica oleracea*, L. Italica group), [151,224] kiwi fruit (*Actinidia chinensis*, Planch.),[97] and spinach (*Spinacia oleracea*, L.)[169]], inhibited and/or uneven ripening of tomatoes after removal from storage,[182] decreased resistance to pathogens,[238] and inhibited wound healing in potatoes.[27]

Low levels of supplemental carbon dioxide result in a number of positive benefits during storage of some crops, and as a consequence, carbon dioxide is an important component of the controlled atmosphere storage of these species (fig. 6-13). In some cases, the elevated carbon dioxide treatment may represent only a short-term exposure. In most instances, however, the product is maintained at the elevated carbon dioxide level during storage.

A central component in extended quality maintenance using carbon dioxide is a decrease in product respiration. It appears to be due primarily to a progressive decline in succinate dehydrogenase activity with increasing carbon dioxide concentration.[82] Inhibited respiratory rates have been demonstrated in a wide cross-section of harvested products (e.g., spinach, avocado, lettuce, cherries, apples). Other positive benefits include a decreased rate of softening of kiwi and apple fruit,[97,145] inhibition of chlorophyll synthesis in potato tubers held in the light,[8,198] and degradation in chlorophyll-containing products such as green beans (*Phaseolus vulgaris*, L.) and broccoli,[211] stimulated loss of astringency in persimmons (*Diospyros kaki*, L.f.),[68] decreased ethylene synthesis,[171,246] and inhibition of some storage pathogens.[23,24,44,45,70] The differential in carbon dioxide concentration between where maximum storage benefits are obtained and the onset of detrimental in-

Figure 6-13. As with oxygen, modified storage carbon dioxide gas environments can be either beneficial or detrimental. A positive effect of high carbon dioxide on a stored product is illustrated by the improved quality of broccoli florets stored in 10% CO_2. *(Photograph courtesy of W. J. Lipton.)*

juries is often quite narrow. As a consequence, the level of carbon dioxide in the storage atmosphere must be very closely monitored and controlled.

Ethylene Stress

Ethylene within the storage environment, whether produced by the stored product, microorganisms, or other sources, is known to cause a significant stress to many harvested products. The hormone affects the rate of metabolism of many succulent plant products and is generally active at very low concentrations. Although the primary mode of action of the molecule is not known, ethylene has been shown to increase the rate of respiration, alter the activity of a number of enzymes, increase membrane permeability, alter cellular compartmentalization, and alter auxin transport and/or metabolism.[196] Cellular changes induced by ethylene result in an acceleration of sensescence and the deteriorative processes that accompany it. Ethylene within the storage environment results in a significant decline in the level of a number of quality attributes and is responsible for the induction or aggravation of a cross-section of physiological disorders.[121]

Primary losses in quality are due to the induction of abscission (flower, leaf, and fruit) and changes in flavor, color, and texture. The exposure of carrots to ethylene increases the concentration of phenolics[212] and induces the formation of isocoumarin,[38] which is very bitter. Likewise, undesirable flavors develop in cabbage[244] and sweetpotatoes[25] and in climacteric fruits general changes in sugars and acids occur upon exposure to ethylene.

In green tissues, ethylene accelerates the degradation of chlorophyll, resulting in undesirable yellowing. It has been shown to occur in cucumber fruit,[194] cabbage,[101] brussels sprouts (*Brassica oleracea,* L. Gemmifera group), broccoli, cauliflower (*Brassica oleracea,* L. Botrytis group),[244] and acorn squash (*Cucurbita pepo,* L.)[187] as well as in the leaves of many ornamentals. Ethylene also greatly accelerates undesirable changes in coloration of many flowers. When carotenoid pigments are prevalent, degradation of the chromoplasts occurs at a much more rapid rate. In flowers where anthocyanins predominate, changes in vacuole pH hastened by ethylene can result in pronounced alterations in color.[236]

Exposure of some postharvest products to ethylene stress results in significant changes in texture. Watermelons (*Citrullus lunatus*, (Thunb.) Mansf.) display a decrease in firmness[203] as also do sweetpotato roots after cooking.[25] In some products, however, there is a marked acceleration of lignin synthesis, which increases fiber formation and toughness. This acceleration has been shown in asparagus spears[92] and rutabaga (*Brassica napus*, L. Napobrassica group) roots.[201]

Ethylene stress can induce the abscission of leaves, flowers, fruits, and in some cases, stems of an extremely wide cross-section of harvested products (fig. 6-14*D*). It occurs not only in intact products such as ornamental species[161] and vegetable transplants[128] but also in decapitated products such as cabbage and cauliflower (see review by Reid[200]).

Ethylene in the transit or storage environment has also been shown to induce a number of physiological disorders that impair quality in harvested products. These include flower petal fading, wilting and closure,[94] leaf epinasty[125,161] (fig. 6-14*C* and *D*), russet spotting of lettuce (fig. 6-14*B*),[112] and gummosis, necrosis, and flower-bud blasting in some flowering bulbs.[124] In addition, many products exposed to ethylene stress display an increased susceptibility to storage pathogens.[70]

Not all of the postharvest effects of ethylene stress are undesirable. Ethylene is used widely to initiate the ripening of several climacteric fruits held in storage (e.g., banana and tomato). This use synchronizes the ripening of all of the fruit as well as increases the uniformity of ripening within individual fruits. Ethylene may also be used to induce the selective abscission of leaves, flowers, and fruits of certain ornamentals. In the citrus industry ethylene is utilized to enhance the degradation of fruit chlorophyll of some cultivars prior to sale (fig. 6-14*A*).[258]

RADIATION STRESS

Visible Light

Plant and plant parts can incur light stress after harvest that significantly alters their quality and marketability. Stress may be imposed due to a deficit or an excess of light depending on the product in question and the conditions to which it is exposed. The injury sustained due to excess light may be attributed to the development of a secondary stress (e.g., heat, drought) or a light-sensitive system within the tissue. Light entering the product increases the object's energy level, resulting in a buildup of heat. This heat, if not adequately dissipated, can cause a secondary heat stress. An alternate secondary stress results from an increase in water loss from the tissue due to the temperature buildup. In fruits, foliage, and young bark of some species, water loss develops into the classical symptoms of sunscald. Both secondary stresses typically occur in plants or plant parts that are grown in the light. Plants that are produced in low-light conditions, for example, shade-tolerant species, are particularly susceptible to excess light during the postharvest

Figure 6-14. Ethylene-mediated responses in harvested products. *A:* A beneficial effect of ethylene after harvest is illustrated by degreening of citrus. *B:* Russet spot in head lettuce (*Lactuca sativa,* L.) is a distinctly undesirable effect of ethylene. *C:* Untreated transplants of tomato (*Lycopersicon esculeutum,* Mill.) and pepper (*Capsicum annuum,* L.). *D:* The same plants after exposure to ethylene during simulated transit. Leaves of the tomato exhibit epinasty (downward curvature of the petioles) while pepper leaves abscise at the base of the petioles.

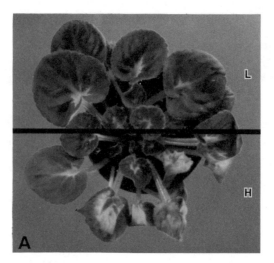

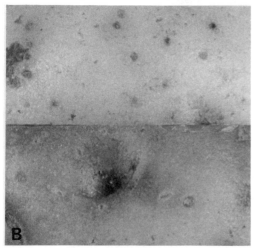

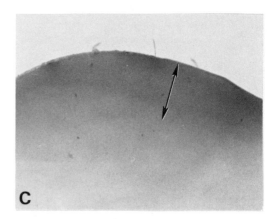

Figure 6-15. The detrimental effect of light stress during the postharvest period is demonstrated by sunscald in African violet (*Saintpaulia ionantha,* Wendl.) during retail sales *(A)* (L, low light; H, high light); greening of potato (*Solanum tuberosum,* L.) tubers during marketing caused by light-induced chlorophyll synthesis *(B, bottom half)*. Unlike the surface pigmentation found in some products that is limited to the surface layer or two of cells, chlorophyll synthesis in the potato can occur some distance into the tuber *(C)*.

period. Injury is unfortunately a common occurrence for ornamental plants such as African violet (*Saintpaulia ionantha,* Wendl.) (fig. 6-15A) and gloxinia (*Sinningia speciosa,* (Lodd.) Hiern.) during retail sales.

In chlorophyll-containing tissues, the symptoms of excess light generally include a decline in chlorophyll concentration within the exposed leaves and an increase in a yellow or whitish coloration. Normally leaf chlorophylls are adequately protected from photooxidation by carotenoids and other pigments. However, when the light intensity exceeds the capacity of the protective molecules, photooxidation and subsequent breakdown of chlorophyll begins. The degree of injury due to light stress depends on species, cultivar, conditions of growth, and light intensity and duration to which the product is exposed. The level of injury can range from slight discoloration to death of the plant.

Another form of excess light stress occurs in products that are grown in the absence of light (e.g., roots and tubers). Even very low-light intensities after harvest (e.g., 25 lux)[5] can induce the formation of chlorophyll (fig. 6-15*B* and *C*). In potato tubers, chlorophyll synthesis is usually accompanied by the accumulation of toxic glycoalkaloids.[116] Root crops may also undergo light stress prior to harvest when the upper portion of the taproot, protruding from the soil, is exposed to light (e.g., green shoulder in carrots). Susceptibility to excess light stress varies widely with species, cultivar, light intensity, duration of exposure, and other factors.

Light-deficiency stress occurs when plants are held under conditions where the daily light exposure is sufficiently low to result in undesirable changes in quality. It is particularly evident when illumination falls below the light compensation point of the plant. An indirect injury occurs due to the plant's inability to synthesize sufficient carbohydrates for essential respiratory and maintenance reactions. Low light is also known to decrease the activity of certain enzymes[155] and mediate significant alterations in chloroplast structure.[184] Therefore, during the postharvest period both light intensity and duration may be critical. Short exposures of light well above the light compensation point may be insufficient to meet the respiratory requirements of the plant over the remainder of the day. Many harvested intact plants, however, are held for prolonged periods in environments devoid of light.

Transferring ornamental plants, such as *Ficus benjamina,* L. produced in conditions of full sunlight, into a low-light postharvest environment generally results in severe leaf abscission (fig. 6-16). If the plants are first acclimatized to low-light conditions, however, leaf shedding is minimized.[48] The acclimatization process results in significant changes in anatomy, morphology,[73] and composition[174] of the leaves. Acclimatization or hardening of plants has also been shown to result in a lower respiratory rate[168] and light compensation point.[118] As a consequence, these plants tend to survive better in low-light interior environments than plants that have not been acclimatized.[49]

Ultraviolet Radiation

Interest in ultraviolet radiation as a source of stress during plant production has increased significantly since the first report of a possible decline in the Earth's protective ozone layer. Ultraviolet radiation can produce significant injury and death to plants if the exposure is sufficiently high.[65,157] Much of the ultraviolet radiation from the sun is filtered out by ozone and oxygen molecules in the upper atmosphere before reaching the Earth's surface. Plants also have evolved organic compounds, for example, anthocyanins, that absorb ultraviolet radiation and act as protectants. At present, there is little evidence to suggest that ultraviolet radiation presents a significant stress (deficit or excess) during production or postharvest phases of agriculture.

Ionizing Radiation

Ionizing radiation represents a form of energy that ionizes or electronically excites the molecules with which it comes in contact when absorbed by an organism. Generally, chemical changes result, and if they are of sufficient

Figure 6-16. Low-light stress during the postharvest period is illustrated by leaf shedding in *Ficus* plants during simulated transportation to market. *(Photograph courtesy of C. A. Connover.)* Symptoms include extensive leaf abscission and stem dieback. Along with light intensity and the duration of light deprivation, preharvest production conditions are known to affect the degree of leaf shedding.

magnitude, damage to the tissue occurs (fig. 6-17). The primary types of ionizing radiation are α-rays, γ-rays, and x-rays. These, however, are found at only very low levels in nature. Thus, exposure of plant products to high levels of ionizing radiation represents an artificial situation imposed on certain harvested products to control microorganisms or insects or to prevent sprouting (e.g., potatoes and onions (*Allium cepa*, L.)).

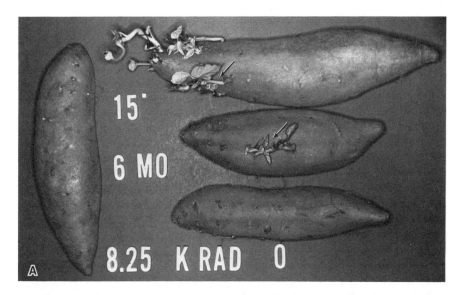

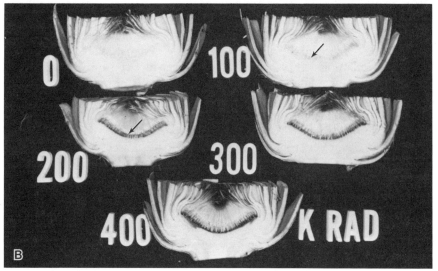

Figure 6-17. *A:* Sweetpotato (*Ipomoea batatas*, (L.) Lam.) storage roots illustrate the beneficial effects of the use of ionizing radiation after harvest to control sprouting. The horizontal root (left) exposed to 8.25 krad displays a pronounced inhibition of sprouting in contrast to control roots. The potential for the successful utilization of postharvest ionizing radiation is dependent on causing more damage to the target organism (e.g., insect, microorganism) than the host plant. *B:* The undesirable effect of radiation on artichoke (*Cynara scolymus*, L.) hearts. Damage is seen as discoloration as the radiation intensity exceeds 100 krad. *(Photographs courtesy of W. J. Lipton.)*

The action of the radiation causes damage to both the cells of the product and those of the target organism, for example, mango anthracnose (*Collectotrichum gloeosporioides*).[22] The objective of a fungicidal irradiation treatment is to cause significant damage or death to a sufficient portion of the microorganism population without causing deleterious side effects to the host product. Thus, radiation is seldom used to sterilize products but rather to diminish the microorganism population as much as possible during the normal postharvest period. One significant advantage of radiation over most chemical treatments is that the rays are of a sufficiently high energy level to pass entirely through the host tissue. As a consequence, microorganisms not only on the surface but also within the tissue can be treated (see Dennison and Ahmed[57]; Maxie and Kader[163]; Romani[207]; and Sommer and Fortlage[232] for detailed reviews).

The dose requirements depend on the sensitivity of the target organism and the ability of the host to withstand undesirable changes. The susceptibility of both the target organism and host vary widely. Since the host product is damaged during exposure, products such as seeds or other propagules that are intended for future growth are not viable candidates for irradiation treatment.

Ionizing radiation has been tested on a wide range of postharvest products but has proven successful primarily on strawberries,[164] mangos,[59] papayas,[64] potatoes,[26] and several grains. Minimum exposure needed for most products to obtain a positive benefit ranges from 8–1,000 krad. This amount, however, is generally greater than the maximum dose the plant tissue will tolerate before injury occurs.[165] At present, irradiation of harvested plant products is little used on a commercial scale outside of Japan and several other countries. Lack of use is due to consumer wariness of products treated by irradiation, the undesirable effects of irradiation on the product, the availability of less expensive methods, and the availability of more effective methods.

Ionizing radiation stress occurs at any level of exposure; however, in postharvest products concern is primarily with levels of radiation that produce significant changes in product quality. Injury may be direct, altering a critical type of molecule, for example, a component of the cell membrane whose alteration leads to a loss of semipermeability.[259] The symptoms of this type of injury occur very rapidly. Injury may also be indirect, such as through the formation of free radicals that cause metabolic disruptions. Sublethal but excessive doses of radiation to the host product have been shown to cause changes in texture, undesirable changes in flavor, formation of lesions, discoloration of the tissue, and other symptoms. Optimum exposures are those sufficiently high to facilitate the desirable effects without causing significant changes in product quality or potential longevity.

CHEMICAL STRESS

Plants and plant parts are subjected to a wide range of chemical stresses that may impair postharvest quality. These include salts, ions, air pollutants, and agricultural chemicals.

Salt Stress

Excess salt represents a potentially significant stress during the postharvest life of products such as ornamental plants. High levels of salt can cause a direct toxic effect on the cells of the plant. More often, however, excess salts result in secondary stresses (e.g., osmotic and/or nutritional)[15] that produce an irreversible plastic strain leading to decreased product quality[195] (fig. 6-18). The addition of salts to the aqueous environment of the plant decreases the osmotic potential of the water. In addition, high salts in the growing media

Figure 6-18. Salt stress during the postharvest period is illustrated by marginal leaf burning on poinsettia (*Euphorbia pulcherrima,* Willd.) *(A)* and necrosis at the tips of *Quercus* spp. *(B). (Photographs courtesy of F. A. Porkney.)*

can suppress the uptake of other nutrients, for example, sodium has been shown to have a competitive effect on potassium uptake.[229]

Secondary osmotic stress represents potentially the most critical concern in the handling, storage, and retail sales of ornamentals. During production, the plants are irrigated on a relatively precise and frequent schedule. However, during the postharvest period the responsibility for this task changes hands, generally to personnel that are less familiar with the requirements of the plants. As conditions change from those of the production zone, water use patterns also change, as does the precision of replenishing water lost due to evapotranspiration. High salts in the potting media greatly increase the osmotic stress as the limited water resources of the container are depleted. This situation causes a decrease in the rate of growth, photosynthesis, and other metabolic events within the plant. Symptoms of salt stress that ensue range from wilting, leaf burn, nutrient deficiency, and leaf drop to death of the plant.

Ion Stress

Unlike salt stress, ions do not result in a significant decrease in water potential. The strain produced by ion stress is due to an alteration in the normal ionic balance within the cell. Spencer[234] divides ions into three groups based on their potential toxicity: nontoxic, toxic at intermediate concentrations, and toxic at low concentrations. Of the rather large number of ions that can cause injury, the most critical that are commonly encountered are Ag, Cd, Co, Hg, Mn, Ni, and Zn. These may be introduced as contaminants in the root media of intact containerized plants from the water to which the product is exposed, or as postharvest treatments.

Ion stress during plant production may result in an elastic strain with no visible symptoms. During this period the plants are grown under near-optimum conditions. However, upon movement to postharvest conditions, other stresses are imposed on the plant (e.g., osmotic) that may increase the plant's susceptibility, resulting in a plastic strain and subsequent damage from the ion stress.

A more direct ion stress and injury occurring during the postharvest period can be seen with silver ions. Silver has been increasingly used on many cut flowers and flowering ornamentals due to its inhibition of ethylene-stimulated senescence and flower abscission. This use, however, has led to an increased incidence of the toxicity of this heavy metal to plant tissues. The most noticeable symptoms are surface discoloration and lesions when applied as an aqueous silver nitrate spray (fig. 6-19). The use of silver thiosulfate has decreased the incidence of injury but has not eliminated the potential for damage, especially if excessively high concentrations are used or treatments are made to susceptible species.

ts cause extensive damage to plants during the production phase o be important during the postharvest period. The most common

Figure 6-19. Heavy metals such as silver can cause extensive injury to plant tissue as illustrated here on carnation (*Dianthus caryophyllus,* L.). *A:* The flower on the left was sprayed with 200 ppm of silver nitrate to delay senescence. *B:* Individual petals display the discoloration and cellular death caused by the silver. Use of silver thiosulfate as a short-exposure vase solution treatment has largely eliminated this problem.

air pollutants encountered are sulfur dioxide, ozone, and ammonia (fig. 6-20).[160,183]

Sulfur dioxide injury varies with species, concentration of the gas, length of exposure, and plant and environmental conditions at the time of exposure.[260] The molecule may either be absorbed on the surfaces of the product or absorbed through the stomates.[71] Upon entry into the plant, sulfur dioxide dissolves to sulfurous acid, which forms sulfite salts. When the concentration is sufficiently low, the plant can oxidize these sulfites to sulfates, which are normal constituents of plants. In fact, some species of plants assimilate sulfur

Figure 6-20. In addition to ethylene, a number of gases can cause quality loss during the postharvest period. *A:* The bottom three kernels illustrate the effect of ammonia vapor on pecans (*Carya illinoensis*, (Wang.) C. Koch.). *B:* Sulfur dioxide under appropriate conditions can cause extensive damage to bedding plants (e.g., *Petunia* × *hybrida*, Hort.) during marketing. (*From Howe and Woltz.*[105])

dioxide with carbon dioxide[74] to be used in various sulfur-containing metabolic products (e.g., the amino acids methionine and cysteine). It is the sulfites that are not converted to sulfates that appear to cause injury to the tissue.

A relatively wide range of agricultural crops are sensitive to sulfur dioxide injury. Typical symptoms are marginal and interveinal chlorosis of leaves[105] (fig. 6-20B). The leaves of some species may develop a water-soaked appearance due to the effect of sulfur dioxide on membrane permeability. While crop response to sulfur dioxide has been studied extensively in the production phase, little information is available on the effects of postharvest exposure.

Ozone, produced through the photochemical action of sunlight on nitrogen oxides and reactive hydrocarbons from fuel combustion emissions, is considered one of the most serious of the gaseous pollutants. Peroxyacetyl nitrate, also an oxidant, causes similar injury. Susceptibility varies widely with species, cultivar, and plant condition.[30] Newly expanded leaves appear to be the most sensitive, and rapid luxuriant growth intensifies sensitivity. Susceptible species develop interveinal chlorosis, blisters, and necrotic areas on the leaves. Blistering, apparently due to swelling of guard and epidermal cells, is followed by tissue collapse and dehydration.[260] Several chemicals such as daminozide[29] and ethylene diurea[30] have been shown to act as anti-air pollution compounds, decreasing the effect of ozone on certain species. As with sulfur dioxide, postharvest effects have been little studied.

Ammonia toxicity to plants has been known for many years. Exposure to the gas during postharvest handling and storage has come primarily from leaks in refrigeration systems in which ammonia was used as the refrigerant. The injury response is a very rapid darkening of the tissue. In pecans, exposure is most dramatic, shifting the surface color from a light brown to black in only a few minutes (fig. 6-20A).[129] Ammonia appears to mediate a change in the oxidation state of iron atoms in the surface testa of the kernels. Strong reducing agents can significantly, although not totally, reverse this color change[254]; however, due to toxicity and/or flavor alterations, these chemicals do not represent a commercially viable method for circumventing injury. Fortunately, the incidence of postharvest exposure to ammonia has declined substantially with the increased utilization of freon as a refrigerant. However, significant losses still occur each year due to exposure of plant products to ammonia fumes.

Other gaseous pollutants, for example, hydrogen fluoride,[90] are also a problem in the production phase of agriculture. Their importance in the postharvest biology of agricultural crops, however, has yet to be established.

Agricultural Chemicals

The number of chemicals used in agriculture has increased dramatically since the 1970s. When used properly, each can have a pronounced effect on yield and/or product quality. Although most are used during the production phase (i.e., herbicides, insecticides, fungicides, growth regulators), some have been found to exhibit distinct advantages when used after harvest. A unifying characteristic of all of these compounds is some form of direct or indirect biological activity. A chemical is selected for a particular type of activity that is advantageous (e.g., the control of one type of insect or group of insects). Its actual biological activity, however, often extends outside of its target role. Plant growth regulators, for example, have been selected for certain beneficial effects they impart on the plant; however, they may also mediate a number of secondary responses, some of which occur after harvest.[7]

The effects of biologically active chemicals can be separated into two categories: the beneficial effects for which application was intended and the unintentional effects mediated. These unintentional effects may be positive and/or negative to the plant or plant part in question. The use of biologically active chemicals, therefore, may result in a stress to the plant (i.e., altering normal metabolism); likewise, they may also prevent or retard the effect of other stresses.

Inadvertent secondary effects of agricultural chemicals on plant processes have been known for many years. Apple trees treated with lime-sulfur fungicidal sprays were shown to undergo a marked reduction in the rate of photosynthesis after application (i.e., 50%) as early as the 1930s.[100,103] Other indirect effects include changes in respiratory rate,[213] fruit shape and color, and physiological disorders.

The time of application of chemicals that may subsequently alter the postharvest biology of a product ranges from very early in the development of the product to just prior to storage. For example, sprays applied to induce parthenocarpy often result in fruits that ripen faster; have higher respiratory rates, lower acidity, less firmness, and less vitamin C; and have an increased susceptibility to postharvest physiological disorders many months later.[87,142,177] Chemicals used to thin fruits early in the season can result in increased fruit color, soluble solids, and titratable acidity.[147] Likewise, postharvest but prestorage fumigation with ethylene dibromide has been shown to decrease the color development of the outer pericarp of tomato fruits upon eventual ripening,[202] increase the respiratory rate of some deciduous fruits,[42] and alter the normal ripening pattern of bananas.[75]

One of the most widely studied chemicals applied preharvest (generally to prevent fruit drop) that results in both positive and detrimental indirect effects during storage is the growth retardant daminozide. In some fruits (especially apple) it significantly retards ripening whereas in stone fruits ripening is accelerated.[76,153,218] Other effects include changes in texture, color, and the susceptibility to certain storage disorders. Recent questions about the safety of secondary effects of daminozide residues to consumers of treated products have led to a marked decline in its use.

The extent of plastic strain mediated by these compounds ranges widely, varying with the chemical in question, its concentration, the plant species, timing of application, and other factors. Generally, the undesirable effects do not result in death of the product or render it totally unacceptable for sale. If this were the case, the chemical's positive benefits would be sufficiently overshadowed by the detrimental effects. More often, undesirable effects are seen as some degree of quality loss, for example, increased susceptibility to a physiological disorder[17] or more rapid softening.[11]

The cause of the strain mediated by chemical stress from biologically active compounds is not well understood. For many compounds, the response is not seen until quite some time after exposure and therefore does not appear to be a direct effect. The response may be via an effect on cell division or assimilate partitioning, or through a more direct means, for example, a metabolic imbalance leading to the formation of lesions.

MECHANICAL STRESS

During growth, plants are subjected to mechanical stresses from a variety of sources. Wind, rain, hail, herbivores, and soil compaction are but a few of the many sources of mechanical stress by which the growth and development of plants or plant parts may be altered. In addition to these forms of mechanical stress, the effect of mechanical perturbation of plants is also important and has been known for many years. Charles Darwin, in a series of experiments first published in 1881,[54] showed that very gentle tactile stimulation of pea (*Pisum sativum,* L.) roots caused a change in both the rate and direction of growth. Subsequently, the effects of various forms of preharvest and postharvest mechanical stress have been studied by a number of scientists. For example, some fruits[223,250] and nuts[233] split while on the tree when subjected to certain conditions.

Mechanical stress is a dominant factor throughout the postharvest handling, transport, and storage phases of agriculture. During this period, mechanical stresses mediate a wide range of physical injuries to harvested products that decrease their value, increase their susceptibility to disease and water loss, and often significantly shorten their potential longevity. Losses due to postharvest mechanical stress can be considerable. For example, retail and consumer losses alone due to mechanical stress for a cross-section of fruits and vegetables sampled in the New York marketing area ranged from 0.2% to over 8% (table 6-4).[31–36]

Mechanical stresses can be separated into two general classes based on the injury produced. The first, mechanical perturbation, originates from a wide range of sources and is characterized by the absence of a direct physical

Table 6-4. Postharvest Losses of Selected Fruits and Vegetables Sampled at the Retail and Consumer Level in the New York Marketing Area

	Cause of Loss		
	Mechanical Injury (%)	Parasitic Disease (%)	Nonparasitic Disorder (%)
Apples, 'Red Delicious'*	1.8	0.5	1.3
Cucumbers	1.2	3.3	3.4
Grapes			
'Emperor'*	4.2	0.4	0.9
'Thompson'*	8.3	0.6	1.6
Lettuce, 'Iceberg'	5.8	2.7	3.2
Oranges			
'Navel'	0.8	3.1	0.3
'Valencia'	0.2	2.6	0.4
Peaches	6.4	6.2	—
Pears			
'Bartlett'	2.1	3.1	0.7
'Bosc'	4.1	3.8	2.2
'd'Anjou'	1.6	1.7	0.8
Peppers, Bell	2.2	4.0	4.4
Potatoes			
'Katahdin'†	2.5	1.4	1.0
'White Rose'†	1.5	2.4	0.4
Strawberries	7.7	15.2	—
Sweetpotatoes	1.7	9.2	4.2

Source: From Ceponis and Butterfield.[31–36]
* Retail losses only.
† Wholesale and consumer losses only.

wounding of the tissue. Mechanical perturbation does, however, result in a decrease in the rate of growth and alters the developmental pattern of many species (termed *thigmomorphogenesis*).[117] The second type of mechanical stress results in a direct physical wounding to the tissue, for example, cuts, bruises, and abrasions.

Generally, some degree of mechanical perturbation accompanies all wounding responses. In addition, the same source of the mechanical stress can result in perturbment at one level and wounding at another, and in some cases, a force causing perturbment may facilitate mechanical wounding, for example, wind-blown sand.[197] Mechanical stress may also arise as a secondary stress induced by a primary stress such as freezing.

Mechanical Perturbment

Mechanical perturbment of plants, whether from stroking,[102] bending,[206] flexing, shaking, rubbing,[176] or other means, exerts a pronounced influence on growth and development (fig. 6-21). Some plants are extremely sensitive to mechanical perturbment, with very light touch (e.g., with a camel hair brush) being sufficient to instigate the response. Plant response to mechanical perturbment varies with species, cultivar, stage of development (young plants being more sensitive), and the type and magnitude of the stress imposed. Shaking of young *Liquidambar styraciflua*, L. trees for 30 seconds each day resulted in a 70–80% reduction in height in comparison with unshaken plants.[185] Likewise, gentle rubbing of the nodes of *Bryonia dioica*, L. three times a day resulted in a 58% reduction in growth. Mechanical perturbment or stimulation has evolved in a few species as part of the function of certain organs (e.g., tendrils) or the trapping movements of certain insectivorous plants (e.g., *Dionaea muscipula*, Ellis.).[222]

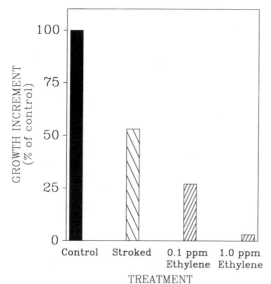

Figure 6-21. The effect of mechanical perturbment on the vertical growth of lily (*Lilium longiflorum*, Thunb.) plants. The leaf blades were lightly stroked five times a day during the course of the experiment (22 days) with a dusting brush. Similar plants were exposed to low levels of ethylene, which also significantly inhibited stem elongation. (*After Hiraki and Ota.*[101])

Mechanical perturbment does not result in physical injury to the tissue; rather it instigates significant changes in the development pattern of the plant or plant part. In addition to an inhibition of elongation and an increase in stem radial diameter, mechanical perturbment has been shown to decrease the number of nodes and leaves, increase greening of new leaves, increase lateral branch development,[176] inhibit flowering (*Mimosa pudica,* L.),[117] and induce epinasty.[176,209]

The mechanism by which mechanical perturbment results in these effects is not yet clear. Reception of the stimulus is quite rapid, while recovery from it is gradual, often requiring several days before normal growth resumes. Perturbment is known to increase the level of ethylene synthesis that has been suggested as the cause for growth inhibition.[108,206] There are also, however, significant changes in the synthesis of gibberellic acid.[241] Both appear to represent sequential responses occurring well after the initial sensory reception and action. It is thought that the early stages of the response to perturbment involve a rapid change in membrane permeability and bioelectrical potential.

An example of a postharvest mechanical perturbment that alters the value of a crop is the effect of sleeving on potted poinsettia plants. To prevent leaf and bract breakage during handling and transport, open-ended plastic sleeves are placed around each individual plant. This procedure, however, often slightly bends the petioles of the leaves and results in a mechanical stress that is sufficient to increase the level of ethylene synthesis by the plant, which in turn causes leaf and bract epinasty.[209] Foliar sprays of an inhibitor of ethylene synthesis or action 24 hours prior to sleeving have been shown to significantly reduce the development of epinasty after the sleeves are removed.[210]

Physical Wounding

Mechanical stresses that cause physical injury (mechanical failure) represent one of the most serious sources of quality loss during the postharvest period. Wounding causes increases in respiration and ethylene production and provides entry sites for decay organisms. Although there are also many preharvest sources of mechanical injury to plant tissue (e.g., hail, growth cracks, insect feeding), of particular interest from a postharvest context are those that occur from harvest until final utilization. Injury may occur at any point during this period. For example, skin punctures of apple fruits increased progressively as the product moved from the orchard (26%), to the packing house (30%), to the retail store (36%), and to the display bin (50%).[37] In some products (e.g., grapes, lettuce, apples, and potatoes), mechanical injury has been shown to be the primary cause of postharvest losses.[31,32] The actual loss may be physical (loss of part of the product), physiological (loss in weight from increased water and respiratory losses), pathological (due to facilitated entry of microorganisms), or qualitative.

Three of the most important types of mechanical stress are friction, impact, and compression. Friction, which results from movement of the product against an adjacent object, can result in surface abrasion. This type of damage occurs on grading belts and during transit where irregularities in the roadbed or wheels of the transport vehicle result in vibrations that are transmitted to the product. Friction-induced injury in fleshy products (e.g., pears)[230,231] is generally limited to the epidermis and a few underlying layers of cells. The surface tissue darkens quickly due to the enzymatic oxidation of the constituents of the injured cells (fig. 6-22 A). It also provides sites for entry of fungi and other microorganisms. In nonfleshy products, such as seeds and nuts,[66] friction results in an abrasive effect on the surface. Generally it does not result in sufficient damage to decrease product quality.

Impact stress occurs when the product is dropped a sufficient distance to cause injury. In very soft fruits, it may be only a few inches, with the injury generally seen as bruising. With bruising the injury is restricted to the interior flesh of the tissue and in many products may be only initially detected as a water-soaked area after peeling. With time, the exposure of the internal contents of the damaged cells to air in the intercellular spaces results in the typical browning symptoms (fig. 6-23). The damaged tissue may eventually become desiccated.[162]

Impact may also cause cleavage failure where the tissue is separated into two pieces (fig. 6-22B) or slip failure where the cells rupture or separate along defined surfaces within the tissue. With slip failure, the tissue along either side of the fracture remains relatively undamaged. Cracking or splitting are common in potato tubers, watermelons, cabbages, tomatoes, and other products.[178,179]

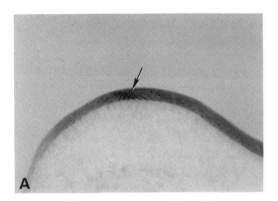

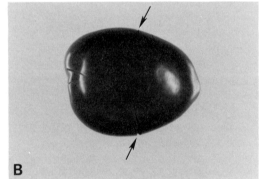

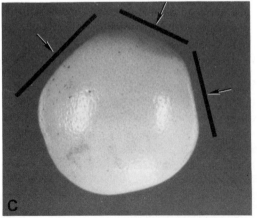

Figure 6-22. The effect of different types of mechanical stress on harvested products. *A:* Friction damage to the surface of a pear (*Pyrus communis,* L.) fruit. In contrast to bruising, the injury incurred with surface abrasions due to friction is normally limited to the surface cells. *B:* Impact stress is illustrated by surface splitting of tomato (*Lycopersicun esculentum,* Mill.) fruit. *C:* Compression stress commonly occurs when products are stacked excessively high. The flat deformed sides of grapefruit (*Citrus × paradisi,* Macfady) are typical of this type of mechanical injury.

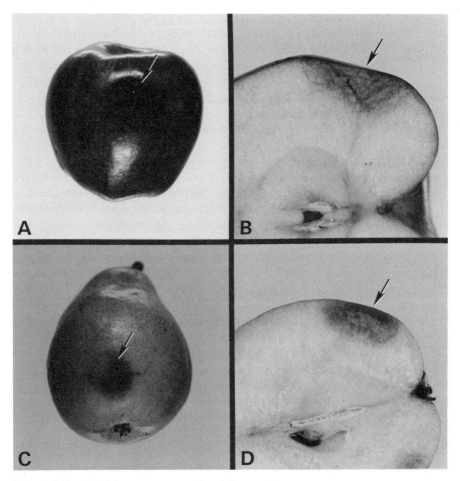

Figure 6-23. Bruising may be readily visible at the surface, as in light-colored pear (*Pyrus communis*, L.) fruit *(C)*, or virtually invisible, as in intensely pigmented 'Red Delicious' apple (*Malus sylvestris*, Mill.) fruit *(A)*. In both cases, the damage to cells below the surface is extensive (*B* and *D*).

Compression-induced injury is also frequently encountered during handling, transport, and storage. Stacking a product at a depth too great is perhaps the most common cause of compression (fig. 6-22C). The injury incurred may be splitting, as in tomatoes where the fruit diameter is expanded to a point of failure; internal shear, as described in potatoes[220]; or a permanent deformation. In one study,[93] 33–60% of the grapefruit shipped to Japan arrived seriously deformed. The most extensive injury was found in fruit located at the base of the load.

The tissues of edible plant products have three primary components that determine product mechanical behavior.[192] These are the parenchyma cells per se, intercellular bonding between neighboring cells, and the intercellular space between the cells. Turgor pressure and cell wall strength control the mechanical properties of parenchyma cells. Turgor pressure within the cell is modulated by the bidirectional movement of water through the plasma membrane. Water movement allows the intercellular and internal water potentials to equilibrate. Restraining the outward stretch of the protoplasm is the cell wall, structurally the most rigid single component of the cell. In parenchyma cells there is only a thin primary wall; no secondary cell wall is formed as in the sclerenchyma cells of more rigid woody products. Stiffness of the primary cell wall is imparted by the cellulose microfibrils; however, since the wall is quite thin, it will bend fairly easily.

To better understand how a mechanical stress damages plant tissues, it is helpful to examine what happens to the cell when a load is applied. Initially the cell begins to change in shape, decreasing in diameter in the direction of the force. Since the contents of the cells do not compress, or at least compress very little, the change in shape alters the surface-to-volume ratio of the cell and increases its turgor pressure. To equilibrate the internal and external water potentials, water must now move out of the cell to compensate for the increased turgor pressure. If the magnitude of the stress is sufficiently low and for a short duration, the deformation will be largely elastic, with the cell recovering most, but not all, of its original shape. Longer periods of compression result in a greater net efflux of water and less recovery. If the magnitude of the stress is great enough, the strength of the wall is exceeded and the cell ruptures.

Intercellular bonding adds another dimension to the mechanical properties of tissue. Neighboring cells are bonded together by a pectin layer called the middle lamella. Physically the middle lamella is plastic in nature, with the pectin bonds allowing the cells to change position slowly during compression. If the shear stress in the middle lamella exceeds its strength, then the cells will separate without necessarily rupturing. When debonding occurs, the separation plane usually proceeds at a 45° angle to the direction of the force applied.

Finally, the intercellular space is also an important component of the tissues' rheological and strength properties. Part of the area between neighboring cells is not middle lamella but represents air space and intercellular fluid. When the intercellular space is sufficiently large, as in peach fruits, it provides room for the cells to reorient when compressed; thus, the tissue volume can change significantly. In very dense products (i.e., potato tubers or sweetpotato storage roots), the intercellular space is low and there is little compression of the tissue.

When a plant tissue is subjected to a sufficient friction, impact, or compression stress it undergoes mechanical failure. Tissue failure may occur by one of four possible modes: cleavage, slip, bruising, or buckling (fig. 6-24A–D).[104] Cleavage occurs when failure is along the line of maximum shear stress, leaving two essentially intact pieces (fig. 6-24A).[62] When a compression force results in failure along planes at approximately a 45° angle to the direction of loading, the two sections of tissue slip or slide relative to each other (fig. 6-24B). This type of failure has been shown in apples[173] and potatoes.[61] However, the most common failure mode in harvested products is bruising (fig. 6-24C). It occurs when the cells rupture, exposing their internal contents to the air and undergoing enzymatic oxidations that cause discoloration. Buckling failure is illustrated by petiole failure in African violet (fig. 6-24D).

Plants and their component parts vary widely in their susceptibility to mechanical damage. Several important factors alter the degree of injury sustained from a given stress, for example, cultivar, degree of cell hydration,

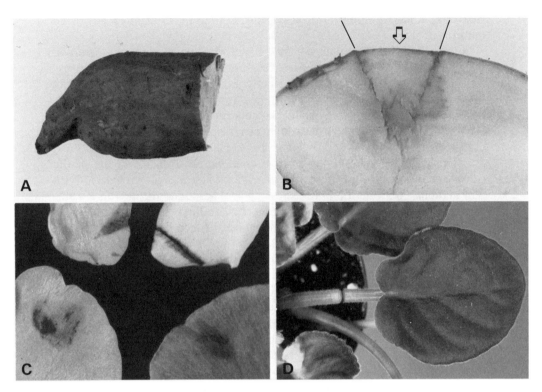

Figure 6-24. Tissue failure occurs via four mechanisms: cleavage *(A)*, illustrated by a sweetpotato *(Ipomoea batatas*, (L.) Lam.) root; slip *(B)*, where the impact force causes the neighboring cells of the potato *(Solanum tuberosum*, L.) tuber to break apart at a 45° angle; bruising *(C)*, illustrated by petal bruising (clockwise from top left: *Impatiens wallerana*, Hook. f.; *Gardenia jasminoides*, Ellis; *Mandevillia laxa*, Woodson; *Hedychium coronarium*, Koenig.); and buckling *(D)*, illustrated by the mechanical failure of an african violet *(Saintpaulia ionantha*, Wendl.) petiole.

stage of maturity, product weight and size, skin characteristics, and environmental conditions (e.g., temperature). Differences in the susceptibility to mechanical damage may be substantial even between individual cultivars. For example, in a damage survey conducted by the Potato Marketing Board in Great Britain, total flesh damage between potato cultivars ranged from 13.3% in the cultivar 'Golden Wonder' to 36.7% in 'Kerrs Pink'.[4] Potato firmness markedly influences the degree of damage within a cultivar. When the tubers are very turgid, the incidence of splitting and cracking increases during harvesting and handling. Conversely, when flaccid they are more susceptible to black spot, a disorder caused by mechanically induced cell rupture and subsequent discoloration due to the oxidation of the amino acid tyrosine. In dried seeds, such as beans,[10] cracking decreases with increasing seed moisture content. Pecans and several other nuts are either soaked in water or steam conditioned prior to shelling to minimize kernel breakage.

Stage of maturity is likewise an important parameter in a number of products. Many fruits undergo distinct textural changes as they ripen, with softening of the flesh greatly increasing their potential for mechanical damage. Thus, many fruits that are to be harvested for commercial sale are gathered before they reach complete ripeness. Peaches harvested at the green ripe stage undergo significantly less bruising than those of more advanced maturity.[110]

Susceptibility to mechanical damage increases with increasing size and weight of the product. In some instances, it may be related partly to maturity differences (e.g., fruits); more importantly, however, is the increased impact force of the individual units when dropped during harvesting and handling.

Skin strength and other mechanical properties are important factors governing differences in damage between types of products and between individual cultivars of a species. For example, tomato skin strength is a critical factor in the resistance of the fruit to cracking.[252] Both the shape of the surface cells and the deposition of cutin appear to affect skin strength.[96] With potato tubers, both periderm thickness and set modulate the degree of abrasion damage sustained during harvesting and handling.[263]

Environmental conditions such as temperature, relative humidity, and oxygen concentration can also significantly alter the degree of damage sustained from a given level of mechanical stress. It may be due to a direct effect on the mechanical properties of the tissue and/or the development of injury symptoms after wounding. Potato tubers are more prone to bruising when cold than when warm.[188] Damage can be reduced, therefore, by starting harvest later in the day when night temperatures have been cool and by warming tubers removed from storage before handling.

The wounding of plant tissue sets into motion a dramatic series of changes that culminate in either rapid deterioration and death of the product or healing of the damaged surfaces. The sequence of events occurring with wound healing in part depends on the nature of the damaged tissue. Healing may involve complete replacement of the damaged tissue (e.g., as in apical meristems), repair through the induction of a secondary meristem (e.g., wound periderm formation in tubers), or changes in the physical and chemical composition of the tissue near the damaged cells.[18,19,123,149] Younger, more meristematic tissues tend to heal wounds more successfully than older tissues.[86] Apple fruit attached to the tree healed mechanically induced wounds completely until mid-July; thereafter wound-healing ability declined rapidly.[223] Harvested apples held in storage displayed virtually no wound-healing ability.

The chronological sequence of anatomical changes occurring in response to wounding includes desiccation of several of the outer layers of parenchyma cells adjacent to the wound; suberization[140] and in some cases lignification[256] of the cells below the desiccated cells; stimulation of cell division in the cells below this level to form a cambium (phellogen); and with continued cell division, the formation of cork (phellem). In some tissue, there may be no phellogen formation, just suberization of the cells adjacent to the wound. In response to mechanical wounding caused by leaf miners, species with thin leaves tended to form callus while species with fleshy or evergreen leaves formed periderm.[267]

Changes in anatomy are mediated through a series of pronounced biochemical alterations in the cell adjacent to the wound. There is an increase in respiration and nucleic acid, RNA, protein and ethylene synthesis; transformation of carbohydrates; and activation of certain enzymes and repression of others.[43,113,122,170] In tomato fruits, the activity of hydroxycinnamate-CoA ligase and O-methyltransferase increases sharply and appears to be operative in the formation of monomeric units that form lignins through the activity of peroxidases.[81]

In addition to age, species, and even cultivar,[85] a number of other factors can influence the response of plants to wounding. Variation due to temperature, relative humidity, light, and oxygen concentration strongly modulate both qualitative and quantitative responses.[149] High relative humidity is conducive to wound healing in many tissues (e.g., 85–95% RH is recommended for sweetpotatoes); however, excess moisture can impede the process. Likewise, moderate to warm temperatures generally facilitate wound healing. In Kalanchoe, an inverse linear relationship was found between temperature (21–36°C) and wound healing.[140]

GRAVITATIONAL STRESS

Plants and many of their parts exhibit a precise gravitationally controlled orientation. Subterranean parts such as primary roots display a positive geotropism, that is, they grow in the direction of gravity. Stems and flower stalks on the other hand are negatively geotropic and develop at a 180° orientation from the pull of gravity. Secondary roots may grow at intermediate angles, for example, 45° (termed *plagiotropism*), and many petioles, stems, rhizomes, roots, and stolons grow more or less horizontally.

When a gravitationally sensitive plant or plant part is altered from the gravitational orientation in which it has grown, it responds by orienting new growth in the direction required to reestablish its original plane. Therefore, a snapdragon raceme laid on its side after harvest (90° to the orientation of gravity) begins to grow in a vertical direction (fig. 6-25A). Potted narcissus also undergo geotropic curvature when their orientation to the gravitational field is altered (fig. 6-25B). Unlike snapdragon, the curvature is a turgor-driven response that is normally reversible upon returning the plant to the proper orientation. This change in orientation results in stress that can be seen as a small decline in the overall growth rate. Much more important, from a postharvest standpoint, is the effect of the plant's new orientation on product quality. Bent stems of gladiolus, asparagus, snapdragon, or narcissus are considered inferior in quality, and as a consequence, care must be taken to keep geotropically sensitive products upright during transit and storage.

The product's response to gravity can be separated into three distinct components: perception of the gravitational stimulus, transduction of the stimulus to the zone of response, and mediation of the growth response.[253,264] Perception is believed to occur in specialized cells called statocytes, found in the apical tip of roots, shoots, and flower stalks. These cells contain geotropically sensitive masses called statoliths that move in response to a change in orientation to the bottom of the cell. Starch-containing amyloplasts are the only subcellular particles of sufficient mass and density that settle in the cytoplasm at a rapid enough rate to account for geoperception, and indeed their movement has been correlated with the plant's perception of an alteration in its gravitational orientation.

The response zone in higher plants is often several cell layers removed from where gravitational perception takes place. The tip of the coleoptiles detect gravity while the cells below respond in an upward growth. If the tip is

Figure 6-25. The effect of gravitational stress on harvested products. *A:* A cut snapdragon (*Antirrhinum majus,* L.) raceme. *B:* A potted narcissus plant. The examples illustrate two different groups of geotropic response. The former *(A)* represents an irreversible curvature due to growth, while the latter *(B)* is a turgor-driven response that will normally reorient in a vertical position once the pot is placed upright.

removed the shoot loses its geoperception even though it continues to elongate. For many years the signal was thought to be transmitted through a differential movement of auxin (Cholodny-Went hypothesis). Auxin accumulated on the lower side of the organ, promoting the growth of those cells in stems while inhibiting them in roots. More recent experimental evidence suggests that auxin does not elicit the response, but more likely a growth inhibitor accumulates on the basal side, at least of roots.[63,120]

The response of the stem or root involves a differential in growth over the cross-section of the organ. Curvature is caused by a growth inhibition; for roots the inhibition is only slight on the upper side but pronounced on the lower. It continues until the organ reestablishes its original gravitational orientation.

The occurrence of geotropic-mediated quality losses is relatively infrequent. While many containerized ornamentals will display curvature if left in a nonvertical position for sufficient time, their response rate is generally slow enough to preclude significant damage during normal handling, transport, and storage periods. Several crops (e.g., seedling cress, asparagus spears, snapdragon racemes, potted flowering bulbs, and gladiolus spikes) do, however, display a high level of sensitivity and relatively short response time. Curvature due to improper stacking of boxes during transit and/or storage can result in significant losses in product quality.

HERBIVORY STRESS

The feeding of an extremely diverse array of herbivores results in extensive damage to many plants and in some cases even death. These range from invertebrate (e.g., protozoa to mites and insects) to vertebrate herbivores (e.g., fish to mammals).[52] In nature, essentially all anatomical parts of plants are subject to consumption by various herbivores (table 6-5). Stems, leaves, roots, buds, flowers, seeds, and other plant parts are consumed or damaged by the various modes of feeding of different herbivores. Significant postharvest losses are incurred, especially due to the action of the Insecta, Arachnida, and Rodentia groups. Obtaining an accurate estimate of worldwide postharvest losses caused by herbivores is difficult; however, losses are known to be extensive. For example, maize storage losses in Honduras (1948–1949) were estimated at 50%.[247] Often the value of the losses sustained via storage herbivores is several orders of magnitude greater than the costs of appropriate control measures.

The feeding of herbivores results in a distinct stress on both preharvest and postharvest plant products. While a substantial volume of research has focused on the herbivore, the short-term and long-term trauma effects of herbivory on the plant have been little studied.[189] Due to the diversity of herbivores consuming plant parts, several potential stresses may be operative. These include mechanical stress, the introduction of biologically toxic or active chemicals (chemical stress), and the development of secondary stresses (e.g., pathogen,[28] water stress).

The action of herbivores can be separated into three general classes based on their effect on the plant or plant part. These include those where the action of the herbivore causes (1) only a small effect on the general growth of the plant or plant part, (2) significant metabolic changes in the plant part attacked, and (3) both metabolic and developmental changes. While these classes are not mutually exclusive, for example, even minor damage results in some localized metabolic changes, they do illustrate gross differences in levels of effect.

The mechanical removal of leaf tissue by phytophagous insects results in short-term effects on growth roughly proportional to the volume of the tissue

Table 6-5. Plant Tissues and the Herbivores That Feed on Them

	Mode of Feeding	*Examples of Feeders*
Leaves	Clipping	Ungulates, slugs, sawflies, butterflies, etc.
	Skeletonizing	Beetles, sawflies, capsid bugs
	Holing	Moths, weevils, pigeons, slugs, etc.
	Rolling	Microlepidoptera, aphids
	Spinning	Lepidoptera, sawflies
	Mining	Microlepidoptera, Diptera
	Rasping	Slugs, snails
	Sucking	Aphids, psyllids, hoppers, whitefly, mites, etc.
Buds	Removal	Finches, browsing ungulates
	Boring	Hymenoptera, Lepidoptera, Diptera
	Deforming	Aphids, moths
Herbaceous stems	Removal	Ungulates, sawflies, etc.
	Boring	Weevils, flies, moths
	Sucking	Aphids, scales, cochineals, bugs
Bark	Tunneling	Beetles, wasps
	Stripping	Squirrels, deer, goats, voles
	Sucking	Scales, bark lice
Wood	Felling	Beavers, large ungulates
	Tunneling	Beetles, wasps
	Chewing	Termites
Flowers	Nectar drinking	Bats, hummingbirds, butterflies, etc.
	Pollen eating	Bees, butterflies, mice
	Receptacle eating	Diptera, Microlepidoptera, thrips
	Spinning	Microlepidoptera
Fruits	Beneficial	Monkey, thrushes, ungulates, elephants
	Destructive	Wasps, moths, rodents, finches, flies, etc.
Seeds	Predation	Deer, squirrels, mice, finches, pigeons
	Boring	Weevils, moths, bruchids
	Sucking	Lygaeid bugs
Sap	Phloem	Aphids, whitefly, hoppers
	Xylem	Spittlebugs, cicadas
	Cell contents	Bugs, hoppers, mites, tardigrades, etc.
Roots	Clipping	Beetles, flies, rodents, ungulates, etc.
	Tunneling	Nematodes, flies
	Sucking	Aphids, cicadas, nematodes, etc.
Galls	Leaves	Hymenoptera, Diptera, aphids, mites
	Fruits	Hymenoptera
	Stems	Hymenoptera, Diptera
	Roots	Aphids, weevils, Hymenoptera

Note: Data from Crawley.[52]

removed.[143] Leaf area loss decreases the potential photosynthetic area of the plant, diminishing the total volume of carbon fixed. If the insect population is small relative to the size of the plant, dry matter losses may represent only a very small portion of the total carbon turnover by the plant. However, when the insect population is large, losses may be extensive, significantly decreasing subsequent growth.[249] During storage, the loss of foliage decreases the quality of the product, and with intact plants or plant propagules, can significantly reduce subsequent growth.

The action of some herbivores also instigates significant metabolic changes within the affected tissue. Aphid feeding has been shown to increase the respiratory rate of leaf tissue while decreasing photosynthetic activity.[138] Leaf miners (e.g., *Stigmella argentipedella,* A.) in birch (*Betula pendula,* Roth) leaves increase the cytokinin and protein concentrations in the tissue surrounding their tunnels (i.e., up to 20 times the cytokinin concentration of normal leaves). Thus, the affected tissue remains greener much longer in the fall.[72] When larvae of the sweetpotato weevil (*Cylas* spp.) tunnel through storage roots (preharvest and postharvest), they release a terpene-inducing factor that triggers the synthesis of the toxic furanoterpenoid, ipomoeamerone, by the roots. Although toxic to animals,[20] it does not appear to affect the insect. The precise biological function of the chemical is not yet known. The terpene-inducing factor consists of a glycoprotein, a protein, and a heat-stable low molecular weight compound.[214,215]

Other herbivores are known to cause both metabolic and developmental changes in the plant part attacked. Perhaps one of the best examples is the effect of gall-forming insects. Various gall wasps deposit their eggs in buds, stems, leaves, or roots of the host species. The ensuing larvae alter the metabolism and development of the tissue, producing a malformed growth. These changes appear to be due to chemicals released by the larvae. In the Oriental chestnut gall wasp (*Dryocosmus kuriphilus,* Yasumatsu), the larvae instigate a preferential movement of carbohydrates and other requisites to the infested bud in lieu of the normal buds on the stem. The resulting distorted growth prevents normal leaf and stem formation, decreasing the photosynthetic area of the plant. When infestations are sufficiently high, tree death often occurs within several seasons.

Losses arising from the action of herbivores on postharvest products can be viewed at several levels. Feeding decreases the net volume of the stored product. For example, under appropriate conditions storage weevils (e.g., *Oryzaephilus surinamenis,* L.) can cause tremendous losses. Herbivore activity also has a major effect on product quality. The external and/or internal appearance of the product may be impaired due to tissue damage (e.g., tunneling insects in fleshy fruits and vegetables) (fig. 6-26). In addition to the physical damage to the product, the presence of excreta, dead insects, and webbing decreases the quality and cleanliness of the product. Quality may be impaired by the mere presence of the insect prior to actual damage to the product. The action of herbivores may also decrease the nutrient value of the product. The production of toxic furanoterpenoids by sweetpotato storage roots in response to the activity of sweetpotato weevil larvae minimizes the sweetpotato's desirability as a food source. Nutrient losses in plant propagules likewise can reduce vigor and subsequent growth.

PATHOLOGICAL STRESS

Harvested plants and plant parts undergo substantial losses due to the stress imposed by invading organisms (fig. 6-27). A wide range of microorganisms (fungi, bacteria, viruses, mycoplasms, and nematodes) is known to affect plant products. In some cases, as in quiescent infections, inoculation occurs prior to harvest with the subsequent development of the disease not occurring until during the postharvest period.

Infection of plants by microorganisms is a complex process. The interaction between pathogen and host can, however, be separated into two general processes: the infection sequence of the organism and the type and degree of stress imposed on the host by the pathogen. While the former is of critical importance in the overall pathogenicity of the organisms, it is the stresses imposed on the host by the invading organisms that result in product losses.

Plants and plant parts exhibit a wide range of structural and biochemical barriers to infection. These barriers may be preexisting or formed in response to invasion by the pathogen. Preexisting structural defenses include substances such as waxes, the cuticle, and other compounds found in or on the epidermal cell walls. In response to infection, the tissue may also form cork layers, deposit gums or gels, form abscission zones, or limit the infection through the formation of necrotic areas. Likewise, plants may exhibit a number of preexisting biochemical inhibitors to infection. For example, the level of phenolics has been correlated with susceptibility of certain plants to invasion and colonization by some pathogens. Onion cultivars resistant to *Colletotrichum circinans,* (Berk) Vogl. accumulate phenolics in the outer scales of the bulb that inhibit germination and penetration of the fungi.[255] Plants may also respond to cellular infection by accumulating antimicrobial stress metabolites called phytoalexins.[239] These compounds are usually of a low molecular weight and are general rather than specific in their activity. Other induced biochemical defenses include mechanisms for the detoxification of pathogen toxins, altered respiration and metabolic pathways, and the induction of hypersensitive reactions.[2,14]

While the invasion sequence is of critical importance in the control of many postharvest pathogens, the focus here is on the stress imposed by the organisms that results in the deterioration of the product. Development of the pathogen within the tissue mediates marked metabolic changes in the host plant. When contrasted with what is known about induced metabolic changes in the host due to herbivory, the effect of pathogens has received considerable attention. It is due in part to the low mobility of plant pathogens after invasion relative to that of most herbivores. As a plant is altered by the action of a herbivore and becomes a less desirable food source, most herbivores can simply move to adjacent plants or a different part of the same plant. Mobility of an individual pathogen is rare, and, as a consequence, there has evolved a very close interrelationship between the host and the disease organism.

When plants are infected by either obligate or facultative pathogens, the rate of respiration generally increases. This increase in respiration begins shortly after invasion and further increases as the organism multiplies.[175] The elevated respiration represents a very general reaction in that it occurs with the invasion of a wide range of microorganisms and in both susceptible and resistant cultivars upon entry of the pathogen. Susceptible and resistant lines differ, however, in that respiratory increases with the entry of the pathogen

Figure 6-26 *(at left).* Herbivory stress on harvested products. *A:* Cigarette beetle (*Lasioderma serricorne*) damage to stored soybean (*Glycine max,* (L.) Merrill.). *B:* Sweetpotato weevil (*Cylas formicarius elegantulus,* Summers) damage to stored sweetpotato (*Ipomoea batatas,* (L.) Lam.) roots. *C:* Preharvest damage to sweetpotato roots caused by wireworms (*Conoderus falli,* Lane). (*Photograph A courtesy of J. W. Todd.*)

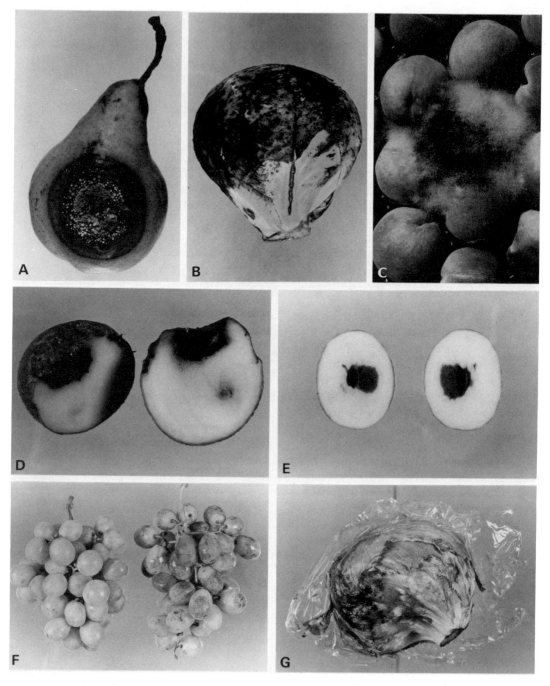

Figure 6-27. Illustrated are a cross-section of postharvest diseases. *A:* Blue mold rot on pear, *Penicillium expansium,* Link; *B:* black rot on brussels sprouts, *Xanthomonas campestris,* (Pam.) Dows.; *C:* Rhizopus rot on peach, caused by both *Rhizopus oryzae,* Went & Prinsen Geerligs and *R. stolonifer,* (Ehreub. ex Fr.) Lind.; *D:* gangrene on potato, *Phoma exigua,* Desm. var. *foveata,* (Foister) Boerema; *E:* black heart of potato, a physiological disorder thought to be caused by low oxygen; *F:* gray mold rot on grapes, *Botryotinia fucheliana,* (de Bary) Whetzel; and *G:* bacterial soft rot on head lettuce, caused by both *Erwinia carotovora,* (Jones) Holland and *Pseudomonas marginalis,* (Brown) Stevens. *(Photographs courtesy of A. Snowdon[228]).*

into a resistant line are generally more rapid and pronounced than in comparable susceptible lines.[2] Therefore, it is thought that much of the increase in respiration is due to an increased energy demand for the synthesis of compounds needed to combat the infection.

Some plant pathogens reduce the rate of photosynthesis in intact plants through (1) the destruction of leaf tissue, (2) the induction of processes leading to chlorosis, or (3) alteration of the actual rate of the photosynthetic process.[154,269] Chlorosis may develop through an effect on either chlorophyll synthesis or degradation or due to the degeneration of the chloroplasts per se. Most virus infections that reduce photosynthesis are operative in actively growing tissues and thus appear to alter chlorophyll synthesis. The importance of invasion of pathogens during the postharvest period that reduces photosynthesis is normally minimal since photosynthesis is generally either much reduced or completely inhibited due to the low level or absence of light. Infection prior to harvest may, however, increase the level of leaf shedding of infected leaves during storage.

The action of some plant pathogens results in blockage of the translocation system within the plant. Blockage may involve an inhibition of the acropetal movement of water and inorganic nutrients from the root system or interference in the translocation of organic compounds basipetally in the phloem.[67] Xylem vessels may be blocked by bacterial bodies or water transport impaired by compounds produced by the microorganism (e.g., slime molds such as *Pseudomonas solanacearum*).[111] Stem blockage is often cited as the cause for decreased vase life of cut flowers and the water does, in fact, generally contain a broad cross-section of microorganisms. Of the 25 microorganisms found in the vase solution of carnation (*Dianthus caryophyllus,* L. cv. 'Improved White Sim'), only three reduced the vase life of the flowers.[272] Several of these microorganisms, when transferred to the vase solution of other flower species, also reduced vase life. It is not yet certain, however, whether the effect is on water transport through the stem or via some other mechanism.

Stresses caused by plant pathogens are ultimately due to the action of the microorganism on the protoplasm of the cells. As a consequence, the parasite must first penetrate the cell wall covering the host cell, which may represent both a physical and chemical barrier. To accomplish this, some pathogens secrete enzymes necessary to degrade the cell wall. Soft rots of stored fruits (e.g., *Penicillium expansum* and *Sclerotinia fructigena*) are a classic example. *S. fructigena* secretes polygalacturonase enzymes that degrade the pectic substances of the cell wall, allowing the organism to enter the protoplasm, thus leading to the death of the cell. For some pathogens, the synthesis of enzymes is an essential requisite for initial invasion from the surface of the tissue.

Plant pathogens are also known to alter the metabolism of nitrogenous compounds, phenols, and hormones within the cells of the infected plant.[1,56,83,89] Many of the effects of plant pathogens on cellular metabolism are in response to damage sustained. Some of these changes, however, represent apparent defense mechanisms of the plant. Studies of the nature, metabolism, and mode of action of these defense compounds represent a rapidly expanding area of research.[6]

One important aspect of the stress mediated by plant pathogens after harvest is that maturation and harvest often result in an increase in susceptibility of the tissue. This increase in susceptibility is illustrated by the resistance of immature fruits to colonization after initial infection by quiescent pathogens. Although present, the organism is not able to proceed with further development. As the fruit matures, this inhibition appears to be lost and colonization proceeds. Several possible reasons have been shown for this greater resistance of immature tissue to pathogens:

1. Young tissues may contain toxic compounds that are not present in older tissues.
2. As most products mature, their composition changes and may thus now meet the energy and nutritional requirements of the pathogen.
3. Young tissues may exhibit a greater propensity for the production of phytoalexins than mature tissues.
4. The enzyme potential of the fungus may not be sufficient for colonization of immature tissues.[242]

Likewise, as the length of time a product is in storage increases, its ability to resist invasion decreases. Carrot roots inoculated with *Botrytis cinerea* at varying intervals after harvest displayed an increase in pathogen development with greater intervals between harvest and inoculation.[88] These changes in resistance appear to be largely associated with changes in the effectiveness of the plant's defense mechanisms.

REFERENCES

1. Akai, S., and S. Ouchi (eds.). 1971. *Morphological and Biochemical Events in Plant-Parasite Interaction*. Phytopathology Society of Japan, Tokyo, 415p.
2. Allen, P. J. 1954. Physiological aspects of fungus diseases of plants. *Ann. Rev. Plant Physiol*. **5**:225–248.
3. Anderson, R. E., and R. W. Penney. 1975. Intermittent warming of peaches and nectarines stored in controlled atmosphere of air. *J. Am. Soc. Hort. Sci*. **100:** 151–153.
4. Anon. 1974. *National Damage Survey, 1973*. U.K. Potato Marketing Board, London.
5. Baerug, R. 1962. Influence of different rates and intensities of light on solanine content and cooking quality of potato tubers. *Eur. Potato J*. **5**:242–251.
6. Bailey, J. A., and B. J. Deverall. 1983. *The Dynamics of Host Defense*. Academic Press, New York, 233p.
7. Bangerth, F. 1983. Hormonal and chemical preharvest treatments which influence postharvest quality, maturity and storability of fruit. In *Post-Harvest Physiology and Crop Preservation*, M. Lieberman (ed.). Plenum Press, New York, pp. 331–354.
8. Banks, N. H. 1985. Coating and modified atmosphere effects on potato tuber greening. *J. Agric. Sci*. **105**:59–62.
9. Barker, J., and L. W. Mapson. 1955. Studies in the respiratory and carbohydrate metabolism of plant tissues. VII. Experimental studies with potato tubers of an inhibition of the respiration and of a "block" in the tricarboxylic acid cycle induced by "oxygen poisoning." *Roy Soc. (London) Proc*. **143B**:523–549.

10. Barriga, C. 1961. The effects of mechanical abuse of Navy bean seed at various moisture levels. *Agron. J.* **53:**250–251.

11. Baumgardner, R. A., G. E. Stembridge, L. O. Van Blaricom, and C. E. Gambrell, Jr. 1972. Effects of succinic acid-2,2-dimethylhydrazide on the color, firmness, and uniformity of processing peaches. *J. Am. Soc. Hort. Sci.* **97:**485–488.

12. Beaudry, R. M., J. A. Payne, and S. J. Kays, 1985. Variation in the respiration of harvested pecans due to genotype and kernel moisture level. *HortScience* **20:**752–754.

13. Becquerel, P. 1932. L'anhydrobiose des tubercules des Renoncules dans l'azote liquid. *Acad. Sci. Paris C.R.* **194:**1974–1976.

14. Bell, A. A. 1981. Biochemical mechanisms of disease resistance. *Ann. Rev. Plant Physiol.* **32:**21–81.

15. Bernstein, L. 1964. Effects of salinity on mineral composition and growth of plants. *Plant Anal. Fert. Probl. Colloq.* **4:**25–45.

16. Bjorkman, O., M. Badger, and P. Armond. 1980. Adaptation of plants to water and high temperature stress. In *Response and Adaptation of Photosynthesis to High Temperature,* N. C. Turner and P. J. Kramer (eds.). Wiley-Interscience, New York, pp. 233–249.

17. Blanpied, G. D. 1978. The soft McIntosh problem. *New York State Hort. Soc. Proc.* **123:**122–124.

18. Bloch, R. 1941. Wound healing in higher plants. *Bot. Rev.* **7:**110–146.

19. Bloch, R. 1952. Wound healing in higher plants. II. *Bot. Rev.* **18:**655–679.

20. Boyd, M. R., L. T. Burka, T. M. Harris, and B. J. Wilson. 1973. Lung toxic furanoterpenoids produced by sweet potatoes. *Biochem. Biophys. Acta* **337:**184–195.

21. Brock, T. D., and G. K. Darland. 1970. Limits of microbial existence: Temperature and pH. *Science* **169:**1,316–1,318.

22. Brodrick, H. T., and R. Thord-Gray. 1982. Irradiation in perspective—the significance for the mango industry. *S. Afr. Mango Grow. Assoc. Res. Rep.* **2:**23–26.

23. Brooks, C., E. V. Miller, C. O. Bratley, J. S. Cooley, P. V. Mook, and H. B. Johnson. 1932. Effect of solid and gaseous carbon dioxide upon transit diseases of certain fruits and vegetables. *USDA Tech. Bull. 318*, 59p.

24. Brown, W. 1922. On the germination and growth of fungi at various temperatures and in various concentrations of oxygen and of carbon dioxide. *Ann. Bot.* **36:**257–283.

25. Buescher, R. W., W. A. Sistrunk, and P. L. Brady. 1975. Effects of ethylene on metabolic and quality attributes in sweet potato roots. *J. Food Sci.* **40:**1,018–1,020.

26. Burton, W. G. 1978. Post-harvest behavior and storage of potatoes. *Appl. Biol.* **3:**86–228.

27. Butchbaker, A. F., D. C. Nelson, and R. Shaw. 1967. Controlled atmosphere storage of potatoes. *Am. Soc. Agric. Eng. Trans.* **10:**534–538.

28. Carter, W. 1973. *Insects in Relation to Plant Disease.* John Wiley and Sons, New York, 759p.

29. Cathey, H. M., J. Halperin, and A. A. Piringer. 1965. Relations of N-dimethylamino-succinamic acid to photoperiod, kind of supplemental light, and night temperature, and its effects on the growth and flowering of garden annuals. *Hort. Res.* **5:**1–12.

30. Cathey, H. M., and H. E. Heggestad. 1982. Ozone and sulfur dioxide sensitivity of petunia: Modification by ethylene diurea. *J. Am. Soc. Hort. Sci.* **107:**1,028–1,035.

31. Ceponis, M. J., and J. E. Butterfield. 1973. The nature and extent of retail and consumer losses in apples, oranges, lettuce, peaches, strawberries, and potatoes marketed in greater New York. *USDA Marketing Res. Rep. 996*, 23p.

32. Ceponis, M. J., and J. E. Butterfield. 1974. Causes of cullage of Florida bell peppers in New York wholesale and retail markets. *Plant Dis. Rep.* **58**:367–369.
33. Ceponis, M. J., and J. E. Butterfield. 1974. Market losses in Florida cucumbers and bell peppers in metropolitan New York. *Plant Dis. Rep.* **58**:558–560.
34. Ceponis, M. J., and J. E. Butterfield. 1974. Retail and consumer losses in sweet potatoes in metropolitan New York. *HortScience* **9**:393–394.
35. Ceponis, M. J., and J. E. Butterfield. 1974. Retail and consumer losses of western pears in metropolitan New York. *HortScience* **9**:447–448.
36. Ceponis, M. J., and J. E. Butterfield. 1974. Retail losses in California grapes marketed in metropolitan New York. *USDA, ARS-NE-53*, 7p.
37. Ceponis, M. J., J. Kaufman, and S. M. Ringel. 1962. Quality of prepackaged apples in New York City retail stores. *USDA, AMS-461*, 12p.
38. Chalutz, E., J. E. DeVay, and E. C. Maxie. 1969. Ethylene-induced isocoumarin formation in carrot root tissue. *Plant Physiol.* **44**:235–241.
39. Chang, L. A., L. K. Hammett, and D. M. Pharr. 1982. Ethanol, alcohol dehydrogenase, and pyruvate decarboxylase in storage roots of four sweet potato cultivars during simulated flood-damage and storage. *J. Am. Soc. Hort. Sci.* **107**:674–677.
40. Chang, T.-S., and R. L. Shewfelt. 1988. Effect of chilling exposure of tomatoes during subsequent ripening. *J. Food Sci.* **53**:1,160–1,162.
41. Christensen, C. M. (ed.). 1982. *Storage of Cereal Grains and Their Products.* American Association of Cereal Chemists, St. Paul, Minn., 549p.
42. Claypool, L. L., and H. M. Vines. 1956. Commodity tolerance studies of deciduous fruits to moist heat and fumigants. *Hilgardia* **24**:297–355.
43. Click, R. E., and D. P. Hackett. 1963. The role of protein and nucleic acid synthesis in the development of respiration in potato tuber slices. *Nat. Acad. Sci. (U.S.A.) Proc.* **50**:243–250.
44. Cochrane, J. C., V. W. Cochrane, F. G. Simon, and J. Spaeth. 1963. Spore germination and carbon metabolism in *Fusarium solani.* I. Requirements for spore germination. *Phytopathology* **53**:1,155–1,106.
45. Cochrane, V. W. 1958. *Physiology of Fungi.* John Wiley & Sons, New York, 524p.
46. Collander, R. 1924. Beobachtungen über die quantitativen Beziehungen zwischen Tötungsgeschwindigkeit und Temperatur beim Wärmetod pflanzlicher Zellen. *Commentat. Biol. Soc. Sci. Fenn.,* pp. 11–12.
47. Collins, P. C., and T. M. Blessington. 1983. Postharvest effects of shipping temperatures and subsequent interior keeping quality of *Ficus benjamina. HortScience* **18**:757–758.
48. Conover, C. A., and R. T. Poole. 1975. Acclimatization of tropical trees for interior use. *HortScience* **10**:600–681.
49. Conover, C. A., and R. T. Poole. 1977. Effects of cultural practices on acclimatization of *Ficus benjamina,* L. *J. Am. Soc. Hort. Sci.* **102**:529–531.
50. Couey, H. M., and J. M. Wells. 1970. Low oxygen and high carbon dioxide atmospheres to control post-harvest decay of strawberries. *Phytopathology* **60**:47–49.
51. Crane, W. M., and D. Martin. 1938. Apple investigation in Tasmania: Miscellaneous notes. *Council Sci. Ind. Res. Aust. J.* **11**:47–60.
52. Crawley, M. J. 1983. *Herbivory: The Dynamics of Animal–Plant Interactions.* Blackwell Scientific Pub., Oxford, England, 437p.
53. Curtis, O. F. 1936. Comparative effects of altering leaf temperatures and air humidities on vapor pressure gradients. *Plant Physiol.* **11**:595–603.
54. Darwin, C. 1881. *The Power of Movement in Plants.* John Murry, London, 592p.
55. Davies, D. D., S. Grego, and P. Kenworthy. 1974. The control of the production of lactate and ethanol by higher plants. *Planta* **118**:297–310.

56. Dennis, C. (ed.). 1983. *Post-Harvest Pathology of Fruits and Vegetables.* Academic Press, New York, 264p.

57. Dennison, R. A., and E. M. Ahmed. 1975. Irradiation treatment of fruits and vegetables. In *Symposium: Postharvest Biology and Handling of Fruits and Vegetables,* N. F. Haard and D. K. Salunkhe (eds.). Westport, Conn., pp. 118–129.

58. Dewey, D. H. 1962. Factors affecting the quality of Jonathan apples stored in controlled atmospheres. *16th Int. Hort. Congr. Brussels Proc.* **1:**278.

59. Dharker, S. D., K. A. Savagaon, A. N. Spirangarajan, and A. Sreenivasan. 1966. Irradiation of mangoes. I. Radiation-induced delay in ripening of Alphonso mangoes. *J. Food. Sci.* **31:** 863–869.

60. Dickens, B. F., and G. A. Thompson, Jr. 1982. Phospholipid molecular species alterations in microsomal membranes as an initial key step during cellular acclimation to low temperature. *Biochemistry* **21:**3,604–3,611.

61. Diehl, K. C., and D. D. Hamann. 1980. Relationships between sensory profile parameters and fundamental mechanical parameters for potatoes, melons and apples. *J. Texture Stud.* **10:**401–420.

62. Diehl, K. C., D. D. Hamann, and J. K. Whitfield. 1980. Structural failure in selected raw fruit and vegetables. *J. Texture Stud.* **10:**371–400.

63. Digby, J., and R. D. Firn. 1976. A critical assessment of the Cholodny-Went theory of shoot geotropism. *Curr. Adv. Plant Sci.* **8:**953–960.

64. Dollar, A., M. M. Hanaoka, G. A. McClish, and J. H. Moy. 1970. Semicommercial scale studies on irradiated papaya. *U.S. Atomic Energy Comm. Rep. 1970.* Contract No. AT-(26-1)-374, 414p.

65. Dubrov, A. P. 1968. *The Genetic and Physiological Effect of the Action of Ultraviolet Radiation on Higher Plants.* Nauka, Moscow, 250p.

66. Dull, G. G., and S. J. Kays. 1988. Quality and mechanical stability of pecan kernels with different packaging protocols. *J. Food Sci.* **53:**565–567.

67. Durbin, R. D. 1967. Obligate parasites: Effect on the movement of solutes and water. In *The Dynamic Role of Molecular Constituents in Plant-Parasite Interaction,* C. J. Mirocha and I. Uritani (eds.). Bruce, St. Paul, Minn., pp. 80–99.

68. Eaks, I. L. 1967. Ripening and astringency removal in persimmon fruits. *Proc. Am. Soc. Hort. Sci.* **91:**868–875.

69. Eaks, I. L., and L. L. Morris. 1957. Deterioration of cucumbers at chilling and non-chilling temperatures. *Proc. Am. Soc. Hort. Sci.* **69:**388–399.

70. El-Goorani, M. A., and N. F. Sommer. 1981. Effects of modified atmospheres on postharvest pathogens of fruits and vegetables. *Hort. Rev.* **3:**412–461.

71. Elkiey, T., D. P. Ormrod, and B. Marie. 1982. Foliar sorption of sulfur dioxide, nitrogen dioxide, and ozone by ornamental woody plants. *HortScience* **17:**358–360.

72. Engelbrecht, L., U. Arban, and W. Heese. 1969. Leaf-miner caterpillars and cytokinins in the "green islands" of autumn leaves. *Nature* **223:**319–321.

73. Fails, B. S., A. J. Lewis, and J. A. Barden. 1982. Anatomy and morphology of sun- and shade-grown *Ficus benjamina,* L. *J. Am. Soc. Hort. Sci.* **104:**410–413.

74. Faller, N. 1976. Simultaneous assimilation of sulfur and carbon dioxides by some plants. *Acta. Bot. Croat.* **35:**87–95.

75. Farooqi, W. A., and E. G. Hall. 1972. Effects of ethylene dibromide on the respiration and ripening of bananas (*Musa cavendishii,* L.). *The Nucleus* **9:** 22–28.

76. Faust, M. 1973. Effect of growth regulators on firmness and red color of fruit. *Acta Hort.* **34:**407–420.

77. Fidler, J. C. 1968. Low temperature injury to fruits and vegetables. *Recent Adv. Food Sci.* **4:**271–283.

78. Fidler, J. C., and G. Mann. 1972. *Refrigerated Storage of Apples and Pears—A Practical Guide.* Commonwealth Agricultural Bureau, Slough, England, 65p.

79. Fidler, J. C., and C. J. North. 1971. The effect of periods of anaerobiosis on the storage of apples. *J. Hort. Sci.* **46**:213–221.

80. Fidler, J. C., B. G. Wilkinson, K. L. Edney, and R. O. Sharples. 1973. *The Biology of Apple and Pear Storage*. Commonwealth Agricultural Bureau, Slough, England, 235p.

81. Fleuriet, A., and J. J. Macheix. 1984. Orientation nouvelle du métabolisme des acides hydroxycinnamiques dans les fruits de tomates blessés (*Lycopersicon esculentum*). *Physiol. Plant* **61**:64–68.

82. Frenkel, C., and M. E. Patterson. 1973. Effect of carbon dioxide on activity of succinic dehydrogenase in 'Bartlett' pears during cold storage. *HortScience* **8**:395–396.

83. Friend, J., and D. R. Threlfall (eds.). 1976. *Biochemical Aspects of Plant-Parasite Relationships*. Academic Press, New York, 354p.

84. Fukushima, H., C. E. Martin, H. Iida, Y. Kitajima, G. A. Thompson, Jr., and Y. Nozawa. 1976. Changes in membrane lipid composition during temperature adaptation by a thermotolerant strain of *Tetrahymena pyriformis*. *Biochem. Biophys. Acta* **431**:165–179.

85. Gallaghen, P. W., and T. D. Sydnor. 1983. Variation in wound response among cultivars of red maple. *J. Am. Soc. Hort. Sci.* **108**:744–746.

86. Garms, H. 1933. Untersuchungen über wundhieling an früchten. *Beih. Bot. Centralbl. I.* **51**:437–516.

87. Gil, G. F., W. H. Griggs, and G. C. Martin. 1972. Gibberellin-induced parthenocarpy in 'Winter Nelis' pear. *HortScience* **7**:559–561.

88. Goodliffe, J. P., and J. B. Heale. 1977. Factors affecting the resistance of stored carrot roots to *Botrytis cinerea*. *Ann. Appl. Biol.* **85**:163.

89. Goodman, R. N., Z. Kiraly, and M. Zaithin. 1967. *The Biochemistry and Physiology of Infectious Plant Disease*. Van Nostrand, New York, 354p.

90. Granett, A. L. 1982. Pictorial keys to evaluate foliar injury caused by hydrogen fluoride. *HortScience* **17**:587–588.

91. Groves, J. F. 1917. Temperature and life duration of seeds. *Bot. Gaz.* **63**:169–189.

92. Haard, N. F., S. C. Sharma, R. Wolfe, and C. Frenkel. 1974. Ethylene induced isoperoxidase changes during fiber formation in postharvest asparagus. *J. Food Sci.* **39**:452–456.

93. Hale, P. W., and J. J. Smoot. 1973. Exporting Florida grapefruit to Japan. An evaluation of new shipping cartons and decay control treatments. *Citrus Veg. Mag.* **37**(3):20–23, 45.

94. Halevy, A. H., and S. Mayak. 1981. Senescence and postharvest physiology of cut flowers—part 2. *Hort. Rev.* **3**:59–143.

95. Hall, C. B. 1964. The effects of short periods of high temperature on the ripening of detached tomato fruits. *Proc. Am. Soc. Hort. Sci.* **84**:501–506.

96. Hankinson, B., and V. N. M. Rao. 1979. Histological and physical behavior of tomato skins susceptible to cracking. *J. Am. Soc. Hort. Sci.* **104**:577–581.

97. Harman, J. E., and B. McDonald. 1983. Controlled atmosphere storage of kiwi fruit: Effects on storage life and fruit quality. *Acta Hort.* **138**:195–201.

98. Harrington, G. T., and W. Crocker. 1918. Resistance of seeds to desiccation. *J. Agric. Res.* **14**:525–532.

99. Havis, J. R. 1976. Root hardiness of woody ornamentals. *HortScience* **11**:385–386.

100. Heiniche, A. J. 1937. How lime/sulphur spray affects the photosynthesis of an entire ten year old apple tree. *Proc. Am. Soc. Hort. Sci.* **35**:256–259.

101. Hicks, J. R., and P. M. Ludford. 1981. Effects of low ethylene levels on storage of cabbage. *Acta Hort.* **116**:65–73.

102. Hiraki, Y., and Y. Ota. 1975. The relationships between growth inhibition and

ethylene production by mechanical stimulation in *Lillium longiflorum*. *Plant Cell Physiol.* **16:**185–189.

103. Hoffman, M. B. 1933. The effect of certain spray materials on the carbon dioxide assimilation by McIntosh apple leaves. *Proc. Am. Soc. Hort. Sci.* **29:**389–398.

104. Holt, J. E., and D. Schoorl. 1982. Mechanics of failure in fruits and vegetables. *J. Texture Stud.* **13:**83–97.

105. Howe, T. K., and S. S. Woltz. 1981. Symptomology and relative susceptibility of various ornamental plants to acute airborne sulfur dioxide exposure. *Proc. Fla. St. Hort. Soc.* **94:**121–123.

106. Hsiao, T. C. 1973. Plant responses to water stress. *Ann. Rev. Plant Physiol.* **24:**519–570.

107. Huber, H. 1935. Der Wärmehaushalt der Pflanzen. *Naturwiss. Landwirtsch.* **17:**148–152.

108. Huelin, F. E., and G. B. Tindale. 1942. Investigations on the gas storage of Victorian pears. *J. Agric. Vict.* **40:**594–606.

109. Hulme, A. C. 1956. Carbon dioxide injury and the presence of succinic acid in apples. *Nature* **178:**218–219.

110. Hung, Y.-C., and S. E. Prussia. 1989. Effect of maturity and storage on the bruise susceptibility of peaches (cv. Red Globe). *Am. Soc. Agric. Eng. Trans.* **32:**1,377–1,382.

111. Husain, A., and A. Kelman. 1958. Relation of slime production to mechanism of wilting and pathogenicity of *Pseudomonas solanacearum*. *Phytopathology* **48:**155–165.

112. Hyodo, H., H. Kuroda, and S. F. Yang. 1978. Induction of phenylalanine ammonia-lyase and increase in phenolics in lettuce leaves in relation to the development of russet spotting caused by ethylene. *Plant Physiol.* **62:**31–35.

113. Hyodo, H., K. Tanaka, and K. Watanake. 1983. Wound-induced ethylene production and 1-aminocyclopropane-1-carboxylic acid synthase in mesocarp tissue of winter squash fruit. *Plant and Cell Physiol.* **24:**963–969.

114. Ilker, Y. 1976. Physiological manifestations of chilling injury and its alleviation in okra plants (*Abelmoschus esculentus* (L.) Moench). Ph.D. thesis, University of California, Davis, 207p.

115. Isenberg, F. M., and R. M. Sayles. 1969. Modified atmosphere storage of Danish cabbage. *J. Am. Soc. Hort. Sci.* **94:**447–449.

116. Jadhav, S. J., and D. K. Salunkhe. 1975. Formation and control of chlorophyll and glycoalkaloids in tubers of *Solanum tuberosum*, L. and evaluation of glycoalkaloid toxicity. *Adv. Food. Res.* **21:**307–354.

117. Jaffe, M. L. 1973. Thigmomorphogenesis: The response of plant growth and development of mechanical stimulation, with special reference to Bryonia dioica. *Planta* **114:**143–157.

118. Johnson, C. R., J. K. Krantz, J. N. Joiner, and C. A. Conover. 1979. Light compensation point and leaf distribution of *Ficus benjamina* as affected by light intensity and nitrogen-potassium nutrition. *J. Am. Soc. Hort. Sci.* **104:**335–338.

119. Jones, R. L., H. T. Freebairn, and J. F. McConnell. 1978. The prevention of chilling injury, weight loss reduction, and ripening retardation in banana. *J. Am. Soc. Hort. Sci.* **103:**129–221.

120. Juniper, B. E. 1976. Geotropism. *Ann. Rev. Plant Physiol.* **27:**385–406.

121. Kader, A. A. 1985. Ethylene-induced senescence and physiological disorders in harvested horticultural crops. *HortScience* **20:**54–57.

122. Kahl, G. 1974. Metabolism in plant storage tissue slices. *Bot. Rev.* **40:**263–314.

123. Kahl, G. (ed.). 1978. *Biochemistry of Wounded Plant Tissues*. Walter de Gruyter, New York, 680p.

124. Kamerbeek, G. A., and W. J. DeMunk. 1976. A review of ethylene effects in bulbous plants. *Scientia Hort.* **4:**101–115.

125. Kang, B. G. 1979. Epinasty. In *Physiology of Movements,* W. Haupt and M. E. Feinleik (eds.). Encylopedia of Plant Physiology, vol. 7. Springer-Verlag, Berlin, pp. 657–667.

126. Kaufman, J., R. E. Hardenburg, and J. M. Lutz. 1956. Weight losses and decay of Florida and California oranges in mesh and perforated polyethylene consumer bags. *Proc. Am. Soc. Hort. Sci.* **67**:244–250.

127. Kaufmann, M. R. 1975. Leaf water stress in Engelmann spruce: Influence of the root and shoot environments. *Plant Physiol.* **56**:841–844.

128. Kays, S. J., C. A. Jaworski, and H. C. Price. 1976. Defoliation of pepper transplants in transit by endogenously evolved ethylene. *J. Am. Soc. Hort. Sci.* **101**:449–451.

129. Kays, S. J., and D. M. Wilson. 1977. Alteration of pecan kernel color. *J. Food Sci.* **42**:982–988.

130. Key, J. L., C. Y. Lin, E. Ceglarz, and F. Schöffl. 1982. The heat shock response in soybean seedlings. In *Structure and Function of Plant Genomes,* L. Dure (ed). NATO Advanced Studies Series, vol. 63. Plenum Press, New York.

131. Key, J. L., C. Y. Lin, and Y. M. Chen. 1981. Heat shock proteins of higher plants. *Nat. Acad. Sci. (USA) Proc.* **78**:3,526–3,530.

132. Kidd, F. 1919. Laboratory experiments on the sprouting of potatoes in various gas mixtures (nitrogen, oxygen and carbon dioxide). *New Phytol.* **18**:248–252.

133. Kidd, F., and C. West. 1923. Brown heart a functional disease of apples and pears. *Great Brit. Dept. Sci. Ind. Res. Food Invest. Board Spec. Rep. No. 12.*

134. Kidd, F., and C. West. 1927. A relation between the concentration of oxygen and carbon dioxide in the atmosphere, rate of respiration and length of storage life in apples. *Great Brit. Dept. Sci. Ind. Res. Food Invest. Board Rep. 1925, 1926,* pp. 41–42.

135. Kidd, F., and C. West. 1939. The gas-storage of Cox's Orange Pippin apples on a commercial scale. *Great Brit. Dept. Sci. Ind. Res. Food Invest. Board Rep. 1938,* pp. 153–156.

136. Kiesselbach, T. A., and J. A. Ratcliff. 1920. Freezing injury of seed corn. *Univ. Neb. Agric. Exp. Sta. Res. Bull.* **16**:1–96.

137. Kiyosawa, K. 1975. Studies on the effects of alcohols on membrane water permeability of *Nitella. Protoplasma* **86**:243–252.

138. Kloft, W., and P. Ehrhardt. 1959. Untersuchungen uber Saugtatigkeit und Schadwirkung der Sitkafichtenlaus *Liosomaphis abietina* (Walk.) (*Neomyzaphis abietina,* Walk.). *Phytopathol. Z.* **35**:401–410.

139. Knee, M. 1973. Effects of controlled atmosphere storage on respiratory metabolism of apple fruit tissue. *J. Sci. Food Agric.* **24**:1,289–1,298.

140. Kolattukudy, P. E., and V. P. Agrawal. 1974. Structure and composition of aliphatic components of potato tuber skin (suberin). *Lipids* **9**:682–691.

141. Kosiyachinda, S., and R. E. Young. 1976. Chilling sensitivity of avocado fruit at different stages of respiratory climacteric. *J. Am. Soc. Hort. Sci.* **101**:665–667.

142. Kotob, M. A., and W. W. Schwake. 1975. Respiration rate and acidity in parthenocarpic and seeded Conference pears. *J. Hort. Sci.* **50**:435–445.

143. Kulman, H. M. 1971. Effects of insect defoliation on growth and mortality of trees. *Ann. Rev. Entomol.* **16**:289–324.

144. Kuraishi, S., and N. Nito. 1980. The maximum leaf surface temperatures of higher plants observed in the Inland Sea area. *Bot. Mag.* **93**:209–228.

145. Lau, O. L. 1983. Effects of storage procedures and low oxygen and carbon dioxide atmospheres on storage quality of 'Spartan' apples. *J. Am. Soc. Hort. Sci.* **108**:953–957.

146. Levitt, J. 1980. *Responses of Plants to Environmental Stresses,* 2 vols. Academic Press, New York.

147. Link, H. 1967. Der Einfuss der Ausdünnung auf Fruchtqualität und Erntemenge bei der Apfelsorte 'Golden Delicious'. *Gartenbauwiss* **32**:423–444.

148. Lipetz, J. 1966. Crown gall tumorigenesis II. Relations between wound healing and the tumorigenic response. *Cancer Res.* **26**:1,597–1,605.

149. Lipetz, J. 1970. Wound-healing in higher plants. *Int. Rev. Cytol.* **27**:1–28.

150. Lipton, W. J. 1977. Toward an explanation of disorders of vegetables induced by high CO_2 or low O_2? *2nd Nat. CA [Controlled Atmosphere] Res. Conf. Proc., Mich. St. Univ. Hort. Rep.* **28**:137–141.

151. Lipton, W. J., and C. M. Harris. 1974. Controlled atmosphere effects on the market quality of stored broccoli (*Brassica oleracea*, L. Italica group). *J. Am. Soc. Hort. Sci.* **99**:200–205.

152. Little, C. R., J. D. Faragher, and H. J. Taylor. 1982. Effects of initial oxygen stress treatments in low oxygen modified atmosphere storage of 'Granny Smith' apples. *J. Am. Soc. Hort. Sci.* **107**:320–323.

153. Liu, F. W. 1979. Interaction of daminozide, harvesting date, and ethylene in CA storage on 'McIntosh' apple quality. *J. Am. Soc. Hort. Sci.* **104**:599–601.

154. Livne, A. 1964. Photosynthesis in healthy and rust-affected plants. *Plant Physiol.* **39**:614–621.

155. Lloyd, E. J. 1976. The influence of shading on enzyme activity in seedling leaves of barley. *Z. Pflanzenphysiol.* **78**:1–12.

156. Lockett, M. C., and B. J. Luyet. 1951. Survival of frozen seed of various water contents. *Biodynamica* **7**(134):67–76.

157. Lockhart, J. A., and U. Brodführer-Franzgrote. 1961. *The Effects of Ultraviolet Radiation on Plants.* Encyclopedia of Plant Physiology, vol. 16. Springer-Verlag, Berlin, pp. 532–554.

158. Lutz, J. M., and R. E. Hardenburg. 1968. The commercial storage of fruits, vegetables, and florist and nursery stocks. *U.S. Dept. Agric. Handb. 66*, 94p.

159. Luyet, B. J. 1937. The vitrification of organic colloids of protoplasm. *Biodynamica* **29**:1–14.

160. Mansfield, T. A. (ed.). 1976. *Effects of Air Pollutants on Plants.* Society of Experimental Biology Seminar Series, vol. 1. Cambridge University Press, Cambridge, England.

161. Marousky, F. J., and B. K. Harbaugh. 1982. Responses of certain flowering and foliage plants to exogenous ethylene. *Proc. Fla. St. Hort. Soc.* **95**:159–162.

162. Mattus, G. E., L. E. Scott, and L. L. Claypool. 1959. Brown spot of Bartlett pears. *Calif. Agric.* **13**(7):8, 13.

163. Maxie, E. C., and A. Abel-Kader. 1966. Food irradiation—physiology of fruits as related to feasibility of the technology. *Adv. Food Res.* **15**:105–145.

164. Maxie, E. C., and N. F. Sommer. 1971. Radiation technology in conjunction with postharvest procedures as a means of extending the shelf-life of fruits and vegetables. *U.S. Atomic Energy Comm. Rep. 1971,* Contract No. AT-(11-1)-34.

165. Maxie, E. C., N. F. Sommer, and F. G. Mitchell. 1971. Infeasibility of irradiating fresh fruits and vegetables. *HortScience* **6**:202–204.

166. Mazur, P. 1969. Freezing injury in plants. *Ann. Rev. Plant Physiol.* **20**:419–448.

167. Mazur, P. 1970. Cryobiology: The freezing of biological systems. *Science* **168**:939–949.

168. McCree, K. J., and J. H. Troughton. 1966. Prediction of growth rate at different light levels from measured photosynthesis and respiration rates. *Plant Physiol.* **41**:559–566.

169. McGill, J. N., A. I. Nelson, and M. P. Steinberg. 1966. Effects of modified storage atmosphere on ascorbic acid and other quality characteristics of spinach. *J. Food Sci.* **31**:510–517.

170. McGlasson, W. B., and H. K. Pratt. 1964. Effects of wounding on respiration and ethylene production by cantaloupe fruit tissue. *Plant Physiol.* **39**:128–132.

171. Meigh, D. F. 1960. Use of gas chromatography in measuring the ethylene production of stored apples. *J. Sci. Food Agric.* **11**:381–385.

172. Meryman, H. T. 1966. *Cryobiology.* Academic Press, New York, 775p.

173. Miles, J. A., and G. E. Rehkugler. 1973. A failure criterion for apple flesh. *Am. Soc. Agric. Eng. Trans.* **16:**1,148–1,153.

174. Milks, R. R., J. N. Joiner, L. A. Garard, C. A. Conover, and B. Tjia. 1979. Influence of acclimatization on carbohydrate production and translocation of *Ficus benjamina, L. J. Am. Hort. Sci.* **104:**410–413.

175. Millerd, A., and K. J. Scott. 1962. Respiration of the diseased plant. *Ann. Rev. Plant Physiol.* **13:**550–574.

176. Mitchell, C. A., C. J. Severson, J. A. Wott, and P. A. Hammer. 1975. Seismorphogenetic regulation of plant growth. *J. Am. Soc. Hort. Sci.* **100:**161–165.

177. Modlibowska, I. 1966. Inducing precocious cropping on young Dr. Jules Guyot pear trees with gibberellic acid. *J. Hort. Sci.* **41:**137–144.

178. Mohsenin, N. N. 1970. *Physical Properties of Plant and Animal Materials,* 2 vols. Gordon and Breach, New York.

179. Mohsenin, N. N. 1977. Characterization and failure in solid food with particular reference to fruits and vegetables. *J. Texture Stud.* **8:**169–193.

180. Molisch, H. 1896. Das erfrieren von pflanzen bei temperaturen über dem eispuckt. *Sitzber. Akad. Wiss. Wien. Math. Naturwis. Kl.* **105:**1–14.

181. Monning, A. 1983. Studies on the reaction of Krebs cycle enzymes from apple tissue (cv. Cox Orange) to increased levels of CO_2. *Acta Hort.* **138:**113–119.

182. Morris, L. L., and A. A. Kader. 1977. Physiological disorders of certain vegetables in relation to modified atmosphere. *2nd Nat. CA Res. Conf. Proc., Mich. St. Univ. Hort. Rep.* **28:**266–267.

183. Mudd, J. B., and T. T. Kozlowski (eds.). 1975. *Responses of Plants to Air Pollution.* Academic Press, New York, 383p.

184. Nagarojah, S. 1976. The effects of increased illumination and shading on the low-light-induced decline in photosynthesis in cotton leaves. *Physiol. Plant.* **36:**338–342.

185. Neel, P. L., and R. W. Harris. 1971. Motion-induced inhibition of elongation and induction of dormancy in Liquidambar. *Science* **173:**58–59.

186. Ogura, N., N. Nakayawa, and H. E. Takehana. 1975. Studies on the storage temperature of tomato fruit. I. Effect of high temperature–short term storage of mature green tomato fruits on changes of their chemical constituents after ripening at room temperature. *Agric. Soc. Japan J.* **49:**189–196.

187. Olorunda, A. O., and N. E. Looney. 1977. Response of squash to ethylene and chilling injury. *Ann. Appl. Biol.* **87:**465–469.

188. Ophuis, B. G., J. C. Hesen, and E. Kroesbergen. 1958. The influence of temperature during handling on the occurrence of blue discolorations inside potato tubers. *Eur. Pot. J.* **1:**48–65.

189. Osborne, D. J. 1973. Mutual regulation of growth and development in plants and insects. In *Insect/Plant Relationships,* H. F. van Emden (ed.). Blackwell Scientific Pub., London, pp. 33–42.

190. Paull, R. E. 1990. Chilling injury of crops of tropical and subtropical origin. In *Chilling Injury of Horticultural Crops,* C.-Y. Wang (ed.). CRC Press, Boca Raton, Fla., pp. 17–36.

191. Pieniazeh, S. A. 1942. External factors affecting water loss from apples in cold storage. *Refrig. Eng.* **44:**171–173.

192. Pitt, R. E. 1982. Models for the rheology and statistical strength of uniformly stressed vegetative tissue. *Am. Soc. Agric. Eng. Trans.* **25:**1,776–1,784.

193. Platt-Aloia, K. A., and W. W. Thomson. 1987. Freeze fracture evidence for lateral phase separations in the plasmalemma of chilling-injured avocado fruit. *Protoplasma* **136**(2/3):71–80.

194. Poenicke, E. F., S. J. Kays, D. A. Smittle, and R. E. Williamson. 1977. Ethylene in relation to postharvest quality deterioration in processing cucumbers. *J. Am. Soc. Hort. Sci.* **102:**303–306.

195. Poljakoff-Mayber, A., and J. Gale (eds.). 1975. *Plants in Saline Environments.* Ecological Studies, vol. 15. Springer-Verlag, New York, 213p.

196. Pratt, H. K., and J. D. Goeschl. 1969. Physiological roles of ethylene in plants. *Ann. Rev. Plant Physiol.* **20:**541–584.

197. Precheur, R., J. K. Greig, and D. V. Armbrust. 1978. The effects of wind and wind-plus-sand on tomato plants. *J. Am. Soc. Hort. Sci.* **103:**351–355.

198. Proapst, P. A., and F. R. Forsyth. 1973. The role of internally produced carbon dioxide in the prevention of greening in potato tubers. *Acta Hort.* **38:**277–290.

199. Raison, J. K., J. M. Lyons, R. J. Mahlhoin, and A. D. Keith. 1971. Temperature-induced phase changes in mitochondrial membranes detected by spin labeling. *J. Biol. Chem.* **246:**4,036–4,040.

200. Reid, M. S. 1985. Ethylene and abscission. *HortScience* **20:**45–50.

201. Rhodes, M. J. C., and L. S. C. Wooltorton. 1973. Changes in phenolic acid and lignin biosynthesis in response to treatment of root tissue of the Swedish turnip (*Brassica napo-brassica*) with ethylene. *Qual. Plant* **23:**145–155.

202. Rigney, C. J., D. Graham, and T. H. Lee. 1978. Changes in tomato fruit ripening caused by ethylene dibromide fumigation. *J. Am. Soc. Hort. Sci.* **103:**402–423.

203. Risse, L. A., and T. T. Hatton. 1982. Sensitivity of watermelons to ethylene during storage. *HortScience* **17:**946–948.

204. Robbins, W. J., and K. F. Petsch. 1932. Moisture content and high temperature in relation to the germination of corn and wheat grains. *Bot. Gaz.* **93:**85–92.

205. Robinson, J. E., K. M. Browne, and W. G. Burton. 1975. Storage characteristics of some vegetables and soft fruits. *Ann. Appl. Biol.* **81:**399–408.

206. Robitaille, H. A., and A. C. Leopold. 1974. Ethylene and regulation of apple stem growth under stress. *Physiol. Plant.* **32:**301–304.

207. Romani, R. J. 1966. Radiobiological parameters in the irradiation of fruits and vegetables. *Adv. Food Res.* **15:**57–103.

208. Rosa, J. R. 1926. Ripening and storage of tomatoes. *Proc. Am. Soc. Hort. Sci.* **23:**233–242.

209. Sacalis, J. N. 1978. Ethylene evolution by petioles of sleeved poinsettia plants. *HortScience* **13:**594–596.

210. Saltveit, M. E., Jr., and R. A. Larson. 1981. Reduced leaf epinasty in mechanically stressed poinsettia (*Euphorbia pulcherrima* cultivar Annette Hegg Diva) plants. *J. Am. Soc. Hort. Sci.* **106:**156–159.

211. Salunkhe, D. K. (ed.). 1974. *Storage, Processing and Nutritional Quality of Fruits and Vegetables.* CRC Press, Cleveland, 166p.

212. Sarkar, S. K., and C. T. Phan. 1979. Naturally-occurring and ethylene-induced compounds in the carrot root. *J. Food Prod.* **42:**526–534.

213. Sasaki, S., and T. T. Kozlowski. 1968. Effects of herbicides on respiration of red pine (seedlings) II. Monuron, Diuron, DCPA, Dalapon, CDEC, CDAA, EPTC, and NPA. *Bot. Gaz.* **129:**286–293.

214. Sato, K., I. Uritani, and T. Saito. 1981. Characterization of the terpene-inducing factor isolated from larvae of the sweet potato weevil, *Cylas formicarius fabricicus* (Coleoptera: Brenthidae). *Appl. Entomol. Zool.* **16:**103–112.

215. Sato, K., I. Uritani, and T. Saito. 1982. Properties of terpene-inducing factor extracted from adults of the sweet potato weevil, *Cylas formicarius fabricius* (Coleoptera: Brenthidae). *Appl. Entomol. Zool.* **17:**368–374.

216. Scott, K. J., and E. A. Roberts. 1968. The importance of weight loss in reducing breakdown of 'Jonathan' apples. *Aust. J. Exp. Agric. Animal Husb.* **8:**377–380.

217. Scott, K. J., and R. B. H. Wills 1975. Postharvest application of calcium as control for storage breakdown of apples. *HortScience* **10:**75–76.

218. Sharples, R. O. 1973. Orchard sprays. In *The Biology of Apple and Pear Storage,* J. C. Fidler, B. G. Wilkinson, K. L. Edney, and R. O. Sharples (eds.). Commonwealth Agricultural Bureau, London, pp. 194–203.

219. Sharples, R. O. 1980. The influence of orchard nutrition on the storage quality of apples and pears grown in the United Kingdom. In *Mineral Nutrition of Fruit Trees,* D. Atkinson, J. E. Jackson, R. O. Sharples, and W. M. Waller (eds.). Butterworths, London, pp. 17–28.

220. Sherif, S. M. 1976. The quasi-static contact problem for nearly incompressible agricultural products. Ph.D. thesis, Michigan State University, East Lansing.

221. Shipway, M. R., and W. J. Bramlage. 1973. Effects of carbon dioxide on activity of apple mitochondria. *Plant Physiol.* **51**:1,095–1,098.

222. Sibaoka, T. 1969. Physiology of rapid movements in higher plants. *Ann. Rev. Plant Physiol.* **20**:165–184.

223. Skene, D. S. 1980. Growth stresses during fruit development in Cox's Orange Pippin Apples. *J. Hort Sci.* **55**:27–32.

224. Smith, W. H. 1938. The storage of broccoli. *Fed. Invest. Board London Rep. for 1937,* pp. 185–187.

225. Smith, W. H. 1958. Reduction of low-temperature injury to stored apples by modulation of environmental conditions. *Nature* **191**:275–276.

226. Smith, W. H. 1965. Storage of mushrooms. *Ditton Lab. Rep. 1964–1965,* p. 25.

227. Smock, R. M. 1977. Nomenclature for internal storage disorders of apples. *HortScience* **12**:306–308.

228. Snowden, A. L. 1990. *A Color Atlas of Post-Harvest Diseases & Disorders of Fruits & Vegetables,* vol. 1: *General Introduction & Fruits.* CRC Press, Boca Raton, 302p.

229. Solov'ev, V. A. 1969. Plant growth and water and mineral nutrient supply under NaCl salinization conditions. *Fiziol. Rast.* **16**:870–876.

230. Sommer, N. F. 1957. Pear transit simulated in test. *Calif. Agric.* **11**(9):3–5,16.

231. Sommer, N. F. 1957. Surface discoloration of pears. *Calif. Agric.* **11**(1):3–4.

232. Sommer, N. F., and R. J. Fortlage. 1966. Ionizing radiation for control of postharvest diseases of fruits and vegetables. *Adv. Food Res.* **15**:147–193.

233. Sparks, D. 1986. Pecan. In *Handbook of Fruit Set and Development,* S. P. Monselise (ed.). CRC Press, Boca Raton, Fla., pp. 323–339.

234. Spencer, E. L. 1937. Frenching of tobacco and thallium toxicity. *Am. J. Bot.* **24**:16–24.

235. Steponkus, P. L. 1984. Role of the plasma membrane in freezing injury and cold acclimation. *Ann. Rev. Plant Physiol.* **35**:543–584.

236. Steward, R. N., K. H. Norris, and S. Asen. 1975. Microspectrophotometric measurement of pH and pH effect on color of petal epidermal cells. *Phytochemistry* **14**:937–942.

237. Stewart, J. K., and M. Uota. 1971. Carbon dioxide injury and market quality of lettuce held in controlled atmospheres. *J. Am. Soc. Hort. Sci.* **96**:27–30.

238. Stoll, K. 1971. Lagerung von Früchten und Gemüsen in kontrollierter Atmosphäre. *Mitt. Eidg. Forsch. Anst. Obst. Wein. Gartenbau, Wädenswil, Flugschrift Schweiz Obst. Weinbau* **107**:614–623, 648–652, 711–714, 741–745.

239. Strobel, G. A. 1974. Phytotoxins produced by plant parasites. *Ann. Rev. Plant Physiol.* **25**:541–566.

240. Studer, E. J., P. L. Steponkus, G. L. Good, and S. C. Wiest. 1978. Root hardiness of container-grown ornamentals. *HortScience* **13**:172–174.

241. Suge, H. 1978. Growth and gibberellin production in *Phaseolus vulgaris* as affected by mechanical stress. *Plant and Cell Physiol.* **19**:1,557–1,560.

242. Swinburne, T. R. 1983. Quiescent infections in post-harvest diseases. In *Post-Harvest Pathology of Fruits and Vegetables,* C. Dennis (ed.). Academic Press, New York.

243. Thomas, M. 1925. The controlling influence of carbon dioxide. V. A quantitative study of the production of ethyl alcohol and acetaldehyde by cells of the higher plants in relation to the concentration of oxygen and carbon dioxide. *Biochem. J.* **19**:927–947.

244. Toivonen, P., J. Walsh, E. C. Lougheed, and D. P. Murr. 1982. Ethylene relationships in storage of some vegetables. In *Controlled Atmospheres for Storage and Transport of Perishable Agricultural Commodities,* D. G. Richardson and M. Meheriuk (eds.). Timber Press, Beaverton, Ore., pp. 299–307.

245. Tomkins, R. G. 1965. The storage of broad beans. *Ditton and Covent Garden Lab. Ann. Rep.,* 1964–65, East Malling, England, p. 24.

246. Tomkins, R. G., and D. F. Meigh. 1968. The concentration of ethylene found in controlled atmosphere stores. *Ditton and Covent Garden Lab. Ann. Rep.,* 1967–68, East Malling, England, pp. 33–36.

247. United Nations, Dept. Economic Affairs. 1951. *Proc. U.N. Scientific Conf. Conservation and Utilization of Resources, 1947.* Lake Success, N.Y., 623 p.

248. Van der Plank, J. E., and R. Davies. 1937. Temperature-cold injury curves of fruit. *J. Pomol Hort. Soc.* **15**:226–247.

249. Varley, G. C., and G. R. Gradwell. 1962. The effect of partial defoliation by caterpillars on the timber production of oak trees in England. *11th Int. Cong. Entomol. Proc., Vienna 1960* **2**:211–214.

250. Verner, L., and E. C. Blodgett. 1931. Physiological studies of the cracking of sweet cherries. *Idaho Agric. Exp. Sta. Bull. 184,* 15p.

251. Vogele, A. C. 1937. Effect of environmental factors upon the color of the tomato and the watermelon. *Plant Physiol.* **12**:929–955.

252. Voisey, P. W., L. H. Lyall, and M. Klock. 1970. Tomato skin strength—its measurement and relation to cracking. *J. Am. Soc. Hort. Sci.* **95**:485–488.

253. Volkmann, D., and A. Sievers. 1979. Graviperception in multicellular organs. In *Physiology of Movements,* W. Haupt and M. E. Feinleik (eds.). Encyclopedia of Plant Physiology, vol. 7. Springer-Verlag, Berlin, pp. 353–367.

254. Von Wandruszka, R. M. A., C. A. Smith, and S. J. Kays. 1980. The role of iron in pecan kernel color. *Lebensm. Wiss. U. Technol.* **13**:38–39.

255. Walker, J. C., and K. P. Link. 1935. Toxicity of phenolic compounds to certain onion bulb parasites. *Bot. Gaz.* **96**:468–484.

256. Walter, W. M., Jr., and W. E. Schadel. 1983. Structure and composition of normal skin (periderm) and wound tissue from cured sweet potatoes. *J. Am. Soc. Hort. Sci.* **108**:909–914.

257. Wang, C. Y. 1982. Physiological and biochemical responses of plants to chilling stress. *HortScience* **17**:173–186.

258. Wardowski, W. F., and A. A. McCormack. 1973. Degreening Florida citrus fruits. *Fla. Agric. Exp. Sta. Ext. Ser. Cir. 389.*

259. Wattendorf, J. 1970. Permeabilität für Wasser, Malonsäurediamid und Harnstoff nach α-Bestrahlung von Convallaria-Zellen. *Ber. Deut. Bot. Ges.* **83**:3–17.

260. Webster, C. C. 1967. *The Effects of Air Pollution on Plant and Soil.* Agricultural Research Council, London, 53 p.

261. Weiser, C. J. 1970. Cold resistance and injury in woody plants. *Science* **169**:1,269–1,278.

262. Whiteman, T. M. 1957. Freezing points of fruits, vegetables and florist stocks. *U.S. Dept. Agric. Market Res. Rep. 196,* 32p.

263. Wilcockson, S. J., R. L. Griffith, and E. J. Allen. 1980. Effects of maturity on susceptibility to damage. *Ann. Appl. Biol.* **96**:349–353.

264. Wilkin, M. B. 1979. Growth-control mechanisms in gravitropism. In *Physiology of Movements,* W. Haupt and M. E. Feinleib (eds.). Encyclopedia of Plant Physiology, vol. 7. Springer-Verlag, Berlin, pp. 601–621.

265. Wilkinson, B. G. 1970. Physiological disorder of fruits. In *The Biochemistry of Fruits and Their Products,* A. C. Hulme (ed.). Academic Press, New York, pp. 537–554.

266. Wills, R. B. H., and K. J. Scott. 1971. Chemical induction of low temperature breakdown in apples. *Phytochemistry* **10**:1,783–1,785.

267. Woit, M. 1925. Über wundreaktionen an blättern und den anatomischen ban der klattminen. *Mitt. Dent. Dendr. Ges.* **35**:163–187.
268. Wright, R. C. 1942. The freezing temperatures of some fruits, vegetables and florists' stocks. *USDA Circ. 447,* 12p.
269. Wynn, W. K., Jr. 1963. Photosynthetic phosphorylation by chloroplasts isolated from rust-infected oats. *Phytopathology* **53**:1,376–1,377.
270. Yoshioka, H., Y. Ueda, and K. Chachin. 1980. Physiological studies of fruit ripening in relation to heat injury II: Effect of high temperature (40°) on the change of acid-phosphatase activity during banana fruit ripening. *Jpn. Soc. Food Sci. and Tech. J.* **27**:511–516.
271. Yoshioka, H., Y. Ueda, and T. Iwata. 1982. Physiological studies of fruit ripening in relation to heat injury. IV: Development of isoamyl acetate biosynthetic pathway in banana fruit during ripening and suppression of its development at high temperature. *Jpn. Soc. Food Sci. and Tech. J.* **29**:333–339.
272. Zagory, D., and M. S. Reid. 1985. Evaluation of the role of vase microorganisms in the postharvest life of cut flowers. In *Third International Symposium Post-Harvest Physiology of Ornamentals,* H. C. M. de Styers (ed.). Drukkerig, Netherlands, pp. 207–217.

ADDITIONAL READINGS

Abeles, F. B. 1973. *Ethylene in Plant Biology.* Academic Press, New York, 302p.

Addicott, F. T. 1981. *Abscission.* University of California Press, Berkeley, 369p.

Berry, J., and O. Björkman. 1980. Photosynthetic response and adaptation to temperature in higher plants. *Ann. Rev. Plant Physiol.* **31**:491–543.

Cooper-Driver, G. A., T. Swain, and E. E. Conn (eds.). 1985. *Chemically Mediated Interactions Between Plants and Other Organisms.* Plenum, New York, 246p.

Cottrell, H. J. (ed.). 1987. *Pesticides on Plant Surfaces.* Wiley, New York, 86p.

Darvill, A. G., and P. Albersheim. 1984. Phytoalexins and their elicitors: A defense against microbial infection in plants. *Ann. Rev. Plant Physiol.* **35**:243–275.

Firm, R. D., and J. Digby. 1980. The establishment of topic cuvatures in plants. *Ann. Rev. Plant Physiol.* **31**:131–148.

Fuch, Y., and E. Chalutz (eds). 1984. *Ethylene: Biochemical, Physiological, and Applied Aspects: An International Symposium.* Nijhoff/Junk, The Hague, 348p.

Graham, D., and B. D. Patterson. 1982. Responses of plants to low, nonfreezing temperatures: Proteins, metabolism, and acclimation. *Ann. Rev. Plant Physiol.* **33**:347–372.

Hart, J. W. 1988. *Light and Plant Growth.* Allen and Unwin, Boston, 204p.

Heath, R. L. 1980. Pathology and initial events in injury to plants by air pollutants. *Ann. Rev. Plant Physiol.* **31**:395–431.

Huner, N. P. A. 1988. Low-temperature-induced alterations in photosynthetic membranes. *Crit. Rev. Plant Sciences* **7**:257–278.

Jackson, M. B. 1985. Ethylene and responses of plants to soil waterlogging and submergence. *Ann. Rev. Plant Physiol.* **36**:145–174.

Kader, A. A. 1985. Ethylene-induced senescence and physiological disorders in harvested horticultural crops. *HortScience* **20**:54–57.

Koziol, M. J., and F. R. Whatley (eds.). 1984. *Gaseous Air Pollutants and Plant Metabolism.* Butterworths, Boston, 466p.

Kozlowski, T. T. 1968–1983. *Water Deficit and Plant Growth,* 7 vols. Academic Press, New York.

Kramer, P. J. 1983. *Water Relations of Plants.* Academic Press, New York, 489p.

Lemon, E. R. (ed.). 1983. CO_2 *and Plants: The Response of Plants to Rising Levels of Atmospheric Carbon Dioxide.* Westview Press, Boulder, Colo., 280p.

Li, P. H. (ed.). 1987. *Plant Cold Hardiness*. A. R. Liss, New York, 381p.

Li, P. H. (ed.). 1989. *Low Temperature Stress Physiology in Crops*. CRC Press, Boca Raton, Fla., 203p.

Little, C. R., and I. D. Peggie. 1984. Storage injury of pome fruit caused by stress levels of oxygen, carbon dioxide, temperature, and ethylene. *HortScience* **22:**783–790.

Longheed, E. C. 1987. Interactions of oxygen, carbon dioxide, temperature, and ethylene that may induce injuries in vegetables. *HortScience* **22:**791–794.

Morgan, J. M. 1984. Osmoregulation and water stress in higher plants. *Ann. Rev. Plant Physiol.* **35:**299–319.

Paleg, L. G., and D. Aspinall (eds.). 1981. *The Physiology and Biochemistry of Drought Resistance in Plants*. Academic Press, New York, 492p.

Pickard, B. G. 1985. Early events in geotropism of seedling shoots. *Ann. Rev. Plant Physiol.* **36:**55–75.

Poovaiah, B. W., J. J. McFadden, and A. S. N. Reddy. 1987. The role of calcium ions in gravity signal perception and transduction. *Physiol. Plant.* **7:**401–409.

Rodricks, J. V. (ed.). 1976. Mycotoxins and other fungal related food problems. *Adv. Chem.* **149:**1–409.

Sachs, M. M., and T. H. D. Ho. 1986. Alteration of gene expression during environmental stress in plants. *Ann. Rev. Plant Physiol.* **37:**363–376.

Saunders, J. A., L. Kosak-Channing, and E. E. Conn. 1987. *Phytochemical Effects of Environmental Compounds*. Plenum Press, New York, 269p.

Schulte-Hostede, S. 1987. *International Symposium on Air Pollution and Plant Metabolism*. Elsevier Applied Science, New York, 381p.

Schulte-Hostede, S., N. M. Darrall, L. M. Blank, and A. R. Wellburn (eds.). 1988. *Air Pollution and Plant Metabolism*. Elsevier Applied Science, New York, 381p.

Schulze, E. D. 1986. Carbon dioxide and water vapor exchange in response to drought in the atmosphere and in the soil. *Ann. Rev. Plant Physiol.* **37:**247–274.

Sexton, R., and J. A. Roberts. 1982. Cell biology of abscission. *Ann. Rev. Plant Physiol.* **33:**133–162.

Simpson, G. M. 1981. *Water Stress on Plants*. Praeger, New York, 324p.

Snowden, A. L. 1990. *Color Atlas of Post-Harvest Diseases & Disorders of Fruits & Vegetables,* vol. 1: *General Introduction & Fruits*. CRC Press, Boca Raton, Fla., 302p.

Steponkus, P. L. 1984. Role of the plasma membrane in freezing injury and cold acclimation. *Ann. Rev. Plant Physiol.* **35:**543–584.

Stern, A. (ed.). 1976–77. *Air Pollution,* 3rd ed. Academic Press, New York.

Taiz, L. 1984. Plant cell expansion: Regulation of cell wall mechanical properties. *Ann. Rev. Plant Physiol.* **35:**585–657.

Timmermann, B. N., C. Steelink, and F. A. Loewus (eds.). 1984. *Phytochemical Adaptations to Stress*. Plenum Press, New York, 334p.

Treshow, M. (ed.). 1984. *Air Pollution and Plant Life*. Wiley, New York, 468p.

Turner, N. C., and P. J. Kramer (eds.). 1980. *Adaptation of Plants to Water and High Temperature Stress*. Wiley, New York, 482p.

Unsworth, M. H., and D. P. Ormrod (eds.). 1982. *Effects of Gaseous Air Pollution in Agriculture and Horticulture*. Butterworths, London, 532p.

7

MOVEMENT OF GASES, SOLVENTS, AND SOLUTES WITHIN HARVESTED PRODUCTS AND THEIR EXCHANGE BETWEEN THE PRODUCT AND ITS EXTERNAL ENVIRONMENT

Plants are composed of a diverse cross-section of organic and inorganic compounds found as solids, liquids, and gases. During the growth and development of a plant or a specific organ, chemical acquisition and distribution processes of tremendous scale are operative. Molecules and ions taken up by the plant must be moved from the site of acquisition to the site of utilization. For example, pumpkin fruits* achieve a final mass of approximately 800 g (dry weight), all of which is transported into the fruit through peduncle sieve tubes[7] having an average total diameter of only 0.05 cm^2. Knowing the concentration of organic material within the sieve tubes, Crafts and Lorenz[7] estimated that an average translocation velocity of 110 cm · hr^{-1} would have to be maintained to account for the tremendous accumulation of substances within a mature fruit. Plants, therefore, have evolved in their xylem and phloem tissues elaborate systems for the movement of chemicals throughout the plant. This rapid, long-distance transport in the xylem and phloem of growing plants is, however, dependent on the movement of water; thus, the ability of the plant to replenish water lost through transpiration is an essential requisite for proper functioning of the system.

Most plant products are severed from the parent plant at harvest, which generally eliminates the product's water replenishing potential, blocking the normal means for long-distance movement of solutes and water throughout the product. In some cases, water can be replenished (e.g., cut flowers held in aqueous solutions) and the transport system remains partly functional for a time.

While the normal transport system is inoperative in most harvested products, the redistribution of organic and inorganic compounds continues.

---

* Cultivar 'Connecticut Field'.

Redistribution is especially prevalent in products that are high in moisture. Transport of solutes and water is now characterized by a much slower rate of movement and generally over shorter distances. The newly dominant forms of movement are driven by different mechanisms than those operative in xylem and phloem transport of intact plants.

Redistribution of constituents within harvested products can be viewed at two levels: (1) a redistribution of compounds within the tissue or product and (2) a loss to or absorption from the external environment. In the latter case, losses normally occur as molecules in the gas phase (e.g., carbon dioxide, benzyl alcohol, esters of organic acids). A tremendous diversity of volatile molecules are given off by plants (chapter 4), and the uptake of gaseous molecules may also be extensive. In addition to oxygen, which is required for respiration, a cross-section of other molecules is absorbed. In dried agronomic crops, water vapor is absorbed and in some cases uptake is extensive. Water may also enter as a liquid, for example, when harvested products are washed, hydrocooled, or inadvertently exposed to rain. With improper temperature management, water vapor may also condense on the surface of the product and be readily absorbed.

Many harvested products not only undergo extensive metabolic and compositional changes after harvest but also undergo a significant movement and redistribution of cellular constituents within the product. Both the forces driving movement and the factors that impede movement are critical components in the redistribution process. To better understand the driving forces that result in the movement of ions and molecules, it is essential to first understand the sources of energy that drive movement.

FORCES DRIVING MOVEMENT

Energy is defined simply as the ability to do work (work = force × distance) and can be separated into two general classes: kinetic and potential. Kinetic energy is the energy of motion. Heat energy is a form of kinetic energy. The addition of heat increases the kinetic energy level and therefore the motion of molecules. Potential energy is the energy of position. An example would be the energy in chemical bonds of a sucrose molecule. For potential energy to do work it must be released. Within the plant this release occurs through chemical reactions. Respiration releases potential energy, some of which ends up as one form of kinetic energy (e.g., heat) or another (e.g., mechanical). Photosynthesis converts kinetic light energy into the potential energy of the carbon bond. Plants require a continual input of free energy or else they drift energetically downward, which eventually results in the cessation of life. The ultimate source of free energy is the sun; for most postharvest products, however, additional free energy inputs are no longer possible. Therefore, the theoretical survival potential of a product is governed to a large extent by its supply of free energy, its biological requirements for the energy, and the thermodynamic laws governing the expenditure of the free energy.

Not all of the total energy is available to do work. Part of it, the entropy factor, is lost. The remaining energy, or free energy, can be converted to

work. Free energy, therefore, in its various forms provides the force necessary to drive the movement of gases, solutes, and solvent within harvested
products. In many cases, differences in free energy between two locations
within the product result in spontaneous movement. For example, as a rice
(*Oryza sativa,* L.) grain desiccates, water moves from areas of high concentration or energy (the interior) to areas of low concentration or energy (the
surface). Both the rate and direction of movement are dependent on gradients
in energy. Movement proceeds from areas of high energy (high potential) to
low energy. Therefore, as the substance moves, it proceeds in an energetically downhill direction.

Many of the changes that occur within plants, however, must proceed in an
energetically uphill direction. These changes cannot occur spontaneously in
that to proceed they require an input of free energy. An example would be the
ATP-driven active transport of substances across the plasma membrane.
Generally the substance is moving from an area of low concentration to an
area of high concentration, against a gradient in energy potential; hence the
need for a supplemental energy input.

Free energy, or Gibb's free energy after the scientist who first described it
in the 1870s, is a measure of the energy within a system that can be used to do
work. Free energy can be viewed on a system or a location level, where the
free energy of each chemical component is added together to obtain a total for
the site. Likewise, it can also be expressed as the energy of a chemical
species, for example, the energy of a single type of substance (such as water)
is monitored. Since the amount of energy available is a function of some unit
quantity of the substance, free energy is usually expressed as free energy per
mole—the chemical potential. Chemical potential, therefore, is a function of
concentration.

Chemical potential also depends on pressure, electrical potential, and
gravitational forces. The existence of a pressure gradient can cause the
movement of liquids. The uptake of water into the cell due to osmosis
(diffusion across a differentially permeable membrane from a region of high
potential to one of lower potential) causes an increase in hydrostatic pressure
within the cell. This outward force influences the chemical potential. Likewise, many substances found within the plant are not electrically neutral. The
presence of a charge, whether positive or negative, influences the behavior of
that substance, and, as a consequence, its chemical potential. For charged
particles, potential is expressed as electrochemical potential to account for
charge effects. Lastly, gravitational effects are operative in very tall intact
trees; however, in harvested products, gravity has a negligible effect on
chemical potential.

Many of the chemical constituents of plants can be found in more than one
physical state (e.g., water is present both as a vapor and as a liquid). Since
one state often diffuses more readily than another, net movement may be
greatly influenced by the relative ratio between the two states. Many of the
principles governing the movement of gases differ from those governing the
movement of liquids. Thus, the movement of gases has been treated separately from that of solvent and solutes. Solute and water movement are
discussed together since water provides the avenue for solute transport and
the solutes in turn alter the movement of the solvent.

GASES: MOVEMENT AND EXCHANGE

While gases make up only a small portion of the Earth's biosphere,* they are of paramount importance to life. Their general lack of visibility tends to minimize awareness of them. Two of the most important gases are carbon dioxide, given off during respiration and consumed during photosynthesis, and oxygen, consumed during respiration and given off during photosynthesis. Because of their central roles in the plant's energy acquisition and utilization systems, they are critical to the overall metabolism of plants.

Water, unlike carbon dioxide and oxygen, is found as both a liquid and a gas at temperatures normally encountered during the postharvest period. Most plant products lose or gain water during storage due to the movement of gaseous water vapor into or out of the product. Alterations in the water status of the cells due to the movement of water vapor can, if of sufficient magnitude, exert a pronounced effect on metabolism and quality. Many other gaseous molecules are also important. Ethylene, a gaseous hormone that stimulates ripening, abscission, and senescence; aromatic compounds that enhance perception of a product; volatile insect behavior modulating chemicals; and air pollutants can have a significant impact on the postharvest behavior and quality of a plant or plant part.

Gaseous molecules may be beneficial or detrimental to the product; they may be produced by the plant or may be from external sources; and they may function either within the product (e.g., oxygen) or externally (e.g., aromatic compounds[24]). In some cases, a specific gas may be beneficial to one product but undesirable to another. The permeation of honeydew melons in storage with volatiles from adjacent onions is decidedly detrimental.

The effect of most gases that modify plants is concentration dependent. Below a certain threshold concentration, there is little or no response by the tissue. Each additional increment in concentration above this threshold level results in a progressively greater response. In some cases, the response saturates (e.g., ethylene-induced leaf abscission) in a relatively narrow range and elevation of the concentration above this range does not result in an increased response.

Because of this relationship between gas concentration and the magnitude of the tissue's response, understanding what controls the concentration of a gas at the site of action is essential. The physical properties of the gas, its site of origin (external versus internal), and factors affecting the supply (e.g., rate of synthesis) and movement into or out of the product modulate the concentration at this site.

One critical characteristic of volatile compounds is boiling point. Compounds such as oxygen, carbon dioxide, and ethylene have very low boiling points (i.e., $-183°C$, $-78°C$, and $-169°C$, respectively), and as a consequence, they are always gaseous at biological temperatures. On the other hand, most volatile molecules of plant origin are found in both a liquid and gaseous state [e.g., H_2O (bp 100°C), benzyl alcohol (bp 204.7°C)]. Since part of the movement of these compounds within the plant is by diffusion of their liquid phase, their movement and exchange differ from compounds present

---

* Total mass of the Earth's atmosphere is estimated to be only 5×10^{18} kg.

only as a gas. These two general classes of molecules (i.e., gases vs. gas/liquid) will be discussed separately.

Properties of Gases

Gaseous molecules are in constant motion due to their kinetic energy. The velocity of this motion increases with increasing temperature and decreases with increasing size of the molecule. Carbon dioxide molecules (molecular weight = 44) have an average velocity of approximately 1,368 km $\cdot$ hr^{-1} at room temperature while smaller hydrogen molecules (molecular weight = 2.01) approach 6,437 km $\cdot$ hr^{-1}. This value represents an average speed, with some of each type of molecule traveling faster and others slower than their respective means. The high-speed (energy) particles tend to be more reactive. For example, with water molecules, the high-energy particles are more likely to cause melting, evaporation, or chemical reactions while the low-energy particles tend to be the first to condense or freeze. Therefore, the energy level of a gas (i.e., temperature) is going to affect its rate of movement into and out of harvested products.

This high-speed, random movement of gases causes their molecules to expand outward, filling all the space available to them and taking the shape of whatever contains them.* This random motion also causes individual gas molecules to collide with neighboring molecules and the walls of the container. Due to the proximity of neighboring molecules and their rate of speed, one molecule may collide billions of times per second. When neighboring gas molecules do collide, typically one molecule will gain energy while the other loses it. Although the energy of each molecule is altered, the sum of the energies of the two molecules remains the same. Therefore, there is no net loss of energy due to collision between neighboring molecules.

Gases differ from liquids in that the distances between molecules are sufficiently great so that no cohesive forces exist. Although oxygen, carbon dioxide, and ethylene are not found as liquids at biological temperatures (fig. 7-1), many of the other gaseous molecules important during the postharvest period are found in both a gaseous and liquid state. Water and essentially all of the aromatic compounds fall into this latter group.

Generally, the higher the molecular weight of a molecule, the more likely it will be found predominantly as a liquid at room temperature (fig. 7-1). Larger molecules require more energy (heat) to break the cohesive forces between neighboring molecules and to escape into the atmosphere. If energy is added, the molecules are accelerated and a greater number of the molecules near the surface of the liquid are able to escape the attractive forces of their neighbors and enter the gas phase. An example in nature of a programmed increase in energy is found in many of the *Araceae*.[23] At flowering, a part of the spathe shifts a major portion of its respiration to the alternate electron transport

---

* The outward expansion of gases is eventually limited by gravitational forces. Fifty percent of the Earth's atmosphere is held within the first 5.5 km of the troposphere; 99% within 30 km of the surface (sea level). As the elevation increases, the density of molecules decreases. The gravitational effect on gases is insignificant, however, under normal postharvest conditions.

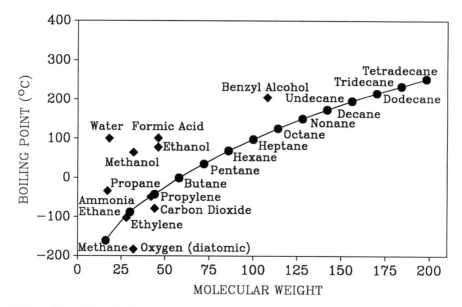

Figure 7-1. The relationship between boiling point and molecular weight of a series of saturated hydrocarbons (connected points) and other compounds of biological interest.

pathway. The alternate pathway's much lower energy-trapping efficiency results in a buildup of heat. The additional heat increases the energy level of several insect attractants that are found predominantly in a liquid state, causing more of the molecules to escape the attractive forces of the liquid and enter the atmosphere where they can be perceived by insects that facilitate pollination.

If energy is added in the form of heat to a gas, the speed of the individual molecules increases as does their outward movement. The volume of a given mass of gas, therefore, is altered, with the change in volume being proportional to the change in temperature (Charles' Law). Likewise, by cooling a gas from temperature T_1 to temperature T_2, the volume decreases from V_1 to V_2:

$$\frac{V_1 - V_2}{T_1 - T_2} = \frac{\Delta V}{\Delta T}$$

The ratio of volume to temperature ($\Delta V / \Delta T$) remains the same for a given mass of gas regardless of changes in temperature. If the volume of the gas is restricted while the temperature increases (e.g., a sealed cold storage room in which the refrigeration system has malfunctioned), the pressure increases. The pressure can be held constant only if the volume the gas occupies increases.* For a given mass of gas held at a constant temperature, the pressure and volume of the gas are inversely proportional (Boyle's Law). The

---

* In controlled atmosphere storage rooms this is accomplished by using "breather" bags open to the outside that change the volume of the room to compensate for changes in barometric pressure.

product of the volume times the pressure for a given mass of gas is the same even though the conditions (pressure or volume) change:

$$P_1V_1 = P_2V_2$$

The laws of Charles and Boyle can be combined into a more general gas law where volume equals temperature divided by pressure times a constant (k).

$$V = k\,(T/P)$$

The constant (k) depends on the amount and kind of gas being considered and may be found in standard chemical tables for most gases. If the same gas is being considered under two sets of conditions, k can be ignored. Then:

$$\frac{P_1V_1}{T_1} = \frac{P_2V_2}{T_2} \text{ (for a fixed mass of gas)}$$

How does this equation relate to harvested products? If the temperature (refrigerated storage), the pressure (hypobaric or low-pressure storage), or the volume of gas within or around a harvested product is altered, the way the gas interacts with the product will be altered. When the gas in question has a significant effect on the product, these changes in conditions are translated into an altered response by the product. For example, decreasing the pressure within a sealed hypobaric storage system decreases the number of molecules present to interact with the product. When the gas is produced by the product (e.g., ethylene, aromatic compounds, carbon dioxide), a reduced pressure also increases the steepness of the concentration gradient from the site of origin of the gas to the exterior, accelerating the rate at which the gas molecules diffuse out of the tissue.

The gas atmosphere surrounding the surface of the Earth exerts a pressure that at sea level is equal to 760 mmHg (1.03 kg $\cdot$ cm^{-2}). This pressure is due to the composite of all of the gases present; however, since scientists are often not interested in all gases, it is useful to determine the partial (part of the total) pressure exerted by one type of gas. A gas's partial pressure is a measure of what part of the total pressure it exerts relative to all of the other gases present. An atmosphere containing 21% oxygen at sea level (760 mmHg) has an oxygen partial pressure (pO$_2$) of 0.21 (21% of the whole). If a partial vacuum is created within a container holding the gases, and the pressure is reduced to one tenth of an atmosphere (i.e., 760 mmHg $\times$ .10 = 76 mmHg), the partial pressure of oxygen decreases proportionally.

$$\frac{0.21 \text{ pO}_2}{760 \text{ mmHg}} = \frac{X \text{ pO}_2}{76 \text{ mmHg}}$$

Therefore $X = 0.021$ pO$_2$. While nine tenths of all of the gas molecules have been removed from the container, the relative ratio of each type of gas within the container does not change, so the percent oxygen remains 21.

Movement and Exchange of Gases

Gases move from areas of high concentration to areas of low concentration by diffusion. This movement from one point in space to another is due to the random movement of the individual molecules caused by their kinetic energy. If a papaya (*Carica papaya*, L.) fruit, in which all of the gases within the tissue have been replaced with an inert gas such as helium, is placed in a sealed jar of nitrogen, molecules of helium begin to diffuse out of the fruit while molecules of nitrogen begin to diffuse into it. Gradually the concentration of helium and nitrogen inside and outside of the fruit become equal, reaching what is called an equilibrium state. While molecules of helium continue to diffuse out of the fruit, an equal number move in the opposite direction back into the fruit. Therefore, while there is a continual exchange, there is no net movement.

If the gas is being continually formed (e.g., carbon dioxide or ethylene) or utilized (oxygen) within the tissue, a more complex situation exists. When the flow of oxygen diffusing into the fruit is equal to the rate of utilization by the tissue, the system is said to have reached steady state. This state differs from an equilibrium state in that there is a continual movement of the gas through the system. Since most gases of biological interest are either being produced or consumed, a true equilibrium state is not reached; however, their exchange often approaches a steady state if the tissue and its environment are not altered.

Gases may also move by bulk flow. Here the movement is caused by a pressure gradient and all of the gases present move together rather than independently as with diffusion. In nature, the bulk flow of gases in plant products is not of major importance. An example, however, can be seen in the seed dispersal mechanism that has evolved in the squirting cucumber (*Ecballium elaterium*, (L.) A. Rich.). As the fruit ripens, the pressure within the central chamber of the fruit builds (i.e., up to 27 atmospheres[29]). Upon dehiscence, the peduncle plug separates from the fruit and the seeds are literally ejected out of the opening in the fruit wall, dispersing over a wide area.[18] A postharvest situation where gas movement is via bulk flow is found with the use of hypobaric (low-pressure) storage. As the atmospheric pressure is reduced around the product, the pressure differential causes an outward flow of gases from the product to equilibrate the internal and external pressures.

Conductance Versus Resistance

The rate of movement of a gas across a given space is referred to as flux density (i.e., flow and quantity). The flux density of a gas moving into or out of a harvested product can be viewed in two ways: by describing the ability or "ease" by which a gas can move through the material (conductance) or by the opposing forces, presented by the tissue, that impede movement of the gas (resistance). Resistance is the reciprocal of conductance (i.e., $C = 1/R$). It may be written:

$$\text{flux density} = \text{conductance} \times \text{force}$$

where force is the difference in concentration of the gas between the two sites, or:

$$\text{Flux density} = \frac{\text{force}}{\text{resistance}}$$

Therefore, flux density is directly proportional to conductance but inversely proportional to resistance. While conductance terminology is generally more readily conceptualized, resistances are specified more commonly than conductances. Resistances are also useful when studying the movement of a gas across a series of components in sequence, for example, the diffusion of ethylene across the cytosol, plasma membrane, cell wall, intercellular channels, tissue surface, and surface boundary layer. The resistances, being in series, are additive.

Tissue Morphology

There is a tendency to visualize the interior of harvested plant products as a dense mass of tightly packed cells. In fact, the interior contains an extensive system of intercellular gas-filled space. In most products, this system represents an interconnecting series of channels reaching virtually every cell through which gases may readily diffuse. The volume of this air-filled space varies widely between products. The cells in potato tubers are very tightly packed, with the intercellular space representing only about 1.0% of the total volume of the product. Most products are much less dense. The apple (*Malus sylvestris,* Mill.), for example, has approximately 36% intercellular space.[28] In addition, some products (e.g., pepper (*Capsicum annuum,* L. Grossum group), muskmelon (*Cucumis melo,* L. Reticulatus group), lotus root (*Nelumbo nucifera,* Gaertn.)) have a large central cavity or cavities that enhance the ability of gases to move readily throughout the tissue.

Intercellular gas volume, however, is not totally indicative of the potential for a gas to move once it is inside a product. The flux of a gas within a tissue depends on the cross-sectional dimensions and length of the air channels and their continuity and distribution within the tissue. In addition to the general architecture of the air channels, their condition is extremely important. As fleshy plant products begin senescence, membrane integrity declines and fluids may leak from the cytosol into the intercellular air spaces. Fluids found external to the plasma membrane displace an equivalent volume of gas, decreasing the effective volume of intercellular air space. This decline in volume may in turn sufficiently restrict the diffusion of gases so that senescence is accelerated.

Since gases diffuse readily in air, if the architecture and physiological condition of the intercellular air spaces are adequate, there will be little restriction (i.e., resistance) of gas movement. Therefore, only small gradients in concentration of gases over reasonable distances will occur within the tissue. However, if the diffusion of gases through the intercellular air channels is restricted, sufficient gradients may develop, resulting in concentrations that may contribute to the deterioration of the tissue.

Plants and their individual parts have achieved a compromise in structure relative to the diffusion of gases. While the movement of gases within the intercellular channels is in many products relatively free (i.e., low resistance), a similar unrestricted exchange between the plant and its environment would result in rapid death due to the loss of water. Early in their evolutionary path to a terrestrial existence, plants developed a waxy epidermal coating, the cuticle, over their aerial portion. In underground plant parts, the cuticle is replaced as the epidermal boundary layer by the periderm. In the periderm, the intercellular air channels are either absent or obstructed. As with the cuticle, the interior air channels are sealed, forming a boundary between the plant and its surrounding environment that greatly restricts the movement of water vapor and other gases. Therefore, unlike the intercellular air channels within most tissues, the outer surface represents a major barrier, an area of high resistance to the diffusion of gases.

Initially it was thought that cuticle thickness was the most important parameter controlling the rate of gas exchange. Cuticle thicknesses are known to vary widely, but in general have not been found to correlate closely with transpirational losses of water vapor. What appears to be of primary importance is the chemical composition of the cuticle and its structure and continuity.

The cuticular surface coating of aerial plant parts is not a continuous, sealed system; rather it is interspersed with localized areas of specialized cells that have a much lower resistance to gas exchange. Stomates and lenticels are the most important of these specialized gas exchange sites; however, other areas, both natural (e.g., hydathodes, stem scars) and accidental (e.g., surface punctures), may also be important.

Diffusion Path

For a gas to diffuse into the interior cells of a harvested product or outward from there, the individual molecules must pass along a diffusion path, from areas of high concentration to areas of low concentration. Different sites along the pathway present varying levels of resistance to the movement of the gas molecules. To understand the net resistance to movement, it is essential to examine each general step in the overall path and the level of resistance presented by it. In the following example, the movement of oxygen molecules from the exterior of the product to the mitochondria of interior cells is described (fig. 7-2). The reverse of the oxygen diffusion pathway is found for CO_2 molecules, originating in the mitochondria and moving outward. Since oxygen and carbon dioxide molecules have different chemical properties, the level of resistance presented by each step in the diffusion path will differ.

BOUNDARY LAYER RESISTANCE

Gas molecules moving from the exterior to the interior of a product must first pass through a boundary layer of air surrounding the product. The boundary layer is a localized area where the air is static and movement of individual molecules is by diffusion forces alone. Exterior to the boundary layer, air turbulence greatly facilitates movement.

The flux of gas molecules across the boundary layer depends on the diffusion coefficient (properties) of the gas in question, the difference in concentration (driving force) of the molecules across the boundary layer, and the thickness (distance) of the layer. Boundary layers represent important barriers to the movement of many gases (e.g., water, carbon dioxide, oxygen). The thickness of the boundary layer depends on the wind speed, size and shape of the product, and surface characteristics of the product (e.g., presence of trichomes). Excessively high air movement in storage rooms decreases the thickness of the boundary layer, which increases the rate at which water vapor is lost from succulent products, accelerating weight and

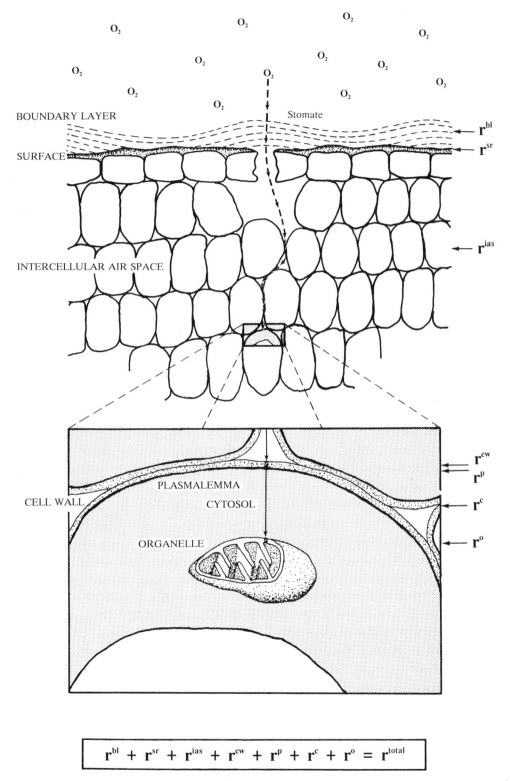

$$r^{bl} + r^{sr} + r^{ias} + r^{cw} + r^p + r^c + r^o = r^{total}$$

Figure 7-2. The pathway and resistances (r^{bl}, boundary layer resistance; r^{sr}, surface resistance; r^{ias}, intercellular air space resistance; r^{cw}, cell wall resistance; r^p, plasmalemma resistance; r^c, cytosol resistance; r^o, organelle resistance; r^{total}, total resistance) encountered by gases diffusing into or out of plant tissues.

quality losses. On the other hand, it increases the rate at which oxygen molecules can diffuse into the product.

The boundary layer of a stored product is found adjacent to its surface; however, other boundary layers may be present and important. For example, if the product is placed within a shipping container, the container's surface and its boundary layer represent two additional resistances that the gas molecules must overcome before approaching the product.

PRODUCT SURFACE RESISTANCE

The surface of harvested products is composed of cuticle or periderm that is generally interspersed with stomates and/or lenticels. Both the cuticle and periderm represent major barriers to the movement of gas molecules. For example, the resistance to the diffusion of water vapor from leaves is 5 to 300 times greater for the cuticle than for open stomates. Thus, the majority of the exchange of gases between the interior and exterior of a product is through these specialized openings. Other important exchange sites are stem scars, the cut end of peduncles, surface punctures, and other discontinuities of the outer integument.

Unlike the lenticels, the aperture of a stomata can be readily altered under appropriate conditions. In response to low internal carbon dioxide, potassium is transported into the two kidney-shaped guard cells of the stomata, lowering their water potential. To compensate, water moves into the guard cells, causing them to bow outward, forming an elliptical pore through which gases may readily pass. Thus, the resistance to gas diffusion presented by the stomates can change from extremely high (when closed) to relatively low (when open), depending on the conditions of the product and its environment. Both decapitation (e.g., harvest of lettuce (*Lactuca sativa*, L.)) and dark storage cause the stomates to close, greatly decreasing the exchange of gases.

While the porosity of lenticels (resistance) is not subject to the type of rapid and reversible changes seen with stomates, lenticel resistance is not necessarily static. Generally there are significant changes in resistance due to partial occlusion of the pores during storage. Thus, the diffusion resistance of gases into or out of products such as potato (*Solanum tuberosum*, L.) tubers has been shown to increase.

In addition to the resistance presented by an individual stomata or lenticel, the size, density, distribution, and condition of these surface openings are extremely important. The leaves of most dicots have a greater density of stomates on the lower surfaces, with a frequency of 40–300 per mm^2. When open, the combined apertures represent 0.2–2% of the leaf surface.

INTERCELLULAR AIR
SPACE RESISTANCE

Once entering the intercellular air space, gas molecules must overcome the resistances to diffusion found therein. Resistance encountered varies widely between individual types of products and their physiological condition. If the intercellular resistance is high, then a significant drop in concentration over a given distance within the tissue would be expected. These gradients in concentration are accentuated when the gas is being utilized (e.g., oxygen) or produced (e.g., carbon dioxide) by the cells along the diffusion path. Burton[3] calculated estimates of the size of this concentration gradient for potato tubers that have only about 1% air space. Tubers respiring at a relatively normal rate (2 cm^3 O$_2$ · kg^{-1} · hr^{-1}) would have a concentration gradient of only about 0.1% O$_2$ · cm^{-1}. In products that have much higher intercellular gas volumes (e.g., apple fruit), the gradients should be negligible. Therefore, under normal conditions, the resistance to diffusion of a gas presented by the intercellular air channels is small. However, if the intercellular air channels lack sufficient continuity due to their structure or the leakage of cell fluids, cells found only short distances from air spaces (i.e., 5 mm) may be anaerobic.

CELL WALL RESISTANCE

Oxygen molecules diffusing from the exterior of the product to the interior must now move from the intercellular air spaces and transverse the water-filled interstices of the cell wall. This movement is accomplished by the oxygen molecules first dissolving in the water and then diffusing in the aqueous medium. Critical parameters include the rate of partitioning of oxygen into the aqueous phase, the diffusion resistance presented by the medium, and the length of the path the molecules must transverse. A number of physical and chemical factors can affect each of these parameters. For example, the pH of the water held in the cell wall lattice can have a significant effect. This effect is especially true for CO$_2$, which can be found in solution as CO$_2$, H$_2$CO$_3$, or HCO$_3$$^-$, the respective concentrations being pH dependent (especially HCO$_3$$^-$). In general, cell wall resistance is considered to be quite low (i.e., approximately 100 s · m^{-1}).

PLASMALEMMA RESISTANCE

The plasmalemma is the outer membrane surrounding the cell, and its properties of semipermeability largely control the entry and exit of molecules from the cell. Since neighboring cells are connected in a continuous cell-to-cell membrane network (plasmodesmata), molecules do not necessarily need to exit through the plasmalemma to move to a neighboring cell (as will be seen, the plasmodesmata is the major diffusion pathway for volatile molecules that are found predominantly in a liquid state).

The potential for a gas in solution to diffuse through the plasmalemma depends on the properties of the gas in question (size, polarity, etc.), the conditions of the membrane, and the concentration gradient across the membrane. Molecules may move by simple diffusion across the membrane or may be actively transported to the opposite side. The resistance presented by the plasmalemma for the diffusion of dissolved gases such as carbon dioxide and oxygen is considered small to moderate (e.g., CO$_2$ is about 500 s · m^{-1}).

CYTOSOL RESISTANCE

Once within the interior of the cell, the dissolved oxygen molecules must diffuse to the mitochondria or other sites of utilization. Diffusion is not a major obstacle since the diffusion resistance of the cytosol is quite small (e.g., about 10 s · m^{-1} for CO$_2$), although greater than in water, and the mitochondria and other organelles are found relatively close to the periphery of the cell due to the central vacuole.

ORGANELLE RESISTANCE

Once transversing the cytosol, the oxygen molecules must pass through the outer membranes of the mitochondria (or other organelle) and subsequently the internal resistance between the interior and the site of utilization. Precise measurements of these resistances are not available; however, estimates for the diffusion of carbon dioxide into the chloroplasts and across their stroma are in the range of 400–500 s · m^{-1}. If active transport (energy-driven) or facilitated diffusion across the membrane occurred, the resistance would be substantially lower.

Carbon dioxide or ethylene molecules produced by the cells of a harvested product must follow the reverse of the oxygen diffusion pathway. Therefore, carbon dioxide molecules must diffuse across the organelle, its outer membranes, the cytosol, the plasmalemma, the cell wall, through the internal air channels, and across the product surface and boundary layer.

Combined Resistances

The total or net resistance to the movement of a gas is the sum of resistances in a series. Therefore:

$$r^{total} = r^{bl} + r^{sr} + r^{ias} + r^{cw} + r^p + r^c + r^o$$

where

$$
\begin{aligned}
r^{total} &= \text{total resistance} \\
r^{bl} &= \text{resistance of the boundary layer} \\
r^{sr} &= \text{surface resistance (cuticle/periderm and stomata/lenticel)} \\
r^{ias} &= \text{internal air space resistance} \\
r^{cw} &= \text{cell wall resistance} \\
r^p &= \text{plasmalemma resistance} \\
r^c &= \text{cytosol resistance} \\
r^o &= \text{resistance of the organelle}
\end{aligned}
$$

The surface resistance is composed of two or more resistances in parallel (e.g., periderm plus lenticel). Gas molecules can move through surface features. The reciprocal of the total resistance of a group in parallel is the sum of the reciprocals of the individual resistances. In this case:

$$r^{st} = \frac{r^{per} \, r^l}{r^{per} + r^l}$$

where r^{per} is the resistance of the periderm and r^l is the resistance of the lenticels, making up the surface total resistance (r^{st}) of a storage root.

While the total resistance (r^{total}) gives a good estimate of the ease or difficulty of getting a gas into or out of a product, it represents only part of the story. The flux of a gas is a function of force divided by resistance. High resistances may in some cases be compensated for by increasing the concentration gradient (driving force), or conversely, by not allowing the concentration gradient to be diminished (e.g., adequate aeration of the product).

In many cases during the postharvest period, the total resistance to diffusion of a gas is of primary interest. An estimate can be obtained by placing the product in an environment containing a known concentration of a gas, preferably one that is not used or produced by the tissue. With time the concentration of this gas within the product will equilibrate with the surrounding environment. At this time, the product is transferred to a second chamber containing air. Since the internal and external concentration of the gas (concentration gradient) and its diffusion coefficient (diffusion characteristics of the gas in question, obtained from chemical tables) are known, the rate of diffusion of the gas from the interior of the product into the surrounding air can be used to estimate the total resistance to diffusion.

MOVEMENT OF SOLVENT AND SOLUTES

The Aqueous Environment

Life on the planet Earth occurs in an aqueous medium with water representing the single most abundant component of actively metabolizing cells. Postharvest products range widely in their water content. Very young leaves of lettuce are as much as 95% water while many dried seeds are only 5–10%. Products such as seeds that are low in water tend to display a correspondingly low level of metabolic activity. An increase in water concentration is a prerequisite for renewed activity and growth.

The physical and chemical properties of water make it a suitable chemical for a tremendous diversity of functions with the plant.

- Due to the size of the molecule, its polar nature, and its high dielectric constant, water is an excellent solvent.
- Water functions as the transport solution, the medium through which solutes are moved throughout harvested products.
- Water represents the medium in which many of the reactions within the plant occur.
- It is a chemical reactant or product in a cross-section of biochemical reactions, for example, photosynthesis, ATP formation, and respiration.
- The incompressibility of water allows it to be used for turgidity of the cells, providing support and cell elongation during growth.
- The high heat of vaporization and thermal conductivity of water makes it the central temperature regulatory compound.
- The transport of water between cells causes a physical movement of certain plant parts (e.g., stomatal guard cells).

While all of the above functions are essential for plant life, the first two (water's role as a solvent and as a transport medium) are of particular importance in the redistribution of compounds within harvested products and the loss of specific compounds from these products. Most chemicals will dissolve to some extent in water and generally solubility increases with increasing temperature (exceptions would be salts, which are relatively unaffected by temperature, and calcium salts of some acids whose solubilities are actually decreased by increasing temperature). The dissolved substance or solute can be a gas, liquid, or solid.* When more than one solute is present in a solution, each solute generally behaves independently of the other.

---

* Gases may also dissolve in a gas.

Water is not the only solvent within plants. Lipids and certain other molecules also act as solvents. A number of cellular compounds are lipid soluble (e.g., chlorophyll, carotenoids). Although very slightly soluble in water, these molecules partition readily into lipids. A primary distinction between water and other solvents, however, is that water acts as a transport medium, an avenue for movement of solutes between cells, neighboring tissues, and organs. Movement of lipid-soluble molecules, in contrast, tends to be over only minute distances, for example, within the fluid membrane component of a chloroplast. When lipid-soluble compounds need to be transported, they are generally first broken down into water-soluble precursors or subunits. Transport then occurs in an aqueous phase with eventual reassembly upon reaching the destination.

Solutes range widely in their solubility in various solvents. For example, sugars are readily soluble in water whereas chlorophyll is relatively insoluble. A unit volume of solvent can dissolve only a fixed amount of solute. When the upper limit is reached, the solution is said to be saturated. When the concentration exceeds this upper limit, solid solutes will crystallize or precipitate out of the solution, while liquid solutes will form droplets of the compound.

Gases, like liquid and solid solutes, are also soluble in water and their degree of solubility likewise varies widely. Gases such as oxygen and nitrogen are only slightly soluble in water and carbon dioxide is very soluble (approximately 100 times more soluble than oxygen). The solubility of a gas is inversely proportional to temperature. Therefore, as the temperature increases, the solubility of a gas decreases. Solubility is directly proportional, however, to the partial pressure of the gas in the atmosphere over the solvent. As the concentration of carbon dioxide in the atmosphere increases, the concentration in solution increases correspondingly. The very high solubility of carbon dioxide in water is due in part to its reaction with water molecules to form a carbonate:

$$CO_2 + H_2O \leftrightarrow H_2CO_3 \leftrightarrow H^+ + HCO_3^-$$

In addition to increasing the concentration of carbon dioxide in solution, the reaction alters the pH of the solution, which if of sufficient magnitude may alter the metabolism of the cells.

Due to its high solubility, carbon dioxide will also diffuse readily in water, allowing it to move within the aqueous medium of a harvested product.

Potential Pathways for Movement in a Liquid Phase

The cytoplasm of neighboring cells is interconnected, forming a virtually continuous system, the symplast, throughout the interior of a product. Water and dissolved solutes move from cell to cell through the symplast system in response to energy gradients. Water and solutes are also found in the apoplast system external to the plasma membrane and movement may be via this second avenue. To enter the apoplast, the solute must first transverse the plasma membrane of the cell; however, its semipermeability restricts the outward movement of many molecules. Once in the apoplast, water and solutes diffuse toward areas of lower chemical potential.

Water and solutes may diffuse in response to energy gradients in either the symplast or apoplast systems. For molecules that can transverse the plasmalemma, both routes may be operative; quantitatively the most important will be the path that represents the least resistance. Most of the movement of water is thought to occur within the apoplast. For molecules that do not readily transverse the plasma membrane, the symplast represents the dominant route.

Liquid solutes moving through the symplast system must transverse the plasma membrane upon reaching the surface if evaporation (loss from the product) is to occur. In some cases, specialized surface cells have ectoplasmata that allow otherwise nonpermeable molecules to move from the cytosol to the exterior. In other surface cells, the plasma membranes, due to differences in composition, may be more readily transversed by certain molecules than the plasma membranes of interior cells. Once exterior to the plasma membranes of the surface cells, volatilization of liquids or crystallization of solids may occur.

The rate at which liquid molecules can escape into the atmosphere is in part a function of how readily they or their precursors can be moved to the surface. Products with large surface areas and relatively short diffusion paths (e.g., leaves) tend to lose liquids more rapidly than bulky products. For example, spinach (*Spinacia oleracea*, L.) leaves lose water approximately 200 times faster than potato tubers under similar conditions.[27]

Factors Affecting the Diffusion of Solvent and Solutes

Operative Forces

Like gases, solvents and solutes move in the plant in response to concentration differences; however, with liquids there are a number of other factors that are also important. Therefore, the movement of solvent and solutes is described in response to gradients in free energy or chemical potential, not just concentration, which represents only one component of free energy or chemical potential. Both the tendency for diffusion and the direction in which it occurs depend on a gradient in chemical potential. For example, in detached fleshy products, significant differences in water concentration due to evaporation may develop, establishing a concentration or chemical potential gradient. Water moves in response to this gradient from areas of high concentration to areas of low concentration.

The presence of solute particles (e.g., sugars, mineral ions) decreases the chemical potential of the solvent molecules in relation to the mole fraction present (number of solute-to-solvent molecules). If an area of high solute concentration is separated by a semipermeable membrane from an area of low solute concentration, the solvent potential is lowest on the side of the membrane that is high in solute molecules (fig. 7-3). If the membrane is permeable to the solvent, the solvent moves across the membrane from areas of high potential (high solvent concentration) to areas of low potential (high solute concentration).

Matric effects are also important in modulating the movement of a solvent. Many of the molecules within a plant have a high affinity for solvents, especially water. Proteins, cellulose, starch, and other molecules are hydrophilic (water-loving) and possess negative charges on their surfaces that attract the positive side of the polar water molecules. Materials that bind

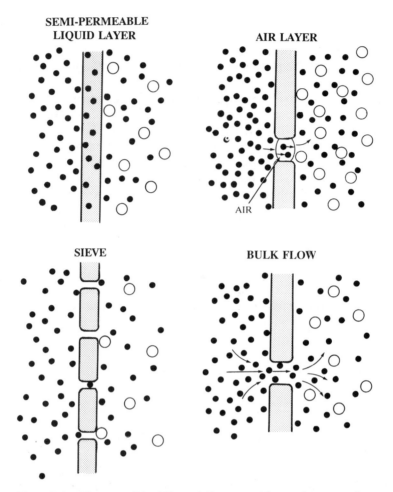

Figure 7-3. Four possible differentially permeable membrane mechanisms that would allow the osmotic diffusion of a solvent across the membrane but prevent the movement of the solute in the opposite direction.

solvent molecules are called a matrix. This affinity of the matrix restrains the movement of the solvent molecules.

Temperature and pressure also influence the movement of liquids. Elevated temperature increases the free energy of a system, increasing movement. Diffusion will occur from points of high temperature to points of low temperature. Since evaporation decreases the temperature of the surface, temperature gradients can be present within a product and affect diffusion. Likewise, increases in pressure increase the level of free energy, which can also enhance solvent movement.

Resistance to the Movement of Solvent and Solutes

From its various sources, energy provides the force required to drive the movement of solvent and solutes. The flux or movement of solute ions or molecules, however, is modulated by more than just these driving forces.

Flux is equal to force divided by resistance. Within both the cell and the apoplast are various resistances that impede the free movement of ions and molecules to varying degrees. The two most important resistances to free diffusion are imposed by membranes and the cell wall.

Individual protoplasts within harvested products are surrounded by a plasma membrane that controls the entry and exit of ions and molecules. Various solutes display marked differences in their ability to transverse this semipermeable barrier. For most solute molecules, the rate-limiting step controlling movement into or out of the cell is diffusion through this membrane. Even when a solute can diffuse readily through the membrane, the rate of diffusion is greatly restricted in comparison with its diffusion within the cytosol or extracellular water. The plasma membrane, therefore, represents a major barrier or resistance to the free movement of most molecules. While encountering some resistance, water molecules, in contrast, move quite readily through the membrane.

In addition to the plasma membrane, most other subcellular organelles are surrounded by membranes. Mitochondria, for example, are enclosed within a double membrane. Dissolved oxygen molecules must diffuse from within the water of the apoplast, through the plasma membrane, across the cytosol, and finally transverse the mitochondrial membrane systems to participate in respiration. Both the plasma membrane and the mitochondrial membranes represent significant resistances when compared with the diffusion of oxygen within the cytosol.

The tonoplast membrane surrounding the vacuole is also of critical importance to cellular metabolism in that it controls the movement of substances into and out of the vacuole. Many of the compounds found within the vacuole would impair or terminate the normal functioning of the cell if they were allowed to enter the cytosol.

What imparts the semipermeable nature to membranes? Lipids represent the largest single component of membranes. Due to their chemical nature, they present an environment within the membrane that differs distinctly from that of the aqueous medium of the cell. Molecules that passively transverse the membrane must first dissolve in this lipid layer before diffusing across it. Therefore, molecules that are lipid soluble partition (move) into the membrane much more readily than those that are highly water soluble. Since the chemical composition varies between membranes (e.g., tonoplast vs. mitochondrial), the permeability of different classes of compounds also varies. Molecules that pass readily through one membrane may be greatly restricted by another.

The ions and molecules that pass through the barrier imposed by a membrane do so by one of two general means: passive or active transport. With passive transport, molecules diffuse across the membrane driven by differences in chemical potential between the respective sides of the membrane. Movement is from the side with the highest chemical potential to the side with the lowest, with the rate of transport being a function of the magnitude of the difference in chemical potential across the membrane. If the molecule carries a positive or negative electrical charge, the charge becomes a part of the total potential (electrochemical potential) that drives diffusion.

In many cases, the transport of ions or molecules across a membrane is against a chemical potential gradient or at a rate that cannot be accounted for by the laws of diffusion or energy potential. Here transport (termed active transport) is in an energetically uphill direction (against the chemical potential gradient) and cannot be achieved without the input of free energy. There have been several model systems proposed to account for the input of energy and the resulting movement. One involves a carrier molecule found within the membrane that attaches to the molecule or ion to be transported. The energy released in the conversion of ATP to ADP drives the movement of the carrier molecule complex to the opposite side of the membrane where the transported substance is released to diffuse away from the membrane. A second ATP molecule is required to move the carrier molecule back to the opposite side of the membrane.

Active transport is essential in plants in that it allows the concentrating of certain substances and the exclusion of others. The loading of sucrose molecules from the apoplast of leaf mesophyll cells into the phloem is an example of an active transport process. If the energy input is blocked, transport ceases.

Factors Affecting the Rate of Diffusion

Knowing the factors that affect the chemical potential of a liquid, what then controls the rate of diffusion? The steeper the chemical potential gradient, the more rapidly diffusion will occur (assuming all other factors are held constant). Likewise, the level of permeability of a membrane to a specific compound significantly alters the rate of diffusion of the molecules. Some compounds move readily through solvents and/or membranes, allowing them to respond more readily to a given gradient in chemical potential. And finally, temperature affects the rate of diffusion. High temperatures increase the average velocity of the solute molecules, increasing their rate of diffusion.

Volatilization of Solvent and Solute Molecules

The volatilization of molecules from the liquid phase occurs when the energy level of individual molecules is high enough to overcome the attractive forces of neighboring molecules. Experimentally these forces are generally measured in pure liquids. In the plant, however, the liquid phase of a compound is exposed to a tremendous cross-section of other molecules (both solid and liquid) that can exert an influence over the liquid in question. The collective effect of these neighboring molecules modulates the general tendency of the compound in question to volatilize. The presence of solutes and solids can greatly affect the rate at which water evaporates (volatilization at a temperature below the liquid's boiling point). Adding sucrose to water increases the amount of energy required for a given amount of water to evaporate. Various liquids differ greatly in their potential for volatilization and this difference is used for the separation of liquids by distillation. Likewise, temperature and pressure alter the rate of volatilization. Evaporation occurs more readily with each increment in elevation of temperature.

Volatilization in turn cools the liquid due to the absorption of energy that occurs during the change of state. The evaporation of 1 g of water from a

1,000-g reservoir lowers the temperature of the remaining liquid by 0.59°C. Likewise, a boiling liquid does not increase in temperature above its boiling point, regardless of the level of energy input, due to the cooling effect of volatilization.

Evaporation also increases as the atmospheric pressure decreases. Therefore, the boiling point of water decreases from 100°C at sea level to only 95°C at an elevation of 1 mile (1.6 km), a common altitude in many mountainous regions of the world. In hypobaric storage systems, the decreased pressure can greatly increase the loss of volatile liquids from the product. If these represent critical components of quality (e.g., volatile flavor compounds), quality can be significantly diminished.* With water, losses can be minimized by keeping the storage room atmosphere as near saturated as possible. Saturation of the air is not presently possible for most other volatile liquids, especially when some air exchange is required in the storage environment.

Evaporation differs from boiling in that evaporation is a surface phenomenon. The change in state occurs only at the surface of the liquid. With boiling, volatilization takes place within the liquid as well as at the surface. Therefore, the surface area of a liquid is an extremely important physical characteristic affecting evaporation. If other factors are held constant, evaporation increases in relation to surface area. Therefore, plant products with high surface-to-volume ratios will lose water or other volatile liquids at a faster rate than those that have a lower surface-to-volume ratio.

The net escape of liquid molecules into the atmosphere is also affected by the relative concentration of these molecules already within the gas phase. In a closed system, as the concentration of these molecules in the gas phase increases, more of the randomly moving molecules return by chance to the liquid phase. When the number of molecules leaving the surface equals those returning, the atmosphere is said to be saturated (i.e., no net loss or gain in either phase).

Molecules that have escaped their liquid surface and exist in the vapor state exert a pressure called the vapor pressure. The vapor pressure is proportional at a given temperature to the net number of molecules escaping into the vapor phase. Therefore, increasing the chemical potential of the liquid (e.g., raising the temperature) increases the vapor pressure of the molecules in the vapor state. The addition of dissolved solutes, on the other hand, decreases the liquid's chemical potential, decreasing the vapor pressure. The rate of evaporation of a liquid such as water from a plant product is proportional to the difference in the vapor pressure for water within the product and the surrounding atmosphere. This differential is called the vapor pressure deficit, and as it increases, the rate of evaporation increases.

Site of Evaporation

Evaporation is a surface phenomenon and plants have two primary surfaces from which it occurs. Surrounding each interior cell is an inner connecting

---

* The subsequent synthesis of the volatile by the tissue after removal from hypobaric storage may also be seriously impaired.

series of air spaces. A portion of the liquids found exterior to the plasma-lemma in the apoplast volatilize within the air space, eventually approaching an equilibrium between the liquid and gaseous phases. In the gas phase, these molecules diffuse through air channels down a concentration gradient toward the surface of the product, eventually escaping to the surrounding atmosphere. Evaporation may also occur at the surface of the product. The predominant site for water evaporation from intact plants is at this surface-exterior interface. For water, most of the evaporation actually occurs on the inner sides of stomatal guard cells and adjacent subsidiary cells. Water also evaporates from nonstomatal areas, for example, lenticels, cuts or breaks in the surface continuum, the cuticle, and the periderm. The amount of cuticular and peridermal transpiration varies widely between species. Generally cuticular evaporation represents a relatively small component of the total water lost from a product.

Which of the two sites, cell surface or product surface, is the most important in the evaporation of liquids and their loss from the plant in a volatile state? Based on an analogy with the movement of oxygen or carbon dioxide, one can imagine molecules of a liquid, such as water, present in the wall matrix of interior cells, evaporating and diffusing down a planar front toward the exterior of the product where the concentration is low. With this scenario, movement of the compound to the product surface is predominantly as a gas rather than as a liquid. This, however, is seldom the case. Although some movement does occur in the vapor state in succulent products, water moves to the surface of the product predominantly in the liquid state.

Due to the presence of water in the wall matrix, the intercellular air spaces are saturated with water vapor. Saturated air is found not only around the interior cells but surrounding the cells as movement is toward the more exterior regions of the product. This high concentration of molecules in the vapor state results in an extremely small drop in vapor state concentration (the diffusion gradient or driving force) between the interior and exterior cells. Since the rate of movement is a function of the steepness of the concentration gradient, diffusion is slow. Most of the movement of liquids between the interior and exterior of the product, therefore, occurs in the liquid state. Exceptions may occur when there is a very localized site of synthesis of a specific compound, for example, only in the cells adjacent to the seeds of a fleshy fruit. Under such a situation, it is conceivable that significant gradients also occur in the vapor state between the site of synthesis and the exterior.

EXCHANGE OF WATER BETWEEN PRODUCT AND ENVIRONMENT

Due to the central role of water in the postharvest biology of plant products, the movement of water between the product and its environment can be of tremendous importance. Depending on the product and the environment in which it is held, movement of water can proceed from the product to the surrounding environment, from the environment into the product, or when in equilibrium, exhibit no net change. Water movement is always toward establishing an equilibrium between the product and its environment, with the individual water molecules diffusing from areas of high chemical potential toward areas of low chemical potential.

Environmental Factors Affecting Water Exchange

During storage the primary concern is to minimize the exchange of water between the product and its environment. With some products, however, prior to storage or during the initial phase of storage, conditions are created to enhance the exchange of water. For example, many grains and pulse crops are exposed to environments that are conducive for drying, that is, causing a net loss of water from the product. In certain other products, it may be desirable to replace the water lost during handling and storage. At the retail level, lettuce is often briefly (10–15 min) submerged in water to rehydrate wilting leaves.

The net flux of water into or out of a detached harvested product is determined by the magnitude of the force's driving movement and the composite of resistances to movement. Environmental factors such as humidity, temperature, pressure, and air movement can have a direct effect on the force's driving movement.

Humidity

The amount of water vapor in the air surrounding a plant product has a pronounced effect on the movement of water into or out of the tissue. Typically humidity is measured using dry and wet bulb thermometer readings from a psychrometer and a psychrometric chart (fig. 7-4). The amount of water vapor present can be expressed in a number of ways, for example, percent relative humidity, absolute humidity, vapor pressure, and dew point.

The most commonly used and misused expression of the moisture content of the air is the percent relative humidity, a ratio of the quantity of water vapor present and the maximum amount possible at that temperature and pressure. Both temperature and pressure have a pronounced effect on the amount of water vapor the environment will hold and seldom during the postharvest period are constant temperature conditions found. For example, there is a small but consistent fluctuation in air temperature in refrigerated storage rooms due to thermostats turning on or off the flow of refrigerant to the coils. In unrefrigerated grain storage bins, temperature also varies with location in the bin. These differences in temperature have a significant effect on the amount of water vapor present in the air without necessarily changing the percent relative humidity. For example, air at 40% RH, 20°C contains approximately 9 g of water vapor $\cdot$ kg^{-1} of dry air. At 30°C and the same percent relative humidity, the air contains 11 g $\cdot$ kg^{-1} of dry air. Although the percent relative humidity is identical, the concentration of water vapor differs greatly.

A more precise measure of humidity is the absolute humidity.* The abso-

---

* Also referred to as specific humidity and humidity ratio when the two weights are in the same units (i.e., kg $\cdot$ kg^{-1}).

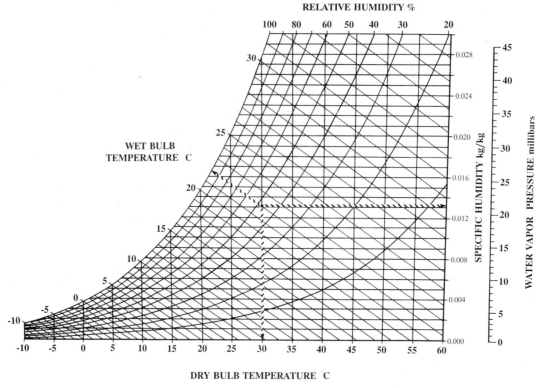

Figure 7-4. Psychrometric charts allow determination of the thermal and moisture characteristics of the atmosphere using any two parameters (e.g., wet bulb temperature, dry bulb temperature, percent relative humidity, specific or absolute humidity, or water vapor pressure). For example, if the dry bulb temperature is 30°C and the wet bulb temperature is 22°C, the point on the chart where the 45° angled wet bulb line intercepts the vertical dry bulb temperature line is the percent relative humidity (curved line), which in this case is 48%. The absolute (specific) humidity and vapor pressure can be read by moving horizontally to the right. The chart is based on a barometric pressure of 1,013.25 mb.

lute humidity is the weight of water in a given weight of dry air (g · kg⁻¹) and is independent of temperature and pressure. Therefore, two storage environments at the same percent relative humidity but at different temperatures will have distinctly different absolute humidities.

Water vapor pressure represents an alternate measure of the amount of water vapor in the atmosphere. Since water vapor is a gas, just as oxygen and nitrogen, it makes up part of the total atmosphere. Each of the gases in the atmosphere exerts a pressure, part of the total pressure. Water vapor pressure is a measure of the pressure exerted by water vapor in the atmosphere. At a given barometric pressure there is a direct relationship between the vapor pressure of water and the absolute humidity.

A useful postharvest measure is the difference in the water vapor pressure between two locations, the water vapor pressure deficit. Since concern is for the movement of water into or out of stored products, the difference in water vapor pressure between the interior of the product and its surrounding environment gives an estimate of how rapidly water will move between the product and its environment. The actual rate is controlled by more than just the vapor pressure deficit, however, since water exchange is also affected by the characteristics of the product itself (e.g., resistance to diffusion–waxy cuticle, amount of water present–chemical potential of the water, etc.).

In succulent plant products, the internal air spaces are considered to be saturated with water vapor.* Thus, it was possible to develop a simple nomogram for water vapor pressure deficit that allows making a direct reading from temperature and the percent relative humidity in the environment (fig. 7-5).[30] The higher the vapor pressure deficit (i.e., the greater the gradient), the more rapid the loss of moisture from the product. Comparison of the water vapor pressure deficit between different storage conditions (temperature and/or percent relative humidity) allows ascertaining under which set of conditions water will be most readily lost from a product.† For example, a product stored in a room with a vapor pressure deficit of 5.0 millibars (mbs) will lose water twice as fast as the same product stored under conditions where the vapor pressure deficit is only 2.5 mbs.

The dew point is the temperature at which the air is saturated with water vapor (100% relative humidity). When lowered below the dew point temperature, condensation occurs since the air can no longer hold as much water. The dew point temperature can be determined from the air temperature (dry bulb) and the relative humidity using a psychrometric chart (fig. 7-4). Dew point is a very important and often neglected parameter in postharvest handling and storage. If a commodity is brought to a temperature in refrigerated storage below the dew point of the air outside of the storage room, water will condense on the surface of the product upon removal from storage. For some crops it can have disastrous consequences.

For example, several years ago a large shipment of Chinese chestnuts (*Castanea mollissima,* Blume.) was being transported from the Orient in a refrigerated vessel. The refrigeration system malfunctioned midway in the voyage. The crew knew that live plant material would begin to heat up in a sealed, insulated room due to its respiration, so to prevent this situation the doors of the refrigeration rooms were opened. Due to the higher temperature and humidity of the air at sea level, and a product temperature below the dew point of the air, water condensed on the surface of the chestnuts. Within 48 hours, the refrigeration system was repaired and the rooms were reclosed; however, upon reaching the port of destination the entire shipment was a solid mass of fungal mycelium. If the shipping line operators had been cognizant of the importance of dew point, a several hundred-thousand-dollar claim could have been avoided.

Temperature

Assuming other factors are held equal, increasing the product temperature increases the free energy of the water molecules, which increases their movement and potential for exchange. Due to the heat given off during

---

* Water saturated internal air spaces are seldom the case in drier products such as grains and pulses.
† Comparisons cannot, however, be made between different types of products.

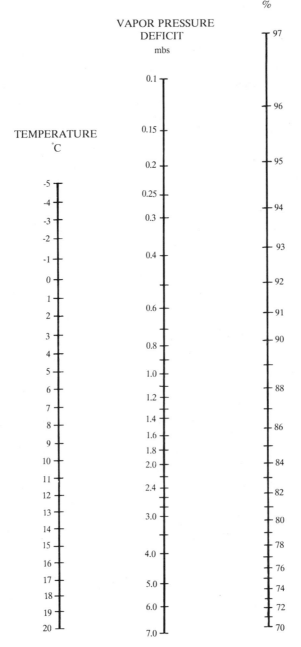

RELATIVE HUMIDITY
%

VAPOR PRESSURE
DEFICIT
mbs

TEMPERATURE
°C

Figure 7-5. A nomogram can be used to determine the water vapor deficit between a fleshy product and its surrounding environment using the air temperature and relative humidity. A line is drawn from the air temperature *(left)* to the relative humidity of the air *(right)*. The intercept on the vapor pressure deficit line *(center)* gives the vapor pressure deficit between the product and its environment; the higher the vapor pressure deficit, the faster water is lost. Accurate use of the nomogram requires that the product and air be at the same temperature and that the product's internal gaseous relative humidity be 100%. The latter is not a valid assumption with low-moisture crops (e.g., grains).

respiration, stored products normally have a slightly higher temperature than the surrounding atmosphere, which enhances water loss.

Temperature also affects the amount of moisture that can be held in the air surrounding the product. As the temperature decreases, the maximum amount of moisture that can be held by the air also decreases. Fluctuations in temperature can result in a much more rapid water loss from stored products than a constant temperature. Temperature deviations such as reaching the dew point have already been described; however, changes in temperature need not be large. For example, small changes due to the thermostat of a direct expansion refrigeration system going on and off can result in significant fluctuations in relative humidity and moisture loss from the product.[14]

In grain storage the movement of moisture from one area of a closed storage bin to another is a common problem. Although the grain moisture levels may have been considered safe when placed in storage, drying of one area due to a slightly elevated temperature results in the transfer of water to cooler grain, causing wetting. It is particularly noticeable in stores of greater than 70 m^3 of product and can result in significant losses.[16]

When the air outside of the storage bin is cold, it cools the grain along the walls (fig. 7-6). Air within the storage chamber moves due to convection, cycling along the outer walls and returning through the center of the grain mass. Since the equilibrium between the moisture content of the grain and the air changes with temperature, warm air holding more moisture rises through the center. As the air reaches the cooler grain at the top, moisture begins to move from the air into the grain, causing an area of moisture accumulation and potential spoilage at the top. When the air outside is warm, the grain in the center is cooler and the airflow is reversed, causing water to accumulate at the base of the bin (fig. 7-6). This redistribution of moisture is a greater problem when (1) the moisture content of the grain is relatively high; (2) the grain is placed in storage in warm weather rather than cold weather; (3) there is a large difference between the temperature of the grain and the external atmosphere; and (4) tall storage structures are used.

Pressure

As with increasing temperature, elevated pressure increases the free energy of water molecules, which in turn increases movement. However, changes in free energy due to increased pressure generally tend to be relatively insignificant in harvested products. In contrast is the effect of reducing the pressure. While reduced pressure decreases the free energy of the water molecules, it also can greatly increase the concentration gradient between liquid phase molecules of water within the tissue and gaseous water molecules in the surrounding atmosphere. This elevation in the steepness of the concentration gradient increases the net movement of water out of the tissue due to the increased differential in the chemical potential of water found within the tissue (liquid phase) and that found in the atmosphere exterior, surrounding the product (gas phase). The rate of evaporation is inversely proportional to pressure; for every 10% decrease in pressure there is a 10% increase in the rate of water loss.

Products placed under a partial vacuum will lose water until an equilibrium

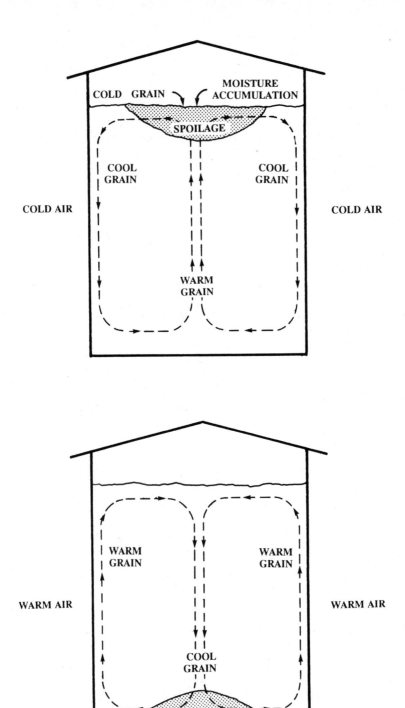

Figure 7-6. Convection currents of air cause a redistribution of moisture in stored grain when *(top)* the outside air is colder than the grain, and *(bottom)* the outside air is warmer than the grain.

is established between the product and its environment. An equilibrium would occur in a small,* sealed chamber at a constant pressure and temperature. In practice, it is seldom this straightforward. Since postharvest products are alive, utilizing oxygen and producing carbon dioxide, fresh air must be continually introduced into the vacuum chamber. If the fresh air does not contain the correct concentration of water vapor (equilibrium concentration), a gradient will be established between the product and the environment. With succulent crops, desiccation occurs. Dryer crops (e.g., rice) either desiccate or rehydrate depending on the direction of the gradient.

The effect of pressure on water loss from harvested products is also important at a high altitude. The cargo areas of airplanes are pressurized to an altitude of about 1,500 m where the air pressure is approximately 130 mmHg less than at sea level. Therefore, there is approximately a 20% differential favoring accelerated water loss.

Air Movement

Air movement over the surface of a commodity decreases the thickness of the boundary layer, decreasing the boundary layer's resistance to exchange of water molecules between the product and its environment. With succulent products this situation increases water loss, the increase at a specific velocity being relative to the magnitude of the vapor pressure deficit between the product and its environment. In dry products, if the water chemical potential gradient is in the opposite direction (i.e., higher in the air than the product), increasing air movement can accelerate rehydration.

The increase in rate of water loss from succulent products does not increase linearly with increasing air velocity. Rather there is a very sharp rise at relatively low air velocities, with the effect decreasing as the velocity increases. The importance of air movement due to convection alone in grain stores has been described in the temperature effects section.

In a closed, refrigerated storage environment, air movement also has an important influence on the vapor pressure deficit between the product and its environment. Under refrigerated conditions, the higher the volume of air circulated per unit time over the cooling coils, the lower the differential in temperature between the return air and the delivery air. A narrow temperature differential decreases the amount of water removed by the refrigeration coils, decreasing the gradient in vapor pressure between the product and air. Thus, increasing the volume of air moved, to a point, can help to maintain high humidity within the chamber. Optimum air movement under refrigerated storage conditions is a compromise between attaining a low-temperature differential between the coils and air and preventing excessive water loss due to reduction of the boundary layer around the product.

Light

Water loss after harvest can be significantly increased by light through its effect on stomatal aperture and/or indirectly by increasing the temperature of

---

* Relative to the volume of the product.

the product. Water losses generally increase with increasing light intensity[19] and duration. Cut roses held under constant light lose five times more water than those in alternating light-dark cycles of 12 hours.[5] When held in dark storage, cut roses did not display the progressive decline in water uptake typically seen with storage in the light after harvest.[8]

Plant Factors

The flux of water from a given surface area and over a fixed time interval into or out of a plant product is equal to the water vapor deficit divided by the cumulative resistances to movement:

$$\text{flux} = \frac{\text{water vapor inside} - \text{water vapor outside}}{\text{resistance to water movement}}$$

Plant characteristics that decrease the water vapor pressure deficit between the interior and exterior and/or increase the resistance to water movement will reduce the rate of exchange of water between the product and its environment. The external water vapor pressure is an environmental parameter; however, the internal water vapor pressure is strongly modulated by the amount of moisture present within the tissue. With succulent plant parts, the interior is generally considered to be at 100% relative humidity. However, with drier, more durable, crops (e.g., grains, pulses, nuts), it is seldom the case. While the water content, or more precisely the chemical potential of the water in the tissue, affects the internal water vapor pressure, the difference in water vapor pressures between the interior and the exterior represents the driving force.

Plant characteristics that increase the resistance to movement of water impede the rate of exchange between the product and the environment. For example, natural or artificial waxes[26] on the surface of the product increase surface resistance. The presence of trichomes often increases boundary layer

Table 7-1. Typical Surface-to-Volume Ratios of Harvested Plant Products

Normal Range in Surface/Volume Ratio $(cm^2 \cdot cm^{-3})$	Plant Material
50–100	Individual leaves (surface exposed); very small seeds
10–15	Cereal grains
5–10	Legume seeds; small fruits (e.g., raspberry, blueberry, currant)
2–5	Legumes (fruits); most nuts; intermediate size fruits (e.g., strawberry); rhubarb; celery; Chinese water chestnut
0.5–1.5	Tubers, storage roots and taproots; apple, pear, peach and citrus fruits; cucumber; squash; banana; plantain; onion bulbs; lotus root
0.2–0.5	Large densely packed cabbages; large turnips, sugar beets and yams; coconut; pumpkin

resistance. The chemical composition and structure of the product can affect how tightly water is held by the tissue. And finally, since evaporation is a surface phenomenon, the ratio of surface area to the volume of the product is of critical importance. Harvested plant products exhibit a tremendous range in surface area to volume ratios (table 7-1). Leafy crops have enormous surface areas relative to their volume, and as a consequence, generally lose moisture quickly when the vapor pressure deficit is conducive to water loss. Leaves exhibit surface area to volume ratios of 50–100 $cm^2 \cdot cm^{-3}$, while large root and tuber crops[3] have ratios of only 0.2–0.5 $cm \cdot cm^{-3}$.

Use of Psychrometric Charts

Psychometrics,[*] the measurement of heat and water vapor properties of air, represents an extremely valuable tool during the postharvest period. The interrelationship between properties of the air such as relative humidity, absolute humidity, dew point, and wet and dry bulb temperatures is graphically illustrated in figure 7-4. Normally, use of the chart involves initially making wet and dry bulb temperature readings.[†] From these two readings, the other thermal and moisture characteristics of the atmosphere can be determined using the psychrometric chart (e.g., relative humidity, dew point, vapor pressure, absolute humidity, and with more elaborate charts, specific heat and other characteristics).

It is not necessary to start with just wet and dry bulb temperature readings; from any two parameters[‡] (e.g., relative humidity and dry bulb temperature) it is possible to determine the remaining characteristics of the air. With any two variables it is possible to find their point of intercept on the chart. From that point each of the other variables can be determined. For example, if the dry bulb temperature (x-axis) is 25°C, and the wet bulb temperature is 18°C (located on the curved line at the upper left-hand portion of the chart and running downward at a 45° angle toward the x-axis), the relative humidity can be read on the curved line where the vertical dry bulb line and the angled wet bulb line intersect (50%). This intercept gives a point on the chart from which the water vapor pressure and absolute (specific) humidity can also be read, that is, by moving horizontally to the right of the intercept point.

The driving force for the movement of moisture is the difference in water vapor pressure between a product and its environment. Since water vapor pressure can be determined from many psychrometric charts, or calculated from relative humidity and temperature data, it is possible to graphically express the relationship between atmospheric relative humidity and tempera-

---

* Derived from the Greek words *psychro* for "cold" and *metron* for "measure."

† The dry bulb temperature is read directly from an ordinary glass thermometer. The wet bulb temperature is read from a similar thermometer in which the reservoir bulb is covered with a thin layer of water (usually accomplished with a wet cotton wick) and is moved through the air at a sufficient velocity and duration until a steady temperature drop occurs due to evaporation. Evaporation cools the bulb to the wet bulb temperature; the difference between the dry and wet bulb temperatures is referred to as the wet bulb depression.

‡ The exception would be the wet bulb temperature and dew point, which are synonymous.

ture conditions and the water vapor pressure deficit between the product and its environment. This calculation is assuming that the internal gas atmosphere of the product is saturated with water vapor, a safe assumption for succulent products, and that the product and environment are at the same temperature. The water vapor pressure deficit is read from the nomogram (fig. 7-5) by simply drawing a straight line from the temperature through the relative humidity for the environment. The water vapor pressure is read at the point of intersection and is given in millibars.

Water vapor pressure alone does not determine the rate of flux of water from a product (since the resistance to water movement is not taken into consideration). Therefore, it is not possible to compare the rate of water loss between two different products (i.e., different resistances) or in many cases even between cultivars of the same product. For a given product, however, it is possible to compare two sets of environmental conditions and determine in which environment the product will lose moisture the fastest (that with the highest water vapor deficit) and how much faster (water vapor pressure of environment A divided by the water vapor pressure of environment B).

Inhibiting the Exchange of Water Between Product and Environment

For most harvested products, once at or near their optimum moisture content, it is desirable to minimize any further change in moisture concentration during handling and storage. This includes not only succulent flowers, fruits, and vegetables but also dried crops such as pulses, grains, and nuts. Inhibiting the exchange of water is accomplished by minimizing the water vapor pressure deficit between the product and its environment and/or by increasing the resistance to exchange. Humidity and temperature are particularly critical in minimizing the difference in water vapor pressure between product and environment.

The humidity of the surrounding environment should be maintained at a level that gives a water vapor pressure as close to that of the internal atmosphere of the product as possible. With succulent products; very high relative humidities, that is, 95–99%, are generally required. Lower humidities may be utilized for products that exhibit a relatively high resistance to water exchange because losses are generally at a much slower rate. With lower-moisture products such as roots, tubers, and corms, much lower storage humidities are required (e.g., 60–70%) or water will be taken up by the product, potentially exceeding the maximum safe concentration.

Whereas the relationship between humidity and water exchange is relatively straightforward, temperature effects are more complex. Three thermal parameters have a pronounced effect on moisture exchange in storage: (1) the actual temperature, (2) the differential in temperature between product and environment, and (3) the fluctuations in storage temperature. Temperature affects both the amount of water a given volume of air will hold and the level of energy of the water molecules.

Lowering the temperature decreases the maximum amount of water the air will hold; if the weight of water vapor in the air is held constant, the relative humidity will increase. The water vapor pressure deficit between a product

Table 7-2. Environmental Conditions and Product Treatments Used to Minimize the Exchange of Water

Environmental Conditions	*Product Treatments*
Lower temperature	Prevent cuts and abrasions during harvest and handling
Maintain a sufficiently high relative humidity	Rapid cooling after harvest
Minimize excessive air movement	Surface coating (e.g., waxes)
Minimize fluctuations in temperature	Packaging (e.g., shrink wraps)
	Decrease surface area (e.g., leaf removal for cut flowers)

and its environment will also decrease at a given relative humidity with decreasing temperature; for example, for apple (*Malus sylvestris,* Mill.) fruit the water vapor pressure deficit at 95% relative humidity at 20°C is 1.15 mbs whereas at 10°C it is only 0.6 mbs (fig. 7-5). Therefore, low temperature decreases the rate of water loss. Likewise, when the product is at a higher temperature than the environment, the gas atmosphere within a succulent product will contain more water than that of the surrounding cooler air, increasing the difference in water vapor pressure between the product and environment and the rate of water loss. As a consequence, it is desirable to cool harvested products that are to be held in refrigerated storage as quickly as possible to minimize water losses. Due to respiratory heat, however, under constant storage temperature conditions many actively metabolizing products will remain at a slightly higher temperature than their surrounding environment. Hence there will normally be a small positive gradient, favoring water loss.

Air movement is essential in refrigerated storage to remove heat generated by the stored product (see the section on air movement). At the same time, air movement decreases the thickness of the boundary layer around the product, decreasing boundary layer resistance and increasing the flux of water vapor from the product. It is essential, therefore, to reach a compromise between obtaining adequate air movement for cooling and minimizing excessive water loss.

There are a number of product treatments that can be utilized to reduce water exchange after harvest (table 7-2). Preventing cuts and abrasions to the surface of the product helps to maintain a high surface resistance to the movement of water vapor. Rapid cooling reduces the temperature differential between the product and the storage environment, decreasing water loss. With some products, specialized surface treatments are advantageous; for example, waxing citrus fruit increases surface resistance, and using wraps and packages impede the movement of water vapor.

Enhancing the Exchange of Water Between Product and Environment

The concentration of water in most harvested crops is of tremendous importance in that it affects the product's physical and chemical properties, its value, and its storage potential. Succulent products, such as lettuce or cut flowers, that lose even a relatively small percent of their total water, often decline markedly in quality and value. Dry products that are less subject to rapid quality changes with moisture concentration are generally sold by weight. Loss of water results in a decline in total weight, and as a consequence, a decline in value.[11] For example, if the corn (100,000 kg at 15.5% moisture) in a small on-the-farm storage bin is allowed to dry to 8% moisture, the total value based on a market price of $0.11 \cdot kg^{-1}$ has declined $825.

It is often desirable, therefore, to adjust the concentration of water within a product after harvest. Moisture alteration may entail accelerating the loss of water (desorption) or accelerating the uptake of water (adsorption and/or absorption).*

The method utilized to alter product moisture content depends on the species in question. Both adsorption and desorption are commonly accomplished by altering the vapor pressure of the environment surrounding the product. Water is generally taken up in a vapor stage; however, with some very high-moisture crops (e.g., cut flowers, leaf lettuce), water may be introduced as a liquid.

Typically, the product is exposed to altered water vapor pressure conditions for a sufficient duration to reach the desired moisture concentration. Environmental conditions can be selected so that when the product reaches an equilibrium with the environment it is at the desired moisture concentration. Considerable time is generally required before an equilibrium is reached, with the rate of change progressively decreasing as the product approaches the point of equilibrium.

An alternative approach that accelerates exchange and decreases the length of time required to reach the desired moisture concentration is to create a very large gradient in vapor pressure between the product and its environment. The product is then either removed or the environment is changed once the product reaches the desired moisture concentration rather than when it reaches an equilibrium with the environment.

The equilibrium moisture concentration (i.e., the amount of moisture in the product when in equilibrium with its environment) depends not only on temperature but also on the previous moisture conditions of the material. Because of the latter, if the equilibrium moisture concentration is plotted versus the percent relative humidity (called a water sorption isotherm), two different moisture content values are obtained† for each relative humidity depending on whether the product is adsorbing (gaining) or desorbing (losing) moisture (fig. 7-7). This difference between the equilibrium moisture content at a given percent relative humidity, called hysteresis, can be as much as 1.5–2.0% moisture. While the reason for hysteresis is not fully understood, the practical implications can be significant. The difference in product mois-

---

* Adsorbed water molecules display a molecular interaction with the adsorbing substance and are more closely bound than molecules of free water. The adsorbent and water molecules both influence the properties of the other. In contrast, absorbed water molecules are held loosely by capillary forces and exhibit the properties of free water. The absorbing substance's role is limited primarily to that of a supporting structure. A combination of adsorption and absorption phenomena often occur in biological materials.

† For most products.

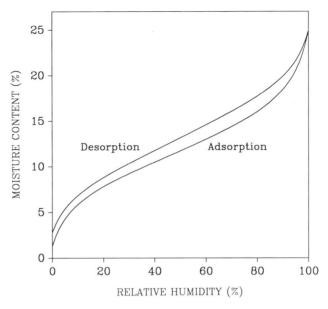

Figure 7-7. Moisture adsorption and desorption isotherms for maize starch.[12]

ture concentration due to adsorption versus desorption of water can, under appropriate conditions, represent the difference between the presence and absence of mold growth during storage.

Enhancing Water Loss

Drying has been the primary preservation process since time in memoria. For field crops, much of the drying occurs prior to harvest. Under favorable weather conditions, seeds lose water, approaching an equilibrium between their moisture content and that of the surrounding environment. Due to inclement weather or other reasons, it is often necessary or desirable to harvest the crop prior to it drying sufficiently in the field for safe storage. When this situation occurs, some additional drying is required and may be accomplished by the use of air, reduced pressure, direct heat, inert gas, or superheated steam. Due to its low cost and convenience, air drying is the most commonly employed technique and will be focused on here. Air drying varies from simply exposing the crop to the sun to the use of elaborate computer-controlled forced air dryers.

RATE OF WATER LOSS

Drying involves two fundamental physical processes: the transfer of heat to evaporate the water and the transfer of mass (water) within and from the product. Mass transfer occurs both as a liquid as it diffuses to the surface and as a gas diffusing outward from the product. The actual rate of water removal from a harvested product is not constant throughout the drying process but declines as the moisture content decreases (fig. 7-8). Drying curves can be

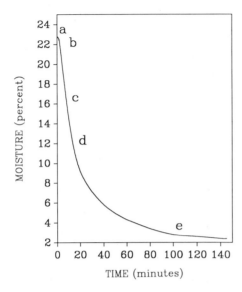

Figure 7-8. A drying rate curve illustrating three general segments: a–b, initial exchange; b–c, the constant rate period; c–e, the falling rate period. The falling rate period is commonly separated into two stages, c–d and d–e, each dominated by a different process. Point c on the curve represents the critical moisture concentration.

separated into three general segments: initial exchange, constant rate period, and falling rate period. The initial rate period (a–b) is typically brief, spanning the time period required for the product to reach the temperature of the surrounding air. During this relatively brief period, some water may actually migrate toward the center of the product due to the temperature differential between the exterior (warm) and interior (cool).

The second segment of the drying curve (b–c) is characterized by a constant rate of water loss (fig. 7-8). During this period, evaporation occurs at the surface of the product and is similar to evaporation from a free water surface. The constant rate period ends when reaching the *critical moisture content:* the minimum concentration of water in the product that will sustain a flow of free water from the interior to the surface at a rate equal to the maximum rate of removal due to evaporation. Upon transversing this point (c), the drying curve enters the falling rate period that is controlled largely by the diffusion of liquid phase moisture to the surface and its removal from the surface. The falling rate portion of the drying curve is often separated into two (occasionally more) stages. The first (c–d) is characterized by unsaturated surface drying and the second (d–e) is controlled by the rate of diffusion of moisture to the surface. Most dried crops are harvested after passing the critical moisture content point of the drying curve.

As a product approaches its equilibrium moisture content, the net exchange of water between the product and its environment nears zero. At the equilibrium moisture content, water molecules continue to move into and out of the product; however, uptake and losses are equal. The actual moisture content of the product does not equal that of the environment; rather the chemical potential of water between the two sites is equal.

The equilibrium moisture content of a product is a function of both the moisture concentration of the air and the characteristics of the individual product. When placed in the same environment, various products exhibit different affinities for water, and as a consequence, display significantly different equilibrium moisture contents (table 7-3). Plant species, cultivar,

Table 7-3. Equilibrium Moisture Contents of Selected Grain Crops at 25°C[15]

	Relative Humidity (%)									
	10	*20*	*30*	*40*	*50*	*60*	*70*	*80*	*90*	*100*
Barley (*Hordeum vulgare,* L.)	4.4	7.0	8.5	9.7	10.8	12.1	13.5	15.8	19.5	26.8
Flaxseed (*Linum ustalissimum,* L.)	3.3	4.9	5.6	6.1	6.8	7.9	9.3	11.4	15.2	21.4
Oats (*Avena sativa,* L.)	4.1	6.6	8.1	9.1	10.3	11.8	13.0	14.9	18.5	24.1
Rice* (*Oryza sativa,* L.)	5.9	8.0	9.5	10.9	12.2	13.3	14.1	15.2	19.1	—
Corn† (*Zea mays,* L.)	5.1	7.0	8.4	9.8	11.2	12.9	14.0	15.6	19.6	23.8
Soybeans (*Glycine max,* (L.) Merrill)	—	5.5	0.5	7.1	8.0	9.3	11.5	14.8	18.8	—
Wheat (*Triticum aestivum,* L.)	5.8	7.6	9.1	10.7	11.6	13.0	14.5	16.8	20.6	—

Note: Data from Hall.[15]
* Whole grain
† Shelled

and maturity may each exert a significant influence on the equilibrium moisture content. In addition, whether water was adsorbed or desorbed to reach the equilibrium moisture content may also affect the precise concentration.

The rate at which a product dries is a function of the difference in the chemical potential of water within the product and that of the surrounding environment and factors that affect the exchange of moisture between the two. Important factors that alter the rate of drying are moisture content of the air, product and air temperature, airflow rate, product surface area, moisture content of the product, and product characteristics that enhance or retard water exchange (e.g., epicuticular waxes). Although growers have little control over most product characteristics, environmental conditions can be modified to enhance or retard the rate of drying.

Simply optimizing environmental conditions for the maximum rate of water removal from a product is seldom a satisfactory approach. Excessively rapid loss of water causes serious quality losses in many products. Seed dried too quickly develop case hardening, a hardening of the outer tissue. Case hardening is caused by the removal of water from the surface of the product at a more rapid rate than water at the interior of the product can diffuse outward. The dry outer tissue tends to seal the surface of the seed, decreasing the subsequent rate of drying. When onion (*Allium cepa,* L.) bulbs are partially dried prior to storage using artificial dryers, the rate of water removal must be carefully controlled. Excessively high rates of water loss can cause splitting of the outer protective surface layers, enhancing an undesirable loss of water later in storage.

Air temperature has a pronounced effect on the rate of drying, with higher temperatures enhancing the rate. Air temperature may also affect the quality of the final product. The temperature at which the product will undergo undesirable changes in quality (e.g., loss of germination potential, color, flavor) is the critical temperature. It varies widely between various products, their intended use, moisture content, type of dryer, and airflow rate. For example, wheat (*Triticum aestivum,* L.) that is to be used as seed and has a moisture content below 24% can be safely dried at 49°C; when above 24% moisture, a lower temperature (43°C) is essential to prevent quality losses.[23] The same grain can be dried at 60°C (>25% moisture) or 66°C (<25% moisture) if it is to be used for milling. Even higher temperatures can be used for wheat that is to be utilized as livestock feed (e.g., 82–104°C). Most commercial drying operations try to maintain the highest possible air temperature that will not impair quality because it increases the capacity of the dryers and decreases costs.

Air drying in the sun with the product placed on drying floors, roofs, courtyards, roadsides, and other suitable areas is a standard practice in many areas of the world where artificial drying is not feasible. For example, in the Guangdong Province of China, under favorable conditions rice (*Oryza sativa*, L.) can be dried from 18% to 13.5% moisture in two days.[13] It requires spreading the rice in relatively thin layers to maximize the surface area exposed (i.e., about 2 cm) and stirring the rice with rakes twice a day. Reabsorption of moisture during the night is minimized by raking the rice into piles and covering them when possible.

LOW- VERSUS HIGH-TEMPERATURE DRYING

Forced air drying depends on the movement of air with a suitable humidity and temperature around the individual units of the product. For some crops it is desirable to decrease the product's moisture content quickly; it is generally accomplished by elevating the temperature of the drying air. Forced-air dryers can be separated into two general classes: low-temperature dryers and high-temperature dryers. Low-temperature dryers utilize air with little or no supplemental heating, relying largely on suitable ambient temperature and humidity conditions. The air is passed over or around the product until a moisture equilibrium is established. The low-temperature air, although requiring a greater drying time, prevents overdrying of the product and potential quality losses. It also allows drying relatively deep layers of product; hence the crop can often be dried in storage, decreasing handling costs by eliminating a separate step.

High-temperature dryers utilize a substantially elevated air temperature to accelerate drying rate. The primary distinction between low- and high-temperature drying is that serious overdrying would occur if the product were allowed to reach an equilibrium moisture content with the air in a high-temperature dryer. Exposure to the elevated-temperature air is continued only until the product has reached the desired moisture content.

High-temperature dryers are separated into two general classes: continuous-flow and batch dryers. The former utilize a continuous movement of product through the dryer, in the latter the product is static. Continuous-flow dryers represent the most widely used of the two classes; however, batch dryers have the advantage of simplicity.

DRYING FRONT AND ZONES

In bulk drying, a relatively large volume of air, the humidity of which is lower than the equilibrium value for the product being dried, is introduced into a bed* of the product. The dryer air lowers the moisture content of the product in which it initially comes in contact, increasing its own moisture content.

---

* A thick layer or stratum of product.

There is also a corresponding decrease in air temperature due to the energy adsorbed with the phase change of water between a liquid and a gas. As the air moves past successive units of the product, it no longer has the same properties (relative humidity and temperature) and drying potential as when first entering the bed of product. Hence the air removes less and less moisture as it proceeds. As a consequence, a series of moisture zones develop within the bed (figure 7-9). At the site of entry, the dry air removes moisture until the equilibrium moisture content is reached, giving a dry product. Within the dry zone there is no additional net exchange of moisture or temperature change as fresh air passes around the product. Drying occurs in the drying zone, which moves progressively forward into the undried product. The forward edge of the drying zone is referred to as the drying front. During the early stages of drying, part of the moisture removed from the product near the air inlet may be reabsorbed by the cooler product near the exhaust. To prevent possible losses, product depth, air temperature, and air velocity can be manipulated to minimize excessive wetting and the length of time required for the drying front to reach the moist zone.

Enhancing Water Uptake

For many harvested products it is necessary to reintroduce water if the internal concentration has dropped below the desired level. Introduction of water is accomplished by using either moisture in the liquid or vapor phase. The physical state of water that can be used depends on the product, rate of uptake required, final concentration desired, intended use of the product, and a number of other factors.

The introduction of vapor phase moisture requires considerably more time

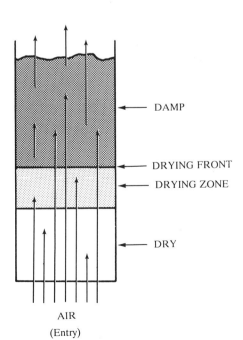

Figure 7-9. Moisture zones are present during deep-layer drying. The drying front progressively moves through the material, leaving a dried product behind.

than water in the liquid phase. The use of water vapor, however, is essential for crops that can not withstand direct wetting. Proceeding from a very dry product toward one of higher moisture level, the relationship between water vapor and product moisture changes. At the low end of the curve (little water), the energy of binding between water molecules and the compounds making up the adsorbing surface is a dominant factor (fig. 7-7). The intermolecular forces involved are thought to be quite strong, accounting for the steepness of the initial segment of the curve. Chemical constituents such as starch and protein provide polar sites with which the water molecules react. The moisture deposited forms a first layer of water molecules. Within the more linear portion of the isotherm, additional water molecules are being deposited on the first layer of water molecules, forming a second layer. In this region the amount of water adsorbed is largely dependent on the water vapor pressure of the atmosphere. In the final segment of the isotherm, the high humidity range, successive layers of water molecules are deposited. The amount of water adsorbed increases rapidly, with the vapor pressure of the air having only a moderate influence.

The rate of uptake of moisture is enhanced by a high water vapor pressure deficit between the atmosphere and product, a positive gradient favoring the movement from the atmosphere to the product. Increasing temperature and air movement also enhances the rate of uptake. For example, the effect of relative humidity and temperature on the uptake of moisture is presented in figure 7-10. The higher the relative humidity, the greater the difference in water vapor pressure between the product and atmosphere, accelerating the rate of moisture uptake.

Liquid phase water may also be used to alter the moisture content of certain crops. Many cut flowers and foliage crops require the reintroduction of liquid phase water after harvest. When severed from the parent plant, water uptake ceases; however, evaporational losses continue. Unless moisture loss is inhibited (e.g., high relative humidity) or water is reintroduced, a

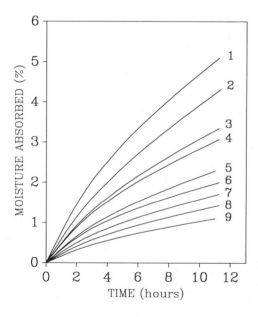

Figure 7-10. The adsorption of water vapor by corn (*Zea mays,* L.) seed with an initial moisture content of 9.26% (at 0 time) under various environmental conditions (1 = 27°C, 80% RH; 2 = 21°C, 80% RH; 3 = 16°C, 80% RH; 4 = 27°C, 70% RH; 5 = 21°C, 70% RH; 6 = 16°C, 70% RH; 7 = 27°C, 60% RH; 8 = 21°C, 60% RH; 9 = 16°C, 60% RH).[25]

water deficit occurs, resulting in reduced turgidity and an accelerated rate of quality loss and senescence. With the loss of turgidity, the stems of some species are no longer able to physically support the weight of the flower, resulting in a disorder called bent neck.

When the stems of cut flowers are placed in containers of water, there is normally a gradual decline in the rate of uptake with time. This decline is in contrast with flowers that remain attached to the parent plant. Here the water conductivity remains essentially constant during aging.[10] This decrease in water conductivity or increase in resistance is caused primarily by two factors: the presence of microorganisms and the introduction of air into the stem. Microorganisms found in the water reservoir enter the base of the stem, causing a direct "stem plugging" due to their physical presence and an indirect blockage due to the production of metabolites by some microbes. The importance of microorganisms in increasing the resistance to water uptake has been shown through correlations between increases in their populations and decreases in water uptake,[21] the effect of chemicals that control microorganisms on water uptake, and use of anatomical studies.

Air entering the base of the stem during storage or shipment can disrupt the continuity of the columns of water within the stem, greatly impeding water uptake. Air may be introduced during cutting and/or dry storage or may represent the movement of dissolved gases out of the liquid phase within the stem, forming bubbles. Degassing the water by boiling or vacuum treatment[10] has been shown to prevent a decrease in conductance of microbial free water through the stem.

Tremendous quantities of intact ornamental floral and foliage plants are marketed each year during which time the reintroduction of water is often essential. Because they are intact, water uptake is seldom a problem; proper quality maintenance centers on balancing the rate of water use by the plant with the reintroduction of water into the root medium. It is especially critical with bedding plants in that they are generally grown in very small-volume containers. Both the volume of the root medium and its water-holding capacity are critical parameters.

Increasing the volume of water available to the plant, whether by increasing the volume of the medium or the physical/chemical properties of the medium (e.g., the use of hydrophilic polymers),* potentially decreases the number of wet-dry cycles the plant goes through after removal from the production zone. Since drying represents a potential stress period when quality can be rapidly lost, decreasing the total number of cycles can decrease the number of times the plant enters a high-risk situation. Conversely, media that display a water-holding capacity that is too great can also cause serious losses due to the occurrence of anaerobic conditions in the root zone. Selection of a medium with proper water-holding and aeration properties is a production decision that can have a critical impact on the postharvest handling and marketing period. Specific media requirements vary widely between individual species.

Cereal grains are often conditioned with water prior to milling. In wheat kernels, absorption represents a heterogenous process.[2] Initially there is a

---

* Such as polyethylene oxide.[1]

very rapid absorption in which the pericarp is saturated by capillary imbibition. The testa of the kernel represents the major barrier to diffusion. The embryo, although small in surface area relative to the testa, has a much lower resistance. Natural capillaries and structural flaws and cracks in the grain increase the rate of diffusion and uptake of water. The gain in product moisture decreases with increasing water temperature.

During retail marketing, water is often directly applied to a number of edible products (e.g., leaf lettuces, green onions, celery (*Apium graveolens* var. *dulce*, Pers.), kale (*Brassica oleracea*, L. Acephala group), kohlrabi (*Brassica oleracea*, L. Gongylodes group), parsley (*Petroselinum crispum*, (Mill.) Nym.), and watercress (*Nasturtium officinale*, R. Br.)). In addition to decreasing product temperature and increasing relative humidity, cold-water sprays allow the uptake of some water by many products. Postharvest products vary widely in their ability to have water applied to their surfaces. Leafy crops that have large surface areas lose water quickly and generally benefit from water sprays. Unsprayed vegetables have been shown to display a 10–20% weight reduction during the first few days under simulated retail conditions.[9] Large surface-to-volume ratios decrease the likelihood of the occurrence of anaerobic conditions within the interior of the product.

The exposure of the surface of many products to free water, however, causes serious problems, generally seen as an increased occurrence of rots. Some species should be kept under high-humidity conditions (e.g., cauliflower (*Brassica oleracea*, L. Botrytis group), the floral parts of cut flowers, raspberries (*Rubus idaeus* var. *strigosus*, (Michx.) Maxim.), and mushrooms) but should never come in contact with water. Dried products (e.g., grains and pulses) should be held under relatively dry atmospheric conditions and should never be wetted.

The uptake of liquid phase water by products during cooling and handling operations can, in some instances, be undesirable in that it may introduce disease organisms into the product. For example, the use of water to float tomatoes (*Lycopersicon esculentum*, Mill.) out of large trucks can introduce water-borne disease organisms into the fruit through the stem scar. Uptake is greatly enhanced by the use of water with a temperature less than that of the fruit. The colder water cools the fruit floating or submerged in it, decreasing the volume of the gas atmosphere within the tissue. A partial vacuum is created that pulls water through the stem scar into the fruit until the internal and external pressures are balanced. To prevent this situation, it is recommended that the temperature of the water be approximately 1–2°C greater than the temperature of the product. Proper sanitation is also important (e.g., the use of chlorine).

Maximum and Minimum Acceptable Product Moisture Content

Succulent products have high moisture contents at harvest, and the loss of even a relatively small amount of water can have, in some species, a serious effect on the physical, physiological, pathological, nutritional, economic, and aesthetic properties of the crop. Table 7-4 shows the maximum permissible

Table 7-4. Maximum Permissible Weight Loss for a Number of Fruits and Vegetables

	Maximum Permissible Weight Loss (%)
Apples (*Malus sylvestris*, Mill.) (4 cultivars)	7.5*
Asparagus (*Asparagus officinalis*, L.)	8.0
Beans	
broad (*Vicia faba*, L.)	6.0
runner (*Phaseolus coccineus*, L.)	5.0
snap (*Phaseolus vulgaris*, L.) (4 cultivars)	41.0*
Beetroot (*Beta vulgaris*, L.)	7.0
Beetroot with tops	5.0
Blackberries (*Rubus* spp.)	6.0
Broccoli (*Brassica oleracea*, L. Italica group)	4.0
Brussels sprouts (*Brassica oleracea*, L. Gemmifera group)	8.0
Cabbage (*Brassica oleracea*, L. Capitata group) (3 cultivars)	8.0, 10.9*
Carrots (*Daucus carota*, L.)	8.0
Carrots with leaves	4.0
Cauliflower (*Brassica oleracea*, L. Botrytis group)	7.0
Celery (*Apium graveolens* var. *dulce*, Pers.)	10.0
Cucumber (*Cucumis sativus*, L.)	5.0
Leek (*Allium ampeloprasum*, L. Porrum group)	7.0
Lettuce (*Lactuca sativa*, L.) (3 cultivars)	3.7
Nectarine (*Prunus persica* var. *nucipersica*, (Suckow) Schneid.)	21.1*
Onion (*Allium cepa*, L.)	10.0
Parsnip (*Pastinaca sativa*, L.)	7.0
Peach (*Prunus persica*, (L.) Batsch.)	16.4*
Pea (*Pisum sativum*, L.)	5.0
Pear (*Pyrus communis*, L.) (3 cultivars)	5.9*
Pepper, green (*Capsicum annuum*, L. Grossum group)	7.0, 12.2*
Persimmons (*Diospyros kaki*, L.f.)	13.3*
Potato (*Solanum tuberosum*, L.)	7.0
Raspberries (*Rubus idaeus* var. *strigosus*, (Michx.) Maxim.)	6.0
Rhubarb, forced (*Rheum rhabarbarum*, L.)	5.0
Spinach (*Spinacia oleracea*, L.)	3.0
Squash, summer (*Curcubita* spp.)	23.9*
Sweet corn (*Zea mays* var. *rugosa*, Bonaf.)	7.0
Tomato (*Lycopersicon esculentum*, Mill.)	7.0, 6.2*
Turnip with leaves (*Brassica rapa*, L. Rapifera group)	5.0
Watercress (*Nasturtium officinale*, R. Br.)	7.0

Note: Data from Robinson[27] and *data from Hruschka.[17]

amounts of water* that can be lost from a cross-section of fruits and vegetables before becoming unsalable. These range from a low of 3% in spinach (*Spinacia oleracea*, L.) leaves to 41% in snapbeans (*Phaseolus vulgaris*, L.), indicating a wide variation between species (fig. 7-11). Deficiency symptoms

---

* Weight loss was measured instead of actual percent moisture since respired substrates generally make up only a very minor portion of the total weight loss.

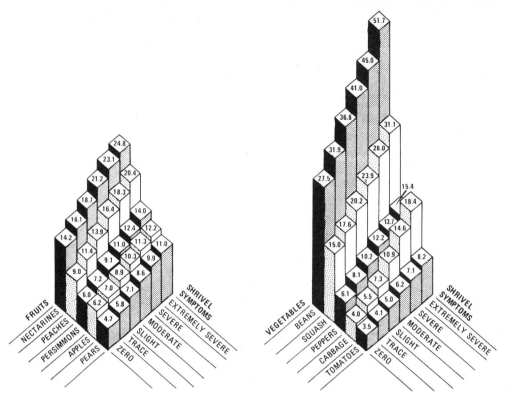

Figure 7-11. Average percent weight loss at the onset of various gradations of shriveling symptoms for selected fruits and vegetables (*from Hruschka*[17]).

Figure 7-12. Shrivel symptoms in peppers (*Capsicum annuum,* L. Grossum group, cv. 'California Wonder'). Top row, left to right: zero (6.1% weight loss), trace (8.1%), slight (10.2%); bottom row, left to right: moderate (12.2%), severe (13.7%), extremely severe (15.4%) (*from Hruschka*[17]).

Table 7-5. Maximum Safe Moisture Content for Stored Grain Held at 25–38°C

	Maximum Acceptable Moisture (%)	
	USA	Canada
Barley (*Hordeum vulgare*, L.)	14.5	14.8
Buckwheat (*Fagopyrum esculentum*, Moench.)	—	16.0
Corn (maize) (*Zea mays*, L.)	14.1	14.0
Oat (*Avena sativa*, L.)	14.0	14.0
Pea (*Pisum sativum*, L.)	15.0	16.0
Rice (*Oryza sativa*, L.)		
rough	14.0	—
milled	15.0	—
Rye (*Secale cereale*, L.)	16.0	14.0
Sorghum (*Sorghum bicolor*, (L.) Moench.)	13.0	—
Soybean (*Glycine max*, (L.) Merrill)	13.0	14.0
Sunflower (*Helianthus annuus*, L.)	—	9.6
Wheat (*Triticum aestivum*, L.)	—	14.5

Source: After Bushek and Lee.[4]

typically increase with increasing water deficit (fig. 7-12). However, not all individual units within a sample display physical desiccation symptoms uniformly relative to weight loss. For example, the actual percent weight loss of individual snapbeans displaying moderate shrivel symptoms ranged from 15% to 68%.[17] Likewise, within a given lot, the onset of dehydration symptoms ranged from zero to extremely severe indicating a tremendous range in the rate of weight loss between individual units within a sample.

Products that are stored dry (e.g., grain) have a maximum moisture concentration that will allow safe storage (table 7-5). Exceeding this moisture level greatly increases the potential for pathogen invasion and storage rots. The actual moisture concentration at which a product can be safely stored depends on storage temperature. Table 7-6 illustrates the relationship between barley (*Hordeum vulgare*, L.) moisture content and storage temperature with regard to the preservation of germination. Both decreasing moisture content and storage temperature prolong the preservation of quality.

Table 7-6. Estimated Maximum Storage Time for Barley (*Hordeum vulgare*, L.) at Various Seed Moisture Contents and Storage Temperatures with Respect to the Preservation of Germination

Storage Temperature °C	Weeks of Storage								
	11	12	13	14	15	16	17	19	23
25	54	39	25	16	9	5	2.5	1	—
20	110	80	50	32	19	10	5	2	0.5
15	240	170	100	65	40	20	10	10.4	1
10	600	400	260	160	90	50	21	8.5	2
5	>1,000	1,000	600	400	200	120	50	17	4

Note: Data from Kreyger.[20]

REFERENCES

1. Anon. 1973. Agricultural hydrogel, concentrate 50G. *Union Carbide Corp. Tech. Bull.* Creative Agricultural Systems, New York.

2. Becker, H. A. 1960. On the absorption of liquid water by the wheat kernel. *Cereal Chem.* **37:**309–323.

3. Burton, W. G. 1982. *Postharvest Physiology of Food Crops.* Longmans, Essex, United Kingdom, 339p.

4. Bushek, W., and J. W. Lee. 1978. Biochemical and functional changes in cereals: Maturation, storage and germination. In *Postharvest Biology and Biotechnology,* H. O. Hultin and M. Milner (eds.). Food and Nutrition Press, Westport, Conn., pp. 1–33.

5. Carpenter, W. J., and H. P. Rasmussen. 1973. Water uptake by cut roses (*Rosa hybrida*) in light and dark. *J. Am. Soc. Hort. Sci.* **98:** 309–313.

6. Colwell, R. N. 1942. Translocation in plants with special reference to the mechanism of phloem transport as indicated by studies on phloem exudation and on the movement of radioactive phosphorus. Ph.D. thesis, University of California, Berkeley.

7. Crafts, A. S., and O. Lorenz. 1944. Fruit growth and food transport in cucurbits. *Plant Physiol.* **19:**131–138.

8. DeStigter, H. C. M. 1980. Water balance of cut and intact 'Sonia' rose plants. *Z. Pflanzenphysiol.* **99:**131–140.

9. Dipman, C. W., J. L. Callahan, A. D. Michaels, and S. R. Barkin. 1936. *How to Sell Fruits and Vegetables.* The Progressive Grocer, New York, 200p.

10. Durkin, D. 1979. Effect of millipore filtration, citric acid, and sucrose on peduncle water potential of cut rose flowers. *J. Am. Soc. Hort. Sci.* **104:**860–863.

11. Duvel, J. W. T., and L. Duval. 1911. The shrinkage of corn in storage. *U. S. Bur. Plant Ind. Circ. 81.*

12. Food and Agriculture Organization of the United Nations. 1970. *Food Storage Manual.* Tropical Stored Products, Ministry Overseas Development, Rome, Italy.

13. Food and Agriculture Organization of the United Nations. 1982. China: Postharvest grain technology. *Agric. Serv. Bull. 50.* Food and Agricultural Organization, United Nations, 60p.

14. Grierson, W., and W. F. Wardowski. 1975. Humidity in horticulture. *HortScience* **10:**356–360.

15. Hall, C. W. 1980. *Drying and Storage of Agricultural Crops.* AVI Pub. Co., Westport, Conn., 381p.

16. Holman, L. E., and D. G. Carter. 1952. Soybean storage in farm-type bins. *Ill. Agric. Exp. Sta. Bull. 552.*

17. Hruschka, H. W. 1977. Postharvest weight loss and shrivel in five fruits and five vegetables. *USDA-ARS Mark. Res. Rep. 1059,* 23p.

18. Jackson, M. B., I. B. Morrow, and D. J. Osborne. 1972. Abscission and dehiscence in the squirting cucumber. *Can. J. Bot.* **50:**1,465–1,471.

19. Kofranek, A. M., and A. H. Halevy. 1972. Conditions for opening cut chrysanthemum flower buds. *J. Am. Soc. Hort. Sci.* **97:**578–584.

20. Kreyger, J. 1964. Investigations on drying and storage of cereals in 1963. *Versl. Tienjarenplan Graanonderz* **10:**157–164 (in Dutch).

21. Larsen, F. E., and M. Frolich. 1969. The influence of 8-hydroxyquinoline citrate, N-dimethylaminosuccinamic acid, and sucrose on respiration and water flow in 'Red Sim' carnations in relation to flower senescence. *J. Am. Soc. Hort. Sci.* **94:**289–291.

22. Meeuse, B. J. D. 1975. Thermogenic respiration in Aroids. *Ann. Rev. Plant Physiol.* **26:**117–126.

23. Nellist, M. E. 1979. Safe temperatures for drying grain. A report to the Home Grown Cereals Authority. *NIAE Rep. No. 29.*

24. Nursten, H. E. 1970. Volatile compounds: The aroma of fruits. In *The Biochemistry of Fruits and Their Products,* A. C. Hulme (ed.). Academic Press, New York, pp. 239–268.

25. Park, S. W., D. S. Chung, and C. A. Watson. 1971. Adsorption kinetics of water vapor by yellow corn. I. Analysis of kinetic data for sound corn. *Cereal Chem.* **48:**14–22.

26. Platenius, H. 1939. Wax emulsions for vegetables. *Cornell Agric. Exp. Sta. Bull. 723,* 43p.

27. Robinson, J. E., K. M. Browne, and W. G. Burton. 1975. Storage characteristics of some vegetables and soft fruits. *Ann. Appl. Biol.* **81:**399–408.

28. Smith, W. H. 1938. Anatomy of the apple fruit. *Fed. Rep. Invest. Board London for 1937,* pp. 127–133.

29. von Guttenberg, H. 1926. Die Bewegungsgewebe-*Ecballium elaterium. Handb. Pflanzenanat.* **5:**117–119.

30. Williams, G. D. V., and J. Brochu. 1969. Vapor pressure deficits vs. relative humidity for expressing atmospheric moisture content. *Naturaliste Can.* **96:**621–636.

ADDITIONAL READINGS

Baker, D. A., and J. L. Hall (eds). 1988. *Solute Transport in Plant Cells and Tissues.* Wiley, New York, 592p.

Boyer, J. S. 1985. Water transport. *Ann. Rev. Plant Physiol.* **36:**473–516.

Gates, D. M., and R. B. Schmerl (eds.). 1975. *Perspectives of Biophysical Ecology.* Springer-Verlag, Berlin, 609p.

Kozlowski, T. T. (ed.). 1968–1983. *Water Deficit and Plant Growth.,* 7 vols. Academic Press, New York.

Kramer, P. J. 1983. *Water Relations of Plants.* Academic Press, New York, 489p.

Kramer, P. J., and T. T. Kozlowski. 1979. *Physiology of Woody Plants.* Academic Press, New York, 811p.

Lange, O. L., L. Kappen, and E. D. Schulze. 1976. *Water and Plant Life.* Ecological Studies, vol. 19. Springer-Verlag, New York, 536p.

Moorby, J. 1981. *Transport Systems in Plants.* Longman, London, 169p.

Passioura, J. B. 1988. Water transport in and to roots. *Ann. Rev. Plant Physiol.* **39:**245–265.

Solomos, T. 1987. Principles of gas exchange in bulky plant tissues. *HortScience* **22**(5):766–771.

Zeiger, E., G. D. Farquhar, and I. R. Cowan (eds). 1987. *Stomatal Function.* Stanford University Press, Stanford, Calif., 503p.

Zimmerman, M. H. 1983. *Xylem Structure and the Ascent of Sap.* Springer-Verlag, Berlin, 143p.

8

HEAT, HEAT TRANSFER, AND COOLING

The flow of energy through substances is one of the underlying requirements for life. In the Earth's biosphere, virtually all of the energy available for biological processes is of solar origin. Heat, a form of kinetic energy, represents one part of the total, and upon entering a substance, accelerates the motion of its atoms. Temperature is simply a measure of the speed of motion of the atoms. Temperature alone, however, does not give the amount of heat present, only the intensity or level of heat. The amount of heat is equal to the object's temperature multiplied by its mass.

The temperature of postharvest products is extremely important since heat alters the physical properties of molecules, which in turn may alter their activity and performance within the tissue. Heat affects the fluidity of membranes, the activity of enzymes, the volatility of aromatic molecules, and numerous other processes. Within limits, increasing temperature increases the rate of change, and during the postharvest period, change is seldom desired. It is possible therefore to accelerate or retard change through the addition or removal of heat. The amount of heat present, in contrast to temperature alone, is extremely important when the objective is the removal of heat to cool an object. Temperature alone does not indicate the heat load present (e.g., a cubic meter of air at 50°C contains only a fraction of the heat found in a cubic meter of mangos (*Mangifera indica,* L.) at the same temperature).

Heat flows from a warmer (higher-energy) substance to a cooler (lower-energy) substance. When heat is transferred, some of the fast-moving atoms give up part of their energy to slower-moving atoms. If you hold one end of a metal rod in a flame and the other in your hand, heat released from the chemical energy of the burning material enters the end of the rod in the fire. The speed of the metal atoms in the flame is accelerated; these in turn interact with their neighbors, causing them to speed up. This transfer of energy moves progressively down the rod until all of the atoms are vibrating rapidly. Eventually the energy is transferred to the molecules within your hand, giving the sensation of hot.

Cold is the lack of heat; objects become cold by the removal of heat. Heat always travels from a warm substance or object to a colder one, never from cold to warm. Therefore, to remove heat, it must either be adsorbed or transferred to a cooler substance.

MEASUREMENT OF TEMPERATURE AND UNITS OF HEAT

Temperature represents the most important single factor in postharvest quality maintenance. As a consequence, accurate measurement of temperature in harvested products is essential. Several different types of temperature-measuring devices are available, each having distinct advantages and disadvantages. Glass thermometers are sealed glass tubes that have a reservoir bulb at the base containing a liquid such as mercury or alcohol. The uniform expansion of the liquid with increasing temperature elevates the level of the liquid within the tube. Therefore, the liquid will rise or fall with the addition or removal of heat.

The thermometer tube is calibrated in degrees using one of several temperature scales. The scale most commonly used has changed with time. In 1706, Daniel Fahrenheit, a German scientist, made a number of improvements on the original thermometer invented around 1592 by Galileo Galilei, an Italian physicist. One improvement was to reference the temperature scale relative to the melting point of ice (32°F) and the boiling point of water (212°F). Fahrenheit's scale gave 180 individual divisions or degrees. In 1742, Anders Celsius, a Swedish scientist, assigned 100 equal divisions between the melting point of ice and the boiling point of water.* The Celsius scale is the metric temperature scale now used fairly universally. Since Fahrenheit and Celsius simply represent a different number of divisions between the melting point of ice and the boiling point of water on identical thermometers, they can be readily converted from one to the other mathematically.†

The Kelvin scale was introduced in the mid-nineteenth century by the English scientist Lord Kelvin and differed from the Celsius scale by making 0 the point at which all motion of molecules should cease—absolute zero. On the Celsius scale this measurement was −273°C (thus the melting point of ice became 273 K and the boiling point of water became 373 K). By utilizing the same units as the Celsius scale, the two can be readily converted (i.e., $T_K = T_C + 273$). The Kelvin scale is especially applicable to chemistry studies; however, the Celsius scale is the most commonly used and is the scale of choice for postharvest work.

For measuring product temperature, glass thermometers tend to be difficult to read and are slow in response time, but they are very inexpensive. Some thermometers utilize metal that will expand or contract to measure changes in temperature. The most commonly used are metal dial thermometers, which are easy to insert into the product and read, but are slow in

---

* Celsius actually selected 100°C for the melting point of ice and 0°C for the boiling point of water. His assistant and successor, Märtin Strømer, quietly reversed the scale in 1750.[23]

† Degrees Celsius = 5/9 (degrees F − 32); degrees Fahrenheit = 9/5 (degrees C) + 32.

response and tend to be the least accurate of the various postharvest temperature-measuring devices.

Thermocouples are also utilized to measure temperature. When two metals have been welded together, upon heating, a voltage develops across the open ends of the junction. By measuring the voltage with a sensitive voltmeter calibrated against temperature, the actual temperature can be determined. Thermocouples have several advantages: They are easy to read and accurate, give fast response times, and can be used for remote sensing; however, they are typically considerably more expensive. Thermistors, in contrast to thermocouples, are solid-state semiconductors that allow fewer electrons to flow through them as the temperature increases. The change in electrical flow through the chemical, typically lithium chloride, is calibrated against temperature.

One final technique is the use of a radiometer. Product temperature is determined by measuring the infrared rays given off by the substance. The amount of infrared radiation changes with product temperature. Although expensive, radiometers are fast, accurate, and measure temperature without actually contacting the product.

When planning a refrigeration system for initial cooling after harvest, storage, transit, or retail use, it is important to determine the amount of refrigeration capacity needed (i.e., how much heat must be removed). Excessive refrigeration capacity greatly increases costs while an insufficient capacity results in slow and/or inadequate cooling of the product. The heat load is the sum of all sources of heat within or moving into the refrigerated area. In addition to product heat, heat from air infiltration, containers, and heat-generating devices (e.g., motors, fans, lights, pumps, forklifts, and people) collectively make up the total heat load. Product heat comprises the primary portion of the total heat load during initial cooling of the product. After removal from the field, product heat depends on the temperature of the product, its cooling rate, the amount of product cooled at a given time, and the specific heat of the product.

For precise heat load determinations, it is necessary to accurately determine the temperature of the product. The air temperature surrounding the product should never be used as an estimate of product temperature. In addition, the temperature is often not uniform throughout the product. Lack of temperature uniformity is especially common during the initial cooling after harvest when there is often a significant temperature gradient from the interior (warm) to the exterior (cool). Normally the pulp at the center of the densest part of the product is the best site to monitor the temperature. For example, the center of an apple fruit or within the stem on celery stalks or lettuce (*Lactuca sativa,* L.) heads are the best indicators of the overall temperature status of the product. Likewise, the temperature of packaged produce should be measured on samples from the center of the package and from center packages when palletized.

Because of the variation in temperature within a given lot of product, a measure called the mass-average temperature is used.[29] The mass-average temperature is a single value from the transient temperature distribution within the tissue, and represents the final uniform temperature that would be reached throughout the product if moved to constant temperature conditions.

In the past, heat has been measured primarily as calories (cal) or kilocalories (1,000 cal or 1 kcal) and British thermal units (BTU). A calorie is the amount of heat required to raise the temperature of 1 g of water 1°C. Both the calorie and the BTU are strictly measures of heat and are still commonly used, although the joule (J) is now the correct unit of measure adopted by Le Systeme International Units. The use of joules is based on the fact that heat can be converted to work, which is measured in joules or foot pounds. One joule is equal to one newtonmeter (i.e., force × distance = work), or for our purposes, 4.187 kJ is the amount of heat required to raise the temperature of 1 kg of water 1°C. Calories, BTUs, and joules can be readily interconverted.*

TYPES OF HEAT

Sensible Heat

When a substance is heated and the temperature rises as heat is added, the increase in heat is called sensible heat, that is, heat that causes a change in temperature of a substance. When heat is removed and the temperature falls, it is also sensible heat. The amount of sensible heat that is required to cause a change in temperature of a substance varies with both the kind and the amount of substance. The specific heat capacity, the amount of heat that must be added to or released from a given mass of substance to change its temperature one degree, has been determined for a cross-section of materials (table 8-1). Specific heat capacity is expressed as $kJ \cdot kg^{-1} \cdot K^{-1}$, $cal \cdot g^{-1} \cdot °C^{-1}$, or $BTU \cdot lb^{-1} \cdot °F^{-1}$. The amount of heat ($Q$) necessary to cause a given change in temperature of a substance is determined by multiplying the mass (m) by the specific heat capacity (c) by the change in temperature (ΔT):

$$Q = mc(\Delta T)$$

These calculations are given as:

$$kJ = (kg)(kJ \cdot kg^{-1} \cdot K^{-1})(\Delta K);$$
$$cal = (g)(cal \cdot g^{-1} \cdot °C^{-1})(\Delta °C);$$
$$BTU = (lb)(BTU \cdot lb^{-1} \cdot °F^{-1})(\Delta °F)$$

and can be readily interconverted.

Latent Heat

One of the most important effects of heat is its ability to change the physical state of pure substances. With the addition of sufficient heat, solids become liquids and liquids eventually become gases. For each substance, these changes in state always occur at the same temperature at a given pressure. To cause a change in state, heat must be added or removed. When a solid (e.g., ice) changes to a liquid (water), heat must be added; when the reaction occurs in the opposite direction, water to ice, heat must be removed. Similarly, when

---

* BTUs × 1,055 = joules; BTUs × 252 = calories; calories × 4.187 = joules; calories × 0.00397 = BTUs.

Table 8-1. Heat Properties of Several Common Substances

	Specific Heat Capacity		Melting Point	Heat of Fusion		Boiling Point	Heat of Vaporization	
	$kJ \cdot kg^{-1} \cdot K^{-1}$	$cal \cdot g^{-1}$	°C	$kJ \cdot kg^{-1}$	$cal \cdot g^{-1}$	°C	$kJ \cdot kg^{-1}$	$cal \cdot g^{-1}$
Water	4.187	1.00	0	335	80	100	2257	540
Methanol	2.512	0.60	−97	67	16	29	272	65
Ethylene glycol	2.345	0.56	−13	188	45	198	800	191
Air	1.005	0.24						
Steam	2.010	0.48						
Ammonia	2.051	0.49	−78			−33	1314	314
Ice	2.094	0.50	0	335	80			
Copper	0.385	0.09	1080	176	42	2310		
Aluminum	0.921	0.22	658	394	94	2057		

a liquid is converted to a gas, heat must be added, and conversely, it must be removed when proceeding from a gas to a liquid.

The change in heat within the substance that occurs during changes in state does not cause a change in temperature. Thus, it is distinct from sensible heat (heat causing a change in temperature) and is referred to as latent (i.e., hidden) heat. Latent heat is defined as heat that brings about a change of state with no change in temperature. Latent heat, illustrated in figure 8-1, can be seen as the heat that is added between points B and C, and points D and E without causing a change in temperature.

When a substance melts, the heat is called the latent heat of melting, or when going from a liquid to a solid upon cooling, the latent heat of fusion. The amount of heat that must be added or removed is the same in both directions for a given substance. With water, 335 kJ · kg^{-1} (80 cal · g^{-1}) is required. When a liquid is converted to a gas, the latent heat is called the latent heat of vaporization, or when cooled, the latent heat of condensation. Again, the amount of heat that must be added or removed is identical. When water (at 100°C) is converted to steam (i.e., points D to E) 2,260 kJ · kg^{-1} (540 cal · g^{-1}) is required. When one assesses the actual amounts of heat exchanged with changes in state, they are quite remarkable. For example, it takes as much heat to change 1 kg of ice to water as it does to raise the temperature of 1 kg of water from 0°C to 80°C. Since the energy change from liquid water to gaseous is approximately 6.75 times that of ice to water, volatilization represents an excellent way to absorb heat. In fact, the latent heat of vaporization is used commercially (vacuum cooling) to cool a number of harvested products.

The temperature at which a change in state occurs depends on pressure. The lower the pressure, the lower the temperature at which water becomes a gas; the amount of heat required to facilitate the change of state, however, remains the same. Likewise, the greater the pressure, the greater the temperature at which the change occurs. The latent heat of vaporization of water would be of little value for cooling harvested products if the temperature of the water had to first reach 100°C. However, by lowering the pressure to 5 mmHg (660 pascals), the pressure used for commercial vacuum cooling water will boil at 1°C.

Since the molecular structure of each substance varies, each will have a

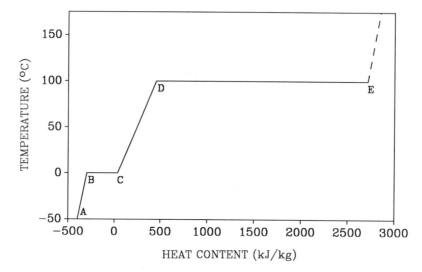

Figure 8-1. The relationship between water temperature and heat content. One kilogram of water at atmospheric pressure (100 kPa) was heated from −50°C to complete vaporization. Between points A and B the temperature of the ice increased with the addition of heat (2 kJ · kg⁻¹ · °C⁻¹ × 50°C = 100 kJ). From points B to C, 335 kJ of heat were added to melt the ice (latent heat of melting) without changing its temperature. Between points C and D, 420 kJ of heat were required to bring 1 kg of water to its boiling point (2.4 kJ · kg⁻¹ · °C⁻¹ × 100°C = 420 kJ). From point D to E, 2,260 kJ of heat were required to convert water to steam (latent heat of vaporization), again with no change in temperature.

different latent heat value for the phase changes. Likewise, the temperature and pressure combination at which the phase change will occur differs widely between substances. For example, methanol has a latent heat of fusion of 67 kJ · kg⁻¹ (16 cal · g⁻¹) at −97°C and water is 335 kJ · kg⁻¹ at 0°C (table 8-1).

The cooling effect of ice comes largely from the absorption of latent heat as the ice melts at 0°C. A range of lower melting point temperatures can be obtained through the addition of salts such as NaCl or $CaCl_2$, a common practice when liquid ice* is used (e.g., for broccoli (*Brassica oleracea*, L. Italica group)). Salt and ice mixtures can be made that reduce the melting point to −18°C, a temperature that is excessively low for direct contact with most harvested plant products. Based on product requirements, the amount of salt added can be adjusted to give the desired melting point.

HEAT TRANSFER

Heat can be transferred from one substance or body to another by three processes: radiation, conduction, and convection. With harvested products, heat transfer generally involves a combination of all three processes.

---

* Liquid ice is a slurry of ground ice and water that has sufficient fluidity to allow it to flow. The mixture is pumped into boxes of product under pressure, filling the space not occupied by the product. The water then drains, leaving the ice dispersed throughout the box.

Radiation

Radiation is the transfer of heat energy by means of electromagnetic waves. These move outward from the surface of an object in all directions. Any object above absolute zero (0 K) emits thermal radiation, the quantity and wavelength of which is dependent on the temperature of the object. The wavelengths involved are typically very long (i.e., longer than 4 μm). Objects in the harvested product's environment that emit thermal radiation include water, carbon dioxide, and other molecules in the air, the sun, the soil, buildings, harvesting containers, and equipment; virtually everything surrounding the product can radiate heat. The quantity of thermal radiation emitted by an object is proportional to the absolute temperature of the object raised to the fourth power (T^4). Therefore, the greater the temperature, the greater the loss of heat as thermal radiation.

Conduction

Conduction is the transfer of heat from one object to another through direct contact or from one part of an object to another. Heat flows readily down an iron rod by conduction, iron being a very good conductor. Not all substances are equally good conductors of heat; for example, wood and air are poor conductors. Several very poor conductors (e.g., urethane foam and fiberglass) are used as insulators for cold rooms and packaging houses. In general, substances that are good conductors of electricity are also good conductors of heat.

Convection

Convection is the transfer of heat through the movement of a liquid or gas. When air is heated, for example, by a stove, the energy level of the air molecules increases, making them move faster and further apart from neighboring molecules. As the distance increases, the actual density of the warm air decreases, making it lighter, which causes it to rise. Heat is transferred from its source to other objects through this rising current of air. Transfer also occurs in liquids where the lighter, warmer molecules rise in currents to the top.

SOURCES OF HEAT

Life within the Earth's biosphere survives due to a continual influx of energy, virtually all of which was at one point emitted by the sun (other nonatomic sources, e.g., geothermal, cosmic, magnetic, are relatively minor contributors). Heat represents but one form of the total energy within the biosphere. Although energy cannot be created or destroyed (the first law of thermodynamics), it can be changed from one form to another, and much of the irradiation striking the Earth eventually becomes heat. Reradiation of energy back into space prevents the biosphere from progressively heating up. During the postharvest period, energy that is in the form of heat is of particular interest.

Energy from the sun may enter the plant directly as absorbed irradiation or indirectly after being first absorbed by other substances in the plant's environment and then transferred to the plant. Most of this transferred energy is in the form of heat that moves into the plant by convection, conduction, and irradiation.

ENERGY BALANCE

To understand heat and heat load in harvested products, it is best to focus first on the inputs and losses of all sources of energy (fig. 8-2). Sunlight, thermal radiation, convection, conduction, and changes in latent heat interact with plants and their individual parts by delivering and removing energy. These can be summarized in what is called an energy balance equation:

$$Q + R + C_D + C_V + L + P + M + S_E = 0 \qquad (8\text{-}1)$$

where

Q = energy absorbed as irradiation;
R = energy lost as thermal radiation;
C_D = heat lost or gained by conduction;
C_V = heat lost or gained by convection;
L = heat lost or gained due to the latent heat of vaporization, condensation, melting, or fusion;
P = energy inputs due to photosynthesis (about 479 kJ is stored per mole of CO_2 fixed and represents a potential source of heat);
M = energy released by exothermic reactions within the plant, for example, respiration and photorespiration;
S_E = storage of heat by the product; when positive, heat storage increases; when negative, it decreases. In some instances, the energy assigned to P and M is lumped together with the storage term.

The terms for conduction (C_D), convection (C_V), and latent heat (L) can be either positive or negative depending on whether there is a net gain or loss of energy via each. The terms for absorbed irradiation (Q) and photosynthesis (P) are positive and their energy contribution ranges from relatively little to substantial for Q and zero to minor for P. Both radiation (R) and exothermic metabolism are negative values in the equation. The storage term, S_E, can also be positive (a net gain in energy) or negative (a net loss); however, its sign and magnitude depend on the composite of all of the other terms in the equation.

The equation can be simplified to:

$$(Q + C_D + C_V + L_C \text{ or } L_F + P) - (R + C_D + C_V + L_E \text{ or } L_M + M) = S_E$$

or

$$(\text{energy in}) - (\text{energy out}) = \text{energy stored} \qquad (8\text{-}2)$$

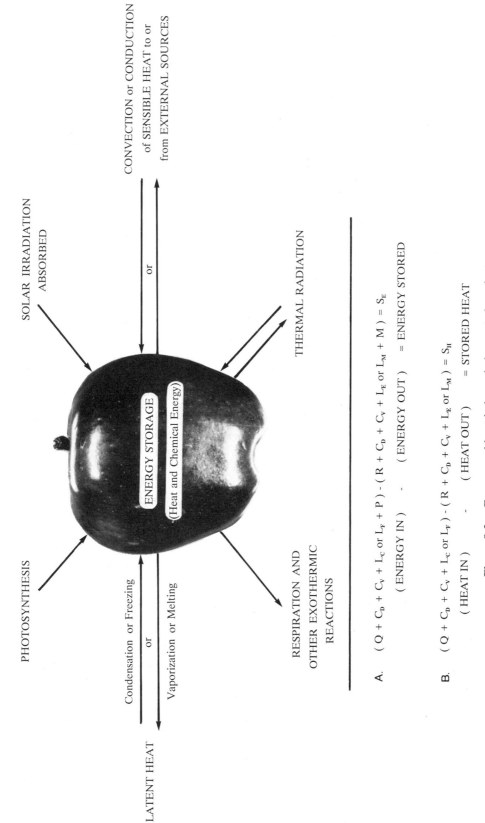

SOLAR IRRADIATION
ABSORBED

CONVECTION or CONDUCTION
of SENSIBLE HEAT to or
from EXTERNAL SOURCES

or

THERMAL RADIATION

PHOTOSYNTHESIS

ENERGY STORAGE

(Heat and Chemical Energy)

Condensation or Freezing
or
Vaporization or Melting

LATENT HEAT

RESPIRATION AND
OTHER EXOTHERMIC
REACTIONS

A. $(Q + C_D + C_V + L_C \text{ or } L_F + P) - (R + C_D + C_V + L_E \text{ or } L_M + M) = S_E$

   (ENERGY IN)   -   (ENERGY OUT)   = ENERGY STORED

B. $(Q + C_D + C_V + L_C \text{ or } L_F) - (R + C_D + C_V + L_E \text{ or } L_M) = S_H$

   (HEAT IN)   -   (HEAT OUT)   = STORED HEAT

Figure 8-2.   Energy and heat balance in harvested products.

HEAT LOAD

Of primary interest during the postharvest period is the heat balance of plants or plant parts rather than the total energy balance. Equation 8-2 can be rewritten to include just the terms that contribute significantly to the overall heat load. Ignoring energy released by exothermic reactions within the plant, which under most circumstances is a minor heat source, and photosynthesis, the heat balance equation is derived. Therefore, equation 8-2 becomes:

$$(Q + C_D + C_V + L_C \text{ or } L_F) - (R + C_D + C_V + L_E \text{ or } L_M) = S_H$$

or

$$\text{(heat in)} - \text{(heat out)} = \text{stored heat} \tag{8-3}$$

The latent heat term becomes L_C for condensation, L_F for fusion, L_E for vaporization, and L_M for melting. Photosynthetic inputs (P) and exothermic losses (M) are typically very minor in harvested products and drop out (exceptions for M are discussed later). The storage term now becomes heat stored (S_H).

Different sources of energy vary in their importance to the overall heat budget of a product and their contribution generally changes markedly from preharvest through final utilization by the consumer. When in the field, both before and after harvest, solar and thermal irradiation represent the primary sources of energy that contribute to the heat load. Solar irradiation may strike the product directly, be scattered by molecules in the atmosphere before striking the product, or be reflected from other objects in the product's environment (e.g., other plants, clouds, the soil) (fig. 8-3). Not all of the solar irradiation is absorbed by the plant; the amount absorbed depends on the wavelength and absorption characteristics of the plant (fig. 8-4). Visible light in the blue (400–500 nm) and red (600–700 nm) regions of the spectrum is efficiently absorbed by green plant parts. Parts of the spectrum in the green range (500–600 nm), however, are absorbed to a much lesser extent as also are regions in the near infrared range (700–1500 nm).

Plants are very efficient absorbers of infrared or thermal irradiation. Infrared irradiation comes from the atmosphere and objects surrounding the plant. Any object with a temperature above 0 K (absolute zero) emits thermal radiation. Thus, temperature can be measured with a radiometer and photographs can be made in complete darkness using infrared film. Aerial infrared photography is occasionally used to detect insect and disease problems in areas of large fields where stress has altered the normal level of infrared emitted by the plants. Most of the thermal irradiation from the sky is emitted from water, carbon dioxide, and other molecules in the air.

Only part of the total solar irradiation striking the plant is absorbed; a significant amount is either transmitted or reflected away. Likewise not all of the absorbed irradiation is converted to heat; in the chlorophyllous tissues of intact plants a small part (e.g., about 1%) of the total absorbed light energy is converted to chemical energy via the photosynthetic process. This chemical energy represents a storage form of potential heat that may eventually be liberated by respiration.

While in the sunlight, solar irradiation represents a primary source of heat for intact plants and harvested plant parts. Placing harvested material in the

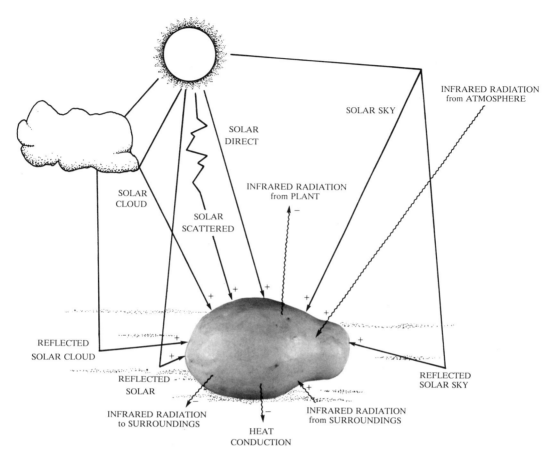

Figure 8-3. The interaction between irradiation from the sun and plant products both before and after harvest.

shade and harvesting at night or during the early-morning hours reduces this input of energy, decreasing the rate of heat buildup.

Thermal radiation of heat (R) represents an important means of dissipating heat from intact plants. Thermal radiation, radiation in the infrared region of the spectrum, is emitted from all sides of the product, moving outward until striking another object. After harvest, plant products are consolidated into bins, boxes, trucks, and other containers where each individual unit is in contact or close proximity with its neighbor. Thermal radiation emitted now tends to largely move into the adjacent product, minimizing its impact on decreasing the heat load of the product.

The importance of the convection and conduction term (C_V and C_D) also shifts after harvest. Convection of heat away from the plant is an important means of heat elimination in intact plants. Convection is especially important for leaves that have large surface areas and small volumes. During convective heat transfer, the distance between individual air molecules* surrounding the product expand due to the energy transferred from the plant. This expansion

---

* Or liquid molecules.

WAVELENGTH (μm)

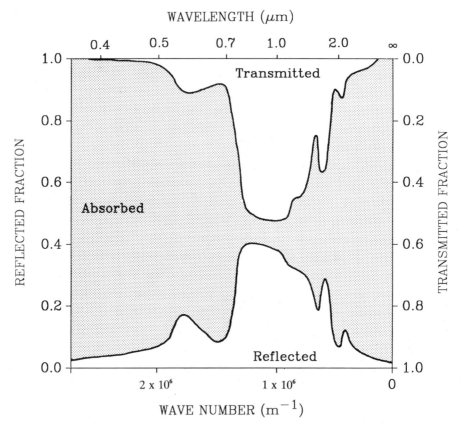

Figure 8-4. Representative fractions of the total solar irradiation at each wavelength that are absorbed, transmitted, and reflected by a leaf *(after Gates[9])*.

in turn decreases the density of the air, causing it to rise upward due to pressure differences, moving the thermal energy away from the plant. Both plant and environmental characteristics have a pronounced influence on the amount of convection. Convective heat exchange is proportional to the difference in absolute temperature between the product and its environment and is inversely proportional to the boundary layer resistance. Plant factors such as the shape and size of the organ in question are critical, as are environmental parameters such as wind speed.

There are two types of convection of heat away from a plant part. Free convection is caused by changes in the density of the surrounding air molecules and forced convection is caused by wind and moves the heated air outside of the boundary layer. Forced convection is the more dominant of the two means of convection and increases with increasing wind speed. Forced convection is responsible for the difference in the rate of temperature reduction between forced air and static refrigeration postharvest precooling techniques.

Conduction, the flow of heat between parts of a substance or neighboring substances that are in contact, can occur between any of the three physical phases (e.g., solid → solid, solid → gas, liquid → solid, gas → liquid, etc.) due to molecular and/or electronic collisions. Conduction in plants occurs in the boundary layer surrounding an organ and within the tissue. In the boundary layer, heat is transferred through the random collisions of gas molecules. The conduction of heat within the boundary layer is much slower than conduction within the plant since the thermal conductivity of air is much lower than water, the primary constituent of most succulent plant parts (i.e., 6.0×10^{-5} cal $\cdot$ s$^{-1} \cdot$ cm^{-1} per °C for air versus 1.43×10^{-3} cal $\cdot$ s$^{-1} \cdot$ cm^{-1} per °C for water).

Conduction is responsible for the movement of heat within harvested products. It is especially important with large and/or dense plant parts (e.g., apple (*Malus sylvestris*, Mill.) fruit or potato (*Solanum tuberosum*, L.) tubers) that can store relatively large amounts of heat. The transfer of heat through plant products is slow and moves as a wave from warm to cooler areas. Thermal waves are caused by variations in heat input that are seldom constant. In the field, heat absorption by intact plants varies markedly over a 24-hour period. Heat inputs during the postharvest period, likewise, are seldom constant. Because of this variation, heat normally moves as a wave, having both magnitude (height) and amplitude or frequency (width). Wave magnitude depends on the temperature differential between the product and the heat source, while the frequency depends on the duration in the oscillation in heat input. The velocity of the wave through a product depends on the product's conductivity, specific heat capacity and density, and the frequency of the thermal wave.[9] Specific heat capacity and thermal conductivity values for several postharvest products are presented in table 8-2. Calculations of the actual heat transfer in a sample depend on the thermal properties and geometry of the product, the temperature of the heat source (or sink) and product, and the temperature differential between the two.

After harvest the duration and magnitude of heat inputs vary widely depending on the source and how the product is handled. For example, heat is often transferred from the rear tires and axles of trucks upward into the load of harvested product. The frequency of the wave will depend on a number of factors, for example, distance traveled and number of stops.

Conduction becomes increasingly important after harvest when the individual harvest units are consolidated, placing them in direct contact with each other. If one uses the difference in thermal conductivity of air and water as a very rough estimate of the difference in the rate of heat transfer between two touching objects versus the object and air, the heat conducted per unit surface area between touching objects is approximately 24 times more effective. This difference in thermal conductivity between water and air is the reason that water or hydrocooling lowers the temperature of a harvested product faster than air cooling at equivalent temperatures.

The state change occurring with the evaporation of water absorbs energy, the latent heat of vaporization, and cools the product. Water evaporates from plants at the air-liquid interface, primarily at the cell walls of mesophyll, epidermal cells, and guard cells. The heat removed due to evaporation can be calculated from the product of the rate of evaporation (g $\cdot$ cm$^{-2} \cdot$ s^{-1}) and latent heat of vaporization of water (2.4 kJ $\cdot$ g^{-1}). With typical intact crop plants growing under moderate conditions,* the latent heat term in the energy balance equation (L) typically represents slightly less than 50% of the daytime

---

* 20°C air temperature, 25°C leaf temperature, 50% RH, net radiation of 370 wm^{-2}.

Table 8-2. Thermal Properties of Selected Fruits and Vegetables

	% H₂O	Density g · cc⁻¹	Specific Heat† J · kg⁻¹ K	Thermal Conductivity w · m⁻¹ K	Latent Heat J · kg⁻¹
Avocado* (*Persea americana*, Mill.)	94.0	1.06	3,810	0.4292	316,300
Banana (*Musa* spp.)	74.8	0.98	3,350	0.4811	251,200
Beet root (*Beta vulgaris*, L.)	87.6	1.53	3,770	0.6006	293,100
Broccoli (*Brassica oleracea*, L. Italica group)	89.9	–	3,850	0.3808	302,400
Cantaloupe (*Cucumis melo*, L. Cantalupenis group)	92.7	0.93	3,940	0.5711	307,000
Carrot (*Daucus carota*, L.)	88.2	1.04	3,770	0.6058	293,000
Cucumber (*Cucumis sativus*, L.)	97.0	0.95	4,103	0.5988	–
Lemon (*Citrus limon*, (L.) Burm. f.)	89.3	–	3,850	1.817	295,400
Nectarines (*Prunus persica* var. *nucipersica*, (Suckow) Schneid.)	82.9	0.99	3,770	0.5850	276,800
Peach (*Prunus persica*, (L.) Batsch.)	86.9	0.93	3,770	0.5815	288,400
Pear (*Pyrus communis*, L.)	–	1.00	–	0.5954	–
Peas, black-eyed (*Vigna unguiculata*, (L.) Walp.)	–	–	–	0.3115	246,600
Pineapple (*Ananas comosus*, (L.) Merrill)	85.3	1.01	3,680	0.5486	283,600
Turnip root (*Brassica rapa*, L. Rapifera group)	90.9	1.00	3,890	0.5625	302,400

Note: Data from ASHRAE,[1] Reidy,[25] Sweat,[33] Turrell and Perry.[37]
* From ASHRAE,[2] with thermal conductivity data from Sweat.[33]
† Specific heat values above freezing.

heat load, the remainder being largely dissipated by radiation, convection, and conduction. If transpiration is stopped, an equivalent amount of heat will be lost due to increased radiation, convection, and conduction, with only a 2–3°C rise in leaf temperature.

The rate of water loss by intact plants via transpiration is generally far greater than with decapitated harvested products. Harvest results in stomatal closure, and with the consolidation of individual plant parts in containers, evaporation is reduced, decreasing the amount of heat that can be dissipated in this manner. By lowering the atmospheric pressure, thereby decreasing the temperature at which water will boil, the rate of evaporation can be greatly increased (i.e., heat removed). This acceleration of evaporation, utilizing the latent heat of vaporization of water, is the thermodynamic principle involved in vacuum cooling of harvested products. Since evaporation is a surface phenomenon, products with large surface-to-volume ratios (e.g., lettuce) are particularly suited to this type of cooling.

The condensation of water on the surface* of a harvested product releases (latent heat of condensation) the same amount of energy ($2.4\,kJ \cdot g^{-1}$) as that absorbed with evaporation. Condensation has two primary effects on the product: It adds heat and wets the surface. The importance of each is dependent on the product in question as well as the conditions to which it is exposed.

The terms for photosynthesis (P), exothermic reactions (M), and storage (S) are often grouped collectively in energy balance equations under the storage term. The rationale is that photosynthesis and metabolic heat are generally very small factors in the overall heat balance of the plant. Since photosynthesis represents irradiation absorbed and converted to chemical energy rather than heat, and metabolic heat represents chemical energy released as heat, their inclusion as a combined stored heat term can be confusing. The actual storage term represents stored heat, with changes in S being seen as changes in temperature. When S is positive, product temperature is increasing; when negative, it is decreasing.

How important is metabolic heat in the overall heat load of a harvested product? Under most situations metabolic heat has a negligible impact on the temperature of the product because of the relatively high rate of heat exchange between the product and its surrounding environment. For example, the temperature excess at the center of an apple fruit caused by metabolic heat was only 0.023°C after 2 hours of cooling.[3] With forced-air cooling of brussels sprouts (*Brassica oleracea*, L. Gemmifera group) (fig. 8-5) metabolic heat also represents an extremely minor heat load.

Under conditions where the product has a very high rate of respiration, and heat exchange is restricted (e.g., tight packing of leafy crops), metabolic heat can be significant. The production of metabolic heat decreases rapidly with decreasing product temperature (table 8-3); therefore, rapid cooling is advantageous. Once cooled to the proper storage temperature, respiratory heat represents a small but significant component in the overall heat load and must be included in estimates of refrigeration requirements.

---

* The formation of dew or frost.

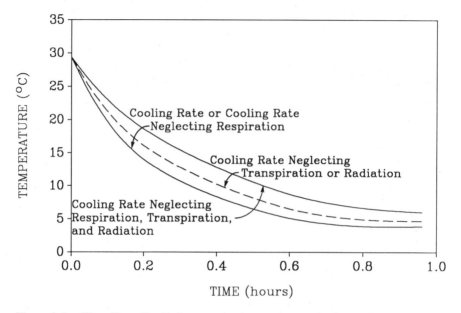

Figure 8-5. The effect of radiation, respiration, and transpiration on forced-air cooling of brussels sprouts (*Brassica oleracea*, L. Gemmifera group) (air velocity of 40 fpm). Metabolic heat (respiration) has a very minor effect during initial product cooling (*after Gaffney et al.[8]*).

Table 8-3. Effect of Product Temperature on the Production of Metabolic Heat by Selected Harvested Plant Products

	Metabolic Heat Produced $kJ \cdot T^{-1} \cdot hr^{-1}$				
	Product Temperature				
	0°C (32°F)	5°C (40–41°F)	15°C (59–60°F)	20°C (68–70°F)	25°C (77–80°F)
Apples (*Malus sylvestris*, Mill.)	24–44	53–78	145–330	179–373	
Asparagus (*Asparagus officinalis*, L.)	300–640	630–1120	1236–2496	1856–2869	3965–5075
Beans, lima (*Phaseolus lunatus*, L.)	112–320	208–383	1066–1328	1415–1910	
Carrots (*Daucus carota*, L.)	102–218	136–281	276–572	490–1013	
Lettuce, head (*Lactuca sativa*, L.)	63–179	141–213	339–480	543–640	780–974
Onions, dry (*Allium cepa*, L.)	112–238	184–727	702–1037	824–1663	1042–2235
Oranges (*Citrus sinensis*, (L.) Osbeck.)	19–53	39–78	136–252	238–364	262–431
Radishes, topped (*Raphanus sativus*,L.)	34–102	63–141	238–451	475–616	645–945
Strawberries (*Fragaria × Ananassa*, Duchesne)	131–189	175–354	756–984	1091–2089	1793–2249
Tomatoes, mature green (*Lycopersicon esculentum*, Mill.)		53–87	175–301	301–441	368–543

Note: Data from Catlin et al.[6]; Haller et al.[14]; Pentzer et al.[24]; Scholz et al.[28]; Smith[30]; Smock and Gross[31]; Tewfik and Scott[34].

ENVIRONMENT AND ENVIRONMENTAL FACTORS AFFECTING HEAT TRANSFER

An intimate relationship exists between a harvested product and its environment. Thermal energy fluxes move into or out of the product depending on the relative properties of the environment and product at any point in time. The conditions of the environment have a pronounced effect on the rate of change in temperature of a product. During the postharvest period, the product is normally in contact with one or more media, whether gaseous, liquid, or solid, which can either add or remove heat. The specific properties of each medium in turn affect the rate and direction of heat transfer.

Gas Environments

The majority of harvested products are stored in gas atmospheres, typically air. Transfer of heat between the product and its gas environment is due to convection, conduction, absorbed irradiation, and thermal radiation losses. Physical parameters that affect heat inputs and losses are irradiation, the temperature differential between the product and its gas environment, movement of the gas, the amount of heat the gas environment will hold (a function of volume and specific heat capacity), and the atmospheric pressure. Therefore, irradiation quality, quantity, and duration; atmospheric humidity; air velocity and direction; atmospheric pressure; and water concentration interact in varying degrees of complexity to modulate heat exchange.

Irradiation

Irradiation from the environment surrounding the product has a major impact on product temperature. Exposure to direct sunlight generally increases product temperature; the higher the intensity (e.g., noon versus early morning), the greater the effect. Moving harvested material out of the sun or direct sunlight greatly reduces irradiation energy inputs.

Solar irradiation also elevates the energy of other objects in the environment surrounding the product (e.g., air, plant, soil, roadways, buildings, harvesting equipment). This, in turn, can contribute to the heat load of the product by reflected, transmitted, and infrared irradiation and the convection and conduction of heat. For example, road temperatures as high as 61°C (air temperature of 40.6°C) are common in the summer months in many areas and part of this heat is radiated upward into the load.[20] Surface friction of the trailer tires also generates heat that is transferred into the product held in the trailer. For example, shelled southern peas (*Vigna unguiculata,* (L.) Walp.) found within areas of the trailer directly over the wheels were 4.4–5.6°C warmer than peas found at the same depth elsewhere in the load.[19]

In intact plants solar irradiation increases the rate of photosynthesis (*P*); thus, part of the absorbed energy that would have otherwise been converted to heat is trapped as chemical energy. This decrease, however, is extremely small (i.e., <1%) in intact plants and is essentially nonexistent in decapitated and nonchlorophyllous tissues.

Humidity

Atmospheric humidity affects the specific heat capacity of the air, the amount of heat removed from the product due to the latent heat of vaporization (L_E or conversely given off with condensation, L_C), the amount of thermal radiation emitted from the air, and the photosynthetic rate of the tissue. Increasing humidity decreases the water vapor pressure deficit between the product and its environment, decreasing the rate of evaporation. As a consequence, less heat is absorbed due to water vaporization. Water molecules in the air also absorb solar irradiation that is reradiated as thermal irradiation. The higher the moisture content of the air, the greater the heat capacity. Finally, humidity can alter the aperture of leaf stomates, thus the rate of photosynthesis. Depressed photosynthesis, therefore, results in a small increase in heat load.

Air Movement

Air movement around the product increases the rate of forced convection, accelerating heat transfer between product and environment. Relatively small increases in air velocity have significant effects. Increases in convective heat transfer make forced-air cooling a much more rapid means of removing field heat than conventional room cooling.

Pressure

Changes in pressure primarily affect the rate of evaporation, hence the latent heat of vaporization. There is an inverse relationship between pressure and the rate of evaporation. Vacuum cooling of harvested products is dependent on the heat absorbed when water vaporizes.

Oxygen and Carbon Dioxide Concentration

Changes in heat balance due to an altered oxygen concentration are largely due to an effect on the rate of respiration of the tissue. While respiration is depressed at very low oxygen levels, under most situations metabolic heat is a very minor component in the overall thermal balance of a tissue.

Carbon dioxide concentration can alter the rate of photosynthesis (P) and respiration (M), but as with oxygen, the effect on the overall thermal status of the product is very small. Atmospheric carbon dioxide also absorbs solar irradiation, emitting thermal irradiation, part of which may be absorbed by the plant. Harvested products held in higher than ambient carbon dioxide concentrations (e.g., during controlled environment storage), however, are seldom exposed to direct solar irradiation.

Liquid Environments

While only a relatively small number of live plant products are stored in liquids, typically water, the surfaces of many are covered to varying degrees and durations with surface water. This presence of water may be intentional, (e.g., during hydrocooling, transfer using water flumes, washing, and display) or unintentional (e.g., condensation on the surface and rain). The use of ice also creates a liquid environment in that ice–product interfaces are typically quite short lived, rapidly becoming an ice–water–product interface.

The transfer of heat to or from a product into a liquid medium differs from air in that conduction and convection account for virtually all of the heat exchange. Critical properties of the exchange therefore are (1) the temperature differential between the product and media, (2) the velocity of the media around the product (forced convection), (3) the surface area of the product in contact with the liquid (e.g., scattered droplets versus complete coverage), (4) the thermal conductivity of the liquid, and (5) the amount of heat the liquid will hold (a function of both the volume of the liquid and the liquid's specific heat capacity). Likewise, heat inputs (e.g., irradiation, conduction) and losses (thermal radiation, conduction) from the medium affect the temperature differential between the product and the liquid. During cooling operations, heat transferred to the liquid medium is removed using some form of refrigeration or liquid exchange (e.g., the flow of cold well water or river water through a cooling chamber). For most cooling rate calculations, the temperature of the liquid is considered to be constant. When heat is not removed from the liquid, the amount of heat in the liquid increases in direct proportion to the amount of heat lost by the product and the medium's surroundings (e.g., container walls).

Solid Environments

A varying percent of a harvested product's surface area may be in contact with a solid object after harvest. The solid may be adjacent product, the harvesting container, or in some cases specialized cooling equipment. The transfer of heat between the product and a solid object in contact with it is by conduction. The rate of heat transfer is modulated by the size of the temperature gradient between the two objects, the size of the contact surface area, and the thermal properties of both objects (e.g., the rate of heat loss or absorption and the specific heat capacity). As mentioned previously, while ice is a solid, it seldom remains in direct contact with the product; rather a film of water forms at the interface as the ice melts. The absence of melting would be due to a product temperature equal to or lower than that of the ice, a detrimental condition for most live plant products.

Under normal conditions, solids are never in contact with the entire surface area of the product; air and/or water generally interface with the remaining area. The solid–product contact area is typically only a relatively small percentage of the total surface area, and as a consequence, product to solid heat transfer is seldom an efficient way to rapidly exchange heat. An exception would be when the temperature differential between the two is quite high, for example, grain on a hot surface. Solids, therefore, generally represent secondary sources of heat inputs or losses for harvested plant products rather than the primary means for heat exchange.

PRODUCT FACTORS AFFECTING HEAT TRANSFER

The transfer of heat into or from a harvested product involves two processes: (1) the movement of heat within the product and (2) the absorption or loss of heat at the product's surface. Both processes are influenced by the physical and chemical characteristics of the plant material.

Internal Movement of Heat

Thermal fluxes move inward as an object gains heat and outward as the object loses heat to its environment. Transfer of heat within solid plant products is by conduction and the rate of thermal conductivity is influenced by the composition and density of the object. The percent water is particularly important in that the thermal conductivity of water is far greater than that of the carbohydrate structural framework or intercellular air. Water also has a much higher specific heat capacity; therefore, much more heat must be removed to cool a saturated rice (*Oryza sativa*, L.) kernel than a dry one.

The relative ratio of solids and liquids to air space and their continuity is also important. Heat will move much more readily when the aqueous phase is continuous than when it must traverse intercellular air spaces.

Heat Transfer at the Surface

The transfer of heat into or out of a product is a surface phenomenon. The larger the surface-to-volume ratio, the greater the potential for transfer. Transfer at the surface is strongly modulated by the product's boundary layer, a transfer zone where the gaseous or liquid medium is in contact with the product. Here, the object influences the temperature, vapor pressure, and velocity of the medium. Generally, the thicker the boundary layer, the slower the convection of heat through it. The physical characteristics of the object can have a pronounced effect on boundary layer thickness. Objects that are relatively small and have large surface-to-volume ratios typically have thinner boundary layers than larger objects. This situation is due to the effect of the movement of the medium on boundary layer thickness. The greater the movement, the lower the thickness and the faster the convective heat transfer. Structural characteristics that alter the topography of the surface (e.g., trichomes, uneven surface structure) decrease the effect of media movement on the boundary layer.

Boundary layer thickness is rarely uniform across the entire object. It is thinnest where the moving medium initially touches the object. Therefore, in addition to product shape and size, the orientation of the product relative to the flow of the medium is also important. Heat transfer will be greatest where the medium first contacts the plant material.

HEAT REMOVAL AFTER HARVEST

Temperature alteration of freshly harvested plant products typically, although not always, involves the removal of heat. Most plant products contain substantially more heat at harvest than is normally acceptable for subsequent handling and storage. In highly perishable fruits, vegetables, and flowers, it is especially true, and it is also often the case in grain and pulse crops if their moisture content is high.

The heat contained by a product at harvest (field heat) largely represents thermal energy obtained from the environment surrounding the plant. To maintain the maximum storage potential of a product, it is desirable to remove this field heat as quickly as possible after harvest. For highly perishable products, the longer the removal of field heat is postponed, the shorter the time interval in which the product remains marketable.

Cooling after harvest can be separated into two processes: (1) the removal of field heat, bringing the product down to or approaching the desired storage temperature, and (2) the maintenance of that temperature through the continued removal of respiratory heat and heat moving into the storage environment (e.g., heat conducted through the walls and floor, air infiltration). For highly perishable products, the rapid removal of heat from any source prior to storage or marketing is termed precooling. The existing definition for precooling is not highly rigid and the term *precooling* is often used to include cooling that occurs prior to shipment or processing.

While it is possible to remove field heat in storage, a common practice for less perishable products (e.g., nuts), the rate of heat removal is relatively slow due to the smaller refrigeration capacity of most storage rooms. The refrigeration capacity needed for rapid precooling is substantially greater than that needed for subsequent storage and the combination of the two is seldom an economically sound choice. As a consequence, for highly perishable crops, precooling and storage are generally accomplished using different refrigeration systems.

Refrigeration

Refrigeration involves the extraction of heat from a substance by lowering its temperature and keeping the temperature below that of its surroundings. Since heat flows from areas of high temperature (energy) to areas of low temperature, the removal of heat from a harvested product requires a sink for this thermal energy. Generally a refrigeration medium is used to lower the temperature of the product. Common cooling media that come in contact with the product are air, water, and ice (fig. 8-6), each differing somewhat in the relative importance of the various means of heat transfer.

There are two basic problems in refrigeration. The first is how to get the medium cold so that it will act as a heat sink. Second, as the medium begins to absorb heat and warm up, the problem becomes how to keep it cold. Until the development of mechanical refrigeration, natural sources of refrigeration were relied upon almost exclusively. For example, ice blocks cut during the winter months in the higher latitudes of temperate zones and stored for use during the summer, cold water from streams or deep wells, and cooler night air and air at high altitudes were used. With air or water from streams and wells, heat is removed from the cooling media by replacing it with cold, new media. There is no attempt to recool the media. As ice melts, losing its cooling capacity, it is likewise replaced with new ice.

Mechanical refrigeration changed this situation in that it allows continual

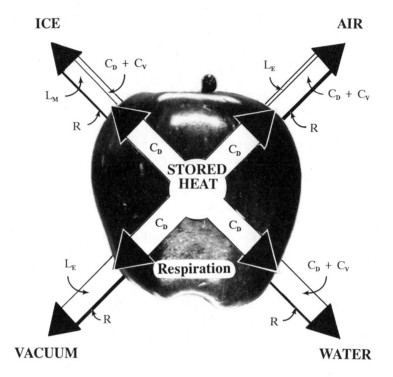

ICE   **AIR**

$C_D + C_V$ L_E

L_M $C_D + C_V$

R   R

C_D C_D

STORED HEAT

C_D C_D

L_E **Respiration** $C_D + C_V$

R   R

VACUUM   **WATER**

$$(Q + R + C_D + C_V + L_C \text{ or } L_F) - (R + C_D + C_V + L_E \text{ or } L_M) = S_H$$

(Minimize Inputs) - (Maximize Outputs) = Decline in Stored Energy

Figure 8-6. Methods of removal of heat from harvested plant products. R, thermal radiation; C_D, conduction; C_V, convection; L_E, latent heat of vaporization; L_M, latent heat of melting; L_c, latent heat of condensation; L_F, latent heat of freezing; S_H, stored heat; Q, solar irradiation absorbed.

removal of the heat absorbed by the media, hence minimizing the need for continous changes and large volumes of the respective media. Heat can be removed using several refrigeration techniques: direct vaporization of a liquid, vapor compression, vapor absorption, air cycle, vapor jet, and thermoelectric cooling (table 8-4). The volatilization of fluids, absorbing energy through the latent heat of vaporization, was first described as a means of mechanical refrigeration in 1755 by William Cullen. It has since become a primary means of direct cooling certain products (e.g., lettuce) in several countries.

The vapor compression technique couples the principle of vaporization with the subsequent reliquification of the refrigerant* using compression. The

---

* A refrigerant differs from the cooling medium. The cooling medium is also referred to as a secondary refrigerant, in that the former is a liquid used for heat transfer in a refrigerating system that absorbs heat at a low temperature and pressure and rejects heat at a higher temperature and pressure, usually involving a change in state. The cooling medium or secondary refrigerant is a nonvolatile substance that absorbs heat from a substance or space and rejects the heat to the evaporator of the refrigeration system.[2]

Table 8-4. Methods of Mechanical Refrigeration

Method	Description
Vapor compression	A closed system where the refrigerant cycles between liquid and vapor phases. For cooling it utilizes the heat absorbed during volatilization of the refrigerant; the gaseous refrigerant is then compressed with removal of the heat away from the cooled area. The now liquified refrigerant moves back to the low-pressure side to begin the cycle again.
Vapor absorption	A closed system where cooling is from vaporization of the refrigerant. Instead of mechanical compression of the gaseous refrigerant, as with the vapor compression technique, the gaseous refrigerant is either absorbed or reacts with a second substance [e.g. water (refrigerant) and ammonia (absorbent)], causing the pressure drop. Both the refrigerant and absorbent are then recharged, using various methods to begin the cycle again.
Air cycle	When air under pressure is allowed to expand, its temperature falls. Air removed from the cold room is compressed, with the heat removed using a water-cooled coil. The air is then allowed to expand in a cylinder against a piston (doing work), causing the air temperature to decline and be returned to the refrigerated area.
Vapor jet	This process is similar to the vapor compression system; however, the gaseous refrigerant is drawn from the evaporator and compressed using a high-pressure vapor, usually steam, that is passed through one or more nozzles (also called a thermocompressor).
Thermoelectric	The cooling effect is produced when an electrical current is passed through a junction of two dissimilar metals; one junction becomes cool and the other warm.

first patent for such a machine was issued in 1834 to Jacob Perkins (fig. 8-7). The essential components of a vapor compression system consist of a compressor, condenser, expansion valve, and evaporator (fig. 8-8). The refrigerant volatilizes in the evaporator, absorbing heat from the cooling medium surrounding the evaporator (i.e., air, water).

Vaporization is caused by a drop in pressure within the closed system generated by the compressor. On the high-pressure side of the compressor, the now gaseous refrigerant is compressed and begins to liquify, with heat being given off (latent heat of condensation). This heat is removed using a condenser situated outside of the refrigerated area, with air or another medium passing over the coils to facilitate heat transfer. The condensed refrigerant then passes through an expansion valve, moving to the low-pressure side of the refrigeration system to begin the cycle again. Therefore, heat is removed from the cooling medium by the evaporator coils utilizing the latent heat of vaporization. The heat is then transferred away from the refrigerated area and dissipated outside using the condenser coils.

Mechanical refrigeration systems are used to cool the medium that will act as a heat sink for the harvested product. In many cases, heat is removed from the medium (e.g., air, water) through its continued circulation over the evaporative coils of the refrigeration unit. In other cases (e.g., ice, liquid carbon dioxide, or liquid nitrogen) an expendable refrigerant is used and is metered into the cooling medium (e.g., liquid nitrogen into air, ice into

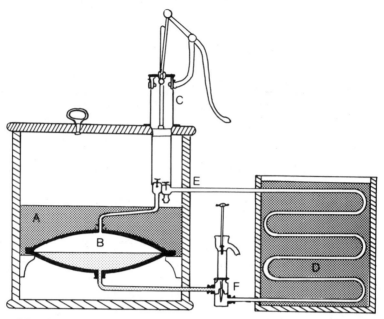

Figure 8-7. The design of Perkin's vapor compression refrigeration system patented in 1834. The refrigerant (a volatile liquid such as ether) boils in the evaporator (*B*), removing heat from the water held in the container (*A*). The volatile refrigerant is removed from the evaporator and compressed using a hand pump (*C*) and passes to the high-pressure side *via* (*E*). As the refrigerant condenses on the high-pressure side of the system, the heat (latent heat of condensation) moves into the water surrounding the condenser coils (*D*). The reliquified refrigerant then moves past a weight-loaded valve (*F*) that maintains the pressure differential between the two sides, and the cycle begins anew (*after Gosney*[10]).

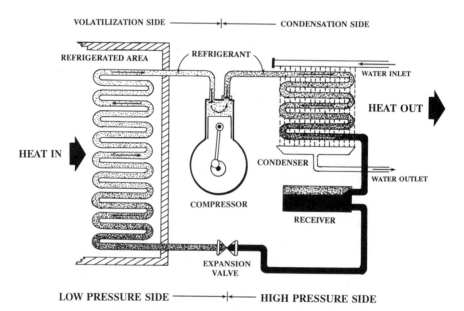

Figure 8-8. A simplified schematic of the essential components of a modern vapor compression refrigeration system. Heat is absorbed with the volatilization of the refrigerant on the low-pressure side and removed with condensation of the refrigerant after being compressed on the high-pressure side.

water). In this situation, the medium does not recycle over the evaporative coils to be recooled.

Expendable refrigerants such as ice, solid or liquid carbon dioxide, and liquid nitrogen have several advantages, the primary one being cooling is not tied to the location of the refrigeration system. Ice is often applied during field harvest of many leafy crops during hot weather. For example, turnip greens (*Brassica rapa,* L. Rapifera group) are layered with crushed ice in the beds of transport trucks for cooling during shipment to the processor. Likewise, in areas where the cost of the electricity used to run the refrigeration system varies with time of day, ice may be produced during nonpeak electrical use periods and stored for use during periods when electrical costs are high.

Ice for precooling, package icing, and transit icing is typically either block ice or one of several forms of fragmentary ice. The latter is being increasingly used by produce packaging plants in that small pieces of ice are formed and are the most desirable for rapid cooling. Block ice is utilized more for field application. Here the ice is generally purchased from a supplier and transported to the field. Block ice has a much greater density than fragmentary ice and requires less space. The low surface-to-volume ratio also decreases melting during transit and holding prior to use. When applied to the product, the ice is mechanically fragmented, increasing the surface area and cooling rate.

The most common fragmentary ice makers produce flake, tube, or plate ice. Each type has certain advantages.[2] For example, flake ice is produced continuously without an intermittent thawing cycle for removal of the ice from the freezing surface, a step that is required by tube and plate ice makers. As a consequence, flake ice is colder, is nonwetted on the surface, and gives a maximum amount of cooling surface. For some uses, however, thicker pieces of ice may be needed, making tube or plate ice more desirable.

Fragmentary ice can be produced on a continuous or a constant number of cycles per hour basis, giving a relatively constant output. Ice usage, on the other hand, is typically on a batch basis. Packing houses seldom are in operation 24 hours a day, and even during periods of operation, usage is seldom constant. Therefore, ice storage systems offer several advantages. First, a smaller ice-making system is required since it can be run 24 hours a day, collecting and storing the ice for use during peak periods. Second, ice can be produced during times of the day when electrical demand is low and power companies offer reduced rates.

Ice storage areas vary depending on whether or not short- or long-term storage is desired and depending on the degree of automation. Ice can be removed from the storage area manually or using screw conveyors, ice rakes, or other systems. Delivery systems to the site of use include screw, belt, and pneumatic conveyors.

Liquid nitrogen and liquid carbon dioxide are also used to a limited extent; however, their very low temperatures ($-195.8°C$ and $-78.5°C$, respectively) in contrast to ice ($0°C$) require much more careful control. Liquid nitrogen is used in some transport-controlled atmosphere systems in that it provides both a heat sink and a source of nitrogen to maintain a low-oxygen environment.

Cryogenics is the name given to the techniques of reaching and using very low temperatures (i.e., $< -150°C$). While the production of liquid nitrogen is an example of cryogenics, its use in postharvest storage is more closely akin to normal refrigeration since ultra-low temperatures are never desired. One interesting property of liquid nitrogen is termed cryoquenching, where hot metal (e.g., cast aluminum) or other material is cooled with this expendable refrigerant. Its advantage is that liquid nitrogen gives a *slower* and more controlled cooling rate than water due to the thin vapor layer that forms between the object and the coolant (heat transfer being much slower through a gas than a liquid). As a consequence, there is less warping and distortion of the object during cooling.

Precooling Methods

Precooling produce to remove field heat as quickly as possible after harvest is essential for slowing the rate of deterioration of highly perishable products. Precooling generally represents a single controlled handling step after harvest and may be accomplished using several different techniques (e.g., room cooling, forced-air cooling, hydrocooling, contact icing, and vacuum cooling), the appropriate choice being determined largely by the type of product in question.

Room Precooling

Harvested produce may be cooled by simply placing it within a refrigerated area. Typically refrigerated air is blown horizontally just below the ceiling, sweeping over and down through the containers of the product below. Upon reaching the floor, it moves horizontally into the return vent to be recycled. Variations of this technique (e.g., ceiling jets) are also used. Air velocities of $200–400 \, \text{ft} \cdot \text{min}^{-1}$ around the containers are required to minimize the length of time required for cooling. After reaching the desired product temperature, airflow is reduced to $10–20 \, \text{ft} \cdot \text{min}^{-1}$, a rate sufficient to maintain product temperature while minimizing water loss. Room cooling also requires well-ventilated containers for the produce and proper container stacking and stack spacing within the room.

The major advantage of room cooling is that the product is cooled and stored in the same place, decreasing the amount of handling required. Primary disadvantages include a relatively slow rate of cooling (fig. 8-9) and greater moisture loss from the product. The increased moisture loss is due to a greater fluctuation in room temperature and a prolonged exposure to high air velocities. The first product placed in the room is subjected to high airflow rates until the last product moved in has cooled. This increased moisture loss can be largely eliminated by using cooling bays within the room (fig. 8-10).

With normal air discharge, refrigerated air is blown outward from one side of the room over the containers of produce. Air velocity, however, decreases with increasing distance from the source, causing produce stacked further from the fans to have less air passing over it. Ceiling jets were designed to increase the uniformity of air turbulence within the room. Air is directed downward from a number of sites in a false ceiling. Typically these discharge sites are metal or plastic nozzles, one being situated over a single pallet-sized stack of produce on the floor. Air moves downward over the sides of the stack, increasing the rate of cooling. Ceiling jets require correct placement of

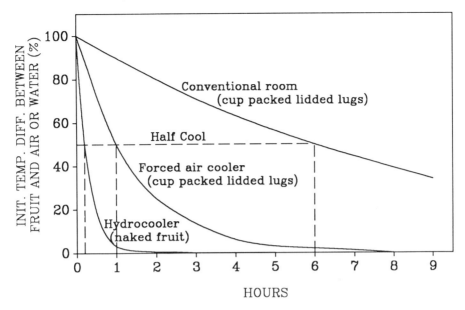

Figure 8-9. Comparison of three methods (room, forced-air, and hydrocooling) for cooling peaches (*Prunus persica,* (L.) Batsch.) *(after Guillou[12]).*

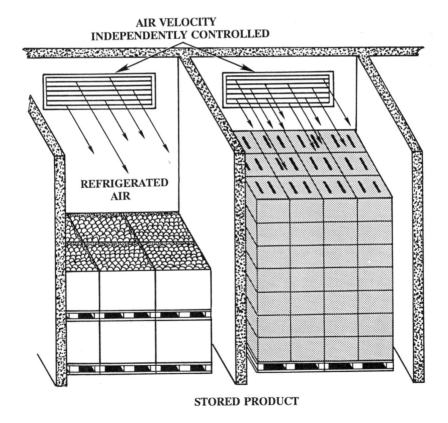

Figure 8-10. Separation of a large refrigerated storage room into cooling bays provides a way to decrease the amount of moisture lost from the product. Air circulation velocity in each bay is independent, allowing a high velocity to be used for new produce that is warm and a low velocity for product that is already cool.

the product under the outlet, greater fan power due to the increased resistance, and higher initial construction costs while reducing the total volume that can be placed within the room.

Heat removal from containerized produce is by forced convection, conduction, radiation, and a relatively small amount of evaporation. The dominant means of heat transfer is via forced convection due to the movement of air around the produce. Thus, air movement into the containers is essential for rapid cooling. It is achieved using containers with holes or slots in the sides. Removal of approximately 5% of the surface area of the container sides decreases the cooling time by approximately 25%.

Large cooling rooms are often separated with internal walls to form individual bays, allowing for segregation of new, warm produce from produce that has already been partially cooled (fig. 8-10). Air circulation in each bay is independent, allowing the air velocity to be reduced in areas with cooled produce but to remain high in areas with warm produce. Heat from the new produce, likewise, does not warm the cooled produce and minimizes the potential for condensation of moisture from the warm air on the already cooled product.

Forced-Air Precooling

Forced-air or pressure cooling utilizes a pressure drop across opposite faces of stacks of vented containers of produce to move air through the internal air spaces of the container.[13] Forced-air cooling differs from room cooling in that forced-air cooling moves the air around the individual product units (e.g., individual fruits) rather than merely around the exterior of the container. It greatly accelerates the removal of heat from the product, typically reducing cooling times to one quarter to one tenth that of conventional room cooling (fig. 8-9). Critical components for forced-air cooling are sufficient refrigeration capacity and air velocity, the use of vented containers, and utilization of a proper stacking pattern for the containers. At present there are three primary variations in how the air is moved through the containers: cold wall cooling, serpentine cooling, and the most commonly used, forced-air tunnel cooling.

With forced-air tunnel cooling, pallets or bins of product are lined up in two adjacent rows perpendicular to a single large fan, leaving an alleyway centered on the fan between the rows (fig. 8-11). The rows are generally one pallet or bin in width with the length being dependent on the fan capacity. The alleyway between the adjacent rows is then covered with a heavy fabric cover, forming an air-return plenum. The fan draws cold air from the surrounding room through vents in the container, across the produce, and into the air-return plenum or alleyway.

Cold wall cooling utilizes a permanently constructed air plenum and fan within one or more walls of the cold room. Stacks of boxes or single pallets are placed against the wall and cold air from the room is drawn through the containers into the return-air plenum (fig. 8-12). One advantage of this system is that the timing for cooling an individual container or stack of containers can be closely controlled, avoiding unnecessary desiccation. With tunnel cooling there may be a considerable time lag between when product at the exterior and

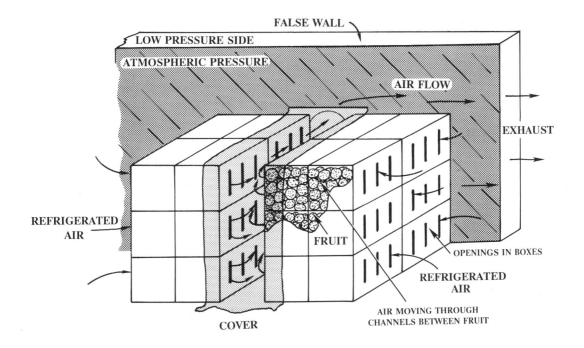

Figure 8-11. Forced-air cooling utilizes a pressure drop between the room and two rows of vented containers. The space between the containers is covered with a baffle made of a heavy but flexible material. The baffle seals the top and one end. An exhaust fan placed on the opposite end (in this case, mounted in the wall) pulls air from the space between the containers, creating a pressure drop. In response to the pressure differential, cold air moves through holes in the boxes, across the product, and into the interior chamber from which it is exhausted. This cools the product much faster than a more static airflow system (i.e., room cooling). *(Photograph courtesy of R. F. Kasmire.)*

Figure 8-12. Cold wall forced-air cooling differs from the standard forced-air cooling system in that it utilizes a permanently constructed air plenum with a built-in exhaust fan. This allows the pallet or boxes to begin cooling immediately since it is not necessary to wait for sufficient product to build a tunnel. Therefore, individual containers can be cooled for differing lengths of time, an advantageous situation when a number of different types of products need to be cooled (e.g., floral crops). *(Photograph courtesy of R. F. Kasmire.)*

interior of the tunnel is sufficiently cool. Cold wall cooling is advantageous for products where desiccation is a problem (e.g., cut flowers) and where a wide range of products with differing cooling times is being handled.

Serpentine cooling is a modification of cold wall forced-air cooling, utilizing bottom rather than side vents in the container. It is designed for produce in pallet bins, with the forklift openings at the base of the pallet being used as the air supply and return plenums (fig. 8-13); cold air is drawn from the room into a forklift opening. By blocking the back of this opening adjacent to the cold wall, the air is forced to move vertically upward and downward through the bins. The return-air plenum or forklift opening on every other vertical bin in the stack is blocked on the exterior but open at the cold wall. Therefore, air enters through one forklift opening, moves upward and downward through the bins, and exists through the forklift opening at the top or bottom of the next row of bins. Serpentine cooling, requiring no space between rows of bins, is particularly desirable for large volumes of product that are handled in bulk.

Adequate container ventilation is essential for forced-air cooling. Sufficient openings to permit a satisfactory volume of airflow through the container with a reasonable pressure drop is needed. Too little venting will restrict the flow of air through the containers while excessive vent openings decrease the strength of the containers. At least 5% of the outside area of the container

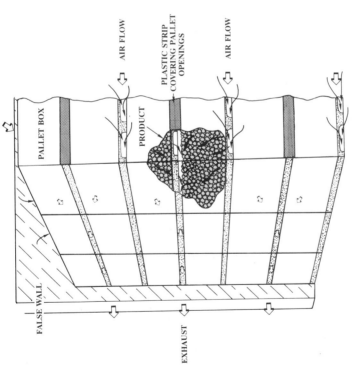

Figure 8-13. Serpentine cooling, a variation of cold wall cooling, is used for pallet boxes of produce. Plastic strips are placed over every other forklift opening on one end, and on the opposite openings on the other end. This forces the air to move either upward or downward through the pallet boxes as it moves to the fan located in the wall. (*Photograph courtesy of F. G. Mitchell.*)

should be open. The size, shape, and arrangement of openings are similar to that used for room cooling. The use of paper wraps around individual product units, plastic bags, or liners are not conducive to forced-air cooling due to the restriction of airflow.

The rate of cooling using forced air is closely related to the volume of airflow through the container per pound of product. Airflow is a function of the fan capacity and the resistance to airflow presented by the container and product. Resistance varies with the type of product, size of the openings in the containers, container stacking pattern, and distance (amount of product) the air must transverse between the site of entry into the outside containers and exit from the boxes adjacent to the fan. The greater the distance, the greater the pressure drop and the longer the cooling time for a given airflow. The optimum airflow rate and duration for proper cooling vary with the type of product being cooled.

Hydrocooling

A wide range of harvested plant products can be rapidly cooled by bringing them in contact with flowing cold water. Hydrocooling is the most rapid means of cooling a wide range of succulent products,* the speed of which is largely due to the much higher heat transfer coefficient of water than air.

Efficient hydrocooling has two basic requirements: (1) The water should move over the product surface, contacting as much of the surface as possible. Thus, hydrocooling is normally utilized for plant material held in bulk bins, and is seldom used after packaging. Packaging restricts water movement, greatly decreasing the cooling efficiency; in addition, special water-tolerant containers are required. (2) The cooling water must be kept as cold as possible without causing chilling damage to the tissue. It is typically 0°C; however, for chill-sensitive products higher temperatures are required.

Hydrocooling is commonly utilized for stem vegetables, many leafy vegetables, and some fruits (e.g., tomatoes (*Lycopersicon esculentum,* Mill.), melons (*Cucumis melo,* L.)). Product requirements include a tolerance to wetting, low susceptibility to physical damage caused by the cooling water striking its surface, and low susceptibility to injury by chemicals used in the water to prevent the spread of disease organisms (e.g., low levels of chlorine).

Exposure of the product to the cooling medium is by one of two methods: (1) showering the water down upon the product, or (2) submerging the product in the water. Water showers are the most commonly used in that they give excellent water movement, a prerequisite for rapid cooling. With showers, however, channeling of the water can occur if the product depth is too great. Channeling takes place due to the water following the path of least resistance, thus moving more rapidly through the largest openings (channels) between the individual product units. Channeling decreases the uniformity of cooling within the bin and increases the cooling time required. Proper product depth and a sufficiently high water application rate can minimize channeling.

Complete submersion of the product is an alternative to using water show-

---

* A possible exception would be vacuum cooling of certain leafy products.

ers. It eliminates the problem with water channeling in that water comes in contact with all of the product surfaces. The primary drawback to submersion is that when the product is held in bulk bins, water movement is greatly restricted. Since movement of the cooling medium over the product surface is a critical component in obtaining rapid cooling, restriction can be a serious disadvantage. Restricted water movement can generally be corrected using pumps or propellers to circulate the water around the product. A second problem with submersion is that many plant products have a density less than water, and as a consequence, they float. To prevent floating, some mechanical means of maintaining the product under water is required.

Hydrocoolers can also be separated into two general designs, conveyer versus batch, based on whether or not the product is stationary within the cooler. The most commonly used is the conveyer hydrocooler (fig. 8-14). Here the bins of product move slowly through the water (both shower and submerged), carried by a conveyer. It allows the product to be continually placed in the cooler, minimizing down time for loading and unloading. The length of the conveyer is critical and depends on the type and volume of product to be cooled and the amount of cooling required (initial versus final product temperature). The conveyer speed can be increased or decreased to adjust for products with different cooling rates and initial temperatures.

In a batch system the product is stationary, being loaded into the cooler and then unloaded when cooled (fig. 8-15). Batch coolers are typically easier and less expensive to construct and generally lend themselves to better insulation (more than 50% of the heat load may be from sources other than the product[36]).

Heat absorbed by the cooling water must be removed for efficient cooling. Removal can be achieved by recooling the water using mechanically refrigerated cooling coils or ice bunkers, or when available, by continually introducing cold, new water into the cooler. In some areas of the world, mountain streams and deep wells have sufficiently cold water for use in cooling.

The amount of refrigeration capacity needed for continuous cooling is presented in table 8-5. Since the greater the mass of product to be cooled, the greater the refrigeration needed, requirements are expressed as the amount of refrigeration per unit weight of produce. Refrigeration capacity requirements also increase with increases in the amount of heat that must be removed and with heat leakage into the cooling medium from the surrounding environment. Typically only about 50% of the total refrigeration is used to absorb product heat; the remaining heat is from the environment. With proper insulation, the percent absorbed from the product can be increased to around 80%.

Refrigeration requirements are still commonly presented as tons of refrigeration, the amount of cooling produced when 1 ton (2,000 pounds) of ice (0°C) melts during a 24-hour period. The metric system does not have a comparable unit of measure; however, one ton converts to approximately 3.54 kJ · sec^{-1} or 3.54 kW.*

---

* The energy absorbed as ice melts is equal to the latent heat of melting (336 kJ · kg^{-1}) times the weight of ice (1 ton = 907 kg), or 305,659 kJ. Dividing this by the time component (24 hr) in seconds (60 sec × 60 min · hr^{-1} × 24 hr · day^{-1} = 86,400) gives 1 ton of refrigeration equal to 3.54 kJ · sec^{-1} or 3.54 kW (1 kW = 1 kJ · sec^{-1}).

Figure 8-14. Conveyer hydrocoolers move containers through a chamber in which ice water is either showered down upon the product or the product is submerged in water. The water is then recooled and recirculated through the system. Conveyer speed can be adjusted to increase or decrease the length of time the product is in the cooler. (*Photographs courtesy of R. F. Kasmire.*)

Figure 8-15. Batch hydrocoolers differ from conveyer hydrocoolers in that the product is stationary. As with shower-type conveyer coolers, ice water cascades down upon the product until it is adequately cooled. While less expensive to build and operate, batch coolers are less efficient in that they cannot be used while the product is being loaded or unloaded. *(Photographs courtesy of R. F. Kasmire.)*

Table 8-5. Refrigeration Capacity Needed for Continuous Precooling*

Change in Product Temperature		% of Total Refrigeration Used to Absorb Heat from the Product					
		50%		65%		80%	
°C	°F	Tons[†]	kW	Tons	kW	Tons	kW
40	70	20.0	70.7	15.4	54.6	12.7	45.0
33	60	17.1	60.4	13.2	46.9	10.9	38.6
28	50	14.2	50.1	11.1	39.2	9.1	32.1
22	40	11.3	39.8	8.9	31.5	7.3	25.7
17	30	8.4	29.6	6.7	23.8	5.4	19.3
11	20	5.4	19.3	4.5	16.1	3.6	12.9

Source: After Guillou.[12]
* The fraction of the refrigeration needed to cool the product will depend on the amount of insulation and heat leakage into the system.
† Tons or kW of refrigeration per 1,000 kg of product per hour.

Icing

Ice has been used to cool harvested produce since pre-Roman times. During the postharvest period it is utilized for temporary cooling during transport from the field (e.g., leafy greens), for package icing during shipment to retail outlets, and in displays of produce at the retail level. The most widely utilized technique, package icing, represents a relatively fast cooling method that can be used for a number of products that are tolerant to contact with water and ice; excluded are most chill-sensitive products. A number of the root and stem vegetables are iced as are some flower-type vegetables (e.g., broccoli), green onions (*Allium cepa*, L.), brussels sprouts, and others.

Icing is a relatively simple operation that in some cases is accomplished in the field. Truck loads of leafy greens for processing are often given alternating layers of ice during hot weather as loading proceeds. Cooling effectiveness increases with increasing contact between the ice and the product. Therefore, techniques that facilitate contact (e.g., small pieces and use of liquid ice) hasten product cooling.

The primary disadvantage of icing products for which ice can be safely used is that the weight of the ice substantially increases the shipping weight. For relatively warm produce (i.e., 35°C), the additional weight may equal 35–40% of the weight of the product.

Several forms of fragmented ice are used: crushed, flake, snow, and liquid.[7] Body icing, using slurry mixes of ice and water (i.e., liquid ice), has increased substantially in popularity in recent years. This technique is especially useful for products that are nonuniform in size and/or configuration (e.g., broccoli). Liquid ice gives a much greater degree of initial contact between the product and the ice and it can be applied after the boxes have been palletized.

Body icing involves pumping an ice-water slurry from an agitated storage tank through the openings in the sides of the boxes of dry-packed product (fig. 8-16). Slurries range in the water-to-ice ratio from 1:1 to 1:4 and often contain a small quantity of salt to lower the melting point. The liquid nature of

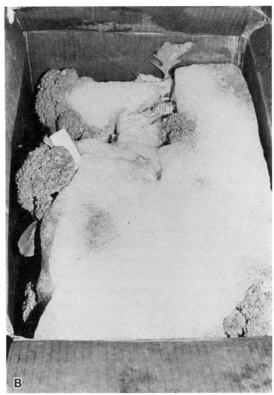

Figure 8-16. Liquid icing (also referred to as body icing) is a form of package icing that utilizes a slurry of ground ice and water that is pumped into palletized boxes (*A*). Moving through the ventilation holes in the containers as a liquid, liquid icing has the advantage over older means of ice application in that it moves around all of the individual units of produce, maximizing contact. The water present drains from the slurry upon filling, leaving a relatively solid mass of crushed ice within the container. *B:* The condition of a box of body-iced broccoli upon reaching the retail market.

the slurry allows the ice to move throughout the box, filling all of the void volume of the container, reaching all the crevices and holes around the individual units of the product. After removal from the body icing machine, the water drains, leaving a relatively solid mass of crushed ice in which the product is embedded. The principal advantage of liquid icing is the much greater contact between the ice and product afforded by this method. When the boxes are palletized prior to application, proper orientation of the openings is required for an unrestricted flow of ice throughout the load.

Ice may also be applied manually using rakes or shovels, or with automatic mechanical package icers. Regardless of the technique of application, special water-tolerant boxes are required.

Vacuum Cooling

Vacuum cooling represents a very rapid and uniform method of cooling used extensively in some production areas for several crops. Vacuum cooling involves decreasing the pressure around the product to a point that lowers the boiling point of water to 0°C (i.e., 4.6 mmHg). The conversion of liquid water to a gas absorbs heat, that is, the latent heat of vaporization (2,260 kJ · kg^{-1} of water). Because evaporation is a surface phenomenon, products with large surface-to-volume ratios are the most effectively cooled (e.g., leafy crops such as lettuce, cabbage (*Brassica oleracea,* L. Capitata group), brussels sprouts). Several crops with lower surface-to-volume ratios (e.g., mushrooms) are also vacuum cooled; however, a greater cooling time is generally required.

Since vacuum cooling requires the volatilization of water, some water is lost from the product. Generally around 1% moisture is lost for every 5–6°C or 9–10°F drop in temperature (fig. 8-17). For most products adequate cooling represents about a 3% loss of water. Losses can be reduced, however, by spraying the surface of the product with water prior to cooling. This procedure, however, requires water-tolerant boxes.

The packaging material must not significantly retard the escape of water vapor since evaporation and, as a consequence, cooling would be greatly impeded. Exposed heads of lettuce, therefore, cool faster than those held in boxes. When sealed in polyethylene bags, no cooling occurs. Perforated polyethylene bags or loosely sealed polystyrene shrink wraps do allow adequate cooling.

During cooling, product temperature is routinely monitored using a temperature probe inserted into a representative sample. Cooling is relatively uniform throughout pallet loads of product; however, individual tissues will vary in their actual temperature. For example, with head lettuce the base of

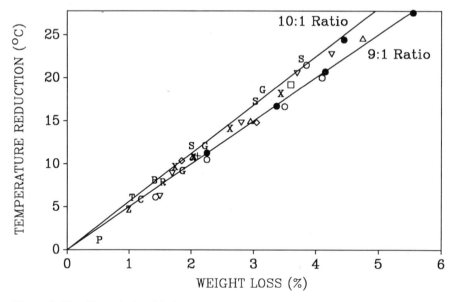

Figure 8-17. The relationship between product temperature reduction and the percent weight loss of selected vegetables when vacuum cooling is used. Most products exhibit a 5–6°C (9–10°F) temperature drop with each percent weight loss of water (△, artichoke; G, asparagus; S, brussels sprouts; C, cabbage; R, carrot; O, cauliflower; X, celery; ▽, corn; •, lettuce; +, mushrooms; □, green onions; ◇, peas; P, potatoes; T, potatoes, skinned; B, snap beans; Z, zucchini squash) *(after Barger[5]).*

the stem is typically several degrees warmer than the leaves due to differences in surface-to-volume ratios and the amount of water lost. Initial product temperature has little influence on the final product temperature, although it does affect the length of time required for cooling to a specific temperature. When individual product units with different initial temperature are cooled together, the warmer product will have lost more moisture than the cooler one when the final temperature is reached.

During cooling the wet bulb temperature within the vacuum chamber decreases rapidly as the initial vacuum is established (fig. 8-18). Four to 6 minutes after the start of evacuation, however, the wet bulb temperature rises dramatically (e.g., about 8.3°C).* The sharp rise in wet bulb temperature is called the flash point and occurs when the product begins to lose moisture rapidly and the tank momentarily becomes saturated with water vapor. After traversing the flash point, product temperature begins to decline rapidly, the rate of which tends to decrease as the actual product temperature declines.

Commercial vacuum cooling operations (fig. 8-19) commonly utilize a pressure of 4.6 mmHg, the point at which water will boil at 0°C. Under these conditions lettuce at typical field temperatures is cooled to 1.1°C in approximately 20–25 minutes. Slightly lower pressures (e.g., 4.0 mmHg, −1.7°C boiling point) can be used to achieve a lower final temperature and/or a faster cooling rate (16–18 minutes for lettuce) with little risk of freezing. When

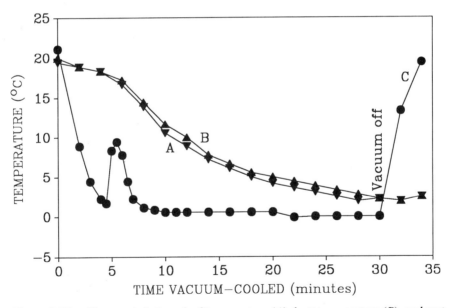

Figure 8-18. Changes in lettuce leaf temperature (*A*), butt temperature (*B*), and wet bulb temperature (*C*) within the vacuum chamber with time during vacuum cooling. The sharp rise in wet bulb temperature at approximately 5 to 6 minutes is due to the rapid volatilization of water, causing a transient water saturation of the chamber environment (*after Barger[4]*).

---

* The precise length of time will depend on the rate at which the vacuum is established.

Figure 8-19. Products such as lettuce (*Lactuca sativa,* L.) are vacuum cooled by rolling pallet loads into the cooler, sealing the chamber and reducing the pressure to 4.6 mmHg for a sufficient duration to give adequate cooling. The product is then rolled out of the opposite end of the chamber (as additional product is being placed in) and moved to conventional cold storage rooms.

higher final temperatures are required, the length of the cooling cycle is shortened rather than selecting a higher pressure.

The principal disadvantages of vacuum cooling are the high initial equipment cost and the need for skilled operators. Nevertheless, in areas where large volumes of amenable crops are produced over extended periods, vacuum cooling is used extensively.

Selection of Precooling Method

A number of factors collectively dictate the precooling method to be used. These include product considerations, packing house size and operating procedures, and market demand. The specific requirements of many products exclude some precooling methods. For example, floral crops are never hydrocooled due to the damage that would be sustained by their typically delicate floral structure. With other crops (e.g., strawberry (*Fragaria* × *Ananassa,* Duchesne)), free water on the surface greatly increases the risk of disease. Some products (e.g., squash (*Cucurbita* spp.), potatoes, fig. 8-20) are not cooled rapidly enough and/or sufficiently using vacuum cooling. When more than one precooling method meets necessary product requirements, the economics of cooling becomes an increasingly major consideration. For example, hydrocooling of palleted boxes of oranges (*Citrus sinensis,* (L.)

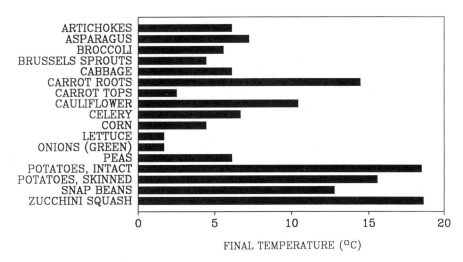

Figure 8-20. A comparison of temperature reduction using vacuum cooling for various types of vegetables under similar conditions [i.e., similar initial product temperature (20–22°C), minimum pressure (4.0–4.6 mmHg), condenser temperature (−1.7 to 0°C), and cooling duration (25–30 min)] *(after Barger[5])*.

Osbeck.) costs \$0.043 per box versus \$0.062 per box with air cooling.[8] Likewise, factors such as personal preference, convenience, and equipment availability may also enter into the decision.

Cooling Rate

For highly perishable products it is desirable to remove the field heat as quickly and economically as possible after harvest since these products may deteriorate significantly during slow cooling. A very rapid temperature reduction does not damage the tissue, providing the lower temperature limit of the product is not exceeded. The actual rate of temperature decrease varies with the cooling method and conditions, the product in question, and other considerations (e.g., packaging). Typically, hydrocooling and vacuum cooling (leafy products) are the most rapid means of cooling. The rate of temperature decline with forced-air cooling can also be quite rapid if a relatively large volume of air is used per unit volume of product. Icing and room cooling are typically slower, requiring substantially longer intervals before adequate cooling is achieved.

For efficient management of a precooling operation, it is desirable to know how long a particular product must be precooled to reach a specific temperature. Knowing the amount of time required for proper precooling gives more control over the flow of produce through the packing house for marketing or storage. With a specific precooling technique, the rate of cooling is determined by the temperature of the product, temperature of the cooling medium, and characteristics of the product. For example, when at the same initial temperature, and exposed to the same precooling conditions, a muskmelon (*Cucumis melo*, L. Reticulatus group) fruit requires a longer cooling time than a pepper (*Capsicum annuum*, L. Grossum group) fruit. Cooling rate equations for specific products can be generated from experimental temperature–time response data or from theoretical relationships for specific products or classes within a product type (e.g., size, packaging method).

One technique for estimating the length of time a product must be pre-cooled is referred to as the half-cooling time. The theoretical concepts governing precooling of harvested plant products were developed in the 1950s[26,35]; however, they were not widely used by the industry until the introduction of simplified charts or nomographs.[32]

Half-cooling time is the amount of time required to reduce the temperature difference between the product and its surroundings (cooling medium) by one-half.* Therefore, if the product temperature is 34°C and the coolant temperature is 0°C, the length of time required to reduce the product temperature 17°C is the half-cooling time. Since the difference in temperature between the product and cooling media is the critical factor, half-cooling time is independent of the initial temperature of the product and remains constant during cooling. Nomographs have been developed for most products that take into account all possible product–cooling medium temperature combinations. In addition, other factors such as cooling method, fruit size, crating, type of container, and trimming must be considered in that each can have a pronounced effect on the half-cooling time. For example, a separate nomograph is used for exposed artichokes (*Cynara scolymus,* L.) versus artichokes in crates with the lids open (fig. 8-21 *A* and *B*).

For most products it is necessary to decrease the temperature to lower than one half the difference between the product and the coolant, the temperature achieved by leaving the product in the coolant for a single half-cooling time. As a consequence, it is necessary to utilize several half-cooling times in succession. For example, the temperature drop from an initial product temperature of 30°C and a coolant temperature of 0°C is 15°C for one half-cooling time. If the product remains in the coolant for a second half-cooling time (*x* minutes) after reaching 15°C, (i.e., a second half-cooling period or cycle), the product temperature now drops from 15 °C to 7.5°C. A third cycle would decrease the temperature to 3.75 °C. While the actual rate of cooling is fastest when the temperature differential between the product and coolant is greatest, the half-cooling time remains constant (fig. 8-22).

Nomographs allow calculations to be bypassed in that the cooling time required can be read directly on the *x*-axis if the desired final product temperature is known. This is illustrated in figure 8-21 *A* with a nomograph for fully exposed globe artichokes. The nomograph is read by placing a straight edge on the initial commodity temperature (left *y*-axis) and the coolant temperature (right *y*-axis). Where the line intersects the desired product temperature (horizontal lines between the left and right *y*-axis), the time is read directly below on the bottom *x*-axis. For example, completely exposed artichokes (fig. 8-21 *A*), with an initial temperature of 26.7°C (80°F), cooled with 4.4°C (40°F) water can be reduced to 10°C (50°F) in 16 minutes.

Having a separate nomograph for every type and size of product, packaging system, cooling method, and other possible variables would entail a

---

* Half-cooling times are not applicable to vacuum cooling in that heat removal is due to the latent heat of vaporization rather than the coolant.

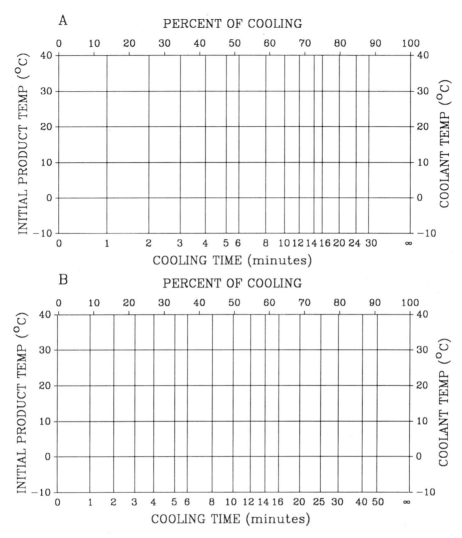

Figure 8-21. Nomographs for hydrocooling globe artichokes, *Cynara scolymus,* L. (size 36) when completely exposed (*A*) and in crates with the lids off and the paper liner open at the top (*B*) *(after Stewart and Covey[31])*.

tremendous number of graphs. However, if the half-cooling time is known for each product under a given set of conditions (table 8-6), then a single universal nomograph (fig. 8-23) can be used, eliminating the large number of graphs otherwise needed. The universal nomograph is read as one would read the nomograph for an individual product (i.e., product temperature, coolant temperature, and desired product temperature) with the exception that the time scale is given in the number of half-cooling periods or cycles rather than minutes. To convert the half-cooling periods to time, the half-cooling time for the commodity is multiplied times the number of half-cooling periods. Therefore, crated broccoli with a half-cooling time (HCT) of 5.8 minutes (table 8-6) is cooled from 26.7°C (80°F) to 10°C (50°F) in 4.4°C (40°F) water in two half-cooling periods (HCP) (fig. 8-23), or 11.6 minutes (2 HCP × 5.8 min, HCT = 11.6 min).

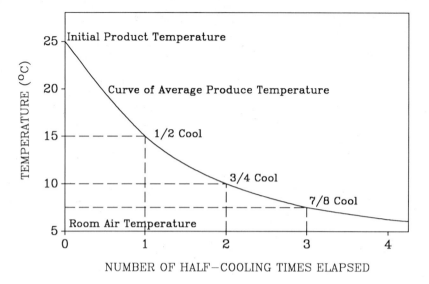

Figure 8-22. Cooling curve showing the drop in temperature from the initial product temperature (25°C) through one (1/2 cool), two (3/4 cool) and three (7/8 cool) half-cooling times. The end of the third half-cooling period represents the seven eighths cooling time. The actual rate of temperature drop and time required to achieve seven-eighths cooling varies with product and cooling conditions.

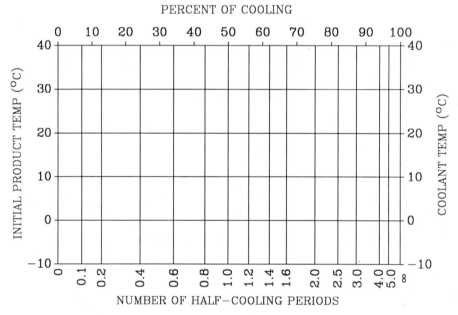

Figure 8-23. General nomograph for cooling harvested products. The nomograph is read as is one for an individual product (e.g., fig. 8-21); however, the reading from the x-axis is in half-cooling periods rather than minutes. The length of time required to cool a product to a specific temperature is determined by multiplying the number of half-cooling periods by the half-cooling time for the particular product (e.g., table 8-6).

Table 8-6. Half-Cooling Times for Selected Hydrocooled Commodities

	Treatment	Half-Cooling Time
Globe artichoke (*Cynara scolymus*, L.)	Exposed	12.8
	Crated	15.5
Asparagus (*Asparagus officinalis*, L.)	Exposed	1.1
	Crated	2.2
Broccoli (*Brassica oleracea*, L. Italica group)	Exposed	5.0
	Crated	5.8
Brussels sprouts (*Brassica oleracea*, L. Gemmifera group)	Exposed	4.4
	Crated	4.8
Cabbage (*Brassica oleracea*, L. Capitata group)	Exposed	69
	Crated	81
Carrots (*Daucus carota*, L.)	Exposed	3.2
	Crated	4.4
Cauliflower (*Brassica oleracea*, L. Botrytis group)	Exposed	7.2
Celery (*Apium graveolens* var. *dulce*, Pers.)	Exposed	5.8
	Crated	9.1
Sweet corn in husk (*Zea mays* var. *rugosa*, Bonaf.)	Exposed	20
	Crated	28
Peas in pod (*Pisum sativum*, L.)	Exposed	1.9
	Basket	2.8
Potatoes (*Solanum tuberosum*, L.)	Exposed	11
	Jumble-stack	11
Radishes (*Raphanus sativus*, L.)	Exposed, bunched	1.1
	Crated, bunched	1.9
	Exposed, topped	1.6
	Crated, topped	2.2
Tomatoes (*Lycopersicon esculentum*, Mill.)	Exposed	10
	Jumble-stack	11
Cantaloupes (*Cucumis melo*, L. Cantalupenis group)	45 size	11
	36 size	20
	27 size	20

Source: After Stewart and Covey.[32]

Seven-eighths cooling time is the length of time the product must be in the coolant to reduce its temperature through seven eighths of the initial difference between the product and coolant (fig. 8-22). This equals three half-cooling times (i.e., for each half-cooling time the product temperature drops by one half; therefore, $1/2 \rightarrow 1/4 \rightarrow 1/8$, or a 7/8 reduction [8/8 − 7/8 = 1/8] in the initial temperature difference). The seven-eighths cooling time for a given product and cooling method is, like the half-cooling time, the same regardless of the initial product and coolant temperature differential. Typically the seven-eighths cooling time is a more practical means of expressing cooling time in the commercial trade in that the expression *half-cooling time* is somewhat ambiguous (i.e., two half-cooling times do not reduce the temperature of the product to the temperature of the coolant).

Another method for determining the cooling rate of a product and therefore the length of time the product must be cooled is the cooling coefficient. A cooling coefficient denotes the change in product temperature per unit change of cooling time for each degree difference between the product and the coolant. The cooling coefficient (C) is equal to R, the change in product temperature per unit change in cooling time ($°C \cdot hr^{-1}$), divided by the average temperature differential between the product and the coolant $(T - T_o)$, or:

$$C = \frac{R}{(T - T_o)}, \quad \text{or} \quad R = C(T - T_o)$$

Therefore, the cooling coefficient can be used to calculate the length of time required to reduce the product temperature to a desired level.

Refrigeration During Storage, Transport, Sales, and Consumer Holding

Refrigeration is used not only during precooling and storage but also during transit, during retail sales, and by the consumer. Storage refrigeration is accomplished primarily using mechanical refrigeration in industrialized areas. While mechanical refrigeration is highly visible in the literature on storage, being the most common means utilized in research, the actual percent of the world population that has access to mechanical refrigeration is relatively low. Most countries that are not highly industrialized utilize mechanical refrigeration only in major metropolitan areas, using other means or no refrigeration in smaller outlying towns, villages, and rural districts. In some cases, ice purchased daily represents the primary means of refrigeration of perishable products. For others, caves and cold water from streams are the only sources of refrigeration.

Harvested plant products are moved in volume from sites of production to sites of utilization by truck, railroad, ships, and air freight, each of which may have refrigeration. Within each means of transport is a diverse range of types of carriers. For example, trucks may be designed for short or long hauls, for local delivery, or for operation within a specific temperature range. Transit refrigeration is essential for highly perishable crops during warm weather in that air and road temperatures may be extreme (e.g., 41°C and 61°C, respectively[20]). In recent years, intermodal transport has become increasingly common. Truck trailers are moved part of the journey by railroad, and containerized products are routinely handled using several modes in tandem, for example, truck–rail–truck, truck–ship–truck. Refrigerated containers allow transferring the product from one carrier mode to another without direct handling of the product. Advantages include continuous refrigeration of the product during transit, reduced damage due to handling, and lower losses.

Refrigeration at the retail level involves the use of display refrigerators for merchandising the produce and walk-in coolers used for the storage of produce not in the sales area. Display refrigeration systems are designed for an attractive presentation of the product, ease of access for self-service, and sufficient refrigeration for short-term protection of the product. Generally, highly perishable produce should be sold within 24 to 48 hours from the time it is placed in the display area. Typically display cases are run at 1.5–7.5°C.

Product delivered to the store in refrigerated trucks should be readily placed in a refrigerated storage area. Separate storage areas are required for plant versus animal products, and larger retail outlets typically have at least two refrigerated storage areas for plant products, one at approximately 1.7°C and a second at 15.6°C for chilling sensitive products such as bananas. As a general rule, the volume of the storage rooms should be equivalent to the capacity of display area.

Home refrigerators are used for the storage of produce at the consumer level. Their use tends to be concentrated in the more industrialized areas of the world. Likewise, home refrigeration is generally less important in the more northern and southern regions of the temperate zones due to lower ambient temperatures during much of the year. Mechanical refrigerators are the most common; however, ice boxes are still utilized extensively. The storage volume of typical home refrigerators ranges from as little as 28 liters to 56 liters upward, with typical units being in the 425–565-liter range.

HEAT INPUT AFTER HARVEST

While much of this chapter has focused on the removal of heat from harvested products, there are occasions when many perishable products must be protected against low ambient temperatures or require specialized heat treatments. As a consequence, heat may need to be added to elevate the product temperature.

Addition of Heat to Maintain a Minimum Safe Product Temperature

The addition of heat to maintain a safe storage temperature is especially important in mountainous areas and during the winter months in the temperate zones as one moves progressively further away from the equator. In these areas, most cold storage rooms are situated in a temperature-controlled building. Therefore, there is a controlled and relatively constant ambient temperature surrounding at least part of the storage area. Most mechanical refrigeration systems now have the capacity for some heating. During cold weather, critical times for exposure to low-temperature extremes typically occur during transfer from one site to another (e.g., movement from wholesale storage to retail stores, or movement from the retail store to the purchaser's home). Most trucks, trailers, railroad cars, and refrigerated containers utilized in these areas have the potential for supplemental heating and can maintain any desired temperature between 0°C and 21°C.

Heating is accomplished using (1) reverse cycle, or hot gas operation of mechanical refrigeration units; (2) fuel-burning heaters that utilize alcohol, kerosene, butane, propane, or charcoal; (3) electrical resistance heaters powered by generators run from any of a number of sources (e.g., wheel-driven generators on railroad cars). The reverse cycle and fuel-burning heaters are the most common. When electrical heaters are utilized, some of the evaporator coil tubes contain tubular electrical heating elements that are used for both heating and defrosting of the evaporator coils. Reverse cycle or hot gas systems are the most common in modern refrigeration systems. In some of the older mechanical refrigerated systems, and when other forms of refrigeration are used (e.g., ice bunker–railroad car), portable thermostatically controlled fuel-burning heaters are employed.

Portable heaters have been used in railroad cars for more than 50 years, with earlier heaters utilizing charcoal as a fuel. Newer heaters are alcohol fueled in that alcohol provides adequate heat production and complete combustion (charcoal heaters release carbon monoxide, ethylene, and other gases into the storage area due to incomplete combustion). Alcohol heaters consist of a fuel tank (typically 5-gal capacity), wick, fuel gauge, thermostat, filler cap, chimney assembly, and spring hooks to secure the heater (fig. 8-24). The fuel moves by capillary action up the wick, with the rate of burning at the tip being modulated by a thermostatically controlled snuffer plate positioned just above the wick. The snuffer plate controls the size of the flame, hence the amount of heat generated. At the lowest level (i.e., the pilot burning) these units release about 600 BTU $\cdot$ hr^{-1} and about 6,000 BTU $\cdot$ hr^{-1} at their maximum with methanol as the fuel. Most systems now utilize a 1 : 1 ratio of methanol and isopropanol, which gives a greater heat yield and heat potential per tank of fuel. Newer high-output alcohol heaters can generate as much as 13,000 BTU $\cdot$ hr^{-1} using this fuel mixture.

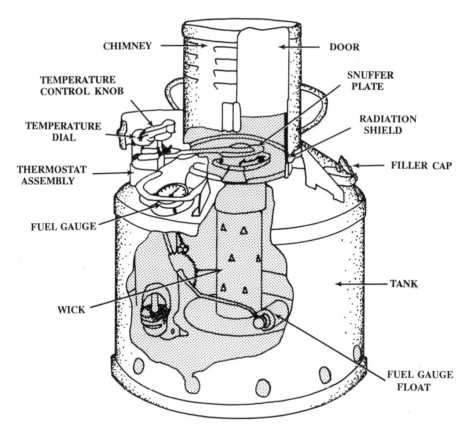

Figure 8-24. Portable thermostatically controlled alcohol heater used in railroad cars. The thermostat controls the height of the flame, and therefore the rate of heat production, by adjusting the height of the snuffer plate over the flame *(after ASHRAE[1]).*

Specialized Heat Treatments

The intentional elevation of product temperature after harvest is used for a cross-section of purposes and crops. Excluded from this group are temperature increases associated with drying (chap. 7). For drying, increased product temperature is not a prerequisite, although in many cases it is the most practical means available to hasten the rate of drying. Short-term exposure to high temperature is used for direct pathogen and insect control, indirect control of pathogens through the healing of surface wounds (invasion sites) prior to storage, inhibition of water loss, decreased chilling injury, and, in some instances, facilitation of sprouting, flowering, or other responses.

Curing of certain crops after harvest is normally with a short-term high-temperature treatment. Freshly harvested sweetpotatoes (*Ipomoea batatas,* (L.) Lam.) are exposed to 29°C (80–90% RH) for 5 to 7 days.[22,38] This exposure stimulates the formation of wound periderm at sites on the roots where the surface is broken, decreasing the incidence of soft rot. In the tropics, due to higher ambient temperatures, the addition of supplemental heat is seldom required for adequate curing. Many bulbs and corms are also cured, that is, gladiolus corms, 10 days at 27–29°C; hyacinth (*Hyacinthus orientalis,* L.) bulbs, several weeks at 25–27°C; Dutch iris (*Iris* spp.) bulbs, 10–15 days at 32°C; narcissus bulbs, 4 days at 30°C; tulip (*Tulipa* spp.) bulbs, 1 week at 26°C.[15]

Harvested grapefruit is increasingly being exposed to a short-term high-temperature treatment that decreases the incidence of chilling injury during subsequent storage. Here, fruit temperature is elevated to 27–29°C for 48 hours[11,17] or held at 21°C for 7 days[16] after harvest, prior to reducing the product temperature to the normal storage level (i.e., 10–16°C).

Brief exposure for some crops to high temperatures is done to control pathogens. Typically hot water is used instead of air due to the rapid heat transfer of water and the absence of drying that occurs with heated air. For example, brief exposure of mango fruit to hot water (55°C) helps to control *Colletotrichum gloesporioides* and *Botryodiplodia theobromal.*[27] Likewise, exposure of gladiolus corms to hot water (56°C) decreases the incidence of *Curvularia trifolii* f. sp. *gladioli.*[21] Poststorage high-temperature treatments are occasionally used to stimulate sprouting and/or flowering of certain crops. Exposure of gladiolus corms and lily bulbs to hot water accelerates sprouting.[18]

The temperature to which the product is raised and the duration of exposure are critical parameters in postharvest high-temperature treatments. With insects and diseases, successful treatments result in significant damage or death to the organism with little or no reduction in the quality of the product. Therefore, the target organism must be substantially more susceptible than the product to the imposed temperature treatment.

REFERENCES

1. ASHRAE. 1967. *Handbook of Fundamentals.* American Society of Heating, Refrigeration and Air Conditioning Engineers. New York.
2. ASHRAE. 1983–1986. *Handbook of Fundamentals.* American Society of Heating, Refrigeration and Air Conditioning Engineers. New York.

3. Awberry, J. H. 1927. The flow of heat in a body generating heat. *Phil. Mag.* **4:**629–638.

4. Barger, W. R. 1961. Factors affecting temperature reduction and weight-loss in vacuum-cooled lettuce. *USDA Mark. Res. Rep. 469*, 20p.

5. Barger, W. R. 1963. Vacuum precooling. A comparison of cooling of different vegetables. *USDA Mark. Res. Rep. 600*, 12p.

6. Catlin, P. B., F. G. Mitchell, and A. S. Greathead. 1959. Studies on strawberry quality. *Calif. Agric.* **13**(2):11, 16.

7. Eddy, D. E. 1965. Manufacture, storage handling and uses of fragmentary ice. *Am. Soc. Heat., Refrig., Air Cond. Eng. J.* **7:**66.

8. Gaffney, J. J., C. D. Baird, and K. V. Chau. 1985. Methods for calculating heat and mass transfer in fruits and vegetables individually and in bulk. *Am. Soc. Heat., Refrig., Air Cond. Eng. Trans.* **91:**333–352.

9. Gates, D. M. 1965. Energy, plants, and ecology. *Ecology* **46:**1–13.

10. Gosney, W. B. 1982. *Principles of Refrigeration*. Cambridge University Press, New York, 606p.

11. Grierson, W. 1974. Chilling injury in tropical and subtropical fruit. V. Effect of harvest date, degreening, delayed storage and peel color on chilling injury of grapefruit. *Proc. Trop. Reg. Am. Soc. Hort. Sci.* **18:**66–72.

12. Guillou, R. 1960. Coolers for fruits and vegetables. *Calif. Agric. Exp. Sta. Bull. 773*, 65p.

13. Guillou, R. 1963. Pressure cooling for fruits and vegetables. *Am. Soc. Heat, Refrig., Air Cond. Eng. J.* **5**(11):45–49.

14. Haller, M. H., P. L. Harding, J. M. Lutz, and D. H. Rose. 1932. The respiration of some fruits in relation to temperature. *Proc. Am. Soc. Hort. Sci.* **28:**583–589.

15. Hardenburg, R. E., A. E. Watada, C. Y. Wang. 1986. The commercial storage of fruits, vegetables and florist and nursery stock. *USDA-ARS Agric. Handb. 66*, 130p.

16. Hatton, T. T., and R. H. Cubbedge. 1982. Conditioning Florida grapefruit to reduce chilling injury during low-temperature storage. *J. Am. Soc. Hort. Sci.* **107:**57–60.

17. Hawkins, L. A., and W. R. Barger. 1926. Cold storage of Florida grapefruit. *Proc. Trop. Reg. Am. Soc. Hort. Sci.* **18:**66–72.

18. Hosoki, T. 1984. Effect of hot water treatment on respiration, endogenous ethanol and ethylene production from gladiolus corms and Easter lily bulbs. *HortScience* **19:**700–701.

19. Hurst, W. 1986. Unpublished data. University of Georgia, Athens.

20. Kasmire, R. F., and R. T. Hinsch. 1982. Factors affecting transit temperatures in truck shipments of fresh produce. *Univ. Calif. Perish. Handling Trans. Suppl. 1*, 10p.

21. Kelling, K., and R.-M. Niebisch. 1985. Nachiveis der Curvularia—Krankheit an Gladiolenpflanzgut. *Gartenbau* **32**(7):218–219.

22. Lutz, J. M. 1943. Factors influencing the relative humidity of the air immediately surrounding sweet potatoes during curing. *Proc. Am. Soc. Hort. Sci.* **43:**255–258.

23. Middleton, W. E. K. 1966. *A History of the Thermometer and Its Use in Meteorology*. Johns Hopkins Press, Baltimore, Maryland, 249p.

24. Pentzer, W. T., R. L. Perry, G. C. Hanna. 1936. Precooling and shipping California asparagus. *Calif. Agric. Exp. Sta. Bull. 600*, 45p.

25. Reidy, G. A. 1968. Values for thermal properties of foods gathered from the literature. M.S. thesis, Michigan State University, 59p.

26. Sainsbury, G. F. 1951. Improved fruit cooling methods. *Refrig. Eng.* **59:**464–469, 506, 508–509.

27. Sampaio, V. R. 1983. Controle em pos-colheita das podridoes da manga Bourbon, conservada em camara fria. *Anais da Escola Superior de Agricultura "Luiz de Queiroz"* **40**(1):519–526.

28. Scholz, E. W., H. B. Johnson, and W. R. Buford. 1963. Heat-evolution rates of some Texas-grown fruits and vegetables. *Rio Grande Valley Hort. Soc. J.* **17:**170–175.

29. Smith, R. E., and A. H. Bennett. 1965. Mass-average temperature of fruits and vegetables during transient cooling. *Am. Soc. Agric. Eng. Trans.* **8:**249–252.

30. Smith, W. H. 1957. The production of carbon dioxide and metabolic heat by horticultural produce. *Mod. Refrig.* **60:**493–496.

31. Smock, R. M., and C. R. Gross. 1950. Studies on respiration of apples. *N.Y. (Cornell) Agric. Exp. Sta. Man. 297,* 47p.

32. Stewart, J. K., and M. H. Covey. 1963. Hydrocooling vegetables: A practical guide to predicting final temperatures and cooling times. *USDA Mark. Res. Rep. 637,* 32p.

33. Sweat, V. E. 1974. Experimental values of thermal conductivity of selected fruits and vegetables. *J. Food Sci.* **39:**1,080–1,083.

34. Tewfik, S., and L. E. Scott. 1954. Respiration of vegetables as affected by postharvest treatment. *J. Agric. Food Chem.* **2:**415–417.

35. Thevenot, R. 1955. Precooling. *9th Int. Cong. Refrig. Paris Proc.* 0.10:0051–0071.

36. Toussaint, W. D., T. T. Hatlow, and G. Abshier. 1955. Hydrocooling peaches in the North Carolina sandhills. *N.C. St. Agric. Exp. Sta. Info. Ser. 39.*

37. Turrell, F. M., and R. L. Perry. 1957. Specific heat and heat conductivity of citrus fruit. *Proc. Am. Soc. Hort. Sci.* **70:**261–265.

38. Weimer, J. R., and L. L. Harter. 1921. Wound cork formation in the sweet potato. *J. Agric. Res.* **21:**637–647.

ADDITIONAL READINGS

ASHRAE. 1983–1986. *Handbook of Fundamentals.* American Society of Heating, Refrigeration and Air Conditioning Engineers, New York.

Dennis, D. T. 1987. *The Biochemistry of Energy Utilization in Plants.* Blackie, Glasgow, 145p.

Gates, D. M., and R. B. Schmerl (eds.). 1975. *Perspectives of Biophysical Ecology.* Springer-Verlag, Berlin, 609p.

Gosney, W. B. 1982. *Principles of Refrigeration.* Cambridge University Press, New York, 606p.

Hanan, J. J. 1984. *Plant Environmental Measurement.* Bookmaker Guide, Longmont, Colo., 326p.

Jones, H. G. 1983. *Plants and Microclimate: A Quantitative Approach to Environmental Plant Physiology.* Cambridge University Press, Cambridge, England, 323p.

Mitchell, F. G., R. Guillou, and R. A. Parsens. 1972. Commercial cooling of fruits and vegetables. *Univ. Calif. Ext. Manual 43,* 44p.

Nobel, P. S. 1983. *Biophysical Plant Physiology and Ecology.* Freeman, San Francisco, Calif., 608p.

Woodward, F. I. 1987. *Climate and Plant Distribution.* Cambridge University Press, Cambridge, England, 174p.

SPECIES INDEX

SUBJECT INDEX